CRYSTAL STRUCTURES

Second Edition

Ralph W. G. Wyckoff, *University of Arizona, Tucson, Arizona*

VOLUME 2
Inorganic Compounds RX_n, R_nMX_2, R_nMX_3

INTERSCIENCE PUBLISHERS

a division of John Wiley & Sons, New York • London • Sydney

addl

Copyright © 1948, 1951, 1953, 1957, 1958,
  1959, 1960 by Interscience Publishers, Inc.
Copyright © 1964 by John Wiley & Sons, Inc.

ALL RIGHTS RESERVED

This book or any part thereof must not be
reproduced in any form without the written
permission of the publisher.

PHYSICS

Library of Congress Catalog Card Number 48-9169
Printed in the United States of America

QD951
W8
1963
v. 2
Physics
Library

Preface

The second volume of *Crystal Structures* closely follows the format adopted for Volume 1. As this attempt to provide a unified statement of the atomic positions in all analyzed crystals progresses, improvements become possible. The principal one introduced with this volume is a more reliable bibliography. The author is indebted to his wife for undertaking to control the accuracy of all the bibliographic entries. As the older references are checked, errors carried over from the earlier compilations are being corrected. It is hoped that with time these can all be eliminated. In the meantime readers will find the bibliography of this volume more accurate than that of the loose-leaf edition.

As the work on compilation proceeds, the author becomes increasingly aware of the incompleteness of the simple statement of atomic positions, principal coordination relationships, and bond lengths that were the basic objectives of this work. For many purposes it would be desirable to have a fuller listing of interatomic distances and a more explicit evaluation of the accuracy of each determination of structure. Thermal parameters have an essential place in a complete description of atomic arrangement as has also a more detailed discussion of the disorder being found in many crystals. Undoubtedly these should in the long run find a place in the kind of summary *Crystal Structures* aims to provide. The author has been tempted to initiate such an expanded description immediately, but the critical evaluation of existing data this would require would very materially add to the task of bringing the compilation up to date. It has accordingly seemed more useful to retain the present limited objectives and to complete the survey as quickly as may be, reserving a more expanded coverage to the future.

An explanation should be given of the system followed in making the illustrations. They include drawings used in both *The Structure of Crystals* and the previous edition of *Crystal Structures;* some, therefore, date back as much as forty years. In earlier days left-hand axes were in common use and the author has continued to employ them even though right-hand systems are now in many circles considered standard. When the time came to prepare this edition, it was apparent that there would be real advantages in adopting this right-hand system but by then so many drawings had been made according to the earlier convention that remaking them was impractical. It was therefore decided to continue with left-hand axes rather than to have an indiscriminate mixture of the two. There are a few exceptions

v

685

where existing right-hand drawings have been used but unless expressly stated, the axial sequence is left-handed with the origin usually lying in the lower left corner of the projected crystal unit. The third axis will therefore generally arise from the plane of the paper. Numbers attached to the atoms are their fractional coordinates along this axis; except when the plane of the other two does not coincide with that of the paper, the larger numbers are proportionately nearer the eye and the packing drawings express this fact. Since the arrangements according to right-hand axes are mirror images of the illustrations, they correspond to the projections if the numbers are taken to indicate fractional coordinates *below* instead of *above* the plane of the paper.

Something further should also be said about the space group symbols used. The second symbol (in parentheses) is of the Herrmann-Mauguin type. Those used in the two editions of the *International Tables* differ somewhat from one another; in this book the older ones have not necessarily been replaced by the new.

The writer is further indebted to his wife and to Helen E. Barnes whose help in preparing the manuscript for the printer and in reading proof has very greatly accelerated the rate at which this and succeeding volumes can appear.

The author wishes to acknowledge financial help in meeting editorial expenses in connection with the preparation of this volume from a grant to the University of Arizona out of National Science Foundation Institutional Grant funds.

Contents

Volume 1 Contents

Chapter V

STRUCTURES OF COMPLEX BINARY COMPOUNDS R_nX_m

This chapter describes the structures found for all binary compounds having compositions more complicated than those of the RX and RX_2 crystals of Chapters III and IV. As in the earlier loose-leaf edition, these have been classified according to their formulas, the grouping being roughly that employed before. They are arranged as follows:

  A.   Compounds R_2X_3
  B.   Compounds RX_3
  C.   Compounds RX_4
  D.   Compounds R_3X_4
  E.   Compounds R_nX_5
  F.   Compounds R_nX_m (where $m = 6$ or more)

In group F, crystals with the same value of m are treated together, $m = 6$ coming first. In group E, crystals with $n = 1$ are described before those with $n = 2, 3, 4$, etc. In all groups and subgroups the compounds have been arranged in order of decreasing electronegativity, i.e., the halides have been followed by the oxides and sulfides, and these by nitrides, phosphides, and carbides.

A. COMPOUNDS R_2X_3

Oxides

V,a1. A number of rare-earth oxides with large metal-to-oxygen separations form hexagonal crystals of which *lanthanum sesquioxide*, La_2O_3, is typical. Assuming that the atoms have the dimensions of their ions, these are compounds with $r(R)/r(O)$ greater than 0.87. There is but one molecule in the unit which for La_2O_3 has the cell edges:

$$a_0 = 3.9373 \text{ A.}, \qquad c_0 = 6.1299 \text{ A.}$$

The space group is D_{3d}^3 ($C\bar{3}m$) and atoms are in the positions:

  La: $(2d)$ $\pm(\frac{1}{3} \, \frac{2}{3} \, u)$ with $u = 0.245$
  as determined by neutron diffraction

O(1): $(1a)$ 000 and O(2): $(2d)$ with $u = 0.645$

1

In this arrangement (Fig. VA,1) each lanthanum atom has four oxygen neighbors at a distance of ca. 2.30 A. and three more at ca. 2.70 A. The oxygen atoms have their usual ionic separations with the closest O–O = ca. 2.75 A.

In addition to the oxides, many oxysulfides and oxyselenides of rare earths have this structure. All these are listed in Table VA,1. One form of Al_2S_3 and the nitride of thorium, Th_2N_3, are isostructural. Magnesium forms an antimonide and a bismuthide, Mg_3Sb_2 and Mg_3Bi_2, that are said to have atomic arrangements "anti" to that of La_2O_3.

Parameters have been determined for only a few of these compounds. For the three oxysulfides, Ce_2O_2S, La_2O_2S, and Pu_2O_2S, they have been chosen as $u = 0.29$ and $v = 0.64$. For Pu_2O_3, $u = 0.235$, $v = 0.63$.

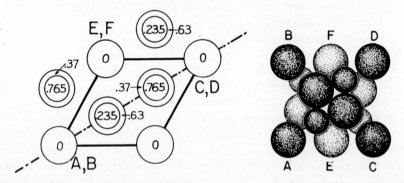

Fig. VA,1a (left). A projection on its base of atoms in the hexagonal unit of La_2O_3. The smaller circles are the metallic atoms.

Fig. VA,1b (right). A packing projection of the La_2O_3 structure on the dot-and-dash diagonal line of Figure VA,1a. Letters refer to the same atoms in the two drawings. The larger spheres are the oxygen atoms.

V,a2. Several of the rare-earth oxides having the hexagonal structure described in **V,a1** have a second cubic modification. Its atomic arrangement was first determined for the mineral *bixbyite* $(Fe,Mn)_2O_3$. The unit cell is large and contains 16 molecules; for bixbyite,

$$a_0 = 9.365 \text{ A.}$$

The space group is T_h^7 ($Ia3$) with the atoms in the positions:

[Fe,Mn](1):   (8a) $^1/_4\,^1/_4\,^1/_4$; $^1/_4\,^3/_4\,^3/_4$; $^3/_4\,^1/_4\,^3/_4$; $^3/_4\,^3/_4\,^1/_4$; B.C.

[Fe,Mn](2):   (24d) $\pm(u\,0\,^1/_4;$ $^1/_4\,u\,0;$ $0\,^1/_4\,u;$
                   $\bar{u}\,^1/_2\,^1/_4;$ $^1/_4\,\bar{u}\,^1/_2;$ $^1/_2\,^1/_4\,\bar{u})$; B.C.

O: (48e) $\pm(xyz;\ x,\bar{y},^1/_2-z;\ ^1/_2-x,y,\bar{z};\ \bar{x},^1/_2-y,z;$
              $zxy;\ ^1/_2-z,x,\bar{y};\ \bar{z},^1/_2-x,y;\ z,\bar{x},^1/_2-y;$
              $yzx;\ \bar{y},^1/_2-z,x;\ y,\bar{z},^1/_2-x;\ ^1/_2-y,z,\bar{x})$; B.C.

TABLE VA,1
Crystals with the Hexagonal La_2O_3 Arrangement

Crystal	a_0, A.	c_0, A.
Ac_2O_3	4.08	6.30
β-Am_2O_3	3.817	5.971 (prep. at 800°C.)
Ce_2O_3	3.888	6.069
Nd_2O_3	3.831	5.991 (26°C.)
Pr_2O_3	3.851	5.996
Pu_2O_3	3.840	5.957
β-Al_2S_3	3.579	5.829
Ce_2O_2S	4.004	6.872
Dy_2O_2S	3.8029	6.603
Er_2O_2S	3.7601	6.5521
Eu_2O_2S	3.8716	6.6856
Gd_2O_2S	3.8514	6.667
Ho_2O_2S	3.7816	6.5800
La_2O_2S	4.0509	6.943
Lu_2O_2S	3.7093	6.486
Nd_2O_2S	3.946	6.790
Pr_2O_2S	3.9737	6.825
Pu_2O_2S	3.927	6.768
Sm_2O_2S	3.8934	6.717
Tb_2O_2S	3.8249	6.6260
Tm_2O_2S	3.747	6.538
Y_2O_2S	3.780	6.563
Yb_2O_2S	3.7233	6.5031
Er_2O_2Se	3.792	6.743
Gd_2O_2Se	3.873	6.854
Ho_2O_2Se	3.807	6.766
La_2O_2Se	4.070	7.124
Nd_2O_2Se	3.975	6.985
Pr_2O_2Se	4.009	7.031
Sm_2O_2Se	3.916	6.912
Yb_2O_2Se	3.761	6.697
Th_2N_3	3.883	6.287
Mg_3Bi_2	4.666	7.401
Mg_3Sb_2	4.573	7.229

The parameters originally found were $u = -0.030$, $x = 0.385$, $y = 0.145$, $z = 0.380$. The most recent values have been given as: $u = -0.034$, $x = 0.375$, $y = 0.162$, $z = 0.400$.

The resulting structure is shown in Figure VA,2. It is most easily thought of as an incomplete cubic close-packing of the oxygen ions. The metallic atoms are distributed as are the calcium atoms in CaF_2 (**IV,a1**); the oxygen atoms occupy three-fourths of the fluorine positions in the larger cell made necessary by this relative deficit in negative atoms. Each metallic atom has around it four closest oxygen atoms, with (Fe,Mn)–O = ca. 2.00 A.; the closest approach of oxygen atoms to one another is ca. 2.51 A.

The various oxides with this arrangement have a radius ratio of their ions which is, on the average, less than that found for the oxides described in **V,a1**: $r(R)/r(O)$ exceeds ca. 0.60 but is generally not greater than 0.87.

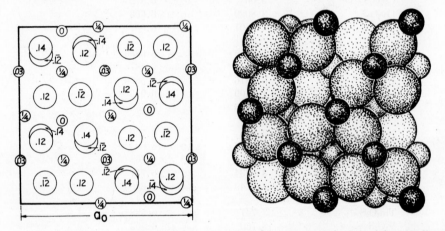

Fig. VA,2a (left). A projection on a cube face of half the contents of the unit of the complicated bixbyite, or Tl_2O_3, structure. The larger circles are the oxygen atoms.
Fig. VA,2b (right). A packing drawing of the atoms of Figure VA,2a when given their usual ionic sizes.

Neutron diffraction studies have been carried out on several of the rare-earth oxides having this structure. Their cell dimensions and those of a number of other isostructural compounds are listed in Table VA,2. Their parameters are as follows:

Compound	u	x	y	z
Er_2O_3	−0.0330	0.394	0.149	0.380
Ho_2O_3	−0.0270	0.388	0.152	0.382
Mn_2O_3	−0.0347	0.378	0.167	0.397
Pr_2O_3	−0.0290	0.385	0.155	0.382
Y_2O_3	−0.0314	0.389	0.150	0.377
Yb_2O_3	−0.0336	0.391	0.151	0.380

TABLE VA,2
Crystals with the Cubic Bixbyite Structure

Crystal	a_0, A.
α-Am_2O_3[a]	11.03 (prep. at 600°C.)
Cm_2O_3	11.00
Dy_2O_3	10.667
Er_2O_3	10.547
Eu_2O_3	10.866
β-Fe_2O_3	9.40
Gd_2O_3	10.813
Ho_2O_3	10.607
In_2O_3	10.118 (26°C.)
La_2O_3	11.38
Lu_2O_3	10.391
β-Mn_2O_3	9.408
Nd_2O_3	11.048
Pr_2O_3	11.136
Pu_2O_3	11.04
Sc_2O_3	9.845
Sm_2O_3	10.932
Tb_2O_3	10.728
Tl_2O_3	10.543 (26°C.)
Tm_2O_3	10.488
Y_2O_3	10.604 (27°C.)
Yb_2O_3	10.439
Be_3N_2	8.13
Ca_3N_2	10.40
Cd_3N_2	10.79
Mg_3N_2	9.95
U_2N_3	10.678
Zn_3N_2	9.743
Be_3P_2	10.15
Mg_3P_2	12.01
Mg_3As_2	12.33

[a] $u = -0.030.$

In paragraph V,a1 it was pointed out that the compounds of magnesium with antimony and bismuth have structures that are "anti" to the bixbyite arrangement. Compounds of magnesium with the lighter fifth column elements—arsenic, phosphorus, and nitrogen—might therefore be expected to have structures that are "anti" to the bixbyite arrangement. As the data of Table VA,2 indicate, this is the case for them and for some of the related compounds of beryllium, calcium, and zinc.

As far as most of these compounds are concerned, they appear to have the fixed composition R_2O_3, but it has been reported that thallic oxide can be prepared with a relative deficit in thallium atoms. For such a material with the formula $Tl_2O_{3.08}$, $a_0 = 10.534$ A.

V,a3. With smaller metallic atoms which make $r(R)/r(O)$ less than 0.60, oxygen ions can approach nearer to a perfect close-packing than is the case with the preceding two structures, and such oxides are often found with an arrangement typified by that of *chromium sesquioxide*, Cr_2O_3. Its symmetry is rhombohedral with a unit containing two molecules and having the dimensions:

$$a_0 = 5.350 \text{ A.}, \qquad \alpha = 55°9'$$

The space group is D_{3d}^6 $(R\bar{3}c)$ and atoms are in the special positions:

Cr: (4c) $\pm(uuu)$; B.C. with $u = 0.3475$
O: (6e) $\pm(u,^1/_2-u,^1/_4; \, ^1/_2-u,^1/_4,u; \, ^1/_4,u,^1/_2-u)$ with $u = 0.556$

For this description the origin lies in a center of symmetry; together with the observed parameters, it is thus displaced one-fourth the length of the body diagonal of the unit rhombohedron from the origin used in many earlier descriptions.

The dimensions of the corresponding hexamolecular cell referred to hexagonal axes are

$$a_0' = 4.954 \text{ A.}, \qquad c_0' = 13.584 \text{ A.}$$

In this cell the atoms have the positions:

Cr: (12c) $\pm(00u'; \, 0,0,u'+^1/_2)$; rh with $u' = 0.3475$
O: (18e) $\pm(v \, 0 \, ^1/_4; \, 0 \, v \, ^1/_4; \, \bar{v} \, \bar{v} \, ^1/_4)$; rh with $v = 0.306$

This is a structure (Fig. VA,3) which is perhaps best considered as a slightly distorted hexagonal close-packing of oxygen ions with small metallic ions lying in some of the interstices. The packing would be perfect if α for the unit rhombohedron were $53°47'$ instead of $55°9'$ and if $u(Cr)$ were $^1/_3$ and $u(O)$ were $^7/_{12}$ [0.583]. In this arrangement each chromium atom has around it six oxygen atoms, three at a distance of 1.97 A. and three more with Cr–O = 2.02 A.

This structure differs from that found for the double oxide ilmenite, $FeTiO_3$ (**VII,a32**) only in the latter's lower symmetry required for an

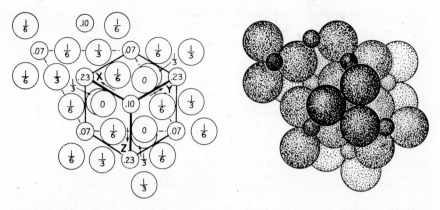

Fig. VA,3a (left). A projection of the portion of the Cr_2O_3 arrangement on a plane normal to the threefold axis and passing through the apex of the unit rhombohedron. The small circles are the metal atoms. Heights refer to an origin displaced by $1/4\ c_0$ from that used in the text.

Fig. VA,3b (right). A packing drawing of the atoms of Figure VA,3a if they are given their usual ionic sizes.

ordered arrangement of its two kinds of metallic atoms. The arrangement within the hexagonal cell of Cr_2O_3 is therefore illustrated by the ilmenite figure, Figure VIIA,35, if its two types of metallic atom are taken as identical.

The unit cells of other compounds having this structure are listed in Table VA,3. Of these, parameter determinations have been made on the following:

Compound	$u = v'(R)$	$u(O)$	$v(O)$
Al_2O_3	0.3520	0.556	0.306
α-Fe_2O_3	0.355	0.550	0.300
Ti_2O_3	0.3450	0.567	0.317
V_2O_3	0.3463	0.565	0.315

In terms of these parameters, the six oxygen atoms around each metal atom are at the distances: Ti–O = 2.01 and 2.08 A. for Ti_2O_3; V–O = 1.96 and 2.06 A. for V_2O_3; Al–O = 1.86 and 1.97 A. for Al_2O_3; Fe–O = ca. 1.91 and 2.06 A. for α-Fe_2O_3. A neutron diffraction study of α-Fe_2O_3 made at room temperature and at 4.2°K. showed no detectable change in parameters with lowered temperature.

TABLE VA,3. Crystals with the Rhombohedral Cr_2O_3 Structure

Crystal	a_0, A.	α	a_0', A.	c_0', A.
Al_2O_3 (corundum)	5.128	55°20′	4.76280	13.00320 (31°C.)
$(Cr_{1.90},V_{0.09},Fe_{0.01})_2O_3$				
(eskolaite)	5.361	55°5′	4.958	13.60
α-Fe_2O_3 (hematite)	5.4135	55°17′	5.035	13.72
Ga_2O_3	5.320	55°48′	4.9793	13.429 (24°C.)
Rh_2O_3	5.47	55°40′	5.11	13.82
Ti_2O_3	5.431	56°36′	5.148	13.636
V_2O_3	5.647	53°45′	5.105	14.449
γ-Al_2S_3	6.86	56°16′	6.47	12.27
Co_2As_3	6.24	59°6′	6.16	15.40

V,a4. The compound *niobium manganese oxide* $(Nb_2Mn_4)O_9$, has a structure closely resembling that of Cr_2O_3 (**V,a3**). The niobium and manganese atoms together occupy positions analogous to those of the chromium atoms, but the oxygens are somewhat differently distributed.

The structure is truly hexagonal rather than rhombohedral, and the two $(Nb_2Mn_4)O_9$ molecules are in a cell of the dimensions:

$$a_0 = 5.335 \text{ A.,} \qquad c_0 = 14.320 \text{ A.}$$

Employing neutron diffraction data, the atoms have been placed in the following positions of D_{3d}^4 ($P\bar{3}c$):

Nb: (4c) $\pm(00u;\ 0,0,u+{}^1/_2)$ with $u = 0.3575$

Mn(1):   (4d) $\pm({}^1/_3\ {}^2/_3\ v;\ {}^1/_3,{}^2/_3,v+{}^1/_2)$ with $v = 0.0181$

Mn(2):   (4d)   with $v' = 0.297$

O(1):   (6f) $\pm(w\ 0\ {}^1/_4;\ 0\ w\ {}^1/_4;\ \bar{w}\ \bar{w}\ {}^1/_4)$ with $w = 0.300$

O(2):   (12g) $\pm(xyz;\qquad \bar{y},x-y,z;\qquad y-x,\bar{x},z;$
$\bar{y},\bar{x},z+{}^1/_2;\ x,x-y,z+{}^1/_2;\ y-x,y,z+{}^1/_2)$
with $x = 0.334,\ y = 0.295,\ z = 0.0884$

The separation Mn(1)–O = 2.08 and 2.26 A.; Mn(2)–O = 1.98 and 2.25 A.

This structure (Fig. VA,4) can be compared with that of Cr_2O_3 by referring this figure to Figures VA,3 and VIIA,35 (for ilmenite).

The analogous cobalt compound $Nb_2Co_4O_9$ has the same structure, with

$$a_0 = 5.177 \text{ A.,} \qquad c_0 = 14.168 \text{ A.}$$

and the parameters $u(Nb) = 0.360$, $v(Co,1) = 0.027$, $v'(Co,2) = 0.307$, $w(O,1) = 0.305$, $x(O,2) = 0.333$, $y(O,2) = 0.295$, $z(O,2) = 0.0833$.

The distance Co(1)–O = 2.22 and 2.27 A.; Co(2)–O = 1.99 and 2.35 A.

At the temperature of liquid nitrogen $(Nb_2Mn_4)O_9$ became antiferromagnetic, but this has not occurred with the cobalt compound.

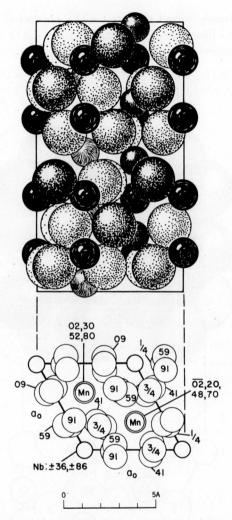

Fig. VA,4. Two projections of the hexagonal unit of $(NbMn_2)_2O_9$.   In the upper projection the niobium atoms are black and the oxygen atoms dotted.

Other examples of this structure are the following:

$$(Nb_2Fe_4)O_9: \quad a_0 = 5.233 \text{ A.,} \quad c_0 = 14.236 \text{ A.}$$
$$(Nb_2Mg_4)O_9: \quad a_0 = 5.171 \text{ A.,} \quad c_0 = 14.173 \text{ A.}$$
$$(Ta_2Co_4)O_9: \quad a_0 = 5.181 \text{ A.,} \quad c_0 = 14.174 \text{ A.}$$
$$(Ta_2Mg_4)O_9: \quad a_0 = 5.170 \text{ A.,} \quad c_0 = 14.143 \text{ A.}$$
$$(Ta_2Mn_4)O_9: \quad a_0 = 5.337 \text{ A.,} \quad c_0 = 14.333 \text{ A.}$$

V,a5. The ternary mixed oxides $(Cr,Mo)_2O_3$ and $(Cr,W)_2O_3$, have a structure different from those of the pure oxides. The symmetry is hexagonal with unimolecular rhombohedral units of the dimensions:

$$(Cr,Mo)_2O_3: \quad a_0 = 3.643 \text{ A.}, \qquad \alpha = 85°36'$$
$$(Cr,W)_2O_3: \quad a_0 = 3.640 \text{ A.}, \qquad \alpha = 85°26'$$

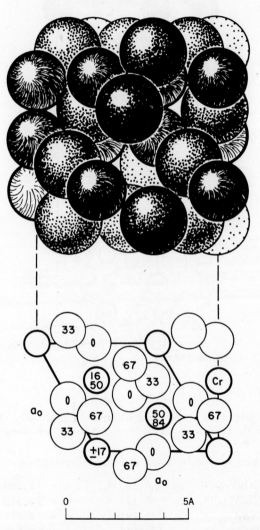

Fig. VA,5. Two projections of the hexagonal structure of $(Cr,Mo)_2O_3$. In the upper projection the oxygen atoms are dotted.

Atoms have been placed in the following special positions of $D_3{}^7$ ($R32$):

(Cr,Mo or W): (2c) $\pm$ (uuu) with u = ca. 0.17

O: (3d) $0u\bar{u}$; $\bar{u}0u$; $u\bar{u}0$ with u = ca. 0.47

For the corresponding trimolecular hexagonal cells:

$(Cr,Mo)_2O_3$: $a_0' = 4.937$ A., $c_0' = 6.763$ A.

$(Cr,W)_2O_3$: $a_0' = 4.938$ A., $c_0' = 6.787$ A.

The atomic positions within these cells are:

(Cr,Mo or W): (6c) $\pm(00u)$; rh with u = ca. 0.17

O: (9d) $u00$; $0u0$; $\bar{u}\bar{u}0$; rh with u = ca. 0.47

This simple atomic arrangement is shown in Figure VA,5.

V,a6. The so-called B-form of *samarium sesquioxide*, Sm_2O_3, is monoclinic with a hexamolecular unit of the following dimensions:

$$a_0 = 14.177 \text{ A.}; \quad b_0 = 3.638 \text{ A.}; \quad c_0 = 8.847 \text{ A.}; \quad \beta = 99°59'$$

The space group is $C_{2h}{}^3$ ($C2/m$). Two oxygen atoms are in the special positions

$$(2b) \quad {}^1/_2\,0\,0; 0\,{}^1/_2\,0$$

All other atoms are in

$$(4i) \quad \pm(u0v; u+{}^1/_2,{}^1/_2,v)$$

with the parameters listed in Table VA,4.

This is a structure (Fig. VA,6) in which all atoms lie in two layers succeeding one another at a separation of ${}^1/_2\,b_0$. Each samarium atom has six oxygen atoms at distances between 2.28 and 2.57 A. and one or two others at distances between 2.71 and 3.12 A.

TABLE VA,4
Parameters of Atoms in B-Sm_2O_3

Atom	u	v
Sm(1)	0.6349	0.4905
Sm(2)	0.6897	0.1380
Sm(3)	0.9663	0.1881
O(1)	0.128	0.286
O(2)	0.824	0.027
O(3)	0.799	0.374
O(4)	0.469	0.344

The B-form of *gadolinium sesquioxide*, Gd_2O_3, has the same structure with:

$$a_0 = 14.061 \text{ A.}; \quad b_0 = 3.566 \text{ A.}; \quad c_0 = 8.760 \text{ A.}; \quad \beta = 100°6'$$

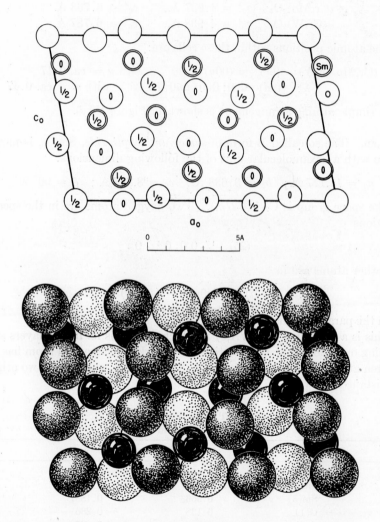

Fig. VA,6a (top). The monoclinic structure of B-Sm_2O_3 viewed along its b_0 axis.   Origin in lower left.

Fig. VA,6b (bottom). A packing drawing of the B-Sm_2O_3 structure viewed along its b_0 axis. The samarium atoms are black.

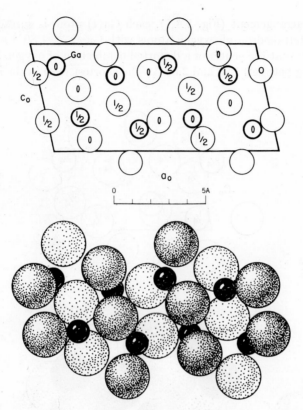

Fig. VA,7a (top). The monoclinic structure of β-Ga$_2$O$_3$ seen along its b_0 axis. Origin in lower left.

Fig. VA,7b (bottom). A packing drawing of the β-Ga$_2$O$_3$ arrangement seen along its b_0 axis. The gallium atoms are black.

V,a7. The β-form of *gallium sesquioxide*, Ga$_2$O$_3$, is monoclinic with a tetramolecular cell having the dimensions:

$$a_0 = 12.23 \text{ A.;} \quad b_0 = 3.04 \text{ A.;} \quad c_0 = 5.80 \text{ A.;} \quad \beta = 103°42'$$

All atoms have been placed in special positions:

$$(4i) \quad \pm(u0v; u+^1/_2,{}^1/_2,v)$$

of C$_{2h}^3$ ($C2/m$) with the parameters:

Atom	u	v
Ga(1)	0.0904	−0.2052
Ga(2)	0.3414	−0.3143
O(1)	0.1674	0.1011
O(2)	0.4957	0.2553
O(3)	0.8279	0.4365

In this arrangement (Fig. VA,7), each Ga(1) atom is surrounded by a distorted tetrahedron of oxygen atoms with Ga–O = 1.80–1.85 A., while each Ga(2) is at the centér of a distorted octahedron of oxygen atoms with Ga–O lying between 1.95 and 2.08 A. The shortest O–O in the structure is 2.67 A.

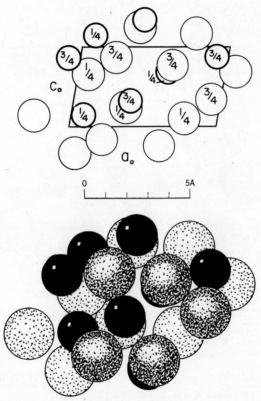

Fig. VA,8a (top). A projection along the b_0 axis of the monoclinic structure of Pb_2O_3
Oxygen are the larger circles.   Origin in lower left.
Fig. VA,8b (bottom). A packing drawing of the monoclinic Pb_2O_3 structure viewed along
the b_0 axis.   The dotted atoms are oxygen.

V,a8.   A structure has been determined for the monoclinic *lead sesqui-oxide*, Pb_2O_3. Its bimolecular cell has the dimensions:

$$a_0 = 7.050 \text{ A.;} \quad b_0 = 5.616 \text{ A.;} \quad c_0 = 3.865 \text{ A.;} \quad \beta = 80°6'$$

All atoms have been placed in the following special positions of C_{2h}^2 $(P2_1/m)$:

$$(2e) \quad \pm (u \: ^1/_4 \: v)$$

with the parameters listed below

Atom	u	v
Pb(1)	0.083	0.144
Pb(2)	−0.383	−0.328
O(1)	−0.228	0.122
O(2)	0.380	0.247
O(3)	0.069	−0.350

In the structure that results (Fig. VA,8), one set of lead atoms has four oxygen atoms at distances lying between 1.94 and 2.22 A. The other lead atoms have two oxygen atoms at 2.24 A., two at 2.54 A., and two more at 2.81 A. The oxygen atoms about the first lead atoms can be thought of as lying at the corners of a distorted square, those about the second type of lead atom as being at the corners of a distorted octahedron.

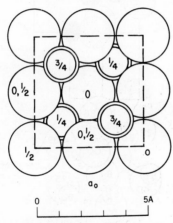

Fig. VA,9. The cubic structure proposed for Ag_2O_3 seen along a cube axis. The silver atoms are smaller and doubly ringed.

V,a9. A very simple cubic structure has been proposed for the *silver sesquioxide*, Ag_2O_3. There are two molecules in a unit with an edge length

$$a_0 = 4.904–4.963 \text{ A.}$$

that depends on the amount of oxygen in the preparation. The space group has been chosen as O_h^4 (*Pn3m*) with atoms in the positions:

Ag: (4b) $1/4\ 1/4\ 1/4$; $1/4\ 3/4\ 3/4$; $3/4\ 1/4\ 3/4$; $3/4\ 3/4\ 1/4$
O: (6d) $0\ 1/2\ 1/2$; $1/2\ 0\ 1/2$; $1/2\ 1/2\ 0$; $1/2\ 0\ 0$; $0\ 1/2\ 0$; $0\ 0\ 1/2$

In this arrangement (Fig. VA,9) the O–O distance is 2.45 A. and each silver atom is surrounded by six oxygens with Ag–O = 2.12 A.

V,a10. *Boron sesquioxide*, B_2O_3, forms hexagonal crystals whose tri‑ molecular cell has the dimensions.

$$a_0 = 4.325 \text{ A.,} \qquad c_0 = 8.317 \text{ A.}$$

Atoms have been placed in general positions of C_3^2 $(C3_1)$:

$$(3a) \quad xyz; \; \bar{y}, x-y, z+\tfrac{1}{3}; \; y-x, \bar{x}, z+\tfrac{2}{3}$$

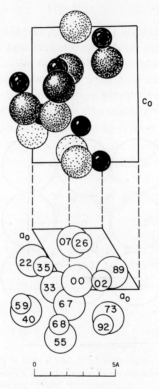

Fig. VA,10. Two projections of the hexagonal cell of B_2O_3.   In the upper part the oxygen atoms are large and dotted.

with the following parameters:

Atom	x	y	z
B(1)	0.54	0.15	0.02
B(2)	0.59	0.77	0.26
O(1)	0.20	0.15	0.00
O(2)	0.46	0.79	0.07
O(3)	0.51	0.23	0.56

This is a structure (Fig. VA,10) made up of a three-dimensional network of distorted BO_4 tetrahedra that share corners with one another. Three of the corners are shared with three other and the fourth with two other tetrahedra. In the two kinds of tetrahedra, B–O varies between 1.31 and 2.12 A.

V,a11. An *arsenious oxide*, As_2O_3, and the modification of Sb_2O_3 expressed by the mineral senarmontite form molecular crystals. Their symmetry is cubic with a unit containing 16 molecules.

$$As_2O_3: \quad a_0 = 11.0745 \text{ A.,}$$
$$Sb_2O_3: \quad a_0 = 11.152 \text{ A.} \quad (26°\text{C.})$$

Atoms are in the following special positions of O_h^4 ($Fd3m$):

R: (32b) $uuu; u\bar{u}\bar{u}; \ ^1/_4 - u, ^1/_4 - u, ^1/_4 - u; \ ^1/_4 - u, u + ^1/_4, u + ^1/_4;$
  $\bar{u}u\bar{u}; \ \bar{u}\bar{u}u; \ u + ^1/_4, ^1/_4 - u, u + ^1/_4; \ u + ^1/_4, u + ^1/_4, ^1/_4 - u;$ F.C.

X: (48c) $v00; \bar{v}00; \ ^1/_4 - v, ^1/_4, ^1/_4; \ v + ^1/_4, ^1/_4, ^1/_4;$
  $0v0; 0\bar{v}0; \ ^1/_4, ^1/_4 - v, ^1/_4; \ ^1/_4, v + ^1/_4, ^1/_4;$
  $00v; 00\bar{v}; \ ^1/_4, ^1/_4, ^1/_4 - v; \ ^1/_4, ^1/_4, v + ^1/_4;$ F.C.

For As_2O_3, u has been determined as 0.885, v as 0.235; for Sb_2O_3 these parameters are $u = 0.883$, $v = 0.235$.

The structure that results (Fig. VA,11) consists of R_4O_6 molecules packed together in the unit cube in the same manner as are the carbon atoms in the diamond. Within the molecule, each R atom has three oxygen neighbors at a distance that is 1.80 A. in As_2O_3 and 2.27 A. in Sb_2O_3; each oxygen atom is bound to two R atoms. The different molecules approach one another through O–O contacts, the shortest being 2.83 A. in As_2O_3.

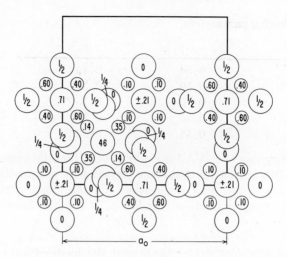

Fig. VA,11a. The structure of As$_2$O$_3$ viewed along a cubic axis. The larger circles are
oxygen. Origin in lower right.

V,a12. The monoclinic form of As$_2$O$_3$ which occurs naturally as the
mineral *claudetite* has a tetramolecular cell of the dimensions:

$$a_0 = 5.25 \text{ A.}; \quad b_0 = 12.87 \text{ A.}; \quad c_0 = 4.54 \text{ A.}; \quad \beta = 93°49'$$

The space group is C_{2h}^5 ($P2_1/n$) with all atoms in general positions:

$$(4e) \quad \pm(xyz; \; x+{}^1\!/_2, {}^1\!/_2-y, z+{}^1\!/_2)$$

Their parameters have been assigned the values listed below:

Atom	x	y	z
As(1)	0.258	0.102	0.040
As(2)	0.363	0.352	0.007
O(1)	0.45	0.22	0.03
O(2)	0.62	0.41	0.18
O(3)	0.95	0.16	0.13

This structure (Fig. VA,12) does not contain the As$_4$O$_6$ molecules present
in the cubic form. Instead, it is built up of sheets of composition As$_2$O$_3$
stacked normal to the b_0 axis. Each arsenic atom has about it three oxygens
at distances between 1.74 and 1.82 A. The closest O–O = 2.4 A. and the
closest As–As = 3.02 A.

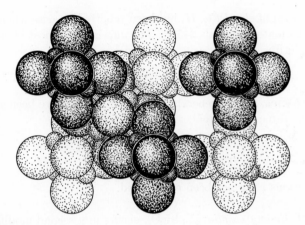

Fig. VA,11b. A packing drawing of the As_2O_3 structure viewed along a cubic axis.   The smaller atoms are arsenic.   The As_4O_6 molecular groupings are apparent.

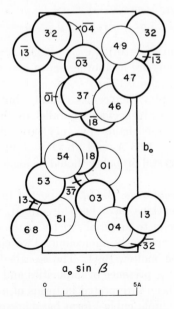

Fig. VA,12. A projection along the c_0 axis of the monoclinic structure of claudetite, As_2O_3. Arsenic atoms are the larger circles.   Origin in lower right.

A second form of *claudetite*, *II*, has been reported to arise when the temperature of formation exceeds 245°C. The unit cell is almost identical with that of claudetite:

$$a_0 = 5.25 \text{ A.}; \quad b_0 = 12.90 \text{ A.}; \quad c_0 = 4.53 \text{ A.}; \quad \beta = 93°53'$$

and arsenic atoms, also in general positions of $P2_1/n$, have been given substantially the same parameters:

$$x(1) = 0.335; \quad y(1) = 0.351; \quad z(1) = 0.00$$
$$x(2) = 0.250; \quad y(2) = 0.101; \quad z(2) = 0.00$$

Oxygen positions have not been established.

V,a13. *Antimony sesquioxide*, Sb_2O_3, occurs in a second modification as the mineral *valentinite*. It is orthorhombic with a four-molecule unit having the dimensions:

$$a_0 = 4.92 \text{ A.}; \quad b_0 = 12.46 \text{ A.}; \quad c_0 = 5.42 \text{ A.}$$

An arrangement has been described with atoms in the following positions of V_h^{10} (*Pccn*):

Sb: $(8e)$ $\pm(xyz; 1/2-x,1/2-y,z; x+1/2,\bar{y},1/2-z; \bar{x},y+1/2,1/2-z)$
O(1): $(4c)$ $\pm(1/4\ 1/4\ u; 1/4,1/4,u+1/2)$

For the antimony atoms, $x = 0.1215$, $y = 0.207$, $z = 0.175$; for O(1), $u = -0.075$. The O(2) atoms are in another set of general positions $(8e)$ with $x = -0.11$, $y = 0.156$, $z = 0.175$.

This does not yield Sb_4O_6 molecules (**V,a11**), but produces a grouping (Fig. VA,13) of endless Sb_2O_3 chains parallel to the c_0 axis. The chosen parameters separate these chains by very large distances, while those within chains (Sb–Sb = 1.66 A. and Sb–O = 1.30 A.) are very short. A further study of this crystal probably should be made.

V,a14. *Bismuth sesquioxide*, Bi_2O_3, is reported to be very pleomorphic. Four different modifications have been described, though it is probable that these do not all refer to material strictly of the composition Bi_2O_3.

The alpha form is apparently monoclinic. Two cubic phases have been obtained by quenching molten Bi_2O_3. The so-called body-centered phase was prepared only in the presence of impurities and may not be Bi_2O_3; its diffraction lines have been interpreted in terms of a 12-molecule cell with $a_0 = 10.08$ A. The "simple cubic" form has lines corresponding to a bimolecular unit with $a_0 = 5.25$ A. It was given an atomic arrangement that

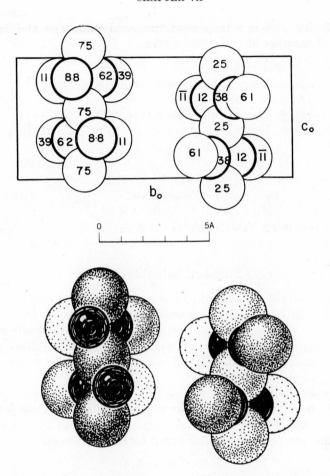

Fig. VA,13a (top). A projection along the a_0 axis of the orthorhombic structure of valentinite, Sb_2O_3. The larger circles are oxygen. Origin in lower right corner.
Fig. VA,13b (bottom). A packing drawing of the valentinite atoms shown in Figure VA,-
13a. The antimony atoms are black.

is a slight distortion of the one originally, and mistakenly, assigned to
Mg_3P_2. It cannot be considered as well established.

The fourth modification of Bi_2O_3, designated as beta-Bi_2O_3, is supposed
to be a tetragonal distortion of the "simple cubic" form. Its eight-molecular
prism has the dimensions:

$$a_0 = 10.93 \text{ A.}, \quad c_0 = 5.62 \text{ A.}$$

The following atomic arrangement, proposed years ago and based on V_d^7 ($C\overline{4}2b$), requires further confirmation:

Bi: (16i) xyz; $\bar{x},y+^1/_2,\bar{z}$; $x+^1/_2,y+^1/_2,z$; $^1/_2-x,y,\bar{z}$;
$\bar{x}\bar{y}z$; $x,^1/_2-y,\bar{z}$; $^1/_2-x,^1/_2-y,z$; $x+^1/_2,\bar{y},\bar{z}$;
$\bar{y}x\bar{z}$; $y,x+^1/_2,z$; $^1/_2-y,x+^1/_2,\bar{z}$; $y+^1/_2,x,z$;
$y\bar{x}\bar{z}$; $\bar{y},^1/_2-x,z$; $y+^1/_2,^1/_2-x,\bar{z}$; $^1/_2-y,\bar{x},z$

with $x = 0.135$, $y = 0.115$, $z = 0.250$

O(1): (8g) $\pm(u\ ^1/_4\ 0;\ ^3/_4\ u\ 0;\ u+^1/_2,^3/_4,0;\ ^1/_4,u+^1/_2,0)$,

with $u = 0.02$

O(2): (8h) $\pm(u\ ^1/_4\ ^1/_2;\ ^3/_4\ u\ ^1/_2;\ u+^1/_2,^3/_4,^1/_2;\ ^1/_4,u+^1/_2,^1/_2)$,

with $u = 0.02$

O(3): (4c) $^1/_4\ ^1/_4\ 0;\ ^1/_4\ ^3/_4\ 0;\ ^3/_4\ ^3/_4\ 0;\ ^3/_4\ ^1/_4\ 0$
O(4): (4b) $0\ 0\ ^1/_2;\ ^1/_2\ 0\ ^1/_2;\ ^1/_2\ ^1/_2\ ^1/_2;\ 0\ ^1/_2\ ^1/_2$

Sulfides, Selenides, etc.

V,a15. *Gallium sesquisulfide*, Ga_2S_3, as commonly prepared, shows a simple x-ray pattern which at low temperatures corresponds to a cubic zinc blende (**III,c1**) and at high temperatures to a hexagonal wurtzite structure (**III,c2**), the metal positions in both cases not being entirely filled.

The low-temperature cubic form has a cell with the edge length:

$$a_0 = 5.441\ \text{A.}$$

It is to be presumed that the sulfur atoms are in a cubic close-packing as they are in ZnS, but that all the corresponding metallic positions are not filled. Other compounds with this deficit cubic structure are

Ga_2Se_3: $a_0 = 5.418$ A.
Ga_2Te_3: $a_0 = 5.899$ A.
α-In_2S_3: $a_0 = 5.36$ A.
α-In_2Te_3: $a_0 = 6.158$ A.

After annealing for several days at 1000°C., a pattern richer in lines is obtained. The unit of this α-form of Ga_2S_3 is hexagonal with six molecules in a cell of the dimensions:

$$a_0 = 6.389\ \text{A.}, \qquad c_0 = 18.086\ \text{A.}$$

The atoms have been placed in the general positions of C_6^3 ($C6_5$):

(6a) xyz; $\bar{y},x-y,z+^2/_3$; $y-x,\bar{x},z+^1/_3$;
$\bar{x},\bar{y},z+^1/_2$; $y,y-x,z+^1/_6$; $x-y,x,z+^5/_6$

with approximate parameters as listed below:

Atom	x	y	z
Ga(1)	0	$^1/_3$	0.125
Ga(2)	$^2/_3$	$^2/_3$	0.125
S(1)	$^1/_3$	0	0
S(2)	0	$^1/_3$	0
S(3)	$^2/_3$	$^2/_3$	0

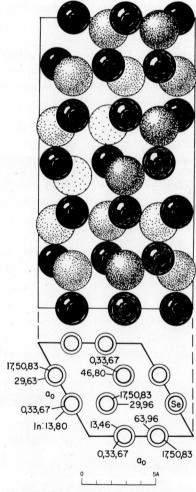

Fig. VA,14. Two projections of the hexagonal In$_2$Se$_3$ arrangement.   In the upper part the indium atoms are black.

In this arrangement the sulfur atoms are hexagonally close-packed. The gallium atoms are in tetrahedral holes of this packing. It thus appears as an ordering of the wurtzite-like form mentioned above.

An electron diffraction study has assigned this structure to the high-temperature form of *indium sesquiselenide*, In_2Se_3. For it,

$$a_0 = 7.11 \text{ A.}, \qquad c_0 = 19.3 \text{ A.}$$

The atomic arrangement is the same as that just described for Ga_2S_3 except that z for the metallic atoms is here 0.130 instead of 0.125. The structure is illustrated by Figure VA,14.

V,a16. *Trinickel disulfide*, Ni_3S_2, has been given a simple structure which is not closely related to that of other known crystals. At first it was thought to be strictly cubic, but more recent data indicate it to be rhombohedral with a unimolecular cell of the dimensions:

$$a_0 = 4.041 \text{ A.}, \qquad \alpha = 90°18'$$

Atoms have been assigned the following special positions of D_3^7 ($R32$):

$$\begin{array}{lll} \text{S:} & (2c) & \pm(uuu) & \text{with } u = {}^1/_4 \\ \text{Ni:} & (3d) & 0v\bar{v};\ \bar{v}0v;\ v\bar{v}0 & \text{with } v = {}^1/_4 \end{array}$$

In terms of hexagonal axes the trimolecular cell has the edges:

$$a_0' = 5.715 \text{ A.}, \qquad c_0' = 6.962 \text{ A.}$$

Atoms are in the positions:

$$\begin{array}{lll} \text{S:} & (6c) & \pm(00u);\ \text{rh} & \text{with } u = {}^1/_4 \\ \text{Ni:} & (9d) & v00;\ 0v0;\ \bar{v}\bar{v}0;\ \text{rh} & \text{with } v = {}^1/_4 \end{array}$$

In this grouping each sulfur atom has about it six nickel atoms at a distance of 2.28 A.

The corresponding *selenide*, Ni_3Se_2, also has this arrangement. For it the unit rhombohedron has:

$$a_0 = 4.2375 \text{ A.}, \qquad \alpha = 90°42'$$

and the trimolecular hexagonal cell:

$$a_0' = 6.029 \text{ A.}, \qquad c_0' = 7.249 \text{ A.}$$

For both, $u(Ni) = 0.26$ and $v(Se) = 0.25$.

In this crystal (Fig. VA,15), each selenium atom has about it six nickel atoms at the corners of a distorted trigonal prism; the Ni–Se distances are 2.36 and 2.38 A. Each nickel atom is surrounded by a distorted tetrahedron of selenium atoms. Nickel atoms approach one another at a distance of 2.57 A.; the closest Se–Se $= 3.48$ A.

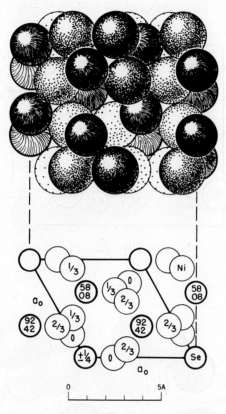

Fig. VA,15. Two projections of the hexagonal Ni_3Se_2 structure. In the shaded part the nickel atoms are dotted.

It is probable that the mineral *heazlewoodite*, Ni_3As_2, also has this structure. For it,

$$a_0 = 4.080 \text{ A.}, \qquad \alpha = 89°25'$$

The trimolecular hexagonal cell has the edge lengths:

$$a_0' = 5.740 \text{ A.}, \qquad c_0' = 7.138 \text{ A.}$$

V,a17. A structure has been described for *orpiment*, As_2S_3. Its tetra-molecular monoclinic unit has the dimensions:

$$a_0 = 11.46 \text{ A.}; \quad b_0 = 9.57 \text{ A.}; \quad c_0 = 4.22 \text{ A.}; \quad \beta = 90°30'$$

All atoms have been placed in general positions of C_{2h}^5 $(P2_1/n)$:

$$(4e) \quad \pm(xyz; \ 1/2-x, y+1/2, 1/2-z)$$

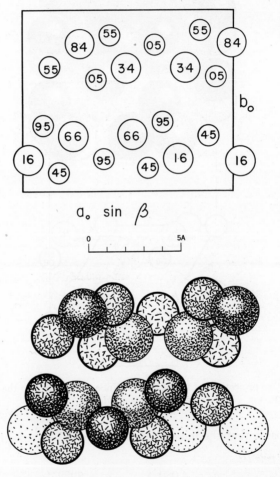

Fig. VA,16a (top). A projection along the c_0 axis of the monoclinic structure of orpiment, As_2S_3. Arsenic atoms are the larger circles. Origin in the lower right.

Fig. VA,16b (bottom). A packing drawing of the monoclinic As_2S_3 structure viewed along the c_0 axis. The sulfur atoms are line shaded.

with, according to a recent redetermination, the following parameters:

Atom	x	y	z
As(1)	0.267	0.190	0.143
As(2)	0.484	0.323	0.643
S(1)	0.395	0.120	0.500
S(2)	0.355	0.397	0.013
S(3)	0.125	0.293	0.410

This is a structure (Fig. VA,16) composed of As_2S_3 layers stacked along the b_0 axis. Each arsenic atom has three sulfur neighbors at distances between 2.21 and 2.28 A., while each sulfur atom is closely bound to two arsenic atoms. The layers are far apart in the b_0 direction.

V,a18. The mineral *stibnite, antimony trisulfide*, Sb_2S_3, has four molecules in an orthorhombic unit of the dimensions:

$$a_0 = 11.299 \text{ A.}; \quad b_0 = 11.310 \text{ A.}; \quad c_0 = 3.8389 \text{ A. } (25°\text{C.})$$

All atoms have been found to be in the following special positions of V_h^{16} (*Pbnm*):

$$(4c) \quad \pm (u,v \; {}^1/_4; \; {}^1/_2 - u,v + {}^1/_2, {}^1/_4)$$

The parameters as recently redetermined are:

Atom	u	v
Sb(1)	0.326	0.030
Sb(2)	0.536	0.351
S(1)	0.880	0.055
S(2)	0.559	0.869
S(3)	0.189	0.214

This structure (Fig. VA,17) consists of sheets of Sb_2S_3 indefinitely prolonged in the plane parallel to the c_0 axis and roughly diagonal to the a_0 and b_0 axes. In a sheet, each antimony and each sulfur atom is surrounded by three atoms of the opposite kind at distances that are approximately the sum of the neutral radii.

Antimony triselenide, Sb_2Se_3, has this structure, with

$$a_0 = 11.62 \text{ A.}; \quad b_0 = 11.77 \text{ A.}; \quad c_0 = 3.962 \text{ A.}$$

Parameters found for its atoms are

Atom	u	v
Sb(1)	0.3280	0.0305
Sb(2)	0.5397	0.3522
Se(1)	0.8732	0.0534
Se(2)	0.5566	0.8698
Se(3)	0.1935	0.2132

Fig. VA,17a (left). A projection along its c_0 axis of the orthorhombic structure of stibnite, Sb_2S_3. The small circles are the sulfur atoms.

Fig. VA,17b (right). A packing drawing of the atoms of stibnite shown in Figure VA,17a when given their neutral radii. The larger spheres are the antimony atoms.

The same arrangement has also been found for *thorium trisulfide*, Th_2S_3. Its unit has the edge lengths:

$$a_0 = 10.85 \text{ A.}; \quad b_0 = 10.99 \text{ A.}; \quad c_0 = 3.96 \text{A.}$$

The parameters are:

Atom	u	v
Th(1)	0.314	−0.022
Th(2)	0.519	0.300
S(1)	0.818	0.053
S(2)	0.561	0.871
S(3)	0.206	0.230

For *uranium trisulfide*, U_2S_3, with the same atomic arrangement, the unit has the edges:

$$a_0 = 10.41 \text{ A.}; \quad b_0 = 10.65 \text{ A.}; \quad c_0 = 3.89 \text{ A.}$$

The sulfur parameters are said to be the same as for the thorium compound and the parameters for uranium to have the values:

$$u(U,1) = 0.311; v(U,1) = -0.014; u(U,2) = 0.508; v(U,2) = 0.305$$

Cell dimensions, but not parameters, have also been established for the following isostructural substances:

$$Th_2Se_3: \quad a_0 = 11.34 \text{ A.}; \quad b_0 = 11.57 \text{ A.}; \quad c_0 = 4.27 \text{ A.}$$
$$U_2Se_3: \quad a_0 = 10.9 \text{ A.}; \quad b_0 = 11.2 \text{ A.}; \quad c_0 = 4.04 \text{ A.}$$
$$Np_2S_3: \quad a_0 = 10.3 \text{ A.}; \quad b_0 = 10.6 \text{ A.}; \quad c_0 = 3.86 \text{ A.}$$
$$Bi_2(S,Se)_3: \quad a_0 = 11.32 \text{ A.}; \quad b_0 = 11.48 \text{ A.}; \quad c_0 = 4.17 \text{ A.}$$
(guanajuatite)
$$Bi_2S_3: \quad a_0 = 11.150 \text{ A.}; b_0 = 11.300 \text{ A.}; c_0 = 3.981 \text{ A. } (26°C.)$$

According to the original determination for stibnite, the parameters for Bi_2S_3 are essentially the same as for the antimony compound.

V,a19. The mineral *tetradymite*, Bi_2Te_2S, is rhombohedral. Its unit contains one molecule and has the dimensions:

$$a_0 = 10.31 \text{ A.}, \qquad \alpha = 24°10'$$

The sulfur atom is in the origin $(1a)$ 000 of the space group D_{3d}^5 $(R\bar{3}m)$ The other atoms are in $(2c)$ $\pm(uuu)$, with $u(Bi) = 0.392$ and $u(Te) = 0.788$.

The corresponding hexagonal cell containing three molecules has the edge lengths:

$$a_0' = 4.316 \text{ A.}, \quad c_0' = 30.01 \text{ A.}$$

The sulfur atom is in $(3a)$ 000; rh, and the other atoms in $(6c)$ $\pm(00u)$; rh, with the same parameters that apply to the rhombohedral unit. This structure (Fig. VA,18) can be imagined as composed of layers of atoms along the c_0' axis following one another in the succession characteristic of a cubic close-packing and in the order

$$S–Bi–Te–Te–Bi–S–Bi–Te–Te–...$$

The shortest interatomic distances are Bi–S $= 3.05$ A., Bi–Te $= 3.12$ A., and Te–Te $= 3.69$ A.

The mineral *tellurobismuthite*, Bi_2Te_3 has this structure, with a rhombohedral cell of the dimensions:

$$a_0 = 10.473 \text{ A.,} \qquad \alpha = 24°10'$$

The corresponding hexagonal cell has the edges:

$$a_0' = 4.3835 \text{ A.,} \qquad c_0' = 30.487 \text{ A.}$$

One tellurium atom is in the origin; the other atoms are in $(2c)$, or $(6c)$, with $u(Bi) = 0.399$ and $u(Te) = 0.792$ according to the original study, and with $u(Bi) = 0.400$ and $u(Te) = 0.788$ according to a more recent analysis.

Complete structures have been determined for three other compounds with this arrangement. They are

Bi_2Se_3:	$a_0 =$	9.841 A.,	$\alpha =$	24°16'
	$a_0' =$	4.138 A.,	$c_0' =$	28.64 A.
	$u(Bi) =$	0.399,	$u(Se) =$	0.794
Sb_2Te_3:	$a_0 =$	10.426 A.,	$\alpha =$	23°31'
	$a_0' =$	4.25 A.,	$c_0' =$	30.4 A.
	$u(Sb) =$	0.400,	$u(Te) =$	0.789
Bi_2Te_2Se:	$a_0 =$	10.255 A.,	$\alpha =$	24°5'
	$a_0' =$	4.28 A.,	$c_0' =$	29.86 A.
	$u(Bi) =$	0.3961,	$u(Te) =$	0.7883

A structure has been described, based on electron diffraction, for a related selenide for which the formula Bi_3Se_4 has been given. Its rhombohedral unit has the dimensions:

$$a_0 = 13.719 \text{ A.,} \qquad \alpha = 17°44'$$

The hexagonal cell has the edges:

$$a_0' = 4.23 \text{ A.,} \qquad c_0' = 40.5 \text{ A.}$$

All atoms have been placed in the positions of D_{3d}^5 stated above, with $u(Bi) = 0.428$, $u(Se,1) = 0.139$, and $u(Se,2) = 0.286$, the third bismuth atom being in the origin. The resulting arrangement is of the same type as that of tetradymite, differing in having a somewhat more complex stacking of layers along the c_0' axis.

Probably the mineral paraguanajuatite, $Bi_2(Se,S)_3$, also has this basic structure with a still different and more complex stacking of layers. For its rhombohedral unit:

$$a_0 = 18.38 \text{ A.,} \qquad \alpha = 12°44'$$

For the hexagonal cell:

$$a_0' = 4.076 \text{ A.}, \qquad c_0' = 54.7 \text{ A.}$$

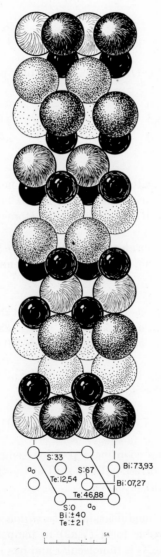

Fig. VA,18. Two projections of the atoms in the hexagonal cell of Bi_2Te_2S.   In the upper projection the bismuth atoms are black and the tellurium atoms dotted.

Borides, Phosphides, etc.

V,a20. A double boride (Incoloy 901) of the composition R_3B_2, where R consists of $Mo_{1.01}$, $Ti_{0.59}$, $Cr_{0.59}$, $Fe_{0.39}$, $Ni_{0.21}$, and $Al_{0.22}$, has been found to be tetragonal with a cell containing two R_3B_2 molecules and having the edges:

$$a_0 = 5.783 \text{ A.}, \qquad c_0 = 3.134 \text{ A.}$$

The space group is D_{4h}^5 ($P4/mbm$), with atoms placed in the positions:

R(1) = (Ni,Fe,Cr): (2b) $0\ 0\ ^1/_2$; $^1/_2\ ^1/_2\ ^1/_2$
R(2) = (Mo,Ti,Al): (4g) $\pm(u,u+^1/_2,0;\ u+^1/_2,\bar{u},0)$ with $u = 0.183$
    B: (4h) $\pm(u,u+^1/_2,^1/_2;\ u+^1/_2,\bar{u},^1/_2)$

with $u = 0.394$

As Figure VA,19 indicates, the boron atoms lie in pairs with B–B = 1.73 A. Each R(1) has four and each R(2) has six boron atoms at distances of 2.36 and 2.33 A. Each R(1) has eight R(2) atoms around it at a distance of 2.63 A.

Three other metallic borides undoubtedly have this structure, though atomic parameters have not been established for any of them. Their units have the dimensions:

$$Nb_3B_2:\quad a_0 = 6.185 \text{ A.}, \quad c_0 = 3.280 \text{ A.}$$
$$Ta_3B_2:\quad a_0 = 6.184 \text{ A.}, \quad c_0 = 3.286 \text{ A.}$$
$$V_3B_2:\quad a_0 = 5.746 \text{ A.}, \quad c_0 = 3.032 \text{ A.}$$

A more thorough study has been made of the alloy Si_2U_3, which likewise has this arrangement. For it,

$$a_0 = 7.3299 \text{ A.}, \qquad c_0 = 3.9004 \text{ A.}$$

Making use of the same coordinate positions employed for Incoloy 901 (instead of values that correspond to a shift of $^1/_2\,c_0$), the uranium parameter is $u = 0.181$ and $u(Si) = 0.389$.

V,a21. A hexagonal modification of *beryllium nitride*, Be_3N_2, is prepared by heating the cubic form (**V,a2**) to temperatures above 1400°C. The unit of this beta-Be_3N_2 is bimolecular and has the edge lengths:

$$a_0 = 2.8413 \text{ A.}, \qquad c_0 = 9.693 \text{ A.}$$

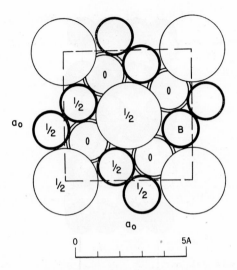

Fig. VA,19. A projection along its c_0 axis of the tetragonal structure of Incoloy 901. The (Ni,Fe,Cr) atoms are represented by the large circles.

Atoms are in the following positions of D_{6h}^4 ($P6_3/mmc$):

N(1):   (2a) $000; 0\ 0\ ^1/_2$
N(2):   (2c) $\pm(^1/_3\ ^2/_3\ ^1/_4)$
Be(1):   (2b) $\pm(0\ 0\ ^1/_4)$
Be(2):   (4f) $\pm(^1/_3\ ^2/_3\ u;\ ^2/_3, ^1/_3, u+^1/_2)$ with $u = 0.075$

This is a layered structure (Fig. VA,20) in which the Be(1) atoms have a triangle of surrounding nitrogen atoms (Be–N = 1.64 A.) and the Be(2) atoms a tetrahedron of nitrogen atoms with Be–N = 1.70 or 1.79 A. The shortest Be–Be = 2.19 A. and the shortest N–N = 2.84 A. The structure can be described as a closest packing of the beryllium atoms with nitrogen in interstices. In this case the packing is mixed cubic and hexagonal in its sequence; using the terminology described in **III,c4,** this is

$$C\ H\ C\ C\ H\ C\ C\ H\ C\ C$$

V,a22. *Zinc phosphide,* Zn_3P_2, is tetragonal with a unit containing eight molecules and having the cell dimensions:

$$a_0 = 8.097\ \text{A.,}\qquad c_0 = 11.45\ \text{A.}$$

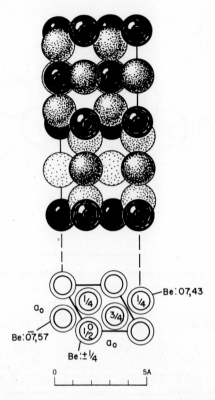

Fig. VA,20. Two projections of the hexagonal β-Be$_3$N$_2$ arrangement. In the upper projection the nitrogen atoms are black.

The space group is D_{4h}^{15} ($P4/nmc$), with atoms in the special positions:

P(1):   (4c)   00u; 00$\bar{u}$; B.C.   with $u = 0.25$

P(2):   (4d)   0 $\frac{1}{2}$ u; $\frac{1}{2}$ 0 $\bar{u}$; B.C.   with $u = 0.239$

P(3):   (8f) $\pm$ (uu0; $u\bar{u}$0); B.C.   with $u = 0.261$

Zn(1):   (8g)   0uv; u0$\bar{v}$; $\frac{1}{2}$,$u+\frac{1}{2}$,$\frac{1}{2}-v$; $u+\frac{1}{2}$,$\frac{1}{2}$,$v+\frac{1}{2}$;

                   0$\bar{u}v$; $\bar{u}$0$\bar{v}$; $\frac{1}{2}$,$\frac{1}{2}-u$,$\frac{1}{2}-v$; $\frac{1}{2}-u$,$\frac{1}{2}$,$v+\frac{1}{2}$

                                      with $u = 0.283$, $v = 0.386$

Zn(2):   (8g)   with $u = 0.217$, $v = 0.103$

Zn(3):   (8g)   with $u = 0.250$, $v = 0.647$

In this structure (Fig. VA,21), each zinc atom is surrounded by a deformed tetrahedron of phosphorus atoms with Zn–P lying between 2.35 and 2.77 A. Each atom of phosphorus has six zinc neighbors which are neither octahedrally nor prismatically distributed.

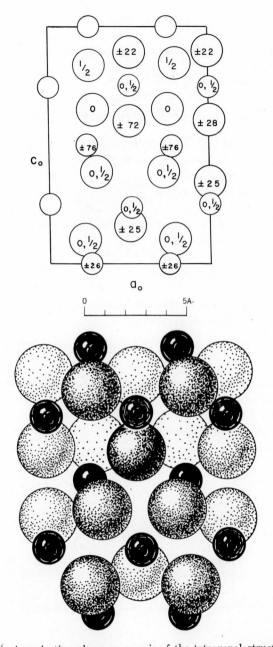

Fig. VA,21 (top). A projection along an a_0 axis of the tetragonal structure of Zn_3P_2. The small circles are phosphorus. Origin in lower left.

Fig. VA,21 (bottom). A packing drawing of atoms in the Zn_3P_2 structure viewed along an a_0 axis. The phosphorus atoms are black.

The following phosphides and arsenides also have this structure:

$$Cd_3As_2: \quad a_0 = 8.945 \text{ A.}, \quad c_0 = 12.65 \text{ A.}$$
$$Cd_3P_2: \quad a_0 = 8.746 \text{ A.}, \quad c_0 = 12.28 \text{ A.}$$
$$Zn_3As_2: \quad a_0 = 8.316 \text{ A.}, \quad c_0 = 11.76 \text{ A.}$$

It should be noted that a larger cell containing 16 molecules and having its a_0' diagonal to the foregoing would be almost exactly cubic in dimensions for each of these compounds; in other words, for the units given above, c_0 is almost exactly $\sqrt{2}\ a_0$.

V,a23.   One of the *carbides* of *chromium* has the composition Cr_3C_2. Its orthorhombic unit has the dimensions:

$$a_0 = 11.46 \text{ A.}; \quad b_0 = 5.52 \text{ A.}; \quad c_0 = 2.82 \text{ A.}$$

There are four molecules in this cell and the space group is V_h^{16} (*Pbnm*). All atoms are in the special positions:

$$(4c) \quad \pm (u\ v\ {}^1\!/_4;\ {}^1\!/_2 - u, v + {}^1\!/_2, {}^1\!/_4).$$

The positions of the chromium atoms were determined many years ago and parameters were suggested for the atoms of carbon. A recent neutron diffraction study has confirmed the previous parameters for chromium and determined experimentally those for carbon. These are close to the positions earlier proposed for C(2), but for C(1) the u value is somewhat different.

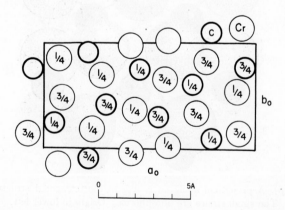

Fig. VA,22a. A projection along its c_0 axis of the orthorhombic structure of Cr_3C_2. Origin in lower right.

They are:

Atom	u	v
Cr(1)	0.406	0.03
Cr(2)	−0.230	0.175
Cr(3)	−0.070	−0.150
C(1)	0.204	0.092
C(2)	−0.048	0.228

The resulting structure is shown in Figure VA,22.

V,a24. *Plutonium* forms the *sesquicarbide*, Pu_2C_3. It is cubic, having a unit of the edge length

$$a_0 = 8.129 \text{ A.}$$

and containing eight molecules. The space group is $T_d{}^6$ ($I\bar{4}3d$) with plutonium atoms in special positions:

$(16c)$ $uuu;$ $u+{}^1/_2, {}^1/_2-u, \bar{u};$
$\bar{u}, u+{}^1/_2, {}^1/_2-u;$ ${}^1/_2-u, \bar{u}, u+{}^1/_2;$
$u+{}^1/_4, u+{}^1/_4, u+{}^1/_4;$ $u+{}^3/_4, {}^1/_4-u, {}^3/_4-u;$
${}^3/_4-u, u+{}^3/_4, {}^1/_4-u;$ ${}^1/_4-u, {}^3/_4-u, u+{}^3/_4;$ B.C.

with $u = 0.050$. Carbon positions could not be determined experimentally, but it was concluded that these atoms should be in $(24d)$ [see below].

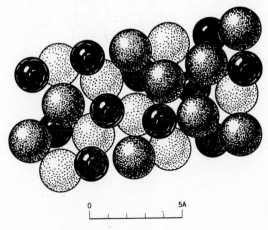

Fig. VA,22b. A packing drawing of the Cr_3C_2 arrangement viewed along its c_0 axis. The carbon atoms are black.

Uranium sesquicarbide, U_2C_3, has this structure with the cube edge:

$$a_0 = 8.0885 \text{ A.}$$

A neutron diffraction study has put the uranium atoms in (16c), above, with $u = 0.050$ and the carbon atoms in

(24d) $u\ 0\ {}^1/_4;\ \bar{u}\ {}^1/_2\ {}^1/_4;\ u+{}^1/_4,{}^1/_2,{}^1/_4;\ {}^3/_4-u,0,{}^1/_4;$ tr;   B.C.
with $u = 0.295$

This leads to a C–C = 1.295 A.   A recalculation of these data led to the slightly altered parameters: $u(U) = 0.0519$, $u(C) = 0.292$. The structure is illustrated in Figure VA,23.

Rare-earth sesquicarbides to which this structure has been assigned are listed in Table VA,5. As the data indicate, there is a considerable spread in the cell dimensions for most of these compounds, presumably due to variations in their contents of carbon.

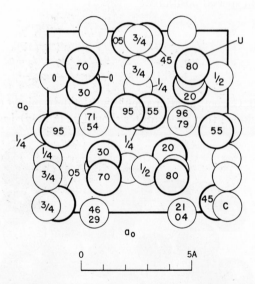

Fig. VA,23a. The cubic U_2C_3 structure projected along a cube axis.

TABLE VA,5
Carbides with the Pu$_2$C$_3$ Structure

Compound	a_0, A.	$u(M)$	$u(C)$	C–C, A.
Ce$_2$C$_3$	8.4476	0.0504	0.2995	1.276
Dy$_2$C$_3$	8.198	—	—	—
Gd$_2$C$_3$	8.3221–8.3407	—	—	—
Ho$_2$C$_3$	8.176	—	—	—
La$_2$C$_3$	8.8034–8.8185	0.0504	0.3049	1.236
Nd$_2$C$_3$	8.5207–8.5478	—	—	—
Pr$_2$C$_3$	8.5731–8.6072	0.0515	0.3029	1.239
Sm$_2$C$_3$	8.3989–8.4257	—	—	—
Tb$_2$C$_3$	8.2434–8.2617	0.0516	0.2999	1.240

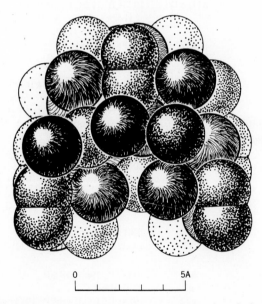

0   5A

Fig. VA,23b. A packing drawing of the U$_2$C$_3$ arrangement viewed along a cube axis.
The carbon atoms are dotted.

BIBLIOGRAPHY TABLE, CHAPTER VA

Compound	Paragraph	Literature
Ac_2O_3	a1	1949: Z
Ac_2S_3	d9	1949: Z
Ag_2O_3	a9	1934: B; 1958: S,W&V
Al_2O_3	a3, [VIII]	1922: B; H; 1923: D; 1924: M; 1925: B&B; G,B&L; P&H; U; 1927: H; 1928: Z; 1930: P; 1932: B; 1935: K; V; 1936: C&DA; 1938: G&V; 1939: A; 1945: T,T&E; 1948: K, 1950: S&A; T; 1953: C; NBS; 1956: S&Y; 1957: S&S; Z&U; 1959: B&N; 1961: S; 1962: N&dH
Al_2S_3	a1, a3	1951: F; 1952: F
Am_2O_3	a1, a2	1953: T&D; 1955: A,E,F&Z
Am_2S_3	d9	1948: Z; 1949: Z
As_2O_3	a11, a12	1923: B; 1928: P; 1932: L; 1937: M,H&D; 1938: H&S; 1942: A&W; 1949: K,P&S; 1951: B,P&S; F; 1952: B,P&S; 1954: B; B,P&S
As_2S_3	a17	1949: M; 1954: M
B_2O_3	a10	1935: C&T; 1952: B; 1953: B
Be_3N_2	a2, a21	1933: vS&P; 1960: E&R
Be_3P_2	a2	1933: vS&P
Bi_2O_3	a14	1937: S; 1940: S; 1941: S; 1945: A&S; 1950: Z
Bi_2S_3 (bismuthinite)	a18	1933: H; 1938: G&F; 1953: NBS
$Bi_2(S,Se)_3$	a18, a19	1948: R
Bi_2Se_3	a19	1951: D; 1953: S&F; 1954: S; 1960: W&M
Bi_2Te_3	a19	1939: F; L; 1951: D; 1954: S; 1958: F; 1960: W&M
Bi_2Te_2S (tetradymite)	a19	1934: H; 1938: G&F; 1939: L
Bi_2Te_2Se	a19	1961: B&B
Ca_3N_2	a2	1933: F,B&H; vS&P; 1934: H&F
Cd_3As_2	a22	1928: P; 1935: vS&P
Cd_3N_2	a2	1940: J&H
Cd_3P_2	a22	1928: P; 1933: vS&P; 1935: vS&P
Ce_2C_3	a24	1958: S,G&D; 1961: A&W
Ce_2O_3	a1	1925: G,B&L; 1926: Z; 1929: P; 1955: B
Ce_2O_2S	a1	1949: Z; 1956: P&P
Ce_2S_3	d9	1949: Z; 1956: P&F
Cm_2O_3	a2	1955: A,E,F&Z
Co_2As_3	a3	1957: V
Co_2S_3	[VIII]	1955: B&R

(continued)

BIBLIOGRAPHY TABLE, CHAPTER VA (*continued*)

Compound	Paragraph	Literature
Cr_3C_2	**a23**	1926: W&P; 1931: W; 1960: M&K
Cr_2O_3	**a3**, [VIII]	1923: D; 1925: G, B&L; 1928: Z; 1930: P; W; 1954: NBS; 1957: S&S; 1958: K&V; 1962: N&dH
$(Cr,Mo)_2O_3$	**a5**	1954: S
$(Cr,W)_2O_3$	**a5**	1954: S
Cr_2S_3	**f6**	1955: J; 1957: J
Cs_4O_6	**d9**	1939: H&K
Dy_2C_3	**a24**	1958: S,G&D
Dy_2O_3	**a2**	1925: G,B&L; 1927: Z; 1928: Z; 1930: P&S; 1939: B; 1954: B&G; T&D
Dy_2O_2S	**a1**	1958: E
Er_2O_3	**a2**	1925: G,B&L; 1927: Z; 1930: P&S; 1939: B; 1954: T&D; 1962: F
Er_2O_2S	**a1**	1958: E
Er_2O_2Se	**a1**	1960: E
Eu_2O_3	**a2**	1925: G,B&L; 1927: Z; 1930: P&S; 1939: B; 1947: I; 1954: T&D
Eu_2O_2S	**a1**	1958: E; 1959: D,F&G
Fe_2O_3	**a2, a3**, [VIII]	1923: D; 1924: M; 1925: B&B; G,B&L; P&H; W&B; 1928: Z; 1930: P; W; 1931: F&G; T; 1932: B; 1933: K&O; 1934: B; 1935: H; K; V&B; 1936: C&DA; V&dB; 1939: G&G; H&S; 1940: M; 1957: S&S; 1958: BS; F&H; vO&R; 1959: S,P&I
$(Fe,Mn)_2O_3$ (bixbyite)	**a2**	1928: Z; 1930: P&S; 1956: D
Ga_2O_3	**a3, a7**	1925: G,B&L; 1928: Z; 1953: NBS; 1960: G
Ga_2S_3	**a15**	1949: H&K; 1952: H; 1955: H&F; 1961: G,D&S
Ga_2Se_3	**a15**	1949: H&K; 1955: G,K&FK
Ga_2Te_3	**a15**	1949: H&K; 1955: G,K&FK
Gd_2C_3	**a24**	1958: S,G&D
Gd_2O_3	**a2, a6**	1925: G,B&L; 1927: Z; 1930: P&S; 1939: B; 1947: I; 1954: B&G; T&D; 1958: G&M
Gd_2O_2S	**a1**	1958: E
Gd_2O_2Se	**a1**	1960: E
Ho_2C_3	**a24**	1958: S,G&D
Ho_2O_3	**a2**	1925: G,B&L; 1927: Z; 1930: P&S; 1954: T&D; 1962: F
Ho_2O_2S	**a1**	1958: E

(*continued*)

BIBLIOGRAPHY TABLE, CHAPTER VA (*continued*)

Compound	Paragraph	Literature
Ho_2O_2Se	a1	1960: E
In_2O_3	a2	1925: G,B&L; 1927: Z; 1928: Z; 1930: P&S; 1954: NBS
In_2S_3	a15, [VIII]	1949: H&K; 1959: R
In_2Se_3	a15	1957: H&F; 1960: S; 1961: S
In_2Te_3	a15	1949: H&K; 1954: I&S
La_2C_3	a24	1958: A,G,D,R&S; S,G&D; 1961: A&W
La_2O_3	a1, a2	1925: G,B&L; 1926: Z; 1929: P; 1947: I; 1952: NBS; 1953: K&W
La_2O_2S	a1	1949: Z; 1956: P&P; 1958: E
La_2O_2Se	a1	1960: E
La_2S_3	d9	1949: Z; 1956: P&F
Lu_2O_3	a2	1925: G,B&L; 1927: Z; 1928: Z; 1930: P&S; 1939: B; 1954: T&D
Lu_2O_2S	a1	1958: E
Mg_3As_2	a2	1928: N&P; 1933: vS&P; Z&H
Mg_3Bi_2	a1	1933: Z&H
Mg_3N_2	a2	1930: H; 1932: H; 1933: vS&P
Mg_3P_2	a2	1928: P; 1933: vS&P; Z&H
Mg_3Sb_2	a1	1933: Z&H; 1935: L
Mn_2O_3	a2, [VIII]	1928: Z; 1930: P&S; W; 1936: V&dB; 1940: M; 1949: M&K; 1957: S&S; 1962: F
$(Mo,Ti,Cr,Fe\ldots)_3B_2$ (Incoloy 901)	a20	1958: B
Nb_3B_2	a20	1958: N&W
$(Nb_2Co_4)O_9$	a3	1960: B,C&F; 1961: B,C,F,A&P
$(Nb_2Fe_4)O_9$	a3	1960: B,C&F; 1961: B,C,F,A&P
$(Nb_2Mg_4)O_9$	a3	1960: B,C&F; 1961: B,C,F,A&P
$(Nb_2Mn_4)O_9$	a3	1960: B,C&F; 1961: B,C,F,A&P
Nd_2C_3	a24	1958: S,G&D
Nd_2O_3	a1, a2	1925: G,B&L; 1926: Z; 1929: P; 1935: L; 1939: B; 1947: I; 1953: NBS; 1954: B&G
Nd_2O_2S	a1	1956: P&P; 1958: E
Nd_2O_2Se	a1	1960: E
Nd_2S_3	d9	1956: P&F
Ni_3As_2 (heazlewoodite)	a16	1946: P
Ni_3S_2	a16	1925: A; 1938: W

(*continued*)

BIBLIOGRAPHY TABLE, CHAPTER VA (*continued*)

Compound	Paragraph	Literature
Ni_3Se_2	a16	1957: A&S; 1960: H&W
Np_2S_3	a18	1949: Z
Pb_2O_3	a8	1937: C,S&Q; 1941: G; 1944: B; 1945: B
Pr_2C_3	a24	1958: S,G&D; 1961: A&W
Pr_2O_3	a1, a2	1925: G,B&L; 1926: Z; 1929: P; 1947: I; 1962: E&B
Pr_2O_2S	a1	1956: P&P; 1958: E
Pr_2O_2Se	a1	1960: E
Pr_2S_3	d9	1956: P&F
Pu_2C_3	a24	1952: Z
Pu_2O_3	a1, a2	1952: T&D; 1955: A,E,F&Z
Pu_2O_2S	a1	1949: Z
Pu_2S_3	d9	1948: Z
Rb_4O_6	d9	1939: H&K
Rh_2O_3	a3	1927: L; 1928: Z
Sb_2O_3	a11, a13	1923: B; 1927: D; D&G; S; 1929: D; 1936: B; 1942: A&W; 1952: NBS
Sb_2S_3 (stibnite)	a18	1926: O; 1927: G; 1929: G&L; 1933: H; 1954: NBS; 1960: S
Sb_2Se_3	a18	1950: D; 1957: T,K&MC
Sb_2Te_3	a19	1951: D; 1956: S
Sc_2O_3	a2	1925: G,B&L; 1927: Z; 1928: Z; 1930: P&S; 1952: NBS; 1961: M,K,S&S
Sm_2C_3	a24	1958: S,G&D
Sm_2O_3	a2, a6	1925: G,B&L; 1927: Z; 1930: P&S; 1939: B; 1947: I; 1954: B&G; T&D; 1957: C
Sm_2O_2S	a1	1956: P&P; 1958: E
Sm_2O_2Se	a1	1960: E
Sm_2S_3	d9	1956: P&F
Ta_3B_2	a20	1958: N&W
$(Ta_2Co_4)O_9$	a3	1960: B,C&F; 1961: B,C,F,A&P
$(Ta_2Mg_4)O_9$	a3	1960: B,C&F; 1961: B,C,F,A&P
$(Ta_2Mn_4)O_9$	a3	1960: B,C&F; 1961: B,C,F,A&P
Tb_2C_3	a24	1958: S,G&D; 1961: A&W
Tb_2O_3	a2	1925: G,B&L; 1927: Z; 1930: P&S; 1939: B; 1961: B,E,S&E
Tb_2O_2S	a1	1958: E
Th_2N_3	a1	1949: Z

(*continued*)

BIBLIOGRAPHY TABLE, CHAPTER VA (*continued*)

Compound	Paragraph	Literature
Th_2S_3	a18	1949: Z
Th_2Se_3	a18	1952: DE,S&M
Ti_2O_3	a3	1925: G,B&L; 1927: L; 1928: Z; 1929: H; 1962: N&dH; S&E
Tl_2O_3	a2	1925: G,B&L; 1927: Z; 1928: Z; 1930: P&S; 1953: NBS; S&T; 1956: S,Z&C
Tm_2O_3	a2	1925: G,B&L; 1927: Z; 1930: P&S; 1939: B; 1954: T&D
Tm_2O_2S	a1	1958: E
U_2C_3	a24	1951: M,G&V; 1959: A
U_2N_3	a2	1948: R,B,W&MD
U_2S_3	a18	1949: Z
U_2Se_3	a18	1957: K
U_3Si_2	a20	1948: Z
V_3B_2	a20	1958: N&W
V_2O_3	a3	1925: G,B&L; 1928: Z; 1953: A; 1959: M,A,K,A,W,H&N; 1960: P&P; 1962: N&dH
Y_2O_3	a2	1925: G,B&L; 1927: Z; 1928: Z; 1930: P&S; 1952: NBS; 1954: B&G; 1962: F
Y_2O_2S	a1	1956: P&P
Yb_2O_3	a2	1925: G,B&L; 1927: Z; 1930: P&S; 1939: B; 1954: B&G; T&D; 1961: PJ,Q&C; 1962: F
Yb_2O_2S	a1	1958: E
Yb_2O_2Se	a1	1960: E
Zn_3As_2	a22	1928: N&P; 1933: vS&P; 1935: vS&P
Zn_3N_2	a2	1940: J&H
Zn_3P_2	a22	1928: P; 1933: vS&P; 1935: vS&P

B. COMPOUNDS OF THE TYPE RX₃

Halides

V,b1. A number of metallic trihalides crystallize with rhombohedral symmetry in cells that do not differ greatly in shape. There has been much uncertainty in the past as to how closely their structures resemble one another and more work still remains to be done. Nevertheless, the structures that should be assigned to many of them have now become clear.

The rhombohedral units found for BiI_3, SbI_3, AsI_3, $FeCl_3$, and $CrBr_3$ are bimolecular and have the dimensions:

$$AsI_3: \quad a_0 = 8.25 \text{ A.}, \qquad \alpha = 51°20'$$
$$BiI_3: \quad a_0 = 8.13 \text{ A.}, \qquad \alpha = 54°50'$$
$$CrBr_3: \quad a_0 = 7.05 \text{ A.}, \qquad \alpha = 52°36'$$
$$FeCl_3: \quad a_0 = 6.758 \text{ A.}, \qquad \alpha = 53°11'$$
$$SbI_3: \quad a_0 = 8.18 \text{ A.}, \qquad \alpha = 54°14'$$

The space group is considered to be $C_{3i}{}^2$ ($R\overline{3}$) with atoms in the following positions:

$$R: \quad (2c) \quad uuu; \ \bar{u}\bar{u}\bar{u}$$
$$Cl, Br, \text{ or } I: \quad (6f) \quad \pm(xyz; \ zxy; \ yzx)$$

For BiI_3, u has been determined as $\frac{1}{3}$ and $x = -0.245$, $y = 0.421$, $z = 0.088$. For AsI_3, the chosen parameters are $u = $ ca. $\frac{1}{3}$, $x = -0.25$, $y = 0.422$, $z = 0.078$; for $FeCl_3$, they are $u = \frac{1}{3}$, $x = -0.256$, $y = 0.410$, $z = 0.077$. In a perfect hexagonal close-packing of the halogen atoms, the parameters would be $x = 0.25$, $y = 0.41$ ($\frac{5}{12}$), and $z = 0.083$ ($\frac{1}{12}$). The slight deviations of the assigned parameters from these ideal values are scarcely less than the uncertainty with which they are known.

The hexagonal cells for these crystals, each containing six molecules, have the edge lengths:

Crystal	a_0', A.	c_0', A.
AsI_3	7.187	21.39
BiI_3	7.498	20.68
$CrBr_3$	6.26	18.20
$FeCl_3$	6.06	17.38
SbI_3	7.466	20.89

In this atomic arrangement (Fig. VB,1), each R atom is at the center of a nearly perfect octahedron of halogen atoms. In BiI_3 the Bi–I distance in such an octahedron is 3.09 A. The closely packed iodine atoms have 12 iodine neighbors: three at 4.10 A., the other nine at 4.33 and 4.42 A., distances which are in good agreement with the assumption that the atoms are charged.

In a different structure based on electron diffraction, BiI_3 has been given a bimolecular hexagonal cell and the space group D_{3d}^1 ($P\bar{3}1m$). For it, $a_0 = 7.50$ A. and $c_0 = 6.9$ A. More work is needed.

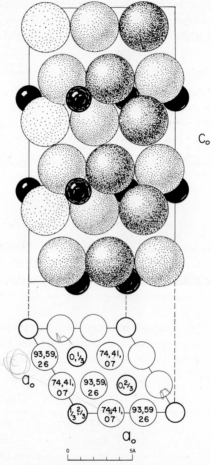

Fig. VB,1. Two projections of the hexagonal cell of BiI_3. The bismuth atoms are the smaller circles in both projections.

V,b2. The cell of *aluminum fluoride*, AlF_3, is similar in shape to that of the BiI_3-like crystals of the preceding paragraph, but it has been given a different atomic arrangement. All the intense reflections can be explained in terms of a unimolecular rhombohedron, but some faint reflections indicating a larger unit have been observed. The structure based on this bimolecular rhombohedron places atoms in the following special positions of D_3^7 ($R32$):

Al: (2c) $uuu;\ \bar{u}\bar{u}\bar{u}$ with $u = 0.237$

F(1):   (3d) $0v\bar{v};\ \bar{v}0v;\ v\bar{v}0$ with $v = 0.430$

F(2):   (3e) $\frac{1}{2}\,w\,\bar{w};\ \bar{w}\,\frac{1}{2}\,w;\ w\,\bar{w}\,\frac{1}{2}$ with $w = 0.070$

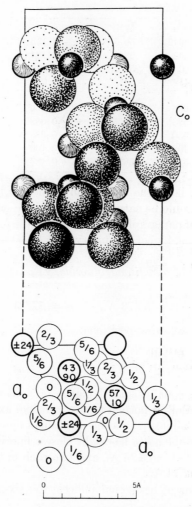

Fig. VB,2. Two projections of the hexagonal unit of AlF₃. The aluminum atoms are heavily ringed in the lower projection, and small and fine-line shaded in the upper.

These parameters place the fluorine atoms in a considerably distorted hexagonal close-packing: for it to be perfect, $v(F)$ would have to be $1/3$ and $w(F) = 1/6$. In the proposed structure (Fig. VB,2), each aluminum has about it three fluorine atoms at a distance of 1.70 A. and three at 1.89 A.; the nearest approach of fluorine to fluorine is 2.53 A.

TABLE VB,1
Some Crystals with the VF_3 Structure

Crystal	a_0, A.	α	u	a_0', A.	c_0', A.
CoF_3	5.279	57°0'	−0.15	5.035	13.218
CrF_3	5.2643	56°34'	−0.136	4.973	12.96
FeF_3	5.362	58°0'	−0.164	5.198	13.331
GaF_3	5.20	57°30'	−0.136	4.972	12.972
IrF_3	5.418	54°8'	−0.083	4.940	13.819
MoF_3	5.666	54°43'	−0.250	5.208	14.409
PdF_3	5.5234	53°55'	−0.083	5.009	14.118
RhF_3	5.330	54°25'	−0.083	4.874	13.579
RuF_3	5.408	54°40'	−0.100	4.967	13.756
TiF_3	5.519	58°53'	−0.183	5.426	13.634

V,b3. The unit cells of the crystals in Table VB,1 so closely resemble the AlF_3 cell that it would be natural to expect these substances to have the atomic distribution of paragraph **V,b2**. The structure found for them differs, however, in having the higher symmetry D_{3d}^6 ($R\bar{3}c$) as space group.

In the typical case of *vanadium trifluoride*, VF_3, the bimolecular rhombohedral cell with

$$a_0 = 5.373 \text{ A.}, \qquad \alpha = 57°31'$$

has its atoms in the following positions:

V: (2b) 000; $1/2\ 1/2\ 1/2$
F: (6e) $\pm(u,1/2-u,1/4;\ 1/2-u,1/4,u;\ 1/4,u,1/2-u)$ with $u = -0.145$

In this structure, equally spaced planes of vanadium and of fluorine atoms succeed one another normal to the principal axis. Each metal atom is at the center of an almost regular octahedron of fluorine atoms, with V–F = 1.94 A. Each fluorine atom has eight closest fluorine neighbors, four in its own plane, with F–F = 2.75 A., and two each in the planes above and below, with F–F = 2.74 A.

Expressed in terms of hexagonal indices, the six-molecule cell has the edges:

$$a_0' = 5.170 \text{ A.}, \qquad c_0' = 13.402 \text{ A.}$$

The coordinates of the atoms in this prism are:

V: (6b)   000; 0 0 $^1/_2$; rh
F: (18e) $\pm(u\ 0\ ^1/_4;\ 0\ u\ ^1/_4;\ \bar{u}\ \bar{u}\ ^1/_4)$; rh   with $u = -0.395$

The arrangement referred to this cell is shown in Figure VB,3.

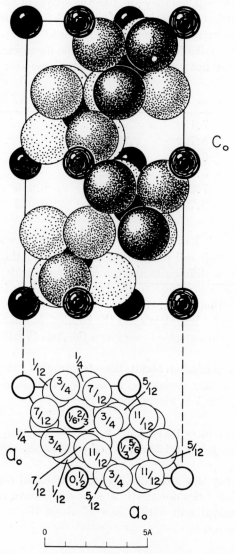

Fig. VB,3. Two projections of the hexagonal unit of VF₃. The vanadium atoms are heavily ringed in the lower projection, small and black in the upper.

A number of other fluorides are now known to have this arrangement based on $D_{3d}{}^6$. Their cells and the parameters u (referred to the rhombohedral axes) are listed in Table VB,1.

These fall into three groups depending on the values of u and the fluorine packing they express. For $u = -0.083$, fluorines are hexagonally close-packed, and for $u = -0.250$ they are, like the anions of the ReO_3 grouping (**V,b5**) in three-fourths of the positions of a cubic close-packing. With parameters between these extremes, the packing is transitional.

Several additional trihalides have cells like the foregoing, but their complete structures have not yet been established. These are:

Compound	a_0, A.	α	a_0', A.	c_0', A.
FeBr$_3$	7.159	53°16′	6.42	18.40
InF$_3$	5.722	56°15′	5.40	14.40
ScCl$_3$	6.972	54°30′	6.384	17.78
ScF$_3$	5.708	59°32′	5.667	14.03
TiBr$_3$	7.26	52°48′	6.456	18.69
VCl$_3$	6.735	53°1′	6.012	17.34

The two compounds RuF$_3$ and IrF$_3$ (Table VB,1) have shown small but real variations in their cell dimensions depending on the way they have been prepared. These differences and the corresponding departures from an exact RX$_3$ composition appear to be due to deviations from a mean trivalency of the metallic atoms.

In view of the structural identity of the halides considered here, it would seem desirable to reinvestigate those given the AsI$_3$ (**V,b1**) and AlF$_3$ (**V,b2**) structures.

There is much confusion about the structure of *titanium trichloride*, TiCl$_3$. One form, usually designated as α, has cell dimensions like those of other salts having the VF$_3$ arrangement:

$$a_0 = 6.814 \text{ A.}, \qquad \alpha = 53°23'$$
$$a_0' = 6.122 \text{ A.}, \qquad c_0' = 17.50 \text{ A.}$$

Very possibly it has this structure, though one based on the lower symmetry $C_3{}^4$ ($R3$) has been described. A second (β) form has been given a bimolecular hexagonal unit with a base of about the same size and one-third the height:

$$\beta\text{-TiCl}_3: \quad a_0 = 6.27 \text{ A.}, \qquad c_0 = 5.82 \text{ A.}$$

Its atoms have been said to be in the following positions of the high symmetry $D_{6h}{}^3$ ($P6_3/mcm$):

Ti: (2b)   000; 0 0 $^1/_2$

Cl: (6g) $\pm(u\,0\,^1/_4;\; 0\,u\,^1/_4;\; \bar{u}\,\bar{u}\,^1/_4)$ with $u = 0.315$

In such an arrangement the Ti–Cl separation is 2.45 A. and the closest Ti–Ti = 2.91 A.

There has also been reported a third (γ) modification having substantially the same cell dimensions as α-TiCl$_3$. For it, $a_0 = 6.14$ A. and $c_0 = 17.40$ A. It has been assigned the low symmetry $D_3{}^5$ ($P3_212$), or its enantiomorphic $D_3{}^3$, found for chromic chloride, CrCl$_3$ (**V,a6**), and will be referred to again under this substance.

V,b4. *Manganese trifluoride*, MnF$_3$, has a structure that is a monoclinic distortion of the VF$_3$ arrangement just discussed (**V,b3**). Its large unit containing 12 molecules has the dimensions:

$$a_0 = 8.904 \text{ A.}; \qquad b_0 = 5.037 \text{ A.}; \qquad c_0 = 13.448 \text{ A.}; \qquad \beta = 92°46'$$

The space group is $C_{2h}{}^6$ ($C2/c$) with atoms in the positions:

(4a)   000; 0 0 $^1/_2$; $^1/_2$ $^1/_2$ 0; $^1/_2$ $^1/_2$ $^1/_2$

(4e) $\pm(0\,u\,^1/_4;\; ^1/_2,u+^1/_2,^1/_4)$

(8f) $\pm(xyz;\; x,\bar{y},z+^1/_2;\; x+^1/_2,y+^1/_2,z;\; x+^1/_2,^1/_2-y,z+^1/_2)$

Assigned positions and parameters are those of Table VB,2

The resulting structure is shown in Figure VB,4. Like those of the other fluorides, it is basically a close-packing of the fluoride ions and, as with VF$_3$, it is a mixed packing between the hexagonal and the cubic. The F–F separations vary between 2.60 and 2.85 A., corresponding to the considerable distortion in the octahedra they form around manganese atoms.

TABLE VB,2
Positions and Parameters of the Atoms in MnF$_3$

Atom	Position	x	y	z
Mn(1)	(4a)	0	0	0
Mn(2)	(8f)	0.167	0.500	0.333
F(1)	(4e)	0	0.617	$^1/_4$
F(2)	(8f)	0.310	0.714	0.244
F(3)	(8f)	0.167	0.117	0.583
F(4)	(8f)	0.477	0.214	0.577
F(5)	(8f)	0.143	0.214	0.911

V,b5. *Scandium trifluoride,* ScF_3, was years ago given a simple rhombo-hedral structure related to those already described. Its unit contains a single molecule and is nearly cubic in its dimensions:

$$a_0 = 4.023 \text{ A.,} \qquad \alpha = 89°34'$$

Atoms were considered to be in the following positions of D_3^7 ($R32$):

Sc: (1a)   000
F: (3e) $^1/_2\, u\, \bar{u};\ \bar{u}\, ^1/_2\, u;\ u\, \bar{u}\, ^1/_2$ with u between 0.025 and 0.030

The structure first given to *rhenium trioxide,* ReO_3, and subsequently to a number of other compounds, differs from this only in being strictly cubic, with u for the negative atoms being exactly zero. The cell dimensions found for these various substances are listed in Table VB,3. It is assumed that for the oxyfluorides having this structure, the oxygen and fluorine atoms are randomly distributed among the positions $^1/_2\, 0\, 0;\ 0\, ^1/_2\, 0;\ 0\, 0\, ^1/_2$.

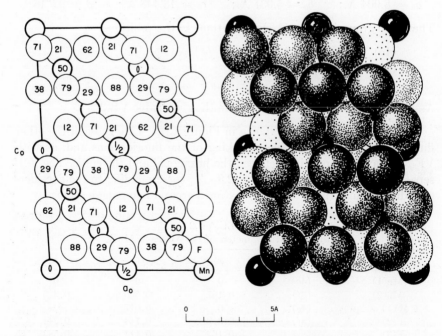

Fig. VB,4a (left). The monoclinic cell of MnF_3 projected along its b_0 axis.   Origin in lower left.
Fig. VB,4b (right). A packing drawing of the monoclinic MnF_3 structure seen along its b_0 axis.   The fluorine atoms are dotted.

TABLE VB,3
Crystals with the Cubic ScF₃-Like Arrangement

Compound	a_0, A.
MoF₃	3.8985
NbF₃	3.903
NbO₂F	3.902
ReO₃	3.734
TaF₃	3.9012
TaO₂F	3.896
TiOF₂	3.798
UO₃	4.156

In this arrangement (Fig. VB,5), the anions are in an approach to a cubic close-packing which is, however, incomplete due to the fact that there are available only three-fourths of the necessary atoms.

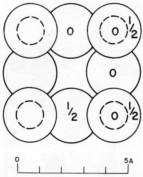

Fig. VB,5. A projection along a cubic axis of the very simple ReO₃ arrangement. Oxygen are the larger circles.

Copper nitride, Cu₃N, has the "anti" form of this grouping. It is described as truly cubic with $a_0 = 3.807$ A. Placing the nitrogen in the origin gives Cu–N = 1.90 A. and Cu–Cu = 2.71 A.

Scandium fluoride has also been described as being truly cubic with $a_0 = 4.03$ A.; presumably this is an error.

V,b6. A different variation on a cubic close-packing of anions is provided by *chromic chloride*, CrCl₃. It is hexagonal with a hexamolecular cell similar in dimensions to the corresponding cells of rhombohedral crystals having the VF₃ structure (**V,b3**):

$$a_0 = 6.00 \text{ A.}, \qquad c_0 = 17.3 \text{ A.}$$

Atoms have been placed in the following positions of D_3^5 ($P3_2 12$), or of the enantiomorphic D_3^3:

$(3b)$ $u\ \bar{u}\ ^2/_3;\ \ 2\bar{u}\ \bar{u}\ ^1/_3;\ u\ 2u\ 0$

$(6c)$ $xyz;\qquad y-x,\bar{x},z+^2/_3;\ \bar{y},x-y,z+^1/_3;$

$\qquad\ \ y-x,y,\bar{z};\ \bar{y},\bar{x},^1/_3-z;\qquad x,x-y,^2/_3-z$

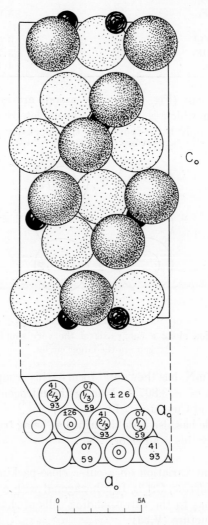

Fig. VB,6. Two projections of the hexagonal cell of CrCl₃. In both projections the chromium atoms are the small circles.

Approximate parameters defining the atomic positions were found to be the following:

Atom	Position	x	y	z
Cr(1)	(3b)	$2/9$	$4/9$	0
Cr(2)	(3b)	$5/9$	$1/9$	0
Cl(1)	(6c)	$2/9$	$4/9$	0.26
Cl(2)	(6c)	$5/9$	$1/9$	0.26
Cl(3)	(6c)	$8/9$	$7/9$	0.26

The cubic packing of the chlorine atoms in this structure would be perfect if z were $1/4$ and if c/a were 2.828 instead of the observed 2.88. This is an arrangement (Fig. VB,6) closely related to that of the dichloride, $CrCl_2$ (**IV,b2**) which has an additional set of chromium atoms in (3b) with $u = 8/9$.

Chromic iodide, CrI_3, is reported to have this arrangement, with

$$a_0 = 6.86 \text{ A.}, \qquad c_0 = 19.88 \text{ A.}$$

Atomic parameters have not been established.

Ruthenium trichloride, $RuCl_3$, is another substance for which this structure has been determined. For it,

$$a_0 = 5.97 \text{ A.}, \qquad c_0 = 17.2 \text{ A.}$$

The chosen x and y parameters are the same as those found for $CrCl_3$, and z for all the chlorine atoms is 0.259.

It seems probable that the gamma form of *titanium trichloride*, $TiCl_3$ (see **V,b3**) also has this structure. For it,

$$a_0 = 6.14 \text{ A.}, \qquad c_0 = 17.40 \text{ A.}$$

The parameter chosen for the z atoms of chlorine is 0.254.

V,b7. *Aluminum chloride*, $AlCl_3$, though actually monoclinic, is almost hexagonal, with a pseudocell having nearly the dimensions of $CrCl_3$ (**V,b6**):

$$a_0'' = 5.91 \text{ A.}, \qquad c_0'' = 17.52 \text{ A.}$$

Initially, it was given a related structure but with the aluminum atoms in highly improbable positions.

More recently, an arrangement has been described which, though monoclinic, leaves the chlorine atoms in much the same cubic close-packing found for the chromium salt. The unit contains four molecules and has the dimensions:

$$a_0 = 5.92 \text{ A.}; \qquad b_0 = 10.22 \text{ A.}; \qquad c_0 = 6.16 \text{ A.}; \qquad \beta = 108°$$

Atoms have been put in the following positions of C_{2h}^3 $(C2/m)$:

Al: (4g) $\pm(0u0; \,^1/_2,u+^1/_2,0)$ with $u = 0.167$
Cl(1): (4i) $\pm(u0v; \,u+^1/_2,^1/_2,v)$ with $u = 0.226$, $v = 0.219$
Cl(2): (8j) $\pm(xyz; \,x\bar{y}z; \,x+^1/_2,y+^1/_2,z; \,x+^1/_2,^1/_2-y,z)$
                          with $x = 0.250$, $y = 0.175$, $z = -0.219$

Though, as already stated, the chlorine atoms have the same kind of close-packing (Fig. VB,7) as in CrCl$_3$, the layers are stacked differently with respect to one another.

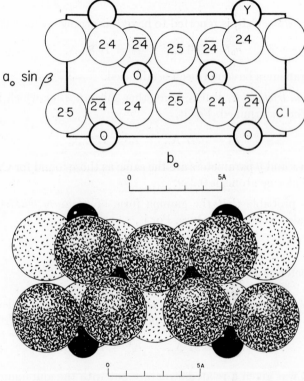

Fig. VB,7a (top). The monoclinic structure of YCl₃ (or AlCl₃) viewed along its c_0 axis. Origin in lower left.
Fig. VB,7b (bottom). A packing drawing of the YCl₃ (AlCl₃) structure seen along its c_0 axis.   The metallic atoms are black.

TABLE VB,4
Cell Dimensions of Crystals with the Monoclinic $AlCl_3$ Structure

Compound	a_0, A.	b_0, A.	c_0, A.	β
$DyCl_3$	6.91	11.97	6.40	111°12′
$ErCl_3$	6.80	11.79	6.39	110°42′
$HoCl_3$	6.85	11.85	6.39	110°48′
$InCl_3$	6.41	11.10	6.31	109°48′
$LuCl_3$	6.72	11.60	6.39	110°24′
$TlCl_3$	6.54	11.33	6.32	110°12′
$TmCl_3$	6.75	11.73	6.39	110°36′
$YbCl_3$	6.73	11.65	6.38	110°24′

This atomic arrangement has also been found for *yttrium trichloride*, YCl_3, and for the several other rare-earth trichlorides listed in Table VB,4. For YCl_3,

$$a_0 = 6.92 \text{ A.}; \quad b_0 = 11.94 \text{ A.}; \quad c_0 = 6.44 \text{ A.}; \quad \beta = 111°$$

With atoms in positions corresponding to those selected for $AlCl_3$,

$$u(Y) = 0.166 \quad u(Cl,1) = 0.211 \quad v(Cl,1) = 0.247$$
$$x(Cl,2) = 0.229 \quad y(Cl,2) = 0.179 \quad z(Cl,2) = -0.240$$

V,b8. The structure found for *aluminum bromide*, $AlBr_3$, also monoclinic, is unlike that of the chloride. Its tetramolecular cell has the dimensions:

$$a_0 = 10.20 \text{ A.}; \quad b_0 = 7.09 \text{ A.}; \quad c_0 = 7.48 \text{ A.}; \quad \beta = 96°$$

Atoms have been placed in general positions of C_{2h}^5 $(P2_1/a)$:

$$(4e) \quad \pm(xyz; \tfrac{1}{2}-x, y+\tfrac{1}{2}, \bar{z})$$

with the parameters that follow:

Atom	x	y	z
Al	0.050	0.095	0.183
Br(1)	0.150	0.075	-0.083
Br(2)	0.169	-0.078	0.411
Br(3)	0.008	0.392	0.252

In contrast to the chloride, which is ionic, this bromide structure (Fig. VB,8) can be imagined as built up of Al_2Br_6 molecules. On this basis, each

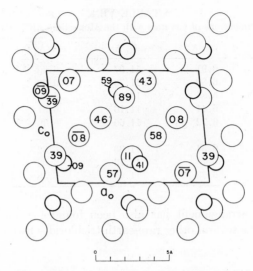

Fig. VB,8a. A projection along the b_0 axis of the monoclinic structure of AlBr₃.   The
large circles are bromine.   Origin in lower left.

aluminum atom is surrounded by a tetrahedron of bromine atoms and a
molecule consists of two of these tetrahedra sharing an edge. The Al–Br
separation within a tetrahedron lies between 2.23 and 2.42 A. Aluminum
atoms are 3.14 A. apart. In spite of this seeming existence of molecules, the
bromine atoms by themselves are in a nearly perfect close-packing throughout the crystal.

V,b9.   *Yttrium trifluoride*, YF₃, is orthorhombic with a tetramolecular
unit of the dimensions:

$$a_0 = 6.353 \text{ A.}; \qquad b_c = 6.850 \text{ A.}; \qquad c_0 = 4.393 \text{ A.}$$

Atoms have been put in the following positions of $V_h{}^{16}$ (*Pnma*):

$$(4c) \quad \pm(u\ ^1/_4\ v;\ u+^1/_2,^1/_4,^1/_2-v)$$
$$(8d) \quad \pm(xyz;\ x,^1/_2-y,z;\ x+^1/_2,y,^1/_2-z;\ x+^1/_2,^1/_2-y,^1/_2-z)$$

with the parameters:

Atom	Position	x	y	z
Y	(4c)	0.367	$^1/_4$	0.058
F(1)	(4c)	0.528	$^1/_4$	0.601
F(2)	(8d)	0.165	0.060	0.363

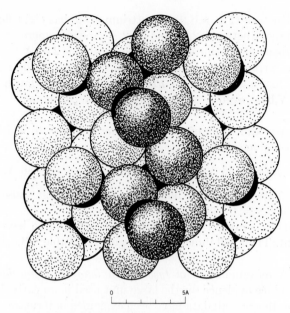

Fig. VB,8b. A packing drawing of the monoclinic structure of AlBr₃ viewed along its b_0 axis. The large bromine atoms are dotted.

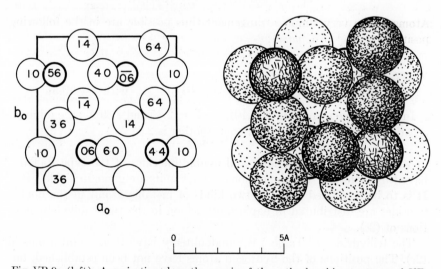

Fig. VB,9a (left). A projection along the c_0 axis of the orthorhombic structure of YF₃. Fluorines are the larger circles. Origin in lower left. Right-hand axes.

Fig. VB,9b (right). A packing drawing of the orthorhombic structure of YF₃ viewed along the c_0 axis. The fluorine atoms are dotted.

In this structure (Fig. VB,9), an yttrium atom has eight fluorine neighbors at distances between 2.25 and 2.32 A. and one more at 2.60 A. The closest approach of fluorine atoms to one another is 2.55 A.

This is an arrangement which after an exchange of the axial sequence from $a_0b_0c_0$ to $c_0a_0b_0$ (*Pnma* to *Pbnm*) becomes very much like that of cementite (**V,b43**) even to values of the atomic parameters.

There is a modification of *bismuth trifluoride*, BiF_3, to which this structure has been given. For it,

$$a_0 = 6.56 \text{ A.}; \qquad b_0 = 7.03 \text{ A.}; \qquad c_0 = 4.86 \text{ A.}$$

Its fluorine parameters have not been established, but those for bismuth are $u = 0.353$, $v = 0.038$.

Numerous rare-earth trifluorides have been found to have this atomic arrangement. Their cell dimensions are given in Table VB,5.

V,b10. A reexamination of *tysonite, lanthanum trifluoride*, LaF_3, has failed to produce evidence for the large unit used in an early determination of structure. Instead, all data have been explained in terms of a bimolecular hexagonal cell of the dimensions:

$$a_0 = 4.148 \text{ A.}, \qquad c_0 = 7.354 \text{ A.}$$

Atoms of the very simple arrangement thus possible are in the following positions of D_{6h}^4 ($P6_3/mmc$):

La: (2c) $\pm(^1/_3 \, ^2/_3 \, ^1/_4)$
F(1): (2b) $\pm(0 \; 0 \; ^1/_4)$
F(2): (4f) $\pm(^1/_3 \, ^2/_3 \, u; \; ^2/_3, ^1/_3, u+^1/_2)$ with $u = 0.57$

In this arrangement (Fig. VB,10), La–F = 2.35 and 2.39 A.; the minimal separation of fluorine atoms from one another is 2.64 A.

A large number of fluorides, and double fluorides of the type $BaThF_6$, have this structure. The dimensions of their units are given in Table VB,6. It is to be presumed that the two kinds of metallic atoms in the double fluorides are distributed in haphazard fashion in the two equivalent positions of (2c).

The trihydrides of Table VB,6 undoubtedly have their metal atoms in (2c). The positions of the hydrogen atoms have not been established, but they may well be distributed as are the fluorine atoms in LaF_3.

When BiF_3 is heated in air for a few minutes at ca. 850°C., an oxyfluoride of the composition $BiO_{0.1}F_{2.8}$ is produced. This has the tysonite structure, with

$$a_0 = 4.098 \text{ A.}, \qquad c_0 = 7.277 \text{ A.}$$

TABLE VB,5
Cell Dimensions of Crystals with the YF_3 Structure

Crystal	a_0, A.	b_0, A.	c_0, A.
DyF_3	6.460	6.906	4.376
ErF_3	6.354	6.846	4.380
EuF_3	6.622	7.019	4.396
GdF_3	6.570	6.984	4.393
HoF_3	6.404	6.875	4.379
LuF_3	6.151	6.758	4.467
SmF_3	6.669	7.059	4.405
TbF_3	6.513	6.949	4.384
TmF_3	6.283	6.811	4.408
YbF_3	6.216	6.786	4.434

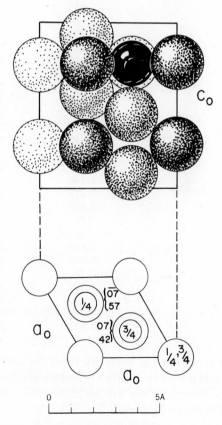

Fig. VB,10. Two projections of the hexagonal cell of LaF_3. In the upper the lanthanum atom is black.

TABLE VB,6
Crystals with the Hexagonal LaF₃ Arrangement

Crystal	a_0, A.	c_0, A.
AcF₃	4.28	7.54
AmF₃	4.067	7.225
BaThF₆	4.288	7.535
BaUF₆	4.273	7.471
CaThF₆	4.033	7.189
CeF₃	4.107	7.273
(Ce,La)F₃ (tysonite)	4.113	7.280
β-EuF₃	3.993	7.091
β-HoF₃	3.945	6.984
NdF₃	4.054	7.196
NpF₃	5.042	7.288
NpI₃	4.116	7.287
PbThF₆	4.200	7.410
PbUF₆	4.183	7.352
PrF₃	4.085	7.238 (26°C.)
PuF₃	4.095	7.254
SmF₃	4.016	7.120
SrThF₆	4.133	7.342
SrUF₆	4.111	7.304
ThOF₂	4.047	7.304
β-TmF₃	3.905	6.927
UF₃	4.146	7.348
DyH₃	3.671	6.615
ErH₃	3.621	6.526
GdH₃	3.73	6.71
HoH₃	3.642	6.560
LuH₃	3.558	6.443
SmH₃	3.782	6.779
TbH₃	3.700	6.658
TmH₃	3.599	6.489
YH₃	3.672	6.659

V,b11. A layer structure has been described for crystals of *plutonium tribromide*, PuBr₃, and a variety of other bromides and iodides of rare-earth and trans-uranium atoms. They are orthorhombic with tetramolecular units. For PuBr₃, the dimensions are

$$a_0 = 12.64 \text{ A.}; \qquad b_0 = 4.10 \text{ A.}; \qquad c_0 = 9.14 \text{ A.}$$

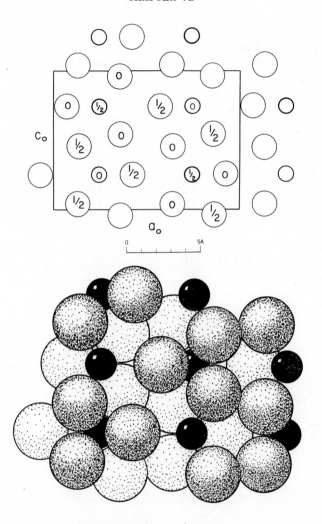

Fig. VB,11a (top). A projection along the b_0 axis of the orthorhombic structure of PuBr₃. The larger circles are bromine. Origin in lower left.

Fig. VB,11b (bottom). A packing drawing of the orthorhombic PuBr₃ structure viewed along its b_0 axis. The large dotted atoms are the bromines.

The space group is given as V_h^{17} (*Ccmm*), with atoms in the special positions:

Pu: (4c) $\pm(u\ 0\ ^1/_4;\ u+^1/_2, ^1/_2, ^1/_4)$ with $u = 0.25$

Br(1): (4c)   with $u = -0.07$

Br(2): (8f) $\pm(u0v;\ u,0,^1/_2-v;\ u+^1/_2, ^1/_2, v;\ u+^1/_2, ^1/_2, ^1/_2-v)$

                                  with $u = 0.36$, $v = -0.05$

As can be seen from Figure VB,11, the atomic layers of this structure are parallel to the b_0c_0 plane.

Other crystals with this arrangement have the cell dimensions listed in Table VB,7.

TABLE VB,7
Crystals with the Orthorhombic PuBr₃ Structure

Crystal	a_0, A.	b_0, A.	c_0, A.
AmBr₃	12.6	4.11	9.11
AmI₃	14.0	4.31	9.9
LaI₃	14.1	4.34	10.06
NdBr₃	12.65	4.11	9.16
β-NpBr₃	12.66	4.12	9.16
NpI₃	14.01	4.30	9.94
PuI₃	14.01	4.30	9.91
SmBr₃	12.63	4.04	9.07
UI₃	13.99	4.32	10.00

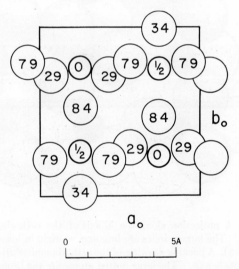

Fig. VB,12a. A projection along the c_0 axis of the orthorhombic structure of SbF₃. Fluorines are the larger circles. Origin in lower right.

V,b12. Crystals of *antimony trifluoride*, SbF₃, are orthorhombic with a tetramolecular unit of the dimensions:

$$a_0 = 7.25 \text{ A.}; \qquad b_0 = 7.49 \text{ A.}; \qquad c_0 = 4.95 \text{ A.}$$

Atoms have been found to be in the following positions based on the space group C_{2v}^{16} (Ama):

Sb: (4b) $^1/_4$ u v; $^3/_4$ $\bar{u}$ v; $^1/_4, u+^1/_2, v+^1/_2$; $^3/_4, ^1/_2-u, v+^1/_2$

with $u = 0.214$, $v = 0.00$

F(1):   (4b)   with $u = 0.467$, $v = -0.161$

F(2):   (8c) xyz; $\bar{x}\bar{y}z$;

$^1/_2-x, y, z$; $x+^1/_2, \bar{y}, z$;

$x, y+^1/_2, z+^1/_2$; $\bar{x}, ^1/_2-y, z+^1/_2$;

$^1/_2-x, y+^1/_2, z+^1/_2$; $x+^1/_2, ^1/_2-y, z+^1/_2$

with $x = 0.069$, $y = z = 0.286$

The choice of these fluorine parameters involves the assumption that the Sb–F distance equals the Sb–O separation in Sb_2O_3 (2.0 A.).

The structure that results (Fig. VB,12) is a packing of SbF_3 molecules for which the closest approach of the fluorine atoms of adjacent molecules is 2.73 A. Atoms of antimony and fluorine belonging to different molecules approach as near as 2.55 A. to one another.

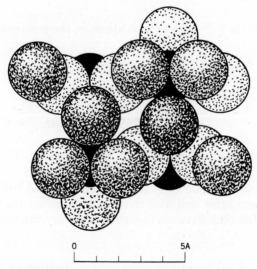

Fig. VB,12b. A packing drawing of the orthorhombic SbF_3 structure viewed along its c_0 axis. The fluorine atoms are dotted.

V,b13.   Crystals of *antimony trichloride*, $SbCl_3$, are orthorhombic with a tetramolecular cell of the dimensions:

$$a_0 = 6.37 \text{ A.}; \qquad b_0 = 8.12 \text{ A.}; \qquad c_0 = 9.47 \text{ A.}$$

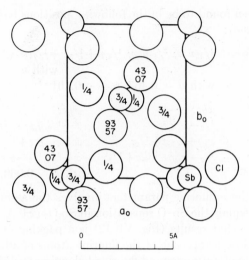

Fig. VB,13a. The orthorhombic SbCl₃ structure projected along its c_0 axis.   Origin in lower right.

The space group is V_h^{16} (*Pbnm*) with atoms in the positions:

Sb: (4c) $\pm (u\,v\,{}^1/_4;\ {}^1/_2-u,v+{}^1/_2,{}^1/_4)$

with $u = 0.025$, $v = 0.995$

Cl(1):   (4c)   with $u' = 0.671$, $v' = 0.077$

Cl(2):   (8d) $\pm (xyz;\ {}^1/_2-x,y+{}^1/_2,{}^1/_2-z;\ x,y,{}^1/_2-z;\ {}^1/_2-x,y+{}^1/_2,z)$

with $x = 0.132$, $y = 0.176$, $z = 0.066$

This is a structure built up of SbCl₃ molecules (Fig. VB, 13) which are trigonal pyramids having antimony atoms at the apices. In each molecule, Sb–Cl = 2.35–2.37 A., Cl–Cl = 3.48 or 3.51 A. and Cl–Sb–Cl = 95°12′. These dimensions agree well with those determined for the vapor molecule by electron diffraction. Between molecules in the crystal, the shortest Sb–Cl = 3.5 A.

The β-form of *antimony tribromide*, SbBr₃, melting at 96°C., has this structure, with

$$a_0 = 6.68\ \text{A.}; \qquad b_0 = 8.25\ \text{A.}; \qquad c_0 = 9.96\ \text{A.}$$

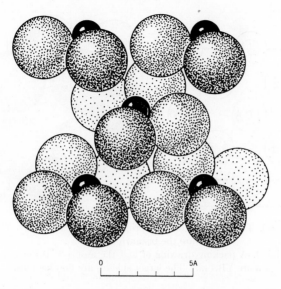

Fig. VB,13b. A packing drawing of the SbCl₃ structure viewed along its c_0 axis. The
antimony atoms are small and black.

Its atomic positions are:

  Sb: (4c)   with $u = 0.0267$, $v = 0.9950$
  Br(1): (4c)   with $u = 0.6710$, $v = 0.0750$
  Br(2): (8d)   with $x = 0.1420$, $y = 0.1840$, $z = 0.0670$

The closest Sb–Br = 2.46 and 2.52 A. and Br–Sb–Br = 92°48′ and 97°37′.

V,b14.   There has been considerable uncertainty concerning the structure of the cubic form of *bismuth trifluoride*, BiF_3. For it,

$$a_0 = 5.853 \text{ A.}$$

This contains four molecules.

Originally it was given a simple arrangement based on O_h^5 (*Fm3m*), with
atoms in the positions:

  Bi: (4a)   000; F.C.
  F(1):   (4b) $^1/_2$ $^1/_2$ $^1/_2$; F.C.
  F(2):   (8c) $\pm(^1/_4$ $^1/_4$ $^1/_4)$; F.C.

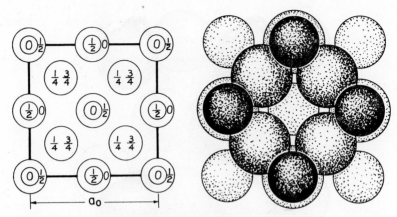

Fig. VB,14a (left). The structure of BiF₃ projected on a cube face. The small circles
are the bismuth atoms.
Fig. VB,14b (right). A packing drawing of half the atoms of BiF₃ in the unit cube of
Figure VB,14a. The larger spheres are fluorine ions.

This is the atomic distribution found to prevail for the NH_4 and FeF_6 ions
in $(NH_4)_3FeF_6$ [Chapter IX]. In it (Fig. VB,14), each atom has eight
nearest neighbors always at a distance of 2.56 A. For bismuth and for the
$F(1)$ atoms, these are $F(2)$ atoms; for $F(2)$ they are four bismuth and four
$F(1)$ atoms.

Subsequently, a less symmetrical arrangement was proposed. It is based
on T_d^1 ($P\bar{4}3m$) with atoms in the positions:

$$
\begin{array}{lll}
\text{Bi:} & (4e) & uuu;\ u\bar{u}\bar{u};\ \bar{u}u\bar{u};\ \bar{u}\bar{u}u \qquad \text{with } u = 0.737 \\
\text{F(1):} & (4e) & \text{with } u = {}^{1}\!/_{4} \\
\text{F(2):} & (1a) & 000 \\
\text{F(3):} & (1b) & {}^{1}\!/_{2}\,{}^{1}\!/_{2}\,{}^{1}\!/_{2} \\
\text{F(4):} & (3c) & 0\,{}^{1}\!/_{2}\,{}^{1}\!/_{2};\ {}^{1}\!/_{2}\,0\,{}^{1}\!/_{2};\ {}^{1}\!/_{2}\,{}^{1}\!/_{2}\,0 \\
\text{F(5):} & (3d) & {}^{1}\!/_{2}\,0\,0;\ 0\,{}^{1}\!/_{2}\,0;\ 0\,0\,{}^{1}\!/_{2}
\end{array}
$$

This revised structure differs from the original in having the bismuth
atoms slightly displaced from their more symmetrical positions.

Other compounds said to be isostructural are:

$$
\begin{array}{ll}
\text{Li}_3\text{Bi:} & a_0 = 6.721 \text{ A.} \\
\beta\text{-Li}_3\text{Sb:} & a_0 = 6.572 \text{ A.}
\end{array}
$$

A suboxide of molybdenum to which the formula Mo_3O has been assigned is reported to have a deficit structure "anti" to the foregoing. Its
cell has been given as cubic with $a_0 = 5.549$ A. and containing three molecules.

V,b15. Crystals of *auric chloride*, AuCl₃, have monoclinic symmetry with a tetramolecular cell of the dimensions:

$$a_0 = 6.57 \text{ A.}; \quad b_0 = 11.04 \text{ A.}; \quad c_0 = 6.44 \text{ A.}; \quad \beta = 113°18'$$

All atoms are in general positions of C_{2h}^5 $(P2_1/c)$:

$$(4e) \quad \pm(xyz; \ x,{}^1\!/_2-y,z+{}^1\!/_2)$$

with the following parameters:

Atom	x	y	z
Au	0.0415	0.0868	0.2337
Cl(1)	0.258	0.003	0.059
Cl(2)	0.335	0.169	0.509
Cl(3)	0.820	0.162	0.395

This is a structure built up of planar Au₂Cl₆ molecules distributed as shown in Figure VB,15. The dimensions of this molecule, which has a center of symmetry, are given in Figure VB,16. Between molecules, the shortest Cl–Cl = 3.54 A.

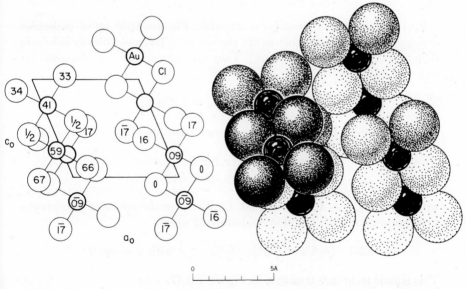

Fig. VB,15a (left). The monoclinic structure of AuCl₃ viewed along its b_0 axis. Origin in lower left.
Fig. VB,15b (right). The AuCl₃ structure seen along its b_0 axis. The gold atoms are small and black.

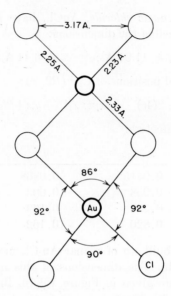

Fig. VB,16. The dimensions of the planar Au_2Cl_6 molecule as it occurs in the crystal. The larger circles are chlorine.

V,b16. Crystals of *phosphorus triiodide*, PI_3, are built up of molecules arranged in the same manner as are the similar CHI_3 molecules in iodoform (Chapter XIII, loose-leaf edition). They are hexagonal with a bimolecular unit having:

$$a_0 = 7.11 \text{ A.}, \qquad c_0 = 7.273 \text{ A.}$$

The iodine atoms have been put in general positions of C_6^6 ($C6_3$):

(6c) $xyz;$ $y-x,\bar{x},z;$ $\bar{y},x-y,z;$
  $\bar{x},\bar{y},z+^1/_2;\ x-y,x,z+^1/_2;\ y,y-x,z+^1/_2$

with $x = 0.30$, $y = 0.35$, $z = 0.00$. Positions for the atoms of phosphorus were not determined, but they undoubtedly are in

(2b) $^1/_3\ ^2/_3\ u;\ ^2/_3,^1/_3,u+^1/_2$ with $u = $ ca. 0

This simple structure is shown in Figure VB,17.

The same arrangement has been found for the following boron halides:

BCl_3: $a_0 = 6.08 \text{ A.}, \qquad c_0 = 6.55 \text{ A.}$

The boron atoms are in (2b) with $u = 0$ (arbitrary) and the chlorine atoms in $(6c)$ with $x = 0.0455$, $y = 0.3763$, $z = 0.00$.

$$BI_3: \quad a_0 = 7.00 \text{ A.}, \quad c_0 = 7.46 \text{ A.}$$

The boron atoms presumably are in $(2b)$ with $u = $ ca. 0; the iodine atoms are in $(6c)$ with $x = 0.03$, $y = 0.37$, $z = 0.00$.

$$BBr_3: a_0 = 6.406 \text{ A.}, \quad c_0 = 6.864 \text{ A.}$$

The atomic parameters have not been established for this compound.

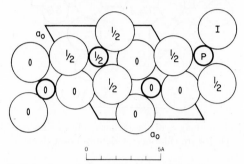

Fig. VB,17. A projection along the c_0 axis of the simple hexagonal structure of PI_3.

V,b17. The form of *chlorine trifluoride*, ClF_3, which is stable below $-82.5°C$. is orthorhombic with a tetramolecular cell of the dimensions:

$$a_0 = 8.825 \text{ A.}; \quad b_0 = 6.09 \text{ A.}; \quad c_0 = 4.52 \text{ A.}$$

The space group is $V_h{}^{16}$ $(Pnma)$ with atoms in the positions:

Cl: $(4c)$ $\pm(u\ ^1/_4\ v;\ u+^1/_2,^1/_4,^1/_2-v)$ with $u = 0.1582$, $v = 0.3790$
F(1): $(4c)$ with $u = 0.0422$, $v = 0.1010$
F(2): $(8d)$ $\pm(xyz;\ x+^1/_2,^1/_2-y,^1/_2-z;\ x,^1/_2-y,z;\ x+^1/_2,y,^1/_2-z)$
    with $x = 0.1517$, $y = 0.5315$, $z = 0.3634$

The resulting structure is illustrated in Figure VB,18. It is an array of planar ClF_3 molecules having the bond dimensions of Figure VB,19. Between molecules the closest F–F $= 2.66$ A. and the closest Cl–F $= 3.06$ A.

V,b18. According to a preliminary determination, *bromine trifluoride*, BrF_3, forms crystals which at $-125°C$. are orthorhombic with the tetramolecular unit:

$$a_0 = 5.34 \text{ A.}; \quad b_0 = 7.35 \text{ A.}; \quad c_0 = 6.61 \text{ A.}$$

The space group has been chosen as $C_{2v}{}^{12}$ $(Cmc2_1)$, and all atoms have been placed in

$(4a)$ $0uv;\ 0,\bar{u},v+^1/_2;\ ^1/_2,u+^1/_2,v;\ ^1/_2,^1/_2-u,v+^1/_2$

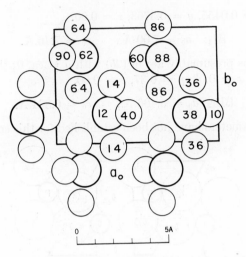

Fig. VB,18a. A projection along c_0 of the orthorhombic structure of ClF_3. The larger circles are chlorine. Origin in lower right.

with the following parameters:

Atom	u	v
Br	0.1587	0.250
F(1)	0.270	0.478
F(2)	0.394	0.152
F(3)	−0.058	0.390

This structure (Fig. VB,20) is built up of molecules similar to those of ClF_3 (**V,b17**) but their arrangement is different. Within the planar molecules, Br–F(1) = 1.72 A., Br–F(2) = 1.85 A., and Br–F(3) = 1.84 A. The significant bond angles are F(1)–Br–F(2) = 82°0′ and F(1)–Br–F(3) = 88°24′.

V,b19. A study of its crystal structure shows that *iodine trichloride* is not ICl_3, but I_2Cl_6. Its triclinic cell, containing one I_2Cl_6, has the dimensions:

$$a_0 = 5.71 \text{ A.}; \quad b_0 = 10.88 \text{ A.}; \quad c_0 = 5.48 \text{ A.};$$
$$\alpha = 130°50′; \quad \beta = 80°50′; \quad \gamma = 108°30′$$

Atoms, in the general positions (2i) $\pm(xyz)$ of C_1^1 ($P\bar{1}$), have been given the parameters:

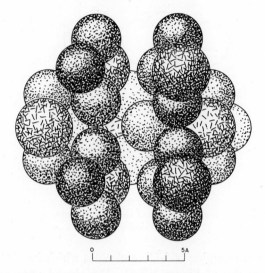

Fig. VB,18b. A packing drawing of the orthorhombic structure of ClF₃ viewed along its
c_0 axis. The chlorine atoms are line shaded.

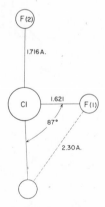

Fig. VB,19. The dimensions of the ClF₃ molecule as it occurs in the crystal.

Atom	x	y	z
I	0	0.186	0
Cl(1)	0.275	0.135	0.263
Cl(2)	0.272	0.464	0.257
Cl(3)	0.272	0.796	0.255

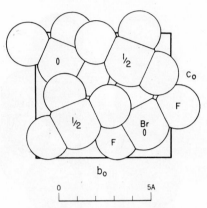

Fig. VB,20. A projection along its a_0 axis of the orthorhombic structure of BrF$_3$. Origin in lower right.

This leads to a grouping of I$_2$Cl$_6$ molecules which have the interatomic distances and angles of Figure VB,21. The closest approach of chlorine atoms in different molecules is 3.46 A.

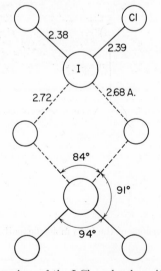

Fig. VB,21. The dimensions of the I$_2$Cl$_6$ molecule as it occurs in the crystal.

V,b20. *Indium hydroxyfluoride*, In(OH)F$_2$, is monoclinic with a bimolecular cell of the dimensions:

$$a_0 = 5.780 \text{ A.}; \qquad b_0 = 5.157 \text{ A.}; \qquad c_0 = 3.874 \text{ A.}; \qquad \beta = 98°46'$$

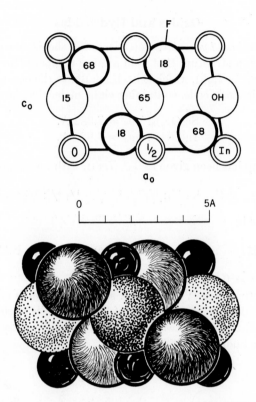

Fig. VB,22a (top). A projection of the monoclinic structure of In(OH)F$_2$ along its b_0 axis.
Fig. VB,22b (bottom). A packing drawing of the simple In(OH)F$_2$ arrangement viewed
along its b_0 axis.   The indium atoms are black, the hydroxyls dotted.

Atoms have been placed in the following positions of C_2^3 ($C2$):

In:	(2a)	$0u0$; $^1/_2,u+^1/_2,0$	with $u = 0$
O:	(2b)	$0\ v\ ^1/_2$; $^1/_2,v+^1/_2,^1/_2$	with $v = 0.149$
F:	(4c)	xyz; $\bar{x}y\bar{z}$; $x+^1/_2,y+^1/_2,z$; $^1/_2-x,y+^1/_2,\bar{z}$	

with $x = 0.328$, $y = 0.180$, $z = 0.154$

In this structure (Fig. VB,22), each indium atom has around it four
fluorine atoms and two OH groups. The In–F separations are 2.06 and 2.11
A. and In–OH = 2.08 A. The shortest F–F = 2.46 A. and O–F = 2.49 A.;
the closest O–O = 3.87 A.

This arrangement can be described in terms of a triclinic cell with $a_0 = b_0 = c_0 = 3.874$ A. and $\alpha = \beta = \gamma = 83°29'$. In this pseudocubic aspect it
does not differ much from the ReO$_3$ arrangement (V,b5).

Oxides and Hydroxides

V,b21. *Scandium hydroxide,* $Sc(OH)_3$, and the isomorphous $In(OH_3)$ have been given a structure that is in a sense a superlattice on the simple ReO_3 arrangement (**V,b5**), which is itself related to the perewskite structure (**VII,a21**). Their cubic cells, containing eight molecules, have the edges:

$$Sc(OH)_3: \quad a_0 = 7.882 \text{ A.}$$
$$In(OH)_3: \quad a_0 = 7.923 \text{ A.}$$

The space group has been given as T_h^5 (*Im3*), with metal atoms in

$$(8c) \quad \pm(^1/_4\, {}^1/_4\, {}^1/_4;\, {}^1/_4\, {}^3/_4\, {}^3/_4;\, {}^3/_4\, {}^1/_4\, {}^3/_4;\, {}^3/_4\, {}^3/_4\, {}^1/_4)$$

and OH groups in

$$(24g) \quad \pm(0uv;\, v0u;\, uv0;\, 0u\bar{v};\, \bar{v}0u;\, u\bar{v}0);\quad \text{B.C.}$$

For $Sc(OH)_3$, $u = 0.307$ and $v = 0.182$.

In this structure (Fig. VB,23), each scandium atom is enveloped by an octahedron of OH ions at a distance of 2.08 A., while each of these (OH) groups is equidistant from two scandium atoms. The closest approach of OH ions to one another is 2.88A.

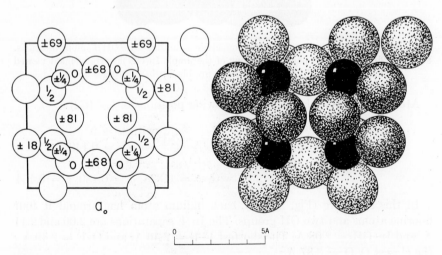

Fig. VB,23a (left). A projection along a cubic axis of the structure of $Sc(OH)_3$. Hydroxyls are the larger circles.

Fig. VB,23b (right). A packing drawing of the structure of $Sc(OH)_3$ viewed along a cubic axis. The hydroxyl oxygens are dotted.

V,b22.   *Yttrium hydroxide,* $Y(OH)_3$, is hexagonal with a bimolecular unit of the dimensions:

$$a_0 = 6.24 \text{ A.,} \qquad c_0 = 3.53 \text{ A.}$$

Atoms have been placed in the following special positions of C_{6h}^2 $(P6_3/m)$:

Y: (2d) $^2/_3 \, ^1/_3 \, ^1/_4; \, ^1/_3 \, ^2/_3 \, ^3/_4$

OH: (6h) $\pm (u \; v \; ^1/_4; \; \bar{v}, u-v, ^1/_4; \; v-u, \bar{u}, ^1/_4)$

with $u = 0.287$, $v = 0.382$

The resulting arrangement (Fig. VB,24) is one that surrounds each metal atom by nine OH radicals; six of these are at a distance of 2.42 A., the other three at 2.54 A. The nearest approach of hydroxyls to one another is 2.78 A.

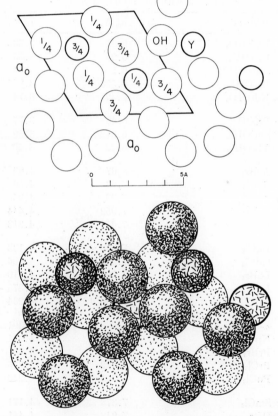

Fig. VB,24a (top). A basal projection of the hexagonal unit of $Y(OH)_3$.
Fig. VB,24b (bottom). A packing drawing of the hexagonal structure of $Y(OH)_3$ viewed along the c_0 axis.   The yttrium atoms are line shaded.

Through a neutron diffraction study of La(OD)$_3$, positions have been found for the deuterium (hydrogen) atoms. For this substance, with substantially the same cell as that of La(OH)$_3$, the oxygen parameters were found to be $u = 0.292$, $v = 0.375$. The deuterium atoms, in the same special positions $(6h)$, were assigned the values $u = 0.152$, $v = 0.225$.

The numerous compounds, both hydroxides and halides, with this atomic arrangement have the cell dimensions of Table VB,8.

V,b23. *Aluminum hydroxide*, Al(OH)$_3$, or as it was often formerly written Al$_2$O$_3 \cdot$ 3H$_2$O, has been given an arrangement in which the OH radicals are distributed with an approach to close-packing. Its symmetry is monoclinic, and measurements on it as the mineral *gibbsite* [*hydrargillite*]

TABLE VB,8
Crystals with the Y(OH)$_3$ Structure

Crystal	a_0, A.	c_0, A.
Dy(OH)$_3$	6.27	3.53
Er(OH)$_3$	6.25	3.53
Gd(OH)$_3$	6.265	3.54
La(OH)$_3$	6.523	3.855
Nd(OH)$_3$	6.421	3.74
Pr(OH)$_3$	6.48	3.77
Sm(OH)$_3$	6.312	3.59
Yb(OH)$_3$	6.22	3.50
AcBr$_3$	8.07	4.69
AcCl$_3$	7.63	4.56
AmCl$_3$	7.38	4.25
CeBr$_3$	7.952	4.444
CeCl$_3$	7.451	4.313
EuCl$_3$	7.369	4.133
GdCl$_3$	7.363	4.105
LaBr$_3$	7.967	4.510
LaCl$_3$	7.483	4.375
NdCl$_3$	7.396	4.234
α-NpBr$_3$	7.933	4.391
NpCl$_3$	7.420	4.281
PrBr$_3$	7.93	4.39
PrCl$_3$	7.422	4.275
PuCl$_3$	7.395	4.246
SmCl$_3$	7.378	4.171
UBr$_3$	7.942	4.441
UCl$_3$	7.443	4.321

TABLE VB,9
Parameters of the Atoms in Al(OH)$_3$

Atom	x	y	z
Al(1)	0.176 (0.166)	0.520 (0.500)	0.005 (0.000)
Al(2)	0.333 (0.333)	0.020 (0.000)	0.005 (0.000)
O(1)	0.181 (0.183)	0.205 (0.202)	0.110 (0.105)
O(2)	0.681 (0.674)	0.671 (0.670)	0.110 (0.104)
O(3)	0.515 (0.498)	0.131 (0.132)	0.110 (0.106)
O(4)	−0.015 (−0.017)	0.631 (0.632)	0.110 (0.108)
O(5)	0.298 (0.293)	0.701 (0.702)	0.100 (0.105)
O(6)	0.838 (0.806)	0.171 (0.170)	0.100 (0.103)

result in a unit containing eight molecules of Al(OH)$_3$ and having the dimensions:

$$a_0 = 8.6236 \text{ A.}; \quad b_0 = 5.0602 \text{ A.}; \quad c_0 = 9.699 \text{ A.}; \quad \beta = 85°26'$$

All atoms have been assigned general positions of C_{2h}^5 ($P2_1/n$):

$$(4e) \quad \pm(xyz; \; {}^1/_2-x, y+{}^1/_2, {}^1/_2-z)$$

with the two sets of moderately agreeing parameters listed in Table VB,9.

This is a layer structure (Fig. VB,25) consisting of linked sheets of the kind of octahedron that occurs in a close-packing, but in the present instance these octahedra, each enclosing an aluminum atom, are stacked one above another along the c_0 axis. This placing of the oxygen atoms one above another in adjacent layers instead of in holes between the oxygen atoms, as is the case with a complete close-packing, is unusual.

A form of Al(OH)$_3$ produced in the laboratory, the so-called *bayerite*, appears to have a simple bimolecular hexagonal cell of the dimensions:

$$a_0 = 5.047 \text{ A.}, \quad c_0 = 4.730 \text{ A.}$$

It has been said that the limited data obtained can be explained by an arrangement based on D_{3d}^1 ($P\bar{3}1m$) which places the atoms as follows:

Al: (2c) ${}^1/_3 \, {}^2/_3 \, 0; \; {}^2/_3 \, {}^1/_3 \, 0$

O: (6k) $\pm(u0v; \; 0uv; \; \bar{u}\bar{u}v)$ with $u = 0.340, v = 0.210$

More work is needed.

There is also said to be a triclinic form which is closely related to gibbsite in cell dimensions (i.e., the a_0 and b_0 axes are approximately twice the corresponding gibbsite axes):

$$a_0 = 17.338 \text{ A.}; \quad b_0 = 10.086 \text{ A.}; \quad c_0 = 9.730 \text{ A.}$$
$$\alpha = 94°10'; \quad \beta = 92°8'; \quad \gamma = 90°0'$$

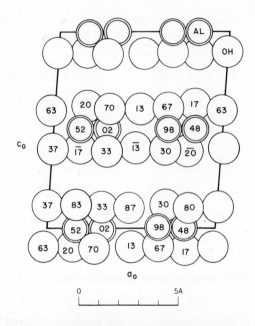

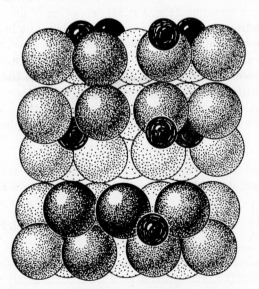

Fig. VB,25a (top). The monoclinic Al(OH)₃ arrangement projected along its b_0 axis. Origin in lower left.

Fig. VB,25b (bottom). A packing drawing of the Al(OH)₃ structure viewed along its b_0 axis. The aluminum atoms are small and black.

V,b24. In *molybdenum trioxide*, MoO_3, the metalloid atoms seemingly are in a distorted cubic close-packed array. There are four molecules in an orthorhombic unit having the dimensions:

$$a_0 = 3.92 \text{ A.}; \qquad b_0 = 13.94 \text{ A.}; \qquad c_0 = 3.66 \text{ A.}$$

Two determinations agree in placing all atoms in special positions

$$(4c) \quad \pm(u\ v\ ^1/_4;\ ^1/_2-u, v+^1/_2, ^1/_4)$$

of V_h^{16} (*Pbnm*) with the parameters:

Atom	u	v
Mo	0.084	0.0998
O(1)	0.015	0.230
O(2)	0.56	0.100
O(3)	0.525	0.435

Fig. VB,26a (left). A projection along the c_0 axis of the orthorhombic structure of MoO_3. The large circles are oxygen. Origin in lower left.

Fig. VB,26b (right). A packing drawing of the orthorhombic MoO_3 structure seen along its c_0 axis. The small, black circles are molybdenum.

The deformation of the oxygen packing in this structure (Fig. VB,26) is such that the distance between a molybdenum atom and its octahedrally distributed oxygen neighbors ranges from 1.88 to 2.45 A.

V,b25. There has been much not altogether conclusive work done on *tungstic oxide*, WO_3, and on a number of complex oxides of both tungsten and molybdenum.

It now appears that the symmetry of WO_3 is really monoclinic, pseudo-cubic, with a tetramolecular cell for which

$$a_0 = 7.274 \text{ A.;} \qquad b_0 = 7.501 \text{ A.;} \qquad c_0 = 3.824 \text{ A.;} \qquad \beta = 89°54'$$

The space group has been chosen as C_{2h}^5 ($P2_1/a$) with all atoms in the general positions:

$$(4e) \quad \pm (xyz; \; x+\frac{1}{2}, \frac{1}{2}-y, z)$$

According to one determination, atoms have the parameters:

Atom	x	y	z
W	0.256	0.229	0.053
O(1)	0.25	0.03	0.00
O(2)	0.00	0.25	0.00
O(3)	0.25	0.28	0.50

This places six oxygen atoms around each tungsten atom at distances between 1.52 and 2.11 A. The oxygen positions were not, however, determined experimentally and it is uncertain how much reliance can be put on the parameters stated above. In another study made at about the same time (1953), somewhat different positions were given the tungsten atoms, while recently it has been asserted that c_0 should have twice the foregoing length, thus resulting in a cell containing eight molecules and having the dimensions:

$$a_0 = 7.30 \text{ A.;} \qquad b_0 = 7.53 \text{ A.;} \qquad c_0 = 7.68 \text{ A.;} \qquad \beta = 90°54'$$

The continuingly unsatisfactory state of our knowledge of this structure is apparent.

Above 740°C., the symmetry of WO_3 becomes tetragonal. According to one determination the unit is tetramolecular, with

$$a_0 = 7.21 \text{ A.,} \qquad c_0 = 3.78 \text{ A.}$$

According to another, the unit is bimolecular, with

$$a_0 = 5.272 \text{ A.,} \qquad c_0 = 3.920 \text{ A.} \quad (950°C.)$$

For this unit a structure based on D_{4h}^7 ($P4/nmm$) has been proposed. The two tungsten atoms are in:

$$(2c) \quad 0 \; {}^1/_2 \; u; \; {}^1/_2 \; 0 \; \bar{u} \qquad \text{with } u = 0.06$$

Four oxygens have been put in $(4d)$: $\pm({}^1/_4 \; {}^1/_4 \; 0; \; {}^1/_4 \; {}^3/_4 \; 0)$ and two more in $(2c)$ with u perhaps near ${}^1/_2$.

The tungsten bronzes have the composition WO_3 plus varying amounts of an alkali metal. Structures based on limited x-ray data have been proposed for some of them. Thus, potassium, rubidium, and cesium tungsten bronzes with the composition $R_{ca \; 0.3}WO_3$ all have hexagonal units of substantially the same size:

$$a_0 = \text{ca. } 7.40 \text{ A.}, \qquad c_0 = \text{ca. } 7.55 \text{ A.}$$

There are six of these molecules in the unit and it has been proposed that their atoms occupy the following positions of D_{6h}^3 ($C6/mcm$):

W: $(6g)$ $\pm(u \; 0 \; {}^1/_4; \; 0 \; u \; {}^1/_4; \; \bar{u} \; \bar{u} \; {}^1/_4)$ with $u = 0.48$

O(1): $(6f)$ ${}^1/_2 \; 0 \; 0; \; 0 \; {}^1/_2 \; 0; \; {}^1/_2 \; {}^1/_2 \; 0; \; {}^1/_2 \; 0 \; {}^1/_2; \; 0 \; {}^1/_2 \; {}^1/_2; \; {}^1/_2 \; {}^1/_2 \; {}^1/_2$

O(2): $(12j)$ $\pm(u \; v \; {}^1/_4; \; \bar{v}, u - v, {}^1/_4; \; v - u, \bar{u}, {}^1/_4;$

$\qquad\qquad\qquad v \; u \; {}^1/_4; \; \bar{u}, v - u, {}^1/_4; \; u - v, \bar{v}, {}^1/_4)$

$$\text{with } u = \text{ca. } 0.42, \; v = \text{ca. } 0.22$$

The alkali atoms, with no more than two per cell, are considered to occupy some of the large holes defined by $(2b)$ $000; \; 0 \; 0 \; {}^1/_2$.

In this structure, tungsten atoms are octahedrally surrounded by oxygens, these octahedra sharing corners with one another.

A neutron diffraction study has now been made of a sodium tungsten bronze of the composition $Na_{0.75}WO_3$. It is cubic with an eight-molecule cell of the approximate dimensions:

$$a_0 = 7.72 \text{ A.}$$

Atoms have been assigned the following positions based on O_h^9 ($Im3m$):

W: $(8c)$ $\pm({}^1/_4 \; {}^1/_4 \; {}^1/_4)$;   F.C.

O(1):   In 12 of the positions of $(48k)$ with $u = 0 \; 235$, $v = 0.011$

O(2):   In 12 of the positions of $(48m)$ with $u = 0.267$, $v = 0.011$

Na: In about 6 of the positions of

$\qquad$ $(6b)$ $0 \; {}^1/_2 \; {}^1/_2; \; {}^1/_2 \; 0 \; {}^1/_2; \; {}^1/_2 \; {}^1/_2 \; 0$;   B.C.

Further work is called for.

An extensive series of molybdenum oxides has been described with complex compositions lying between MoO_2 (**IV,b6**) and MoO_3 (**V,b24**). There are similar oxides of tungsten, some of which have the same and some different formulas. Atomic arrangements have been published for a number of these. The positions of the heavy metallic atoms may be considered to be reasonably well established by the x-ray data, but it is hard to be sure how much confidence to attach to the oxygen parameters which have been mainly chosen from considerations of packing.

The simplest of these oxides (usually designated as the gamma oxide) has the formula Mo_4O_{11}. Its symmetry is orthorhombic and its tetramolecular unit has the edge lengths:

$$a_0 = 24.4 \text{ A.}; \qquad b_0 = 5.45 \text{ A.}; \qquad c_0 = 6.70 \text{ A.}$$

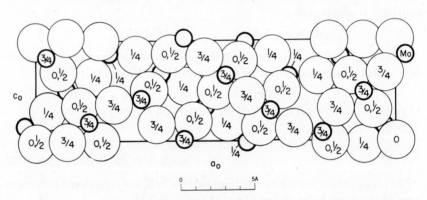

Fig. VB,27. The orthorhombic Mo_4O_{11} structure projected along its b_0 axis. Origin in lower left.

TABLE VB,10

Positions and Parameters of the Atoms in Mo_4O_{11} (γ-Oxide)

Atom	Position	x	y	z
Mo(1)	(4c)	−0.028	$1/4$	0.160
Mo(2)	(4c)	0.087	$1/4$	0.518
Mo(3)	(4c)	0.8535	$1/4$	0.804
Mo(4)	(4c)	0.2055	$1/4$	0.869
O(1)	(4a)	0	0	0
O(2)	(4c)	0.028	$1/4$	0.34
O(3)	(4c)	−0.088	$1/4$	−0.03
O(4)	(4c)	0.163	$1/4$	0.67
O(5)	(4c)	0.778	$1/4$	0.63
O(6)	(8d)	0.063	0.00	0.69
O(7)	(8d)	0.123	0.00	0.35
O(8)	(8d)	0.183	0.00	0.01

Atoms have been placed in the following special and general positions of V_h^{16} ($Pnma$):

(4a) $000; 0\,{}^1/_2\,0; {}^1/_2\,0\,{}^1/_2; {}^1/_2\,{}^1/_2\,{}^1/_2$

(4c) $\pm (u\,{}^1/_4\,v;\ u+{}^1/_2,{}^1/_4,{}^1/_2-v)$

(8d) $\pm (xyz;\ x,{}^1/_2-y,z;\ x+{}^1/_2,y,{}^1/_2-z;\ {}^1/_2-x,y+{}^1/_2,z+{}^1/_2)$

with the parameters listed in Table VB,10. The resulting arrangement is illustrated in Figure VB,27.

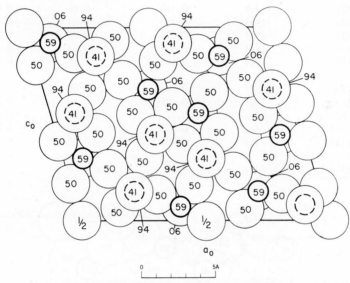

Fig. VB,28. The monoclinic Mo_8O_{23} arrangement projected along its b_0 axis.   The small, heavily ringed circles are molybdenum.

The compound Mo_8O_{23}, commonly called the β-oxide, is monoclinic with a bimolecular cell of the dimensions:

$$a_0 = 16.8\ \text{A.};\qquad b_0 = 4.04\ \text{A.};\qquad c_0 = 13.4\ \text{A.};\qquad \beta = 106°5'$$

The space group is C_{2h}^4 ($P2/a$), with one set of oxygen atoms in the special positions:

(2c) $0\,{}^1/_2\,0; {}^1/_2\,{}^1/_2\,0$

and all other atoms in the general positions:

(4g) $\pm (xyz;\ {}^1/_2-x,y,\bar{z})$

The parameters are those of Table VB,11. The arrangement is shown in Figure VB,28.

TABLE VB,11

Positions and Parameters of the Atoms in Mo_8O_{23} (β-Oxide)

Atom	Position	x	y	z
Mo(1)	(4g)	0.415	0.59	0.063
Mo(2)	(4g)	0.246	0.41	0.188
Mo(3)	(4g)	0.077	0.59	0.316
Mo(4)	(4g)	0.404	0.41	0.445
O(1)	(2c)	0	$^1/_2$	0
O(2)	(4g)	0.415	0.06	0.065
O(3)	(4g)	0.245	0.94	0.19
O(4)	(4g)	0.075	0.06	0.315
O(5)	(4g)	0.405	0.94	0.445
O(6)	(4g)	0.165	0.50	0.065
O(7)	(4g)	0.335	0.50	0.135
O(8)	(4g)	0.495	0.50	0.195
O(9)	(4g)	0.17	0.50	0.27
O(10)	(4g)	0.325	<0.50	0.325
O(11)	(4g)	0.005	ca. 0.50	0.405
O(12)	(4g)	0.16	ca. 0.50	0.46

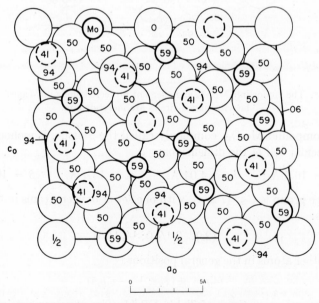

Fig. VB,29. The monoclinic Mo_9O_{26} structure projected along its b_0 axis. The small, heavily ringed circles are molybdenum.

TABLE VB,12
Positions and Parameters of the Atoms in Mo_9O_{26} (β'-oxide)

Atom	Position	x	y	z
Mo(1)	(2e)	$^1/_4$	0.59	0
Mo(2)	(4g)	0.456	0.41	0.113
Mo(3)	(4g)	0.162	0.41	0.221
Mo(4)	(4g)	0.369	0.59	0.334
Mo(5)	(4g)	0.082	0.41	0.450
O(1)	(2e)	$^1/_4$	0.06	0
O(2)	(4g)	0.455	0.94	0.155
O(3)	(4g)	0.16	0.94	0.22
O(4)	(4g)	0.37	0.06	0.335
O(5)	(4g)	0.08	0.94	0.45
O(6)	(4g)	0.355	0.50	0.06
O(7)	(4g)	0.21	0.50	0.115
O(8)	(4g)	0.06	0.50	0.18
O(9)	(4g)	0.42	0.50	0.23
O(10)	(4g)	0.26	0.50	0.285
O(11)	(4g)	0.125	<0.50	0.345
O(12)	(4g)	0.465	ca. 0.50	0.42
O(13)	(4g)	0.32	ca. 0.50	0.46
O(14)	(2c)	0	$^1/_2$	0

The oxide with the next more complicated formula is Mo_9O_{26} (often called the β'-oxide). It is monoclinic, with a bimolecular cell that has the dimensions:

$$a_0 = 16.75 \text{ A.}; \qquad b_0 = 4.03 \text{ A.}; \qquad c_0 = 14.45 \text{ A.}; \qquad \beta = 96°$$

The space group is the same as that chosen for Mo_8O_{23}, C_{2h}^4 ($P2/a$), with atoms in (2c) and (4g) just listed as well as in (2e) $\pm (^1/_4 \ u \ 0)$. The positions determined for the molybdenum atoms and chosen for oxygen are listed in Table VB,12. The structure is shown in Figure VB,29. Like that prevailing in all these oxides, it is one having each metal atom surrounded by six oxygen neighbors.

The oxide containing ten metallic atoms in its molecule is mixed and has the composition $(Mo,W)_{10}O_{29}$. Its bimolecular monoclinic unit has the edge lengths:

$$a_0 = 17.0 \text{ A.}; \qquad b_0 = 4.00 \text{ A.}; \qquad c_0 = 17.5 \text{ A.}; \qquad \beta = 111°$$

The same space group C_{2h}^4 has been selected and atoms have been placed in the same positions used for previous oxides. The positions and parameters are those of Table VB,13.

TABLE VB,13
Parameters Selected for the Oxygen Atoms in $(Mo,W)_{10}O_{29}$[a]

Atom	x	y	z
O(1)	0.117	0.00	0.050
O(2)	0.353	0.00	0.150
O(3)	0.091	0.00	0.252
O(4)	0.325	0.00	0.352
O(5)	0.065	0.00	0.457
O(6)	0	$^{1}/_{2}$	0
O(7)	0.365	0.50	0.050
O(8)	0.235	0.50	0.100
O(9)	0.105	0.50	0.151
O(10)	0.472	0.50	0.201
O(11)	0.340	0.50	0.251
O(12)	0.208	0.50	0.302
O(13)	0.078	0.50	0.354
O(14)	0.440	0.50	0.430
O(15)	0.315	0.50	0.465
[Mo,W](1)	0.117	0.42	0.050
[Mo,W](2)	0.353	0.58	0.150
[Mo,W](3)	0.091	0.58	0.252
[Mo,W](4)	0.325	0.42	0.352
[Mo,W](5)	0.065	0.58	0.457

[a] Oxygen positions not established experimentally.

Another mixed oxide has the composition $(Mo,W)_{11}O_{32}$. Its bimolecular monoclinic unit has the dimensions:

$$a_0 = 16.6 \text{ A.}; \qquad b_0 = 4.00 \text{ A.}; \qquad c_0 = 18.7 \text{ A.}; \qquad \beta = 74°$$

In spite of its differently shaped cell, the atomic distribution is very much like that prevailing in the other oxides. The space group is the same, C_{2h}^4, with atoms in the same special positions. These positions and the parameters used are listed in Table VB,14.

There is a considerable gap in composition between these mixed oxides and the next more complex oxide, the molybdenum compound, $Mo_{17}O_{47}$. It is orthorhombic with two molecules in a cell of the dimensions:

$$a_0 = 21.615 \text{ A.}; \qquad b_0 = 19.632 \text{ A.}; \qquad c_0 = 3.9515 \text{ A.}$$

TABLE VB,14

Parameters Selected for the Oxygen Atoms in $(Mo,W)_{11}O_{32}$[a]

Atom	Position	x	y	z
O(1)	(2e)	$^1/_4$	0.00	0
O(2)	(4g)	0.418	0.00	0.092
O(3)	(4g)	0.088	0.00	0.183
O(4)	(4g)	0.257	0.00	0.277
O(5)	(4g)	0.427	0.00	0.370
O(6)	(4g)	0.096	0.00	0.458
O(7)	(2c)	0	$^1/_2$	0
O(8)	(4g)	0.334	0.50	0.046
O(9)	(4g)	0.169	0.50	0.091
O(10)	(4g)	0.003	0.50	0.137
O(11)	(4g)	0.337	0.50	0.185
O(12)	(4g)	0.173	0.50	0.230
O(13)	(4g)	0.008	0.50	0.276
O(14)	(4g)	0.342	0.50	0.324
O(15)	(4g)	0.177	0.50	0.367
O(16)	(4g)	0.496	0.50	0.432
O(17)	(4g)	0.333	0.50	0.467
[Mo,W](1)	(2e)	$^1/_4$	0.60	0
[Mo,W](2)	(4g)	0.418	0.40	0.092
[Mo,W](3)	(4g)	0.088	0.40	0.183
[Mo,W](4)	(4g)	0.257	0.60	0.277
[Mo,W](5)	(4g)	0.427	0.40	0.370
[Mo,W](6)	(4g)	0.096	0.60	0.458

[a] Oxygen positions not determined experimentally.

Atoms have been placed in the following positions of $C_{2v}{}^8$ (Pba2):

(2a) $00u;\ ^1/_2\ ^1/_2\ u$

(4c) $xyz;\ \bar{x}\bar{y}z;\ ^1/_2-x,y+^1/_2,z;\ x+^1/_2,^1/_2-y,z$

with the parameters listed in Table VB,15.

The resulting structure (Fig. VB,30) can be thought of as built up of polyhedra sharing corners and edges. Those around six of the molybdenum atoms are distorted octahedra, that about Mo(7) is approximately a pentagonal bipyramid. The Mo–O separations are from 1.70 A. upwards.

CRYSTAL STRUCTURES

TABLE VB,15
Positions and Parameters of the Atoms in $Mo_{17}O_{47}$

Atom	Position	x	y	z
Mo(1)	(2a)	0	0	0.579
Mo(2)	(4c)	0.0244	0.2606	0.581
Mo(3)	(4c)	0.1299	0.1189	0.425
Mo(4)	(4c)	0.1355	0.4005	0.434
Mo(5)	(4c)	0.2424	0.2573	0.575
Mo(6)	(4c)	0.2881	0.0653	0.579
Mo(7)	(4c)	0.3824	0.1944	0.428
Mo(8)	(4c)	0.3851	0.3661	0.424
Mo(9)	(4c)	0.4657	0.0556	0.565
O(1)	(2a)	0	0	0.01
O(2)	(4c)	0.020	0.266	0.01
O(3)	(4c)	0.126	0.115	0.98
O(4)	(4c)	0.140	0.396	0.98
O(5)	(4c)	0.246	0.251	0.00
O(6)	(4c)	0.288	0.069	0.01
O(7)	(4c)	0.383	0.191	0.00
O(8)	(4c)	0.392	0.360	0.00
O(9)	(4c)	0.465	0.052	0.05
O(10)	(4c)	0.076	0.045	0.46
O(11)	(4c)	0.060	0.181	0.52
O(12)	(4c)	0.093	0.324	0.44
O(13)	(4c)	0.059	0.462	0.57
O(14)	(4c)	0.206	0.067	0.48
O(15)	(4c)	0.180	0.195	0.46
O(16)	(4c)	0.203	0.335	0.50
O(17)	(4c)	0.191	0.475	0.54
O(18)	(4c)	0.297	0.165	0.48
O(19)	(4c)	0.330	0.280	0.48
O(20)	(4c)	0.325	0.427	0.41
O(21)	(4c)	0.380	0.091	0.56
O(22)	(4c)	0.465	0.160	0.50
O(23)	(4c)	0.434	0.275	0.52
O(24)	(4c)	0.455	0.411	0.52

The next more complicated oxide is of tungsten and has the formula $W_{18}O_{49}$. It is the gamma phase of the tungsten–oxygen system. Its unimolecular monoclinic cell resembles other monoclinic phases in dimensions:

$$a_0 = 18.32 \text{ A.}; \quad b_0 = 3.79 \text{ A.}; \quad c_0 = 14.04 \text{ A.}; \quad \beta = 115°12'$$

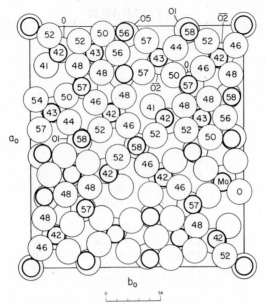

Fig. VB,30. The orthorhombic $Mo_{17}O_{47}$ structure projected along its c_0 axis. The small, heavily ringed circles are molybdenum.

The space group is, however, the different C_{2h}^1 $(P2/m)$, and atoms have been put in the positions:

$$(1h) \quad \tfrac{1}{2}\,\tfrac{1}{2}\,\tfrac{1}{2} \qquad (2m) \quad \pm(u0v) \qquad (2n) \quad \pm(u\,\tfrac{1}{2}\,v)$$

The selected parameters are listed in Table VB,16.

The most complicated oxide for which an arrangement has been proposed is $W_{20}O_{58}$. Its unimolecular monoclinic cell has been given the dimensions:

$$a_0 = 12.05 \text{ A.}; \qquad b_0 = 3.767 \text{ A.}; \qquad c_0 = 23.59 \text{ A.}; \qquad \beta = 85°17'$$

The space group again is C_{2h}^1, with atoms in the positions:

$$(1b) \quad 0\,\tfrac{1}{2}\,0 \qquad (1e) \quad \tfrac{1}{2}\,\tfrac{1}{2}\,0$$
$$(2m) \quad \pm(u0v) \qquad (2n) \quad \pm(u\,\tfrac{1}{2}\,v)$$

and the parameters stated in Table VB,17.

V,b26. A restudy of *chromium trioxide*, CrO_3, has led to a structure different from that originally selected. The orthorhombic unit is, however, the same as before with a tetramolecular cell of the dimensions:

$$a_0 = 5.743 \text{ A.}; \qquad b_0 = 8.557 \text{ A.}; \qquad c_0 = 4.789 \text{ A.}$$

TABLE VB,16
Positions and Parameters of the Atoms in $W_{18}O_{49}$ (γ-Oxide)

Atom	Position	x	y	z
W(1)	(2n)	0.072	$1/_2$	0.003
W(2)	(2n)	0.084	$1/_2$	0.285
W(3)	(2n)	0.127	$1/_2$	0.759
W(4)	(2n)	0.218	$1/_2$	0.575
W(5)	(2n)	0.257	$1/_2$	0.015
W(6)	(2n)	0.277	$1/_2$	0.257
W(7)	(2n)	0.363	$1/_2$	0.869
W(8)	(2n)	0.414	$1/_2$	0.536
W(9)	(2n)	0.453	$1/_2$	0.168
O(1)	(2m)	0.07	0	0.005
O(2)	(2m)	0.085	0	0.285
O(3)	(2m)	0.125	0	0.76
O(4)	(2m)	0.22	0	0.575
O(5)	(2m)	0.255	0	0.015
O(6)	(2m)	0.275	0	0.255
O(7)	(2m)	0.365	0	0.87
O(8)	(2m)	0.415	0	0.535
O(9)	(2m)	0.455	0	0.17
O(10)	(2n)	0.015	$1/_2$	0.715
O(11)	(2n)	0.035	$1/_2$	0.115
O(12)	(2n)	0.125	$1/_2$	0.905
O(13)	(2n)	0.125	$1/_2$	0.43
O(14)	(2n)	0.14	$1/_2$	0.625
O(15)	(2n)	0.185	$1/_2$	0.11
O(16)	(2n)	0.19	$1/_2$	0.30
O(17)	(2n)	0.25	$1/_2$	0.87
O(18)	(2n)	0.29	$1/_2$	0.715
O(19)	(2n)	0.32	$1/_2$	0.555
O(20)	(2n)	0.335	$1/_2$	0.15
O(21)	(2n)	0.35	$1/_2$	0.39
O(22)	(2n)	0.40	$1/_2$	0.015
O(23)	(2n)	0.445	$1/_2$	0.82
O(24)	(2n)	0.495	$1/_2$	0.32
O(25)	(1h)	$1/_2$	$1/_2$	$1/_2$

TABLE VB,17
Parameters Given the Atoms in $W_{20}O_{58}$

Atom	Position	x	y	z
W(1)	(2n)	0.353	$^1/_2$	0.026
W(2)	(2n)	0.063	$^1/_2$	0.077
W(3)	(2n)	0.773	$^1/_2$	0.128
W(4)	(2n)	0.483	$^1/_2$	0.178
W(5)	(2n)	0.192	$^1/_2$	0.229
W(6)	(2n)	0.903	$^1/_2$	0.280
W(7)	(2n)	0.610	$^1/_2$	0.331
W(8)	(2n)	0.322	$^1/_2$	0.382
W(9)	(2n)	0.030	$^1/_2$	0.432
W(10)	(2n)	0.740	$^1/_2$	0.483

O(1)–O(10) in (2m) with $x[O](2m) = x[W](2n)$ and $z[O](2m) = z[W](2n)$

O(11)	(1b)	0	$^1/_2$	0
O(12)	(1e)	$^1/_2$	$^1/_2$	0
O(13)	(2n)	0.208	$^1/_2$	0.052
O(14)	(2n)	0.710	$^1/_2$	0.051
O(15)	(2n)	0.417	$^1/_2$	0.102
O(16)	(2n)	0.128	$^1/_2$	0.153
O(17)	(2n)	0.918	$^1/_2$	0.103
O(18)	(2n)	0.627	$^1/_2$	0.153
O(19)	(2n)	0.337	$^1/_2$	0.204
O(20)	(2n)	0.048	$^1/_2$	0.255
O(21)	(2n)	0.838	$^1/_2$	0.204
O(22)	(2n)	0.546	$^1/_2$	0.255
O(23)	(2n)	0.257	$^1/_2$	0.306
O(24)	(2n)	0.757	$^1/_2$	0.306
O(25)	(2n)	0.466	$^1/_2$	0.357
O(26)	(2n)	0.187	$^1/_2$	0.422
O(27)	(2n)	0.967	$^1/_2$	0.456
O(28)	(2n)	0.675	$^1/_2$	0.407
O(29)	(2n)	0.395	$^1/_2$	0.475
O(30)	(2n)	0.100	$^1/_2$	0.526

The space group is now chosen as C_{2v}^{16} (Ama), with atoms in its special positions:

$(4a)$ $00u;$ $^1/_2\,0\ u;$ $0,^1/_2,u+^1/_2;$ $^1/_2,^1/_2,u+^1/_2$

$(4b)$ $^1/_4\,u\,v;$ $^3/_4\,\bar{u}\,v;$ $^1/_4,u+^1/_2,v+^1/_2;$ $^3/_4,^1/_2-u,v+^1/_2$

The selected parameters are given below:

Atom	Position	x	y	z
Cr	$(4b)$	$1/4$	$0.403\ (.400)$	$0.194\ (.500)$
O(1)	$(4a)$	0	0	$0.556\ (.883)$
O(2)	$(4b)$	$1/4$	$0.222\ (.225)$	$0.000\ (.433)$
O(3)	$(4b)$	$1/4$	$0.278\ (.342)$	$0.500\ (.800)$

This is an arrangement (Fig. VB,31) composed of CrO_3 chains held together only by van der Waals forces. Within the chains, each chromium atom has four oxygen neighbors with Cr–O lying between 1.79 and 1.81 A.

A somewhat different atomic distribution based on the same experimental data has still more recently been proposed. The space group is unchanged, but atoms have been assigned the parameters parenthesized above. The origin is shifted by 0.300 along the c_0 axis compared to that used earlier, but, allowing for this, there still remain significant differences in the oxygen positions.

Additional work is desirable.

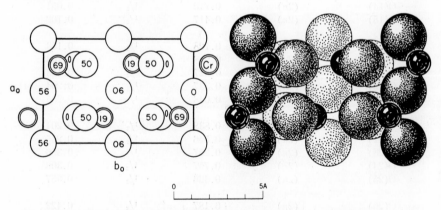

Fig. VB,31a (left). The orthorhombic structure of CrO_3 projected along its c_0 axis.
Fig. VB,31b (right). A packing drawing of the CrO_3 arrangement seen along its c_0 axis. The chromium atoms are small and black.

V,b27. The *cesium suboxide*, Cs_3O, forms hexagonal crystals. Its bimolecular unit has the dimensions:

$$a_0 = 8.78\ \text{A.,} \qquad c_0 = 7.52\ \text{A.}$$

Atoms have been placed in the following positions of D_{6h}^3 $(C6/mcm)$:

O: $(2b)$ $000;\ 0\ 0\ 1/2$

Cs: $(6g)$ $\pm(u\ 0\ 1/4;\ 0\ u\ 1/4;\ \bar{u}\ \bar{u}\ 1/4)$ with $u = 0.250$

In this metal-like compound (Fig. VB,32), Cs–O = 2.89 A., about equal to the sum of the ionic radii. The shortest Cs–Cs separation is 4.34 A.; that between layers involving only metallic contacts is 4.90 A. This is to be compared with the 5.26 A. prevailing in metallic cesium.

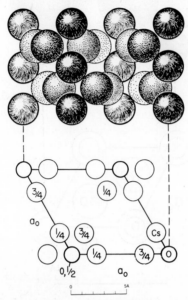

Fig. VB,32. Two projections of the hexagonal Cs_3O structure.   In the upper projection the cesium atoms are dotted.

V,b28.   A structure has been assigned *α-uranium trioxide*, UO_3, which relates it to that of U_3O_8 (**V,f12, 13**). The hexagonal cell containing one molecule has the dimensions:

$$a_0 = 3.971 \text{ A.}, \qquad c_0 = 4.168 \text{ A.}$$

Atomic positions developed from D_{3d}^3 ($C\overline{3}m$) are

$$\begin{array}{lll} \text{U:} & (1a) & 000 \\ \text{O(1):} & (1b) & 0\ 0\ {}^1/_2 \\ \text{O(2):} & (2d) & {}^1/_3\ {}^2/_3\ u;\ {}^2/_3\ {}^1/_3\ \bar{u} & \text{with } u = 0.7 \end{array}$$

In this arrangement (Fig. VB,33), U–O(1) = 2.08 A. and U–O(2) = 2.39 A.

V,b29.   The ice-like, or gamma, form of solid *sulfur trioxide*, SO_3, is a packing of molecules of its trimer S_3O_9. The symmetry is orthorhombic, with a unit containing four of these molecules and having the dimensions:

$$a_0 = 12.3 \text{ A.}; \qquad b_0 = 10.7 \text{ A.}; \qquad c_0 = 5.3 \text{ A.} \quad (\text{ca.} -10°\text{C.})$$

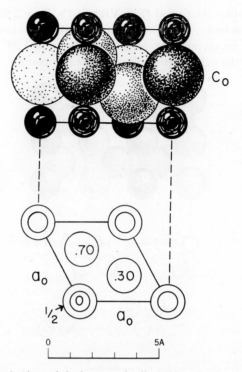

Fig. VB,33. Two projections of the hexagonal cell of UO_3. In the upper projection the uranium atoms are small and black.

The space group is C_{2v}^9 (*Pbn*) and therefore all atoms are in several sets of general positions:

$$(4a) \quad xyz; \ \bar{x},\bar{y},z+{}^1\!/_2; \ {}^1\!/_2-x,y+{}^1\!/_2,z; \ x+{}^1\!/_2,{}^1\!/_2-y,z+{}^1\!/_2$$

The established parameters are listed in Table VB,18.

As can be seen from Figure VB,34, the molecules are puckered rings of SO_2 groups held together by sharing an oxygen atom, the interatomic distances being those given in the figure. To form the crystal, they are packed as indicated in Figure VB,35, with a closest approach through oxygen atoms of 2.9 A.

V,b30. The second, β or asbestos-like, form of *sulfur trioxide*, SO_3, is monoclinic, with a tetramolecular unit of the dimensions:

$$a_0 = 6.20 \text{ A.}; \quad b_0 = 4.06 \text{ A.}; \quad c_0 = 9.31 \text{ A.}; \quad \beta = 109°50'$$

TABLE VB,18
Parameters of the Atoms in γ-SO$_3$

Atom	x	y	z
S(1)	−0.075	0.119	0.191
S(2)	−0.146	0.346	0.017
S(3)	0.078	0.303	0.074
O(1)	−0.086	0.175	0.415
O(2)	−0.154	0.375	0.260
O(3)	0.071	0.350	0.328
O(4)	−0.153	0.194	0
O(5)	−0.090	0.010	0.157
O(6)	0.042	0.163	0.079
O(7)	0.175	0.318	0.066
O(8)	−0.022	0.364	−0.079
O(9)	−0.219	0.397	−0.157

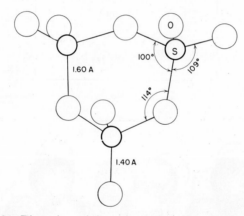

Fig. VB,34. Dimensions of the molecule found for the γ form of SO$_3$.

The space group is C_{2h}^5 ($P2_1/c$) and all atoms are in general positions:

$$(4e) \quad \pm(xyz; \; x,{}^1\!/_2-y,z+{}^1\!/_2)$$

with the parameters:

Atom	x	y	z
S	0.171	0.02	0.294
O(1)	0.001	0.27	0.169
O(2)	0.270	0.20	0.431
O(3)	0.281	−0.15	0.207

As can be seen from Figure VB,36, this leads to a structure in which there are endless $(SO_3)_n$ chains spiraling upwards in the b_0 direction. The S–O distances are 1.41, 1.59, and 1.63 A.; O–O separations within a chain vary from 2.37 to 2.54 A. The O–S–O angles lie between 102° and 128°. Between chains, the shortest O–O distance is 3.09 A. It is pointed out that this nonmolecular modification can be thought of as one in which the oxygen atoms are approximately in a cubic close-packing only slightly distorted by the presence of the atoms of sulfur.

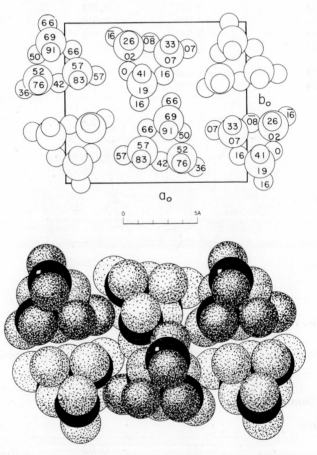

Fig. VB,35a (top). A projection along the c_0 axis of the orthorhombic structure of γ-SO$_3$. The sulfur are the larger circles.   Origin in lower right.

Fig. VB,35b (bottom). A packing drawing of the orthorhombic structure of γ-SO$_3$ viewed along the c_0 axis.   The oxygen atoms are dotted.

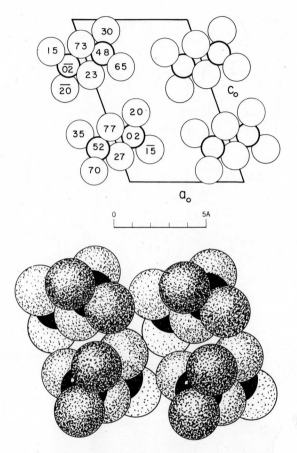

Fig. VB,36a (top). A projection along the b_0 axis of the monoclinic structure of β-SO_3. The larger circles are oxygen. Origin in lower left.

Fig. VB,36b (bottom). A packing drawing of the monoclinic structure of β-SO_3 viewed along the b_0 axis. The oxygen atoms are dotted.

Sulfides and Selenides

V,b31. *Cesium hexasulfide*, Cs_2S_6, forms triclinic crystals, with a bimolecular unit having the dimensions:

$$a_0 = 11.53 \text{ A.}; \qquad b_0 = 9.18 \text{ A.}; \qquad c_0 = 4.67 \text{ A.}$$
$$\alpha = 89°13'; \qquad \beta = 95°12'; \qquad \gamma = 95°8'$$

All atoms are in the general positions of C_i^1 ($P\bar{1}$):

$$(2i) \quad \pm(xyz)$$

with the parameters of Table VB.19.

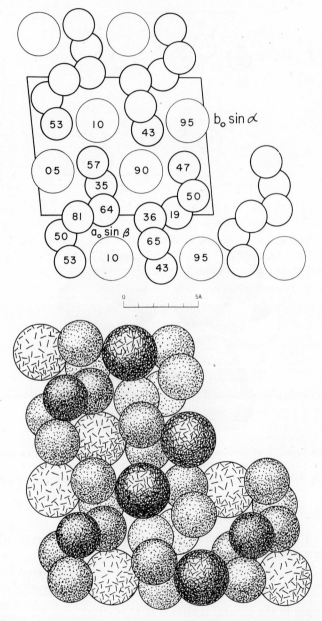

Fig. VB,37a (top). A projection along the c_0 axis of the triclinic structure of Cs_2S_6. The larger circles are cesium. Origin in lower left, c_0 direction downwards.

Fig. VB,37b (bottom). A packing drawing of the triclinic structure of Cs_2S_6 viewed along the c_0 axis. The cesium atoms are line shaded.

TABLE VB,19
Parameters of the Atoms in Cs_2S_6

Atom	x	y	z
Cs(1)	0.0959	0.3237	0.0500
Cs(2)	0.3849	−0.3103	0.1018
S(1)	0.1456	−0.3432	0.5312
S(2)	0.1024	−0.1379	0.5022
S(3)	0.2137	−0.0146	0.8085
S(4)	0.3689	0.0272	0.6355
S(5)	0.3601	0.2086	0.3518
S(6)	0.3295	0.3872	0.5752

This structure (Fig. VB,37) shows the S_6 ion to be a twisted, nonplanar string with the bond dimensions of Figure VB,38. In the crystal, these strings, separated by van der Waals distances of 3.39 A., form helices of sulfur atoms between which the cesium atoms lie embedded. The shortest Cs–S separations range from 3.48 A. upward.

Fig. VB,38. The dimensions of the S_6 ion as found from the structure for Cs_2S_6. The important dihedral angles of this new planar molecule are: $S_1S_2S_3/S_2S_3S_4 = 101°14'$; $S_2S_3S_4/S_3S_4S_5 = 98°4'$; $S_3S_4S_5/S_4S_5S_6 = 118°36'$.

V,b32. A structure which cannot be considered as well established was proposed some time ago for *barium trisulfide*, BaS_3. The powder diffractions on which this study was based could all be accounted for by a tetragonal unit having a single molecule and the dimensions:

$$a_0 = 4.82 \text{ A.}, \qquad c_0 = 4.16 \text{ A.}$$

An orthorhombic cell with

$$a_0' = 8.32 \text{ A.}; \quad b_0' = 9.64 \text{ A.}; \quad c_0' = 4.28 \text{ A.}$$

was, however, chosen instead, because it was concluded that a satisfactory structure could not be built upon the foregoing simple tetragonal unit. The suggested orthorhombic arrangement, based on the space group V^3 ($P2_12_12$) has atoms in the following positions:

$$
\begin{aligned}
&\text{Ba(1):} \quad (2a) \quad 00u; \; {}^1/_2 \, {}^1/_2 \, \bar{u} \qquad \text{with } u = 0 \\
&\text{Ba(2):} \quad (2b) \quad 0 \, {}^1/_2 \, v; \; {}^1/_2 \, 0 \, \bar{v} \qquad \text{with } v = 0 \\
&\text{S(1):} \quad (4c) \quad xyz; \; \bar{x}\bar{y}z; \; x+{}^1/_2, {}^1/_2-y, \bar{z}; \; {}^1/_2-x, y+{}^1/_2, \bar{z} \\
&\qquad\qquad\qquad\qquad\qquad\qquad \text{with } x = {}^1/_4 = y, z = 0 \\
&\text{S(2):} \quad (4c) \quad \text{with } x = {}^1/_4, \; y = z = {}^1/_2 \\
&\text{S(3):} \quad (4c) \quad \text{with } x = 0.124, \; y = 0.309, \; z = 0.382
\end{aligned}
$$

Evidently, a further study should be made of this substance.

V,b33. *Zirconium triselenide*, $ZrSe_3$, is monoclinic with the bimolecular cell:

$$a_0 = 5.41 \text{ A.}; \quad b_0 = 3.77 \text{ A.}; \quad c_0 = 9.45 \text{ A.}; \quad \beta = 97°30'$$

According to a preliminary announcement, atoms are probably in positions of C_2^2 ($P2_1$):

$$(2a) \quad xyz; \; \bar{x}, y+{}^1/_2, \bar{z}$$

The x and z parameters have been determined as:

Atom	x	z
Zr	0.715	0.343
Se(1)	−0.764	−0.553
Se(2)	−0.455	−0.175
Se(3)	−0.888	−0.169

and it is thought that y for all atoms is near to ${}^1/_4$. Additional work is required to complete this determination.

Other substances which appear to have this arrangement have cells of the dimensions:

$$\text{TiS}_3: \quad a_0 = 4.97 \text{ A.}; b_0 = 3.42 \text{ A.}; c_0 = 8.78 \text{ A.}; \beta = 97°10'$$
$$\text{USe}_3: \quad a_0 = 5.68 \text{ A.}; b_0 = 4.06 \text{ A.}; c_0 = 9.64 \text{ A.}; \beta = 99°30'$$

Nitrides, Phosphides, etc.

V,b34. A structure, which appears to need further investigation, has been proposed for *copper trinitride*, CuN_3. It is tetragonal, with

$$a_0 = 8.653 \text{ A.}, \qquad c_0 = 5.594 \text{ A.}$$

The atoms of the eight molecules in this unit have been given the following positions of C_{4h}^6 $(I4_1/a)$:

Cu:	$(8d)$	$0\ ^1/_4\ ^5/_8;\ 0\ ^3/_4\ ^5/_8;\ ^1/_4\ 0\ ^3/_8;\ ^3/_4\ 0\ ^3/_8;$ B.C.
N(1):	$(8c)$	$0\ ^1/_4\ ^1/_8;\ 0\ ^3/_4\ ^1/_8;\ ^1/_4\ 0\ ^7/_8;\ ^3/_4\ 0\ ^7/_8;$ B.C.
N(2):	$(16f)$	$xyz;\ \bar{x}\bar{y}z;\ x,y+^1/_2,^1/_4-z;\ \bar{x},^1/_2-y,^1/_4-z;$
		$\bar{y}x\bar{z};\ y\bar{x}\bar{z};\ \bar{y},x+^1/_2,z+^1/_4;\ y,^1/_2-x,z+^1/_4;$ B.C.

with $x = 0.077,\ y = 0.173,\ z = 0.250$

This arrangement bears no obvious relationship to others thus far described.

V,b35. *Nickel phosphide*, Ni_3P, is tetragonal with a unit containing eight molecules and having the dimensions:

$$a_0 = 8.954 \text{ A.}, \qquad c_0 = 4.386 \text{ A.}$$

The space group is the low symmetry S_4^2 $(I\bar{4})$, with all atoms in the general positions:

$$(8g) \quad xyz;\ \bar{x}\bar{y}z;\ y\bar{x}\bar{z};\ \bar{y}x\bar{z};\quad \text{B.C.}$$

The parameters were found to be:

Atom	x	y	z
Ni(1)	0.0775	0.1117	0.2391
Ni(2)	0.3649	0.0321	0.9765
Ni(3)	0.1689	0.2200	0.7524
P	0.2862	0.0487	0.4807

In this structure (Fig. VB,39), the shortest Ni–P = 2.21 A. and the shortest Ni–Ni = 2.43 A. Each phosphorus atom has nine nickel neighbors at distances between 2.21 and 2.44 A. The three kinds of nickel atoms have different environments, with Ni–Ni distances increasing so gradually that no exact coordination can be specified.

Other compounds for which this structure has been established in detail are as follows:

$$Fe_3P: \quad a_0 = 9.107 \text{ A.},\ c_0 = 4.460 \text{ A.}$$

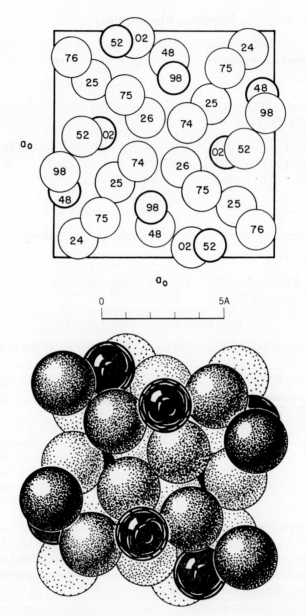

Fig. VB,39a (top). A projection of the tetragonal structure of Ni₃P along its c_0 axis. The phosphorus atoms are the smaller, heavily ringed circles.

Fig. VB,39b (bottom). A packing drawing of the Ni₃P structure viewed along its c_0 axis. The nickel atoms are dotted.

The atomic parameters for this substance are:

Atom	x	y	z
Fe(1)	0.0793	0.1059	0.2338
Fe(2)	0.3605	0.0310	0.9860
Fe(3)	0.1717	0.2195	0.7548
P	0.2921	0.0454	0.4903

Mn_3P: $a_0 = 9.178$ A. (or 9.181 A.), $c_0 = 4.608$ A. (or 4.568 A.)
The atomic parameters are:

Atom	x	y	z
Mn(1)	0.0807	0.1071	0.2279
Mn(2)	0.3567	0.0319	0.9863
Mn(3)	0.1721	0.2192	0.7531
P	0.2935	0.0450	0.4880

Other substances which have been shown to have this arrangement but for which parameters have not been established are:

$$Cr_3P: \quad a_0 = 9.144 \text{ A.}, \quad c_0 = 4.569 \text{ A.}$$
$$Mo_3P: \quad a_0 = 9.729 \text{ A.}, \quad c_0 = 4.923 \text{ A.}$$
$$Pd_3As: \quad a_0 = 9.951 \text{ A.}, \quad c_0 = 4.817 \text{ A.}$$
$$(Fe_1 \text{ Ni, Co})_3 P: \quad a_0 = 9.09 \text{ A.}, \quad c_0 = 4.42 \text{ A.}$$
(schreibersite)

The β (high-temperature) modification of *trivanadium sulfide*, V_3S, has a unit like the foregoing and might be expected to have the same structure, but it has been assigned the higher symmetry space group $V_d{}^{11}$ ($I\bar{4}2m$). The eight-molecule tetragonal unit has the edge lengths:

$$a_0 = 9.470 \text{ A.}, \qquad c_0 = 4.589 \text{ A.}$$

Atoms have been placed in the special positions:

V(1): (8i) $uuv; \bar{u}\bar{u}v; u\bar{u}\bar{v}; \bar{u}u\bar{v};$ B.C.   with $u = 0.0932, v = 0.250$
V(2): (8i) with $u = 0.2000, v = 0.750$
V(3): (8f) $\pm(u00; 0u0)$; B.C.   with $u = 0.3550$
  S: (8g) $\pm(u \ 0 \ ^1/_2; 0 \ u \ ^1/_2)$; B.C.   with $u = 0.2851$

This arrangement (Fig. VB,40) and that chosen for Ni_3P differ by only minor atomic displacements, but it would be important through further work to be sure that two different structures actually exist.

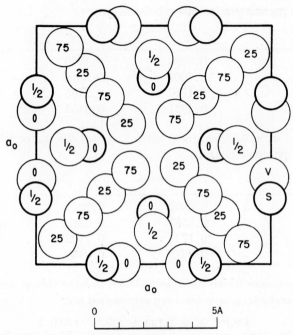

Fig. VB,40. The tetragonal arrangement of high-temperature V_3S projected along its c_0 axis.

The low-temperature, α. form of *trivanadium sulfide*, V_3S, has a tetragonal unit of nearly the same size as that of the β modification. For it,

$$a_0 = 9.381 \text{ A.}, \qquad c_0 = 4.663 \text{ A.}$$

The chosen space group is of still higher symmetry, $D_{4h}{}^{11}$ ($P4_2/nbc$), and atoms are in the following of its special positions:

V(1): (8*j*) $\pm(u,u+{}^1/_2,0;\ u+{}^1/_2,\bar{u},0;\ u,{}^1/_2-u,{}^1/_2;\ u+{}^1/_2,u,{}^1/_2)$

$\qquad\qquad\qquad\qquad\qquad\qquad\qquad\qquad\qquad\qquad$ with $u = 0.4080$

V(2): (8*j*) with $u = 0.2028$

V(3): (8*i*) $u\ 0\ {}^3/_4;\ \bar{u}\ 0\ {}^3/_4;\ u+{}^1/_2,{}^1/_2,{}^3/_4;\ {}^1/_2-u,{}^1/_2,{}^3/_4;$

$\qquad\qquad\quad\ 0\ u\ {}^1/_4;\ 0\ \bar{u}\ {}^1/_4;\ {}^1/_2,\ u+{}^1/_2,{}^1/_4;\ {}^1/_2,{}^1/_2-u,{}^1/_4$

$\qquad\qquad\qquad\qquad\qquad\qquad\qquad\qquad\qquad\qquad$ with $u = 0.1486$

S: (8*h*) $u\ 0\ {}^1/_4;\ \bar{u}\ 0\ {}^1/_4;\ u+{}^1/_2,{}^1/_2,{}^1/_4;\ {}^1/_2-u,{}^1/_2,{}^1/_4;$

$\qquad\qquad\quad 0\ u\ {}^3/_4;\ 0\ \bar{u}\ {}^3/_4;\ {}^1/_2,u+{}^1/_2,{}^3/_4;\ {}^1/_2,{}^1/_2-u,\ {}^3/_4$

$\qquad\qquad\qquad\qquad\qquad\qquad\qquad\qquad\qquad\qquad$ with $u = 0.217$

A comparison of Figure VB,41 with Figure VB,40 shows the great similarity that exists between this structure and that of the β modification. These

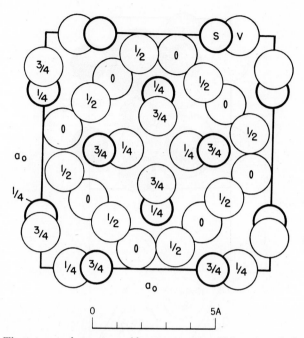

Fig. VB,41. The tetragonal structure of low-temperature V_3S projected along its c_0 axis.

basal projections are almost identical, differences being in the atomic heights along the c_0 axis and a displacement of the origin by $1/2\ a_0$.

In both structures the vanadium atoms have either two or four sulfur neighbors at distances between 2.31 and 2.47 A. The shortest V–V separations range upward from 2.44 A.

V,b36. *Sodium arsenide*, Na_3As, is typical of a number of phosphides, arsenides, and related compounds having the same structure. They are hexagonal with bimolecular cells. For Na_3As the unit has the edges:

$$a_0 = 5.088 \text{ A.,} \qquad c_0 = 8.982 \text{ A.}$$

The proposed arrangement puts its atoms in the following special positions of D_{6h}^4 ($C6/mmc$):

As: (2c) $\pm(1/3\ 2/3\ 1/4)$
Na(1): (2b) $\pm(0\ 0\ 1/4)$
Na(2): (4f) $\pm(1/3\ 2/3\ u;\ 2/3,1/3,u+1/2)$ with $u = 0.583$

In the resulting simple structure (Fig. VB,42), the closest Na–As separations are 2.94 and 2.99 A.

Other crystals known to have this grouping are listed in Table VB.20.

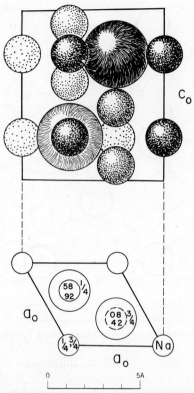

Fig. VB,42. Two projections of the hexagonal unit of Na₃As. In the upper projection
the large arsenic atoms are fine-line shaded.

V,b37. *Cobalt triarsenide*, CoAs₃, which occurs as the mineral skutterudite, is cubic. Its cell, containing eight molecules, has the edge length:

$$a_0 = 8.189 \text{ A.}$$

Atoms have been placed in the following special positions of T_h^5 (*Im3*):

Co: (8*e*) $^1/_4\ ^1/_4\ ^1/_4$; $^1/_4\ ^3/_4\ ^3/_4$; $^3/_4\ ^1/_4\ ^3/_4$; $^3/_4\ ^3/_4\ ^1/_4$; B.C.

As: (24*d*) $\pm(0uv;\ v0u;\ uv0;\ 0u\bar{v};\ \bar{v}0u;\ u\bar{v}0)$; B.C.

with $u = 0.35$ and $v = 0.15$

In this arrangement (Fig. VB,43), each cobalt atom is octahedrally surrounded by six arsenic neighbors at a distance of 2.35 A., while each arsenic
atom at the center of a tetrahedron has two cobalt neighbors at 2.35 A. and
two arsenics at 2.45 A. These are separations to be expected if the atoms
have their usual neutral radii.

TABLE VB,20
Crystals with the Hexagonal Na₃As Structure

Crystal	a_0, A.	c_0, A.
K₃As	5.782	10.222
K₃Bi	6.178	10.933
K₃P	5.691	10.05
K₃Sb	6.025	10.693
Li₃As	4.387	7.810
Li₃P	4.264	7.579
α-Li₃Sb	4.710	8.326
Na₃Bi	5.448	9.655
Na₃P	4.980	8.797
Na₃Sb	5.355	9.496
Rb₃As	6.052	10.73
Rb₃Bi	6.42	11.46
Rb₃Sb	6.283	11.18

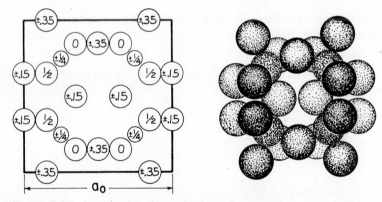

Fig. VB,43a (left). A projection on a cube face of atoms in a unit of CoAs₃. The smaller circles are the cobalt atoms.
Fig. VB,43b (right). A packing drawing of the CoAs₃ atoms shown in Figure VB,43a, giving them their neutral radii. The line-shadowed spheres are the cobalt atoms.

Detailed studies have been made of two other crystals with this arrangement. They are:

$$\text{IrSb}_3: \quad a_0 = 9.2495 \text{ A.}$$

The antimony parameters are $u = 0.343$, $v = 0.157$. These result in the interatomic distances Ir–Sb = 2.613 A., Sb–Sb = 2.904 A.

$$\text{IrAs}_3: \quad a_0 = 8.4691 \text{ A.}$$

TABLE VB,21
Compounds with the Cubic CoAs$_3$ Structure

Compound	a_0, A.
CoSb$_3$	9.034
IrP$_3$	8.015
PdP$_3$	7.705
RhAs$_3$	8.453
RhP$_3$	7.996
RhSb$_3$	9.229

The arsenic parameters are $u = 0.350$, $v = 0.146$. The resulting interatomic distances are Ir–As = 2.445 A., As–As = 2.473 and 2.541 A.

Unit cells of other isostructural compounds are given in Table VB,21.

V,b38. The mineral *domeykite*, Cu$_3$As, is cubic with a unit having

$$a_0 = 9.592 \text{ A.}$$

and containing 16 molecules. It has been assigned an arrangement developed from T_d^6 ($I\bar{4}3d$):

As: (16f) $uuu;$ $u+^1/_2,^1/_2-u,\bar{u};$
  $u+^1/_4,u+^1/_4,u+^1/_4;$ $u+^3/_4,^1/_4-u,^3/_4-u;$ tr; B.C.
  with $u = -0.03$

Cu: (48e) $xyz;$ $x+^1/_2,^1/_2-y,\bar{z};$
  $z+^1/_2,^1/_2-x,\bar{y};$ $y+^1/_2,^1/_2-z,\bar{x};$
  $y+^1/_4,x+^1/_4,z+^1/_4;$ $y+^3/_4,^1/_4-x,^3/_4-z;$
  $^3/_4-y,x+^3/_4,^1/_4-z;$ $^1/_4-y,^3/_4-x,z+^3/_4;$ tr; B.C.
  with $x = -0.03$, $y = 0.12$, $z = 0.20$

These parameters yield a nearest Cu–Cu distance of 2.56 A., which is almost exactly that of the copper atoms in metallic copper; the smallest separation of unlike atoms, Cu–As = 2.63 A., is considerably in excess of the sum of the neutral radii. More work should be done on this crystal.

V,b39. In a structure proposed long ago for Mn$_3$As there are four molecules in an orthorhombic, pseudotetragonal cell of the dimensions:

$$a_0 = b_0 = 3.788 \text{ A.}, \qquad c_0 = 16.29 \text{ A.}$$

Atoms were placed in the following positions of V_h^{13} (*Pmmn*):

Mn(1): (2a) $00u;$ $^1/_2$ $^1/_2$ $\bar{u}$ with $u = 0.1935$
Mn(2): (2a) with $u = -0.1935$

Mn(3):   (2a)   with $u = -0.4345$
Mn(4):   (2b) $0 \; {}^1/_2 \, v; \; {}^1/_2 \, 0 \, \bar{v}$ with $v = 0.3065$
Mn(5):   (2b)   with $v = -0.3065$
Mn(6):   (2b)   with $v = -0.0655$
As(1):   (2a)   with $u = 0.409$
As(2):   (2b)   with $v = 0.091$

A further study is called for.

V,b40. Many years ago a structure was proposed for *trilithium nitride*, Li_3N. It is hexagonal, with a cell said to be unimolecular and to have the dimensions:

$$a_0 = 3.658 \text{ A.}, \qquad c_0 = 3.882 \text{ A.}$$

The lithium atoms were put in the positions $0 \; 0 \; {}^1/_2; \; {}^1/_3 \; {}^2/_3 \; 0; \; {}^2/_3 \; {}^1/_3 \; 0$ and the nitrogen in the origin 000.

Additional work should be carried out.

V,b41. The *bismuthide of cesium*, Cs_3Bi, is cubic, with four molecules in a cell of the edge length:

$$a_0 = 9.305 \text{ A.}$$

Atoms, in partial disorder, have been placed in the following special positions of O_h^7 (*Fd3m*):

Cs(1):   (8a) $000; \; {}^1/_4 \; {}^1/_4 \; {}^1/_4;$ F.C.
Cs(2)+Bi: (8b) ${}^1/_2 \; {}^1/_2 \; {}^1/_2; \; {}^3/_4 \; {}^3/_4 \; {}^3/_4;$ F.C.

Other compounds with this structure are:

Cs_3Sb: $a_0 = 9.128$ A. (9.147 A.)
Rb_3Sb: $a_0 = 8.84$ A.
Na_2KSb: $a_0 = 7.74$ A.

In this crystalline substance, atoms have been distributed as follows in the more ordered array based on O_h^5:

Sb: (4a)   000;   F.C.
K: (4b) ${}^1/_2 \; {}^1/_2 \; {}^1/_2;$ F.C.
Na: (8c) $\pm({}^1/_4 \; {}^1/_4 \; {}^1/_4);$ F.C.

This is an alloy-type structure also possessed by intermetallic compounds such as Na_3Tl.

V,b42. *Trititanium antimonide*, Ti_3Sb, is dimorphous with a cubic and a tetragonal form.

For the tetragonal modification, the unit cell containing eight molecules has the dimensions:

$$a_0 = 10.465 \text{ A.}, \qquad c_0 = 5.2639 \text{ A.}$$

The space group is D_{4h}^{18} ($I4/mcm$), with atoms in the positions:

Ti(1):　(4a)　$\pm(0\ 0\ ^1/_4)$;　B.C.

Ti(2):　(4b)　$0\ ^1/_2\ ^1/_4;\ ^1/_2\ 0\ ^1/_4$;　B.C.

Ti(3):　(16k)　$\pm(uv0;\ v\bar{u}0;\ u\ \bar{v}\ ^1/_2;\ v\ u\ ^1/_2)$;　B.C.
　　　　　　　　　　　　　　　　　with $u = 0.0766$, $v = 0.2228$

　Sb:　(8h)　$\pm(u,\ u+^1/_2,0;\ u+^1/_2,\bar{u}\ 0)$;　B.C.　　with $u = 0.1635$

In this structure (Fig. VB,44) the Ti(1) atoms have only titanium neighbors at distances of 2.632 and 2.794 A. The Ti(2) and Ti(3) have both titanium and antimony neighbors with Ti(2)–4Sb = 2.755 A., Ti(2)–Ti = 2.632 and 3.285 A., Ti(3)–4Sb = 2.780–3.029 A., Ti(3)–Ti = 2.794 A. up and Sb–Sb = 3.671 A.

The cubic modification has the structure of β-tungsten (**II,p**) with the antimony atoms in (2a) and the titanium atoms in (6c) of O_h^3 ($Pm3n$). The bimolecular cell has the edge:

$$a_0 = 5.2186 \text{ A.}$$

Carbides, Borides, Hydrides

V,b43. The important *iron carbide, cementite*, Fe_3C, is orthorhombic, with a unit containing four molecules. Many accurate determinations of cell dimensions made on different samples lead to significantly different values. Two examples are

$a_0 = 4.5230$ A.; $b_0 = 5.0890$ A.; $c_0 = 6.7428$ A.　(slowly cooled)

$a_0 = 4.5165$ A.; $b_0 = 5.0837$ A.; $c_0 = 6.7475$ A.　(quenched from 900°C.)

Atoms are in the following positions of V_h^{16} ($Pbnm$) with parameters that also are somewhat different from investigation to investigation.

Fe(1):　(4c)　$\pm(u\ v\ ^1/_4;\ ^1/_2-u,v+^1/_2,^1/_4)$
　　　　　　　　　　　　　　　　with $u = -0.167$, $v = 0.040$

Fe(2):　(8d)　$\pm(xyz;\ ^1/_2-x,y+^1/_2,^1/_2-z;\ \bar{x},\bar{y},z+^1/_2;\ x+^1/_2,^1/_2-y,\bar{z})$
　　　　　　　　　　　　with $x = 0.333$, $y = 0.175(0.183)$, $z = 0.065$

　C:　(4c)　with $u = 0.43(0.47)$ and $v = -0.13(-0.14)$

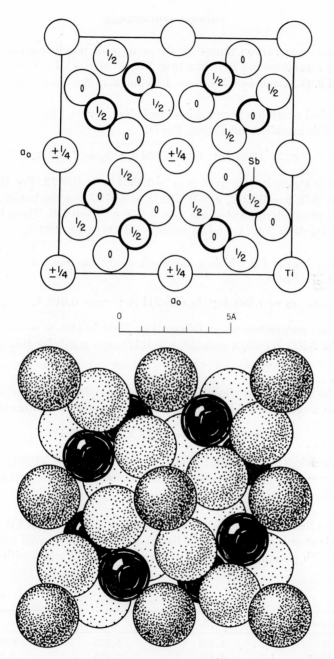

Fig. VB,44a (top). A projection of the tetragonal modification of Ti_3Sb along its c_0 axis. The antimony atoms are heavily ringed.

Fig. VB,44b (bottom). A packing drawing of the tetragonal Ti_3Sb arrangement seen along its c_0 axis. The antimony atoms are black.

This places the carbon atoms at the centers of nearly regular trigonal prisms of iron atoms in a structure (Fig. VB,45) in which Fe–Fe lies between 2.49 and 2.68 A. and Fe–C between 1.85 and 2.15 A.

A detailed study has been made of *palladium boride*, Pd_3B, which proves to have this atomic arrangement. For it,

$$a_0 = 4.852 \text{ A.}; \quad b_0 = 5.463 \text{ A.}; \quad c_0 = 7.567 \text{ A.}$$

The Pd(1) atoms in (4c) have $u = -0.1554$, $v = 0.0372$. For the Pd(2) atoms in (8d), $x = 0.3276$, $y = 0.1798$, $z = 0.0700$. For the boron atoms in (4c), the parameters selected are $u = 0.433$, $v = -0.116$. These lead to a shortest Pd–B = 2.17 A. and an average Pd–Pd = ca. 2.85 A.

Another boride for which a thorough study has been made is *nickel boride*, Ni_3B. For it,

$$a_0 = 4.389 \text{ A.}; \quad b_0 = 5.211 \text{ A.}; \quad c_0 = 6.619 \text{ A.}$$

The chosen parameters are as follows. For Ni(1) in (4c), $u = -0.136$, $v = 0.028$; for Ni(2) in (8d), $x = 0.347$, $y = 0.178$, $z = 0.061$; for B in (4c), $u = 0.433$, $v = -0.111$.

For the isostructural phosphide *palladium phosphide*, Pd_3P, which, however, has a rather broad homogeneity range, the stated cell dimensions are

$$a_0 = 5.170 \text{ A.}; \quad b_0 = 5.947 \text{ A.}; \quad c_0 = 7.451 \text{ A.}$$

The parameters are: for Pd(1) in (4c), $u = -0.1300$, $v = 0.0264$; for Pd(2) in (8d), $x = 0.3373$, $y = 0.1783$, $z = 0.0636$; for P in (4c), $u = 0.4550$, $v = -0.1166$.

Besides the additional borides and carbides listed in Table VB,22 which have this atomic arrangement, it has also been found in solid solutions of borides with phosphides, borides with carbides, and borides with silicides.

TABLE VB,22
Compounds with the Fe_3C Structure

Compound	a_0, A.	b_0, A.	c_0, A.
Co_3B	4.408	5.225	6.629
Co_3C	4.483	5.033	6.731
$(Fe,Si)_3B$	4.458–4.474	5.363–5.299	6.660–6.674
$Fe_3B_{0.9}C_{0.1}$	4.4500	5.4052	6.6685
Mn_3C	4.530	5.080	6.772

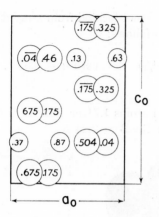

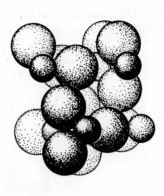

Fig. VB,45a (left). A projection of the atomic contents of the orthorhombic unit of cementite, Fe_3C, on its b_0 face.   Large circles are the iron atoms.   Origin in lower left.
Fig. VB,45b (right). A packing drawing of the structure of Fe_3C if the iron and carbon atoms are given their elementary radii.

V,b44.   An approximate structure has been described for *nickel carbide*, Ni_3C, which is not like cementite. It is rhombohedral, with a bimolecular unit having the dimensions:

$$a_0 = 5.045 \text{ A.}, \qquad \alpha = 53°39'$$

The corresponding hexagonal cell contains six molecules and has the edge lengths:

$$a_0' = 4.553 \text{ A.}, \qquad c_0' = 12.92 \text{ A.}$$

The atoms have been placed in the following positions of D_{3d}^6 ($R\bar{3}c$):

C: (6b) $000; 0\ 0\ 1/2$; rh
Ni: (18e) $\pm(u\ 0\ 1/4; 0\ u\ 1/4; \bar{u}\ \bar{u}\ 1/4)$; rh   with $u = $ ca. $1/3$

The nickel atoms are in a hexagonal close-packing, and it is suggested that there is some disorder in the layering of the atoms of carbon.

V,b45.   A hexagonal structure has been described for *rhenium triboride*. ReB_3. For it there are two molecules in a cell of the dimensions:

$$a_0 = 2.900 \text{ A.}, \qquad c_0 = 7.475 \text{ A.}$$

Rhenium atoms are in twofold positions

$$(2c)   \pm(1/3\ 2/3\ 1/4)$$

and D_{6h}^4 ($P6_3/mmc$). The boron atoms were not located from x-ray data, but a plausible arrangement places them as follows:

B(1):   (4f) $\pm(^1/_3\ ^2/_3\ u;\ ^2/_3,^1/_3,u+^1/_2)$ with $u = 0.55$

B(2).   (2a)   000; 0 0 $^1/_2$

This grouping (Fig. VB,46) makes boron atoms 1.71 or 1.83 A. distant from six other borons and yields a shortest Re–B = 2.24 A.

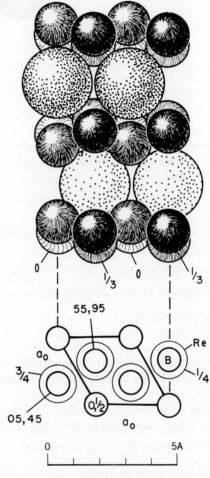

Fig. VB,46. Two projections of the hexagonal structure of ReB₃. In the upper the large rhenium atoms are dotted.

V,b46. *Trirhenium boride*, Re_3B, is orthorhombic with a tetramolecular cell of the dimensions:

$$a_0 = 2.890 \text{ A.}; \quad b_0 = 9.313 \text{ A.}; \quad c_0 = 7.258 \text{ A.}$$

The space group is V_h^{17} (*Cmcm*), with atoms in the positions:

B: (4c) $\pm (0\ u\ ^1/_4;\ ^1/_2, u + ^1/_2, ^1/_4)$ with $u = 0.744$
Re(1): (4c) with $u' = 0.4262$
Re(2): (8f) $\pm (0uv;\ 0, u, ^1/_2 - v;\ ^1/_2, u + ^1/_2, v;\ ^1/_2, u + ^1/_2, ^1/_2 - v)$
                    with $u'' = 0.1345,\ v'' = 0.0620$

In this structure (Fig. VB,47), each boron atom is at the center of a triangular prism of rhenium atoms with Re–B = 2.23 A. The shortest Re–Re = 2.66 A. is not very different from twice the metallic radius. There is no close contact between boron atoms.

V,b47. *Solidified ammonia*, NH_3, is cubic with a unit that contains four molecules. Cell dimensions for it and for the analogous deuterium compound are

NH_3: $a_0 = 5.138$ A. ($-102\,°C.$); 5.084 A. ($-196\,°C.$)
ND_3: $a_0 = 5.091$ A. ($-160\,°C.$); 5.073 A. ($-196\,°C.$)

In an early determination of structure it was found that the nitrogen atoms are in the following positions of T^4 ($P2_13$):

(4a) $uuu;\ u + ^1/_2, ^1/_2 - u, \bar{u};\ \bar{u}, u + ^1/_2, ^1/_2 - u;\ ^1/_2 - u, \bar{u}, u + ^1/_2$

with $u = $ ca. 0.22. Thus it is that these nitrogen atoms, and the molecules of which they are the centers, are in a nearly perfect cubic close-packing; it would be perfect if u were $^1/_4$.

Studies have since been made with x rays to determine the positions of the hydrogen atoms, but a recent neutron diffraction investigation of ND_3 at $-196\,°C.$ probably places the atoms more accurately than is possible with x rays. According to this determination, the nitrogen atoms in (4a) have $u = 0.2127$. The deuterium atoms are in

(12b) $xyz;\ x + ^1/_2, ^1/_2 - y, \bar{z};\ \bar{x}, y + ^1/_2, ^1/_2 - z;\ ^1/_2 - x, \bar{y}, z + ^1/_2;$
       $zxy;\ \bar{z}, x + ^1/_2, ^1/_2 - y;\ ^1/_2 - z, \bar{x}, y + ^1/_2;\ z + ^1/_2, ^1/_2 - x, \bar{y};$
       $yzx;\ ^1/_2 - y, \bar{z}, x + ^1/_2;\ y + ^1/_2, ^1/_2 - z, \bar{x};\ \bar{y}, z + ^1/_2, ^1/_2 - x$
                    with $x = 0.3740,\ y = 0.2632,$ and $z = 0.1094$

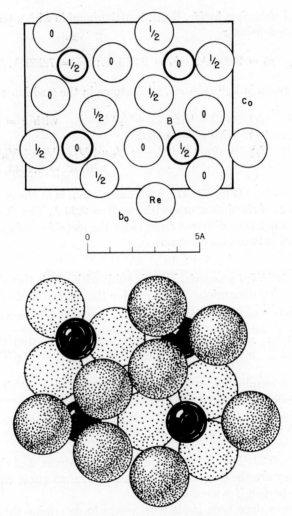

Fig. VB,47a (top). The orthorhombic structure of Re₃B projected along its a_0 axis. Origin in lower right.

Fig. VB,47b (bottom). A packing drawing of the Re₃B structure viewed along its a_0 axis. The boron atoms are black.

The structure is illustrated in Figure VB,48. In its ND₃ molecules, N–D = 1.005 A. and D–N–D = 110°24′. Between molecules N–D–N = 3.352 A. and the angle between DN and NN is 11°18′. Thus the deuterium atom occupies a definite site in the structure and the hydrogen bond is not of the "half-atom" type that prevails in ice.

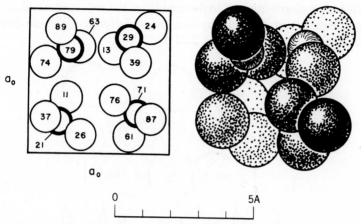

Fig. VB,48a (left). The cubic structure of solid NH_3 projected along a cube axis. The larger circles are hydrogen.

Fig. VB,48b (right). A packing drawing of the NH_3 arrangement seen along a cube axis. The dotted circles are hydrogen.

Arsine, AsH_3, and *phosphine*, PH_3, when solidified give cubic x-ray patterns which, though reportedly face-centered, probably indicate that they have an ammonia-like structure. Their unit cubes have the edges:

$$AsH_3: \quad a_0 = 6.41 \text{ A. } (-170°C.)$$
$$PH_3: \quad a_0 = 6.32 \text{ A. } (-170°C.)$$

V,b48. *Uranium hydride*, UH_3, is cubic, with a unit that has eight molecules in a cell of the edge length:

$$a_0 = 6.6445 \text{ A.}$$

The uranium atoms have the same positions as in β-uranium. Based on O_h^3 ($Pm3n$), they are in:

(2a) $000; \frac{1}{2} \frac{1}{2} \frac{1}{2}$
(6c) $\pm(\frac{1}{4} 0 \frac{1}{2}; \frac{1}{2} \frac{1}{4} 0; 0 \frac{1}{2} \frac{1}{4})$

By neutron diffraction of the corresponding deuteride, it has been decided that the deuterium (and hydrogen) atoms are in:

(24k) $\pm(0uv;$ $v0u;$ $uv0;$
          $0u\bar{v};$ $\bar{v}0u;$ $u\bar{v}0;$
          $\frac{1}{2},v+\frac{1}{2},u+\frac{1}{2}; u+\frac{1}{2},\frac{1}{2},v+\frac{1}{2}; v+\frac{1}{2},u+\frac{1}{2},\frac{1}{2};$
          $\frac{1}{2},\frac{1}{2}-v,u+\frac{1}{2}; u+\frac{1}{2},\frac{1}{2},\frac{1}{2}-v; \frac{1}{2}-v,u+\frac{1}{2},\frac{1}{2})$

with $u = 0.155, v = 0.31$

With such a distribution of hydrogens (Fig. VB,49), each uranium atom is about equidistant from 12 hydrogens, while each hydrogen has four uranium neighbors, with U–H = 2.32 A.

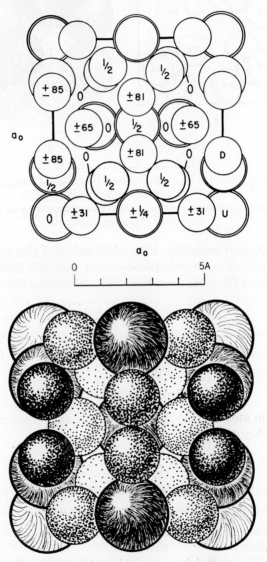

Fig. VB,49a (top). The cubic structure of UD₃ projected along a cube axis.
Fig. VB,49b (bottom). A packing drawing of the UD₃ arrangement seen along a cube axis. The uranium atoms are large and fine-line shaded.

A second form of UH_3 has recently been described. Like the foregoing it is cubic, but its smaller, bimolecular cell has the edge:

$$a_0 = 4.153 \text{ A.}$$

The atomic arrangement has not been established.

BIBLIOGRAPHY TABLE, CHAPTER VB

Compound	Paragraph	Literature
$AcBr_3$	b22	1948: Z
$AcCl_3$	b22	1948: Z
AcF_3	b10	1949: Z
AgN_3	VI,3	—
$AlBr_3$	b8	1945: R&MG
$AlCl_3$	b7	1930: L; 1935: K; 1947: K,MG&R
AlF_3	b2	1929: F&S; 1931: K; 1933: K; 1939: C
$Al(OH)_3$	b23	1934: M; 1942: M; 1956: S; 1958: Y&S; 1960: S
$AmBr_3$	b11	1948: Z
$AmCl_3$	b22	1948: Z
AmF_3	b10	1949: Z; 1953: T&D
AmI_3	b11	1948: Z
AsH_3	b47	1930: N&C
AsI_3	b1	1930: B; H; 1931: H; 1947: V
$AuCl_3$	b15	1958: C,T&MG
BBr_3	b16	1947: R&R
BCl_3	b16	1947: R&R; 1957: A&L
BI_3	b16	1962: R,D&K
B_2H_6	[XIII]	1925: M,B&P; M&P
BaS_3	b32	1936: M&K
$(BaTh)F_6$	b10	1949: Z
$(BaU)F_6$	b10	1949: Z
BiF_3	b9, b10, b14	1929: H&N; 1949: H&F; 1955: A; A&L; 1961: M&C
$BiF_{2.8}O_{0.1}$	b10	1955: A&L
BiI_3	b1	1930: B; 1947: V; 1950: Z; 1952: P
BrF_3	b18	1957: B&B
$(CaTh)F_6$	b10	1949: Z
$CeBr_3$	b22	1948: Z
$CeCl_3$	b22	1948: Z; 1951: K,I&I; 1954: T&D
CeF_3	b10	1929: O

(continued)

BIBLIOGRAPHY TABLE, CHAPTER VB (*continued*)

Compound	Paragraph	Literature
(Ce,La)F$_3$ (tysonite)	b10	1929· O; 1931: O
ClF$_3$	b17	1953: B&B
CoAs$_3$ (skutterudite)	b37	1925: O; 1928: O; 1957: V
Co$_3$B	b43	1958: R
Co$_3$C	b43	1937: M; 1961: N
CoF$_3$	b3	1931: E; 1957: H,J,P&W
CoSb$_3$	b37	1956: Z&Z; 1958: D
CrBr$_3$	b1	1932: B
CrCl$_3$	b6	1927: N; 1930: W
CrF$_3$	b3	1957: J&M; 1960: K; 1961: M&C
CrI$_3$	b6	1952: H&G
CrO$_3$	b26	1931: B; W&W; 1950: B&W; 1960: H&S
Cr$_3$P	b35	1937: A&N; 1954: S
Cs$_3$Bi	b41	1959: Z&S
CsI$_3$	VI,14	—
Cs$_3$O	b27	1956: T,H&L
Cs$_2$S$_6$	b31	1952: A,G&K; 1953: A&G
Cs$_3$Sb	b41	1957: J&W
Cu$_3$As (domeykite)	b38	1929: R; 1930: M; 1938: S; 1949: M
CuN$_3$	b34	1948: W
Cu$_3$N	b5	1938: J&H; 1941: J
DyCl$_3$	b7	1954: T&C
DyF$_3$	b9	1953: Z&T
DyH$_3$	b10	1962: P&W
Dy(OH)$_3$	b22	1947: F&S
ErCl$_3$	b7	1954: T&C
ErF$_3$	b9	1953: Z&T
ErH$_3$	b10	1962: P&W
Er(OH)$_3$	b22	1946: S; 1947: F&S
EuCl$_3$	b22	1954: T&D
EuF$_3$	b9, b10	1953: Z&T
Fe$_3$B	b43	1957: N; 1959; A&L
FeBr$_2$	b3	1951: G
Fe$_3$C	b43	1922: W; W&P; 1924: W&P; 1930: H; S; 1931: O&T; 1932: W; 1940: L&P; 1942: HR,R&L; 1944: P; 1950: A; 1961: L; S&M; 1962: A&R
FeCl$_3$	b1	1932: W; 1949: G; 1951: G

(*continued*)

BIBLIOGRAPHY TABLE, CHAPTER VB (*continued*)

Compound	Paragraph	Literature
FeF_3	b3	1931: E; K; 1933: W; 1957: H,J,P&W
Fe_3P	b35	1928: H; 1929: H; 1962: R
$(Fe,Ni,Co)_3P$ (schreibersite)	b35	1932: H,H&P
GaF_3	b3	1959: B,G&G; 1961: M&C
$GdCl_3$	b22	1954: T&D
GdF_3	b9	1953: Z&T
GdH_3	b10	1962: P&W
$Gd(OH)_3$	b22	1947: F&S
$HoCl_3$	b7	1954: T&C
HoF_3	b9, b10	1953: Z&T
HoH_3	b10	1962: P&W
ICl_3	b19	1954: B&W
$InCl_3$	b7	1954: T&C
InF_3	b3	1961: M&C
$In(OH)_3$	b21	1947: F&S; 1948: P; S&S
$In(OH)F_2$	b20	1957: F
$IrAs_3$	b37	1961: H&C; K&P
IrF_3	b3	1957: H,J,P&W
IrP_3	b37	1960: R
$IrSb_3$	b37	1956: Z&Z; 1961: K&P
K_3As	b36	1937: B&Z
K_3Bi	b36	1937: B&Z
KN_3	VI,1	—
KO_3	VI,1	1951: Z&Z
K_3P	b36	1961: G,D&K
K_3Sb	b36	1937: B&Z
$LaBr_3$	b22	1948: Z
$LaCl_3$	b22	1948: Z
LaF_3 (tysonite)	b10	1929: O; 1952: S
LaI_3	b11	1948: Z
$La(OD)_3$	b22	1959: A&W
$La(OH)_3$	b22	1946: S; 1947: F&S; 1948: Z; 1953: R&MK
Li_3As	b36	1937: B&Z
Li_3Bi	b14	1935: Z&B
Li_3N	b40	1926: B; F; 1935: Z&B
Li_3P	b36	1937: B&Z
Li_3Sb	b36	1937: B&Z

(*continued*)

BIBLIOGRAPHY TABLE, CHAPTER VB (*continued*)

Compound	Paragraph	Literature
LuCl$_3$	b7	1954: T&C
LuF$_3$	b9	1953: Z&T
LuH$_3$	b10	1962: P&W
Mn$_3$As	b39	1951: N,F&P
Mn$_3$C	b43	1957: P&F
MnF$_3$	b4	1957: H&J
Mn$_3$P	b35	1937: A&N; 1962: R
MoF$_3$	b3, b5	1951: G&J; 1960: LV,S,W&Y
MoO$_3$	b24	1931: B; W; 1944: H&M; 1950: A&M; 1952: NBS
MoO$_3$	b14	1954: S
Mo$_3$P	b35	1954: S
NH$_3$(ND$_3$)	b47	1925: M&P; M,B&P; dS; 1942: V&H; 1959: O&T; 1961: R&H
NH$_4$I$_3$	VI,14	—
NH$_4$N$_3$	VI,2	—
Na$_3$As	b36	1937: B&Z
Na$_3$Bi	b36	1937: B&Z
NaN$_3$	VI,12	—
Na$_3$P	b36	1937: B&Z
Na$_2$KSb	b41	1959: S&Z
Na$_3$Sb	b36	1937: B&Z
NbF$_3$	b5	1955: E,P&P; 1961: M&C
NbO$_2$F	b5	1956: F&R
NdBr$_3$	b11	1948: Z
NdCl$_3$	b22	1948: Z
NdF$_3$	b10	1929: O
Nd(OH)$_3$	b22	1947: F&S; 1948: Z; 1953: R&MK
Ni$_3$B	b43	1957: F&M; 1958: R; 1959: R
Ni$_3$C	b44	1958: N
Ni$_3$P	b35	1955: A; 1962: R, H&L
NpBr$_3$	b11, b22	1948: Z
NpCl$_3$	b22	1948: Z
NpF$_3$	b10	1948: Z; 1949: Z
NpI$_3$	b10, b11	1948: Z
PH$_3$	b47	1930: N&C
PI$_3$	b16	1933: B
(PbTh)F$_6$	b10	1949: Z
(PbU)F$_6$	b10	1949: Z
Pd$_3$As	b35	1961: H&C

(*continued*)

BIBLIOGRAPHY TABLE, CHAPTER VB *(continued)*

Compound	Paragraph	Literature
Pd_3B	b43	1961: S
PdF_3	b3	1931: E; 1957: H,J,P&W
PdP_3	b37	1960: R
Pd_3P	b43	1960: R&G
$PrBr_3$	b22	1948: Z
$PrCl_3$	b22	1948: Z; 1954: T&D
PrF_3	b10	1929: O; 1954: NBS
$Pr(OH)_3$	b22	1947: F&S; 1948: Z
$PuBr_3$	b11	1948: Z
$PuCl_3$	b22	1948: Z
PuF_3	b10	1948: Z; 1949: Z
PuI_3	b11	1948: Z
Rb_3As	b36	1961: G,D&K
Rb_3Bi	b36	1960: Z,S&M
Rb_3Sb	b11, b36, b41	1960: Z,S&M; 1961: C,I,T&S; G,D&K
ReB_3	b45	1960: A,S&A
Re_3B	b46	1960: A,B&R
ReO_3	b5	1931: B,L&M; 1932: M; 1933: B
$RhAs_3$	b37	1961: H&C
RhF_3	b3	1931: E; 1957: H,J,P&W
RhP_3	b37	1960: R; R&H
$RhSb_3$	b37	1956: Z&Z
$RuCl_3$	b6	1957: S&O
RuF_3	b3	1957: H,J,P&W
$S_6(NH)_2$	f9	1960:W
SO_3	b29, b30	1941: W&MG; 1954: W&MG
$SbBr_3$	b13	1962: C&H
$SbCl_3$	b13	1956: L&N
SbF_3	b12	1943: B&W
SbI_3	b1	1930: B; 1947: V
$ScCl_3$	b3	1947: K&K; 1954: T&D
ScF_3	b3, b5	1938: N; 1939: N; 1961: M&C
$Sc(OH)_3$	b21	1947: F&S; 1948: S&S
$SmBr_3$	b11	1948: Z
$SmCl_3$	b22	1954: T&D
SmF_3	b9, b10	1929: O; 1953: Z&T
SmH_3	b10	1962: P&W
$Sm(OH)_3$	b22	1947: F&S; 1953: R&MK
$(SrTh)F_6$	b10	1949: Z
$(SrU)F_6$	b10	1949: Z

(continued)

BIBLIOGRAPHY TABLE, CHAPTER VB (*continued*)

Compound	Paragraph	Literature
TaF₃	b5	1951: G&J; 1961: M&C
TaO₂F	b5	1956: F&R
TbF₃	b9	1953: Z&T
TbH₃	b10	1962: P&W
ThOF₂	b10	1949: Z
TiBr₃	b3	1958: N,C,B&P; R&S
TiCl₃	b3, b6	1947: K&K; 1958: N,C,B&P; 1959: N,C&A; 1961: N,C&A; 1962: C
TiF₃	b3	1954: E&P; 1956: S; 1961: M&C
TiOF₂	b5	1955: V&D
TiS₃	b33	1958: J&B
Ti₃Sb	b42	1962: K,G&T
TlCl₃	b7	1954: T&C
TmCl₃	b7	1954: T&C
TmF₃	b9, b10	1953: Z&T
TmH₃	b10	1962: P&W
UBr₃	b22	1948: Z
UCl₃	b22	1948: Z
UD₃	b47	1947: R; 1951: R
UF₃	b10	1949: Z
UH₃	b48	1947: R; 1948: P&E; 1951: R; 1953: C,C&P; 1954: M,E&Z
UI₃	b11	1948: Z
UO₃	b5, b28	1948: Z; 1953: P; 1955: W
USe₃	b33	1957: K&F
VCl₃	b3	1947: K&K
VF₃	b3	1951: J&G
V₃S	b35	1959: P&G
WO₃	b25	1931: B; 1944: H&M; 1950: M; U&I; 1951: M&W; U&I; W&F; 1952: K,H&W; 1953: A; U&K; 1954: H; 1960: T
YCl₃	b7	1954: T&C
YF₃	b9	1923: G&T; 1938: W; 1953: Z&T
YH₃	b10	1962: P&W
Y(OH)₃	b22	1946: S; S&S; 1947: F&S; S&S; 1949: F&D
YbCl₃	b7	1954: T&C
YbF₃	b9	1953: Z&T
Yb(OH)₃	b22	1949: F&D
ZrSe₃	b33	1958: K&P

C. COMPOUNDS OF THE TYPE RX$_4$

Halides

V,c1. Solidified *silicon tetrafluoride*, SiF$_4$, has been given a very simple cubic structure. At $-145°$C. its bimolecular unit has the edge length:

$$a_0 = 5.41 \text{ A.}$$

Atoms have been placed in positions of $T_d{}^3$ ($I\bar{4}3m$):

Si: (2a)   000; $\frac{1}{2}\,\frac{1}{2}\,\frac{1}{2}$
F: (8c)   *uuu; u$\bar{u}\bar{u}$; $\bar{u}$u$\bar{u}$; $\bar{u}\bar{u}$u;* B.C.   with $u = 0.165$

The Si–F separation in this body-centered structure (Fig. VC,1) is 1.56 A.

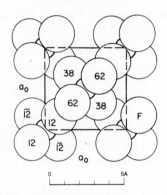

Fig. VC,1. The cubic structure of SiF$_4$ projected along a cube axis.

V,c2. *Zirconium tetrafluoride*, ZrF$_4$, and several fluorides isomorphous with it, are monoclinic with large unit cells containing 12 molecules. In the original studies the axes were chosen to give side-centered cells which for ZrF$_4$ had the dimensions:

$$a_0' = 11.71 \text{ A.}; \quad b_0' = 9.89 \text{ A.}; \quad c_0' = 7.66 \text{ A.}; \quad \beta = 126°9'$$

A structure for UF$_4$ was described in terms of this unit. In it the positions of the metallic atoms were determined and those of the fluorine atoms postulated. In a more recent study of the zirconium compound the axes were chosen so as to make the unit body-centered. The cell dimensions are

$$a_0 = 9.57 \text{ A.}; \quad b_0 = 9.89 \text{ A.}; \quad c_0 = 7.66 \text{ A.}; \quad \beta = 94°28'$$

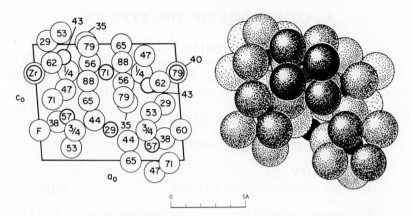

Fig. VC,2a (left). Half the contents of the monoclinic unit of ZrF_4 projected along its b_0 axis. Origin in lower left.

Fig. VC,2b (right). A packing drawing of the atoms of ZrF_4 shown in Fig. VC,2a. The large dotted circles are fluorine.

The space group is C_{2h}^6 which in this body-centered orientation has the symbol $I2/c$. Atoms were found to be in the positions:

$(4d)$ $1/4\ 1/4\ 3/4;\ 3/4\ 1/4\ 3/4;$ B.C.

$(4e)$ $\pm(0\ u\ 1/4);$ B.C.

$(8f)$ $\pm(xyz;\ x,\bar{y},z+1/2);$ B.C.

The parameters, which could be established for all atoms in this compound because of the relatively lower atomic number of zirconium, have the values of Table VC,1.

In this structure (Fig. VC,2), each zirconium atom has eight fluorine neighbors at distances between 2.03 and 2.18 A. Each atom of fluorine is in contact with two atoms of zirconium.

TABLE VC,1
Positions and Parameters of the Atoms in ZrF_4

Atom	Position	x	y	z
Zr(1)	$(4e)$	0	0.2148	$1/4$
Zr(2)	$(8f)$	0.2944	0.9289	0.1278
F(1)	$(4d)$	$1/4$	$1/4$	$3/4$
F(2)	$(4e)$	0	0.598	$1/4$
F(3)	$(8f)$	0.111	0.289	0.040
F(4)	$(8f)$	0.115	0.059	0.161
F(5)	$(8f)$	0.212	0.526	0.104
F(6)	$(8f)$	0.379	0.125	0.162
F(7)	$(8f)$	0.378	0.346	0.028

Other compounds with this structure are listed in Table VC,2.

TABLE VC,2
Crystals with the Monoclinic ZrF_4 Arrangement

Crystal	a_0, A.	b_0, A.	c_0, A.	β
AmF_4	10.10	10.45	8.18	94°55′
CeF_4	10.23	10.6	8.3	95°1′
CmF_4	10.11	10.45	8.16	94°46′
HfF_4	9.48	9.86	7.63	94°29′
NpF_4	10.28	10.64	8.32	94°37′
PuF_4	10.21	10.57	8.28	94°44′
TbF_4	9.82	10.3	7.9	94°36′
ThF_4	10.64	11.0	8.6	94°50′
UF_4	10.38	10.74	8.41	94°41′

V,c3. According to a preliminary description, *tin tetrafluoride*, SnF_4, is tetragonal, with the bimolecular cell:

$$a_0 = 4.048 \text{ A.}, \qquad c_0 = 7.930 \text{ A.}$$

The space group is D_{4h}^{17} ($I4/mmm$), with the atoms in

Sn: (2a) $000; \frac{1}{2}\frac{1}{2}\frac{1}{2}$
F(1): (4c) $0\frac{1}{2}0; \frac{1}{2}00; \frac{1}{2}0\frac{1}{2}; 0\frac{1}{2}\frac{1}{2}$
F(2): (4e) $\pm(00u; \frac{1}{2},\frac{1}{2},u+\frac{1}{2})$ with $u = 0.245$

This simple arrangement is shown in Figure VC,3.

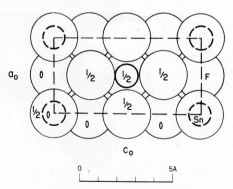

Fig. VC,3. The simple tetragonal structure of SnF_4 projected along an a_0 axis. Origin in lower left.

Lead tetrafluoride, PbF_4, has the same structure, with

$$a_0 = 4.247 \text{ A.}, \qquad c_0 = 8.030 \text{ A.}$$

V,c4.   *Thorium tetrachloride*, $ThCl_4$, and the corresponding uranium chloride, UCl_4, have tetragonal units containing four molecules. Their cells have the dimensions:

$$ThCl_4: \quad a_0 = 8.490 \text{ A.}, \ c_0 = 7.483 \text{ A.}$$
$$UCl_4: \quad a_0 = 8.296 \text{ A.}, \ c_0 = 7.487 \text{ A.}$$

Atoms have been placed in the following special positions of D_{4h}^{19} ($I4/amd$):

Th or U: (4a)   000; 0 $^1/_2$ $^1/_4$; B.C.
  Cl: (16h) $0uv$; $0\bar{u}v$; $u0\bar{v}$; $\bar{u}0\bar{v}$;
    $0,u+^1/_2,^1/_4-v$; $0,^1/_2-u,^1/_4-v$; $u,^1/_2,v+^1/_4$;
                $\bar{u},^1/_2,v+^1/_4$; B.C.

with $u = 0.280$, $v = 0.917$ for both the thorium and the uranium salts.

In the resulting structure for $ThCl_4$ (Fig. VC,4), each thorium atom is surrounded by four chlorine atoms at a distance of 2.46 A. and by four more at the greater distance of 3.11 A. The closest approach of chlorine atoms to one another is 3.2 A.

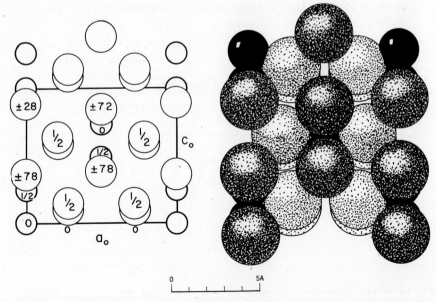

0   5A

Fig. VC,4a (left). A projection along an a_0 axis of the tetragonal structure of $ThCl_4$. The large circles are chlorine.

Fig. VC,4b (right). A packing drawing of the tetragonal structure of $ThCl_4$ viewed along an a_0 axis. The chlorine atoms are dotted.

Thorium tetrabromide, $ThBr_4$, is isostructural, with

$$a_0 = 8.945 \text{ A.}, \qquad c_0 = 7.930 \text{ A.}$$

Its atoms have been given the chlorine parameters: $u = 0.28$, $v = 0.92$.

Two other chlorides of heavy radioactive elements also have this arrangement. They are

$$PaCl_4: \quad a_0 = 8.377 \text{ A.}, \ c_0 = 7.482 \text{ A.}$$
$$NpCl_4: \quad a_0 = 8.27 \text{ A.}, \ c_0 = 7.48 \text{ A.}$$

V,c5. The structure of *stannic iodide*, SnI_4, was determined in the early days of crystal analysis. It is cubic, with eight molecules in a unit having the cell edge:

$$a_0 = 12.273 \text{ A. } (26°\text{C.})$$

Atoms are in the following positions of T_h^6 (*Pa*3):

Sn: (8c) $\pm (uuu; \ u+{}^1/_2,{}^1/_2-u,\bar{u}; \ \bar{u},u+{}^1/_2,{}^1/_2-u; \ {}^1/_2-u,\bar{u},u+{}^1/_2)$

with $u = 0.129$

I(1):   (8c)   with $u = 0.253$

I(2):   (24d) $\pm (xyz; \ x+{}^1/_2,{}^1/_2-y,\bar{z}; \ \bar{x},y+{}^1/_2,{}^1/_2-z; \ {}^1/_2-x,\bar{y},z+{}^1/_2;$

$zxy; \ \bar{z},x+{}^1/_2,{}^1/_2-y; \ {}^1/_2-z,\bar{x},y+{}^1/_2; \ z+{}^1/_2,{}^1/_2-x,\bar{y};$

$yzx; \ {}^1/_2-y,\bar{z},x+{}^1/_2; \ y+{}^1/_2,{}^1/_2-z,\bar{x}; \ \bar{y},z+{}^1/_2,{}^1/_2-x)$

with $x = 0.009$, $y = 0.001$, and $z = 0.253$

Parameters obtained in a recent redetermination differ from the above only in making $u(Sn) = 0.125$, $u(I) = 0.252$, $x = -0.002$, $y = -0.002$, and $z = 0.252$.

In this arrangement (Fig. VC,5), each tin atom is surrounded by a practically regular tetrahedron of iodine atoms. The Sn–I distance of 2.63 A. is slightly less than the sum of the neutral radii of tin and iodine atoms. The observed I–I separation of 4.21 A. is the same as the I–I distance between molecules of solid iodine, and also between the iodide ions in CdI_2 (**IV,c1**).

Platinic chloride, $PtCl_4$, has this structure, with

$$a_0 = 10.45 \text{ A.}$$

Approximate parameters for its atoms are: $u(Pt) = {}^1/_8$, $u(I) = {}^1/_4$, $x = y = 0$, $z = {}^1/_4$.

Though their atomic positions have not been established with accuracy, the compounds listed in Table VC,3 also have this arrangement.

TABLE VC,3.   Crystals with the Cubic SnI₄ Structure

Crystal	a_0, A.
CBr₄ (above 47°C.)	11.34
GeI₄	12.040 (26°C.)
SiI₄	11.986
TiBr₄	11.250
TiI₄	12.002
ZrBr₄	10.94
ZrCl₄	10.32

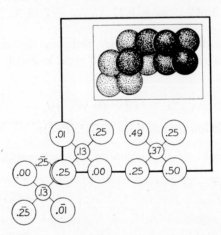

Fig. VC,5. Atoms of three of the eight molecules in the unit cube of SnI₄ projected on a cube face. The insert shows how these molecules pack together if the atoms have their ionic sizes.

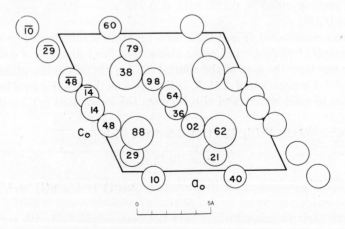

Fig. VC,6a. A projection along the b_0 axis of the monoclinic structure of CsI₄. The larger circles are cesium. Origin in lower left.

V,c6. A determination of the crystal structure of *cesium tetraiodide*, CsI$_4$, shows that it actually contains I$_8$ ions and, therefore, has the composition of Cs$_2$(I$_8$). The symmetry is monoclinic with two Cs$_2$I$_8$ molecules in a unit of the dimensions:

$$a_0 = 11.19 \text{ A.}; \; b_0 = 9.00 \text{ A.}; \; c_0 = 10.23 \text{ A.}; \; \beta = 114°20'$$

All atoms are in general positions of C$_{2h}^5$ ($P2_1/a$):

$$(4e) \quad \pm (xyz; \; x+{}^1/_2, {}^1/_2-y, z)$$

with the following parameters:

Atom	x	y	z
Cs	0.200	0.876	0.277
I(1)	0.006	0.143	0.446
I(2)	0.050	0.485	0.328
I(3)	0.107	0.292	0.121
I(4)	0.174	0.100	0.941

The structure as a whole, shown in Figure VC,6, contains I$_8$ ions that have a nearly planar Z shape and the dimensions given in Figure VC,7. The distance between a cesium ion and eight iodine atoms about it varies from 3.85 to 4.07 A.; all other interatomic separations are greater than this.

This substance is probably identical with the polyiodide measured crystallographically by Penfield in 1893 and at that time given the composition CsI$_5$.

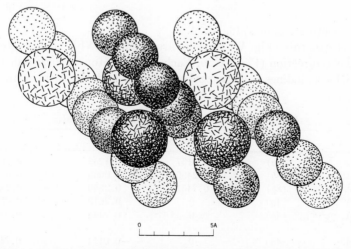

Fig. VC,6b. A packing drawing of the monoclinic structure of CsI$_4$ viewed along its b_0 axis. The cesium atoms are line shaded.

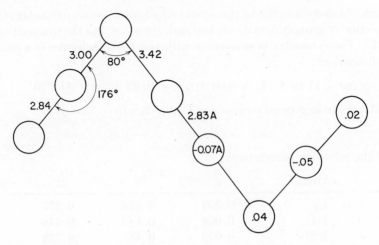

Fig. VC,7. The dimensions of the I_8 ion as found from the structure of Cs_2I_8. The numbers within the circles are atomic distances above or below the mean plane of the molecule.

V,c7. Crystals of the α-form of *niobium tetraiodide*, NbI_4, are orthorhombic, with a unit containing eight molecules. This cell has the edges:

$$a_0 = 7.67 \text{ A.}; \; b_0 = 13.23 \text{ A.}; \; c_0 = 13.93 \text{ A.}$$

Atoms have been placed in the following positions of C_{2v}^{12} ($Cmc2_1$):

(4a) $0uv; \; 0,\bar{u},v+^1/_2; \; ^1/_2,u+^1/_2,v; \; ^1/_2,^1/_2-u,v+^1/_2$

(8b) $xyz; \qquad\qquad \bar{x}yz; \qquad\qquad \bar{x},\bar{y},z+^1/_2; \qquad\qquad x,\bar{y},z+^1/_2;$
  $x+^1/_2,y+^1/_2,z; \; ^1/_2-x,y+^1/_2,z; \; ^1/_2-x,^1/_2-y,z+^1/_2;$
  $\qquad\qquad\qquad\qquad\qquad\qquad\qquad\qquad x+^1/_2,^1/_2-y,z+^1/_2$

with the parameters of Table VC,4.

In this structure (Fig. VC,8) there are indefinite NbI_4 chains extending along the a_0 direction (Fig. VC,9). Each niobium atom is surrounded octahedrally by six iodines at distances between 2.67 and 2.91 A. They are, how-

TABLE VC,4
Positions and Parameters of the Atoms in α-NbI_4

Atom	Position	x	y	z
Nb	(8b)	0.2157	0.1252	0.1202
I(1)	(8b)	0.2460	0.0065	−0.0323
I(2)	(8b)	0.2451	0.2340	0.2834
I(3)	(4a)	0	0.2619	0.0342
I(4)	(4a)	0	0.7443	0.0444
I(5)	(4a)	0	−0.0114	0.2102
I(6)	(4a)	0	0.5077	0.1999

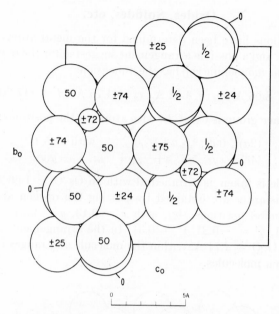

Fig. VC,8. A projection along its a_0 axis of the orthorhombic structure of NbI$_4$. The large circles are the iodine atoms. Origin in lower left.

ever, displaced from the centers of these octahedra towards one another in pairs to give Nb–Nb = 3.31 A. The two iodine atoms bridging these close niobium atoms have Nb–I = 2.74 A., distances shorter than the Nb–I = 2.90 A. involving the more distant niobium atoms in a chain. The shortest I–I between chains is 3.89 A.

In the pentachloride of niobium (**V,e3**), similar double halogen bridges between metallic atoms occur in discrete molecules of the composition Nb$_2$Cl$_{10}$.

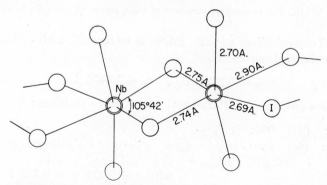

Fig. VC,9. A segment of one of the atomic chains that exist in crystals of NbI$_4$ giving its bond lengths and angles.

Oxides, Sulfides, etc.

V,c8. Positions have been determined for the metal atoms in *osmium tetroxide*, OsO_4, and a possible arrangement suggested for the oxygen atoms. The bimolecular monoclinic cell has the dimensions:

$$a_0 = 8.66 \text{ A.}; \; b_0 = 4.52 \text{ A.}; \; c_0 = 4.75 \text{ A.}; \; \beta = 117°54'$$

The chosen space group is C_2^3 ($C2$), with atoms in the positions:

Os: (2a) $0u0; \; 1/2,u+1/2,0$ with $u = 0$
O: (4c) $xyz; \; \bar{x}y\bar{z}; \; x+1/2,y+1/2,z; \; 1/2-y,y+1/2,\bar{z}$

If the molecule is taken as tetrahedral with an Os–O = 1.66 A., the best interatomic distances are obtained by putting the oxygen atoms in two sets of the general positions (4c), with $x = 0.13$, $y = 0.21$, $z = -0.07$, and $x' = 0.11$, $y' = -0.21$, $z' = 0.31$. In the arrangement thus defined (Fig. VC,10), O–O = 2.71 A. within the molecule and ranges from 2.90 to 3.25 A. between molecules.

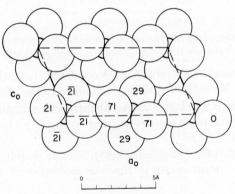

Fig. VC,10. The monoclinic structure of OsO_4 projected along its b_0 axis.

V,c9. *Tetrapalladium selenide*, Pd_4Se, is tetragonal, with a bimolecular cell of the dimensions:

$$a_0 = 5.2324 \text{ A.}, \qquad c_0 = 5.6470 \text{ A.}$$

The space group is V_d^4 ($P\bar{4}2_1c$) with atoms in the positions:

Se: (2a) $000; \; 1/2\,1/2\,1/2$
Pd: (8e) $xyz; \; \bar{x}\bar{y}z; \; 1/2-x,y+1/2,1/2-z; \; x+1/2,1/2-y,1/2-z;$
             $\bar{y}x\bar{z}; \; y\bar{x}\bar{z}; \; y+1/2,x+1/2,z+1/2; \; 1/2-y,1/2-x,z+1/2$
                         with $x = 0.374$, $y = 0.232$, $z = 0.154$

In this arrangement (Fig. VC,11), each palladium atom has two selenium neighbors at 2.46 and 2.49 A.; there are ten palladium atoms within the range from 2.76 to 3.12 A. Each selenium has around it eight palladiums, four at 2.46 A. and four at 2.49 A. In palladium metal the atomic separation is 2.75 A.

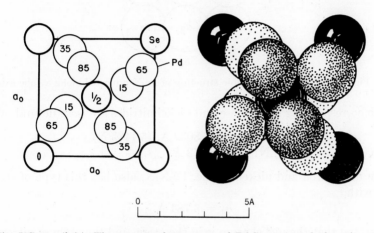

Fig. VC,11a (left). The tetragonal structure of Pd_4Se projected along its c_0 axis. Fig. VC,11b (right). A packing drawing of the Pd_4Se arrangement seen along its c_0 axis. The palladium atoms are dotted.

The sulfide Pd_4S has this structure, with

$$a_0 = 5.1147 \text{ A.}, \qquad c_0 = 5.5903 \text{ A.}$$

Its sulfur parameters have been given as $x = 0.358$, $y = 0.230$, $z = 0.155$, resulting in Pd–S distances of 2.34 and 2.48 A.

Nitrides, Carbides, Borides

V,c10.   The *iron nitride* Fe_4N has been given a very simple structure. Its metallic atoms are in a cubic close-packing with the coordinates:

$$000; \; {}^{1}/_{2}\,{}^{1}/_{2}\,0; \; {}^{1}/_{2}\,0\,{}^{1}/_{2}; \; 0\,{}^{1}/_{2}\,{}^{1}/_{2}$$

The unit containing one molecule has

$$a_0 = 3.7885 \text{ A. } (5.69\% \text{ N})$$

The position of the nitrogen atom has not been established from x-ray data, but it very possibly is in ${}^{1}/_{2}\,{}^{1}/_{2}\,{}^{1}/_{2}$ (Fig. VC,12).

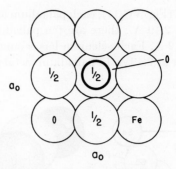

Fig. VC,12. The simple, cubic structure of Fe₄N projected along a cube axis.

The corresponding manganese and nickel nitrides are isostructural, with

$$Mn_4N: \quad a_0 = 3.8032 \text{ A. } (6.14\% \text{ N})$$
$$Ni_4N: \quad a_0 = 3.72 \text{ A.}$$

The mixed iron and nickel nitride, Fe₃NiN, also has this type of structure, with

$$a_0 = 3.784 \text{ A.}$$

It is stated that the nickel atom is in the origin and the atom of nitrogen in $1/2\,1/2\,1/2$.

A second form of Ni₄N has been described as tetragonal with $a_0 = 3.72$ A., $c_0 = 7.28$ A. It would thus appear to be a kind of superlattice on the simpler modification described above.

V,c11. *Boron carbide*, B₄C, forms rhombohedral crystals with a trimolecular unit of the dimensions:

$$a_0 = 5.19 \text{ A.}, \qquad \alpha = 66°18'$$

Corresponding to this is its hexagonal cell containing nine molecules:

$$a_0' = 5.60 \text{ A.}, \qquad c_0' = 12.12 \text{ A.}$$

Atoms have been found to be in special positions of $D_{3d}{}^5$ ($R\bar{3}m$) which, expressed in terms of these hexagonal axes, have the following coordinates:

C(1):   (3b) $0\ 0\ 1/2$; rh
C(2):   (6c) $00u;\ 00\bar{u}$; rh   with $u = 0.385$
B(1):   (18h) $v\bar{v}w:\ v\ 2v\ w;\ 2\bar{v}\ \bar{v}\ w;\ \bar{v}v\bar{w};\ \bar{v}\ 2\bar{v}\ \bar{w};\ 2v\ v\ \bar{w}$; rh
                          with $v = 1/6$, $w = 0.360$
B(2):   (18h)   with $v' = 0.106, w' = 0.113$

There are two ways of viewing this interesting structure (Fig. VC,13). In the first place it can be thought of as a rhombohedrally distorted NaCl-

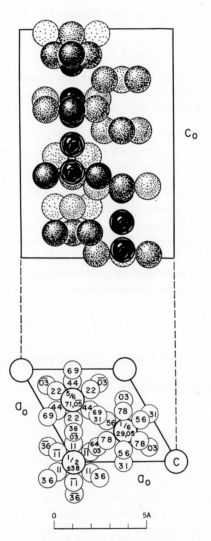

Fig. VC,13. Two projections of the hexagonal unit of B_4C. In the upper projection the
carbon atoms are black.

like grouping of linear C_3 and icosahedral B_{12} groups. At the same time it is
to be noted that the boron atoms in themselves constitute a three-dimensional network extending throughout the crystal; in such a network the
groups of carbon atoms fill the largest holes. Within the linear C_3 chains,
C–C = 1.39 A. The B–B separations are from 1.74 to 1.80 A.; B–C is 1.64 A.

Silicon tetraboride, SiB$_4$, has been reported to have a structure of this type. For it,

$$a_0' = 6.319 \text{ A.}, \qquad c_0' = 12.713 \text{ A.}$$

More recently it has been decided that the composition of this material actually was SiB$_{2.89}$ and that the available atoms were distributed in the following fashion within the B$_4$C arrangement:

$$
\begin{array}{lll}
\text{Si:} & (6c) & \text{with } u = 0.403 \\
\text{B:} & (18h) & \text{with } v = 0.158,\ w = 0.025 \\
(\text{Si,B}): & (18h) & \text{with } v = 0.107,\ w = -0.122
\end{array}
$$

In structurally similar substances, boron can replace carbon to give compounds with such a composition as B$_{13}$C$_2$ (instead of B$_{12}$C$_3$) and with practically no change in cell dimensions. A similar structure has been described for a preparation considered to have the overall composition B$_{14}$C$_3$. In this case it is thought that boron replaces the carbon atom in (3b). Still other substances have been described as solid solutions with excess boron up to B$_{6.75}$C; their cell dimensions are little different from those of B$_4$C itself.

A *boron phosphide* of the composition B$_{13}$P$_2$ seems also to have this structure, with

$$a_0' = 5.984 \text{ A.}, \qquad c_0' = 11.850 \text{ A.}$$

The corresponding arsenide B$_{13}$As$_2$ has likewise been prepared. For it,

$$a_0' = 6.142 \text{ A.}, \qquad c_0' = 11.892 \text{ A.}$$

There is an oxide, originally described as B$_7$O but more probably of the composition B$_{13}$O$_2$, which appears to have this structure. For it,

$$a_0' = 5.37 \text{ A.}, \qquad c_0' = 12.31 \text{ A.}$$

The α modification of boron itself (**II,f2**) has the arrangement of boron atoms described here, the positions occupied by carbon in B$_4$C being empty in the structure of the element.

V,c12. The *tetraboride* of *thorium,* ThB$_4$, is tetragonal with a tetramolecular unit having the edges:

$$a_0 = 7.256 \text{ A.}, \qquad c_0 = 4.113 \text{ A.}$$

Thorium atoms are in the following positions of D$_{4h}^5$ ($P4/mbm$):

$$\text{Th:} \quad (4g) \quad \pm(u,u+{}^1/_2,0;\ {}^1/_2-u,u,0) \qquad \text{with } u = 0.313$$

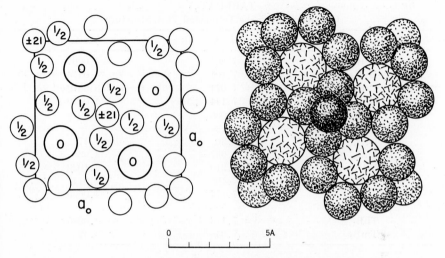

Fig. VC,14a (left). A projection along the c_0 axis of the tetragonal structure of ThB$_4$. The larger circles are thorium.

Fig. VC,14b (right). A packing drawing of the tetragonal ThB$_4$ structure viewed along the c_0 axis. The larger atoms of thorium are line shaded.

If the boron atoms have about the same dimensions as in other borides (B–B = 1.7 – 1.8 A.), it is stated that they can only be in the following positions:

B(1): (4e) $\pm(00u; \, {}^{1}/_{2} \, {}^{1}/_{2} \, u)$ with $u = 0.212$
B(2): (4h) $\pm(u,u+{}^{1}/_{2},{}^{1}/_{2}; \, {}^{1}/_{2}-u,u,{}^{1}/_{2})$ with $u = -0.087$
B(3): (8j) $\pm(u \, v \, {}^{1}/_{2}; \, u+{}^{1}/_{2},{}^{1}/_{2}-v,{}^{1}/_{2}; \, \bar{v} \, u \, {}^{1}/_{2}; \, v+{}^{1}/_{2},u+{}^{1}/_{2},{}^{1}/_{2})$
with $u = 0.170$, $v = 0.042$

This leads to a structure consisting of an uninterrupted network of boron atoms with holes into which the heavy atoms can fit (Fig. VC,14). With the parameters just stated, the B–B distances lie between 1.74 and 1.80 A. Each metal atom has 16 boron neighbors at distances between 2.78 and 2.96 A., and two more at 3.10 A.

Detailed studies have led to this structure for *uranium tetraboride*, UB$_4$. For it,

$$a_0 = 7.075 \text{ A.}, \qquad c_0 = 3.979 \text{ A.}$$

Besides these compounds, there are numerous rare-earth and rare earth-like tetraborides which have been shown to have this ThB$_4$ arrangement. Their cell dimensions are listed in Table VC,5.

TABLE VC,5
Compounds with the Tetragonal ThB$_4$ Structure

Compound	a_0, A.	c_0, A.
CeB$_4$	7.205	4.090
DyB$_4$	7.101	4.0174
ErB$_4$	7.071	3.9972
GdB$_4$	7.144	4.0479
HoB$_4$	7.086	4.0079
LaB$_4$	7.30	4.17
LuB$_4$	6.997	3.938
NdB$_4$	7.219	4.1020
PuB$_4$	7.10	4.014
SmB$_4$	7.174	4.0696
TbB$_4$	7.118	4.0286
TmB$_4$	7.05	3.99

Carbonyls, etc.

V,c13. In the range between their melting point (at $-25°$C.) and $-150°$C., crystals of *nickel carbonyl*, Ni(CO)$_4$, are cubic. There are eight molecules in the unit which at $-55°$C. has the edge:

$$a_0 = 10.84 \text{ A.}$$

The space group is T_h^6 ($Pa3$), with atoms in the following positions:

(8c) $\pm (uuu;\ u+{}^1/_2,{}^1/_2-u,\bar{u};\ {}^1/_2-u,\bar{u},u+{}^1/_2;\ \bar{u},u+{}^1/_2,{}^1/_2-u)$

(24d) $\pm (xyz;\ x+{}^1/_2,{}^1/_2-y,\bar{z};\ \bar{x},y+{}^1/_2,{}^1/_2-z;\ {}^1/_2-x,\bar{y},z+{}^1/_2;$

        $zxy;\ z+{}^1/_2,{}^1/_2-x,\bar{y};\ \bar{z},x+{}^1/_2,{}^1/_2-y;\ {}^1/_2-z,\bar{x},y+{}^1/_2;$

        $yzx;\ y+{}^1/_2,{}^1/_2-z,\bar{x};\ \bar{y},z+{}^1/_2,{}^1/_2-x;\ {}^1/_2-y,\bar{z},x+{}^1/_2)$

The positions occupied by the several atoms and the parameters assigned them are given below.

Atom	Position	x	y	z
Ni	(8c)	0.123	0.123	0.123
C(1)	(8c)	0.220	0.220	0.220
C(2)	(24d)	0.032	0.020	0.220
O(1)	(8c)	0.281	0.281	0.281
O(2)	(24d)	−0.025	−0.044	0.280

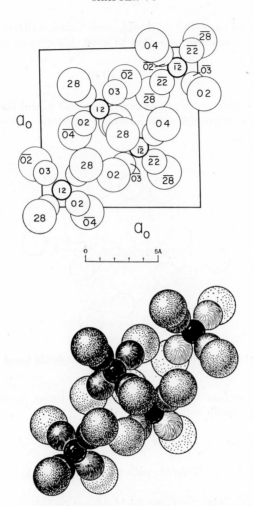

Fig. VC,15a (top). A projection along a cubic axis of some of the atoms in the unit of Ni(CO)$_4$. The smallest heavily ringed circles are nickel, the largest circles are oxygen atoms.

Fig. VC,15b (bottom). A packing drawing of the atoms in the structure of Ni(CO)$_4$ shown in Fig. VC,15a. The nickel atoms are black, the oxygens are dotted.

The molecule consists of a tetrahedral distribution of CO radicals about a nickel atom, with Ni–C = 1.84 A. and C–O = 1.15 A. As Figure VC,15 suggests, these molecules are packed in the crystal much as are the molecules in SnI$_4$ (**V,c5**). Between molecules the closest atomic approach is O–O = 3.24 A.

The corresponding *cobalt carbonyl*, sometimes written $HCo(CO)_4$ instead of $Co(CO)_4$, has been reported to have this structure, with

$$a_0 = 10.70 \text{ A.}$$

According to another study, however, crystals of it and the corresponding iron compound $Fe(CO)_4$ are so disordered that atomic arrangements could not be established.

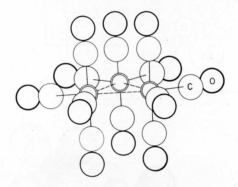

Fig. VC,16. The atomic distribution in the $Os_3(CO)_{12}$ molecule found in solid $Os(CO)_4$.

V,c14. The *osmium tetracarbonyl*, $Os(CO)_4$, forms monoclinic crystals with a large 12-molecule cell of the dimensions:

$$a_0 = 8.10 \text{ A.}; \quad b_0 = 14.79 \text{ A.}; \quad c_0 = 14.64 \text{ A.}; \quad \beta = 100°27'$$

The space group is C_{2h}^5 ($P2_1/n$), with all atoms in the general positions:

$$(4e) \quad \pm (xyz; \ x+\tfrac{1}{2}, \tfrac{1}{2}-y, z+\tfrac{1}{2})$$

The atomic parameters are those of Table VC,6.

This is a structure composed of $Os_3(CO)_{12}$ molecules having the configuration indicated in Figure VC,16. In a molecule the osmium atoms form an approximately equilateral triangle with Os–Os = 2.873–2.886 A. The assigned Os–C = 1.79–2.04 A. and C–O = 1.03–1.20 A., though the carbon and oxygen positions could not be established with great accuracy.

The corresponding *ruthenium carbonyl*, $Ru_3(CO)_{12}$, is reported to be isomorphous, though x-ray data are lacking. These are compounds which had previously been thought to have the composition $R_2(CO)_9$.

TABLE VC,6
Parameters of the Atoms in Os(CO)$_4$

Atom	x	y	z
Os(1)	0.4431	−0.0246	0.2233
Os(2)	0.4412	0.1698	0.2059
Os(3)	0.6962	0.0812	0.3389
C(1)	0.300	−0.011	0.319
O(1)	0.202	−0.012	0.366
C(2)	0.596	−0.032	0.129
O(2)	0.658	−0.040	0.068
C(3)	0.251	−0.044	0.135
O(3)	0.137	−0.069	0.076
C(4)	0.526	−0.145	0.253
O(4)	0.545	−0.215	0.291
C(5)	0.287	0.180	0.296
O(5)	0.239	0.188	0.356
C(6)	0.571	0.153	0.105
O(6)	0.669	0.163	0.062
C(7)	0.502	0.285	0.228
O(7)	0.565	0.353	0.254
C(8)	0.246	0.180	0.109
O(8)	0.120	0.197	0.046
C(9)	0.554	0.088	0.431
O(9)	0.469	0.104	0.487
C(10)	0.836	0.070	0.244
O(10)	0.928	0.064	0.194
C(11)	0.799	−0.026	0.408
O(11)	0.875	−0.081	0.446
C(12)	0.834	0.188	0.398
O(12)	0.897	0.246	0.423

V,c15. Crystals of *platinum tetrathionitrosyl*, Pt(SN)$_4$, are monoclinic, with a tetramolecular unit of the dimensions:

$$a_0 = 8.57 \text{ A.}; \quad b_0 = 7.69 \text{ A.}; \quad c_0 = 11.14 \text{ A.}; \quad \beta = 101°36'$$

The space group is C$_{2h}^5$ ($P2_1/c$), with all atoms in the general positions:

$$(4e) \quad \pm (xyz; \; x, \tfrac{1}{2}-y, z+\tfrac{1}{2})$$

The assigned parameters are those listed in Table VC,7. As Figure VC,17 indicates, the structure is an array of planar molecules which have the bond lengths and angles shown in Figure VC,18.

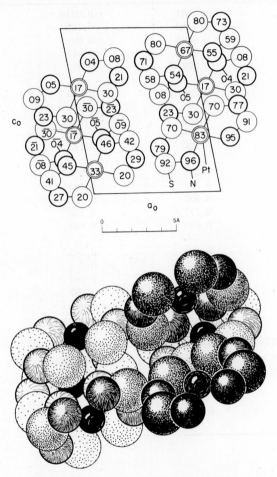

Fig. VC,17a (top). The monoclinic structure of Pt(SN)₄ projected along its b_0 axis.
Fig. VC,17b (bottom). A packing drawing of the Pt(SN)₄ arrangement viewed along its b_0 axis. The platinum atoms are black, the sulfur atoms dotted.

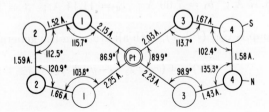

Fig. VC,18. The bond dimensions of the Pt(SN)₄ molecules found in the crystal.

TABLE VC,7
Parameters of the Atoms in Pt(SN)$_4$

Atom	x	y	z
Pt	0.054	0.167	0.636
S(1)	0.904	0.299	0.473
S(2)	0.688	0.092	0.572
S(3)	0.279	0.296	0.611
S(4)	0.371	0.084	0.815
N(1)	0.838	0.046	0.665
N(2)	0.721	0.228	0.473
N(3)	0.176	0.042	0.786
N(4)	0.389	0.206	0.704

BIBLIOGRAPHY TABLE, CHAPTER VC

Compound	Paragraph	Literature
AmF$_4$	c2	1954: A
B$_4$C	c11	1934: R; 1939: S; 1941: Z&S; 1943: C&H; Z&S; 1953: G,M&P; Z,Z&Z; 1954: Z,M,Z&S; 1960: K&S; 1961: LP&P
CBr$_4$	c5	1924: M; 1937: F&H
CeB$_4$	c12	1950: Z&T; 1953: Z&T
CeF$_4$	c2	1949: Z
Co(CO)$_4$	c13	1953: N&C
CsI$_4$	c6	1954: H,B&W
CmF$_4$	c2	1957: A,E,F&Z
DyB$_4$	c12	1957: N&S; 1958: N&S; 1959: E&G
ErB$_4$	c12	1958: N&S; 1959: E&G; S,P&S
Fe(CO)$_4$	c13	1931: B; 1957: D&R; M
Fe$_4$N	c10	1928: B; H; O&I; 1929: H; 1930: H; 1955: B; 1958: D&P; 1960: A&W
Fe$_3$NiN	c10	1960: A&W
GdB$_4$	c12	1958: N&S; S&Z; 1959: E&G
GeI$_4$	c5	1925: J,T&W; 1954: NBS
HfF$_4$	c2	1934: S; 1949: Z
HoB$_4$	c12	1957: N&S; 1958: N&S; S&Z; 1959: E&G
LuB$_4$	c12	1957: N&S; 1958: N&S; S&Z
Mn$_4$N	c10	1929: H; 1946: G&W
NbI$_4$	c7	1962: D&W
NdB$_4$	c12	1959: E&G
Ni(CO)$_4$	c13	1952: L,P&F
Ni$_4$N	c10	1959: T; 1962: T
NpCl$_4$	c4	1949: Z; 1950: E,F,S&Z; 1954: S,F,E&Z
NpF$_4$	c2	1949: Z

(continued)

BIBLIOGRAPHY TABLE, CHAPTER VC (*continued*)

Compound	Paragraph	Literature
NpO_2F_2	[IV,c14]	—
$Os(CO)_4$	c14	1961: C&D; 1962: C&D
OsO_4	c8	1953: Z&T
$PaCl_4$	c4	1950: E,F,S&Z; 1954: S,F,E&Z
PbF_4	c3	1962: H&D
Pd_4S	c9	1962: G&R
Pd_4Se	c9	1962: G&R
$PtCl_4$	c5	1958: F
$Pt(SN)_4$	c15	1958: L&W
PuB_4	c12	1960: MD&S
PuF_4	c2	1949: Z
$Ru(CO)_4$	c14	1961: C&D
SiB_4	c11	1960: B&M; M; 1962: M&B
SiF_4	c1	1930: N; 1954: A&L
SiI_4	c5	1931: H&K
SmB_4	c12	1959: E&G
SnF_4	c3	1962: H&D
SnI_4	c5	1923: D; M&W; 1926: D; O; 1954: NBS; 1955: M&F
TbB_4	c12	1959: E&G; P,S&S; S,P&S
TbF_4	c2	1954: C,F&R
ThB_4	c12	1950: Z&T; 1953: Z&T
$ThBr_4$	c4	1950: DE
$ThCl_4$	c4	1949: M; 1950: E,F,S&Z; 1954: S,F,E&Z
ThF_4	c2	1949: Z
$TiBr_4$	c5	1932: H&K
TiI_4	c5	1932: H&K
TmB_4	c12	1961: P&S
UB_4	c12	1949: B&B; 1950: Z&T; 1953: Z&T; 1954: B&B
UCl_4	c4	1930: H; 1949: M; 1950: E,F,S&Z
UF_4	c2	1949: Z; 1961: M&C
UO_2F_2	[IV,c14]	—
$ZrBr_4$	c5	1962: B,L,V&B
$ZrCl_4$	c5	1930: H
ZrF_4	c2	1934: S; 1949: Z; 1956: B&B

D. COMPOUNDS OF THE TYPE R_3X_4

Oxides, Sulfides, etc.

V,d1. Considerable attention has been given to the structure of *minium* or *red lead*, Pb_3O_4. It is tetragonal, with a tetramolecular cell having

$$a_0 = 8.788 \text{ A.}, \qquad c_0 = 6.551 \text{ A.}$$

The space group is D_{4h}^{13} ($P4/mbc$), with atoms in the positions:

Pb^{III}: (4d) $\pm(0 \; ^1/_2 \; ^1/_4; \; ^1/_2 \; 0 \; ^1/_4)$

Pb^{II}:   (8h) $\pm(uv0; \bar{v} \, u \, ^1/_2; u+^1/_2, ^1/_2-v, 0; v+^1/_2, u+^1/_2, ^1/_2)$
with $u = v = 0.153$

O(1):   (8h)   with $u = 0.114$, $v = 0.614$

O(2):   (8g) $\pm(u, u+^1/_2, ^1/_4; \bar{u}, ^1/_2-u, ^1/_4; u+^1/_2, \bar{u}, ^1/_4; ^1/_2-u, u, ^1/_4)$
with $u = 0.672$

This is the same structure that has been found for Sb_2ZnO_4. In it the zinc atoms are in (4d); the antimony atoms in (8h) have $u = 0.175$, $v = 0.167$; the O(1) in (8h) have $u = 0.114$, $v = 0.614$; and the O(2) in (8g) have $u = 0.669$.

In this arrangement (Fig. VD,1) the zinc atoms are octahedrally surrounded by oxygen atoms, these octahedra being joined by sharing two edges. The antimony atoms have three pyramidally distributed oxygen neighbors at distances of 1.87 and 2.01 A.

Though no other compound of the composition R_3O_4 is known to have this atomic arrangement, there are several compounds R_2MX_4 isostructural with Sb_2ZnO_4. Their cells have the edge lengths listed in Table VD,1.

TABLE VD,1
Crystals with the Tetragonal Pb_3O_4 Structure

Crystal	a_0, A.	c_0, A.
As_2NiO_4	8.22	5.62
Pb_2SnO_4	8.72	6.30
Sb_2CoO_4	8.49	5.91
Sb_2FeO_4	8.592	5.905
Sb_2MgO_4	8.445	5.907
Sb_2MnO_4	8.685	5.980
Sb_2NiO_4	8.35	5.91
Sb_2ZnO_4	8.491	5.920

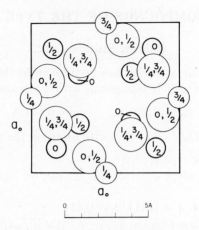

Fig. VD,1. A projection along the c_0 axis of the tetragonal unit of Sb_2ZnO_4 as an example of the Pb_3O_4 structure. The small heavily ringed circles are antimony atoms; the largest circles are oxygen.

V,d2. The *platinum oxide*, Pt_3O_4, is reported to be cubic, with a small unit containing two molecules and having the edge length:

$$a_0 = 6.23 \text{ A.}$$

In the assigned structure, platinum atoms have been put in

(6b) $^1/_2\,0\,0$; $0\,^1/_2\,0$; $0\,0\,^1/_2$; $^1/_2\,^1/_2\,0$; $0\,^1/_2\,^1/_2$; $^1/_2\,0\,^1/_2$

and oxygen atoms in

(8c) $^1/_4\,^1/_4\,^1/_4$; $^3/_4\,^3/_4\,^3/_4$; F.C.

of O_h^9 ($Im3m$).

This very simple atomic distribution yields a Pt–O = 2.2 A. and a Pt–Pt = 3.11 A.

V,d3. Material with the approximate composition of the *chromium sulfide*, Cr_3S_4, is monoclinic, with the bimolecular cell:

$$a_0 = 5.973\text{–}5.960 \text{ A.,} \qquad b_0 = 3.432\text{–}3.427 \text{ A.}$$
$$c_0 = 11.361\text{–}11.253 \text{ A.,} \qquad \beta = 91°9'\text{–}91°38'$$

Atoms have been assigned the following positions of C_{2h}^3 ($I2/m$):

Cr(1):   (2a) 000; $^1/_2\,^1/_2\,^1/_2$
Cr(2):   (4i) $\pm(u0v)$; B.C.   with $u = -0.012$, $v = 0.263$

S(1):   (4i)   with $u = 0.355$, $v = 0.363$
S(2):   (4i)   with $u = 0.320$, $v = 0.876$

This arrangement is illustrated in Figure VD,2. It is related to the NiAs structure (**III,d1**) and like it is one in which the metallic atoms have both metallic and nonmetallic close neighbors.

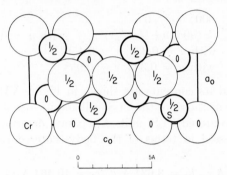

Fig. VD,2. The monoclinic Cr_3S_4 structure projected along its b_0 axis. Origin in lower right.

The corresponding chromium selenide, Cr_3Se_4, has been shown to have this structure, with

$$a_0 = 6.32 \text{ A.}; \quad b_0 = 3.62 \text{ A.}; \quad c_0 = 11.77 \text{ A.}; \quad \beta = 91°28'$$

For Cr(2) the parameters in (4i) are $u = 0.022$, $v = 0.240$. For Se(1), $u = 0.329$, $v = 0.379$; for Se(2), $u = 0.336$, $v = 0.866$.

Other compounds reported to have this arrangement have the cell dimensions listed in Table VD,2.

TABLE VD,2
Cell Dimensions of Crystals with the Cr_3S_4 Structure

Crystal	a_0, A.	b_0, A.	c_0, A.	β
Co_3Se_4	6.14	3.57	10.43	91°43'
Cr_2NiS_4	5.94	3.42	11.14	91°18'
Fe_3Se_4	6.17	3.54	11.17	92°0'
Ti_3Se_4	6.40	3.57	12.04	90°42'
Ti_3Te_4	6.82	3.85	12.66	90°28'
V_3Se_4	6.17	3.44	11.86	91°34'
V_3Te_4	6.736	3.754	12.521	91°26'

A structure has been described for the *nickel selenide*, Ni_3Se_4, which, though expressed in terms of a different set of axes, is undoubtedly the same type of atomic arrangement found for Cr_3S_4. The monoclinic cell employed, also bimolecular, has the dimensions:

$$a_0 = 12.15 \text{ A.}; \quad b_0 = 3.633 \text{ A.}; \quad c_0 = 10.45 \text{ A.}; \quad \beta = 149°22'$$

Atoms were placed in the same special positions of $C_{2h}{}^3$ when given the side-centered orientation $C2/m$:

Ni(1): (2a) $000; \frac{1}{2}\frac{1}{2}0$
Ni(2): (4i) $\pm (u0v; u+\frac{1}{2},\frac{1}{2},v)$ with $u = 0.00, v = 0.25$
Se(1): (4i)   with $u = 0.333, v = 0.458$
Se(2): (4i)   with $u = 0.333, v = 0.958$

This unit cell and its contents are shown in Figure VD,3. The b_0 and c_0 axes are the same as those for the Cr_3S_4 cell and its a_0 axis is the long diagonal of the body-centered Cr_3S_4 unit. A separate determination of the cell dimensions expressed in terms of the body-centered cell for Ni_3Se_4 has given:

$$a_0' = 6.197 \text{ A.}; \quad b_0 = 3.634 \text{ A.}; \quad c_0 = 10.464 \text{ A.}; \quad \beta' = 90°47'$$

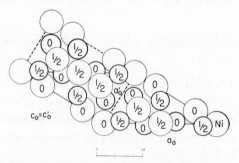

Fig. VD,3. The monoclinic structure of Ni_3Se_4 viewed along its b_0 axis. The axes of its side-centered cell are drawn as full lines; those of the nearly orthogonal cell resembling that of Cr_3S_4 are dotted.

Referred to this cell and making use of the coordinates given under Cr_3S_4, atomic positions in its nearly orthogonal unit are:

Ni(1): (2a)
Ni(2): (4i)   with $u = 0.00, v = 0.25$
Se(1): (4i)   with $u = 0.333, v = 0.128$
Se(2): (4i)   with $u = 0.333, v = 0.628$

As comparison of Figures VD,2 and VD,3 makes immediately evident, these two structures are almost exact reflections of one another in the b_0c_0 plane. The axes of the body-centered unit are so nearly orthogonal that the x-ray data probably are not precise enough to distinguish between the chosen parameters and the same values applicable to the nearly identical arrangement having the signs of v reversed. The important atomic separations are Ni–Ni = 2.61 A., Ni–Se = 2.47 A., and Se–Se = 3.33 A.

V,d4. A titanium phase exists which has been studied with x rays over the composition range between ca. $Ti_{2.2}S_4$ and Ti_3S_4. It shows little change in cell dimensions with composition:

$$\text{For } Ti_{2.2}S_4: \quad a_0 = 3.413 \text{ A.}, \quad c_0 = 11.46 \text{ A.}$$
$$\text{For } Ti_3S_4: \quad a_0 = 3.448 \text{ A.}, \quad c_0 = 11.45 \text{ A.}$$

In this hexagonal structure, atoms have been distributed according to the following special positions of C_{6v}^4 $(P6_3/mc)$:

S(1): (2b) $^1/_3\,^2/_3\,u;\; ^2/_3,^1/_3,u+^1/_2$ with $u = {}^3/_8$
S(2): (2a) $00v;\, 0,0,v+^1/_2$ with $v = {}^1/_8$
Ti(1): (2b) with $u = {}^3/_4$
Ti(2): atoms in some of the positions of (2b) with $u = 0$

In such an arrangement as this, the sulfur atoms would be in a mixed cubic and hexagonal close-packing.

V,d5. Crystals of *tetraphosphorus trisulfide*, P_4S_3, are orthorhombic, with an eight-molecule cell of the dimensions:

$$a_0 = 9.660 \text{ A.}; \quad b_0 = 10.597 \text{ A.}; \quad c_0 = 13.671 \text{ A.}$$

All atoms are in the following positions of V_h^{16} $(Pmnb)$:

(4c) $\pm(^1/_4\,u\,v;\; ^1/_4,u+^1/_2,^1/_2-v)$
(8d) $\pm(xyz;\; ^1/_2-x,y+^1/_2,^1/_2-z;\; x+^1/_2,\bar{y},\bar{z};\; \bar{x},^1/_2-y,z+^1/_2)$

with the parameters, as determined by a three-dimensional least squares analysis, listed in Table VD,3.

TABLE VD,3
Parameters of the Atoms in P_4S_3

Atom	Position	x	y	z
S(1)	(8d)	0.5853	0.1474	0.9711
S(2)	(4c)	$^3/_4$	−0.0699	0.8580
S(1')	(8d)	0.0851	0.5613	0.8728
S(2')	(4c)	$^1/_4$	0.3347	0.9709
P(1)	(8d)	0.6341	0.2324	0.8376
P(2)	(4c)	$^3/_4$	0.0801	0.7580
P(3)	(4c)	$^3/_4$	0.0219	0.9934
P(1')	(8d)	0.1345	0.4456	0.7536
P(2')	(4c)	$^1/_4$	0.2873	0.8223
P(3')	(4c)	$^1/_4$	0.5301	0.9683

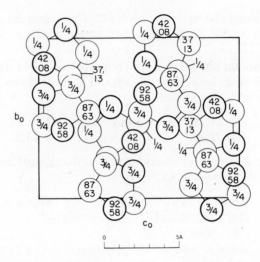

Fig. VD,4a. The orthorhombic structure of P_4S_3 projected along its a_0 axis. The sulfur atoms are the more heavily outlined circles. Origin in lower left.

The resulting structure (Fig. VD,4) is an aggregate of P_4S_3 molecules which have essentially the same configuration as has been found in the vapor by electron diffraction.

There are two sets of these molecules in the crystal. Their bond lengths and angles (Fig. VD,5), however, indicate that the crystallographically unlike molecules are not significantly different in size and shape. The molecules themselves possess a plane of symmetry, and between them the shortest atomic separations are about 3.6 A.

V,d6. Though the molecules of *tetraphosphorus triselenide*, P_4Se_3, resemble those of the analogous sulfur compound (**V,d5**), the two crystal structures are different. Both are orthorhombic with similar a_0 and b_0 axes, but the c_0 axis for the selenide is about twice as long as for the sulfide. Its 16-molecule cell has the dimensions:

$$a_0 = 9.739 \text{ A.}; \quad b_0 = 11.797 \text{ A.}; \quad c_0 = 26.270 \text{ A.}$$

Atoms are nevertheless in the same positions of V_h^{16} (*Pmnb*) as for the sulfide:

$$(4c) \quad \pm (1/4 \, u \, v; \; 1/4, u+1/2, 1/2-v)$$
$$(8d) \quad \pm (xyz; \; 1/2-x, y+1/2, 1/2-z; \; x+1/2, \bar{y}, \bar{z}; \; \bar{x}, 1/2-y, z+1/2)$$

with the parameters of Table VD,4.

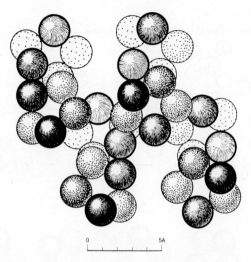

Fig. VD,4b. A packing drawing of the P_4S_3 arrangement viewed along its a_0 axis. The phosphorus atoms are dotted.

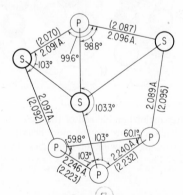

Fig. VD,5. The bond lengths and angles of the P_4S_3 molecule as it is found in its crystals.

The similarity between the two structures is apparent from a comparison of Figure VD,6 with Figure VD,4. The four crystallographically different molecules in the cell of the selenide have a plane of symmetry and on the average have the bond lengths and angles of Figure VD,7. The shortest intermolecular distances, between phosphorus atoms, is ca. 3.6 A.

In both this structure and that of the simpler sulfide, the molecules are approximately in a hexagonal close-packing.

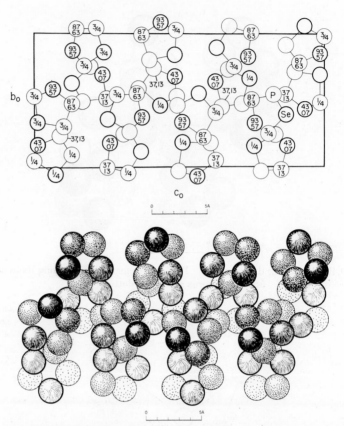

Fig. VD,6a (top). The large orthorhombic cell of the P_4Se_3 structure projected along its a_0 axis. The selenium atoms are the more heavily ringed circles.

Fig. VD,6b (bottom). A packing drawing of the P_4Se_3 structure viewed along its a_0 axis. The phosphorus atoms are dotted.

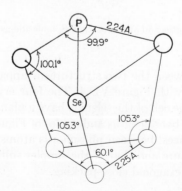

Fig. VD,7. The bond lengths and angles of the P_4Se_3 molecule.

TABLE VD,4
Positions and Parameters of the Atoms in P_4Se_3

Atom	Position	x	y	z
P(1)	(4c)	$1/4$	0.086	0.127
P(2)	(8d)	0.366	0.227	0.0875
P(3)	(4c)	$1/4$	0.041	0.000
Se(1)	(4c)	$1/4$	−0.057	0.0710
Se(2)	(8d)	0.428	0.1592	0.0139
P(1′)	(4c)	$1/4$	−0.034	−0.218
P(2′)	(8d)	0.866	0.034	0.1443
P(3′)	(4c)	$1/4$	0.238	−0.169
Se(1′)	(4c)	$1/4$	0.148	−0.2433
Se(2′)	(8d)	0.928	−0.1473	0.1298
P(1″)	(4c)	$1/4$	−0.048	0.325
P(2″)	(8d)	0.366	−0.010	0.2527
P(3″)	(4c)	$1/4$	0.239	0.307
Se(1″)	(4c)	$1/4$	0.107	0.3700
Se(2″)	(8d)	0.428	0.1723	0.2612
P(1‴)	(4c)	$1/4$	0.192	−0.396
P(2‴)	(8d)	0.866	−0.039	0.3663
P(3‴)	(4c)	$1/4$	−0.032	−0.485
Se(1‴)	(4c)	$1/4$	0.163	−0.4822
Se(2‴)	(8d)	0.928	0.0602	0.4355

Nitrides and Phosphides

V,d7. The *nitrides* of *silicon* and *germanium*, Si_3N_4 and Ge_3N_4, are dimorphous.

An approximate structure has been described for the α modification. It is hexagonal, with a tetramolecular cell that for α-Si_3N_4 has the dimensions:

$$a_6 = 7.753 \text{ A.,} \qquad c_0 = 5.618 \text{ A.}$$

The space group has been chosen as C_{3v}^4 ($P31c$), with atoms in the positions:

N(1): (2a) $00u;\ 0,0,u+^1/_2$ with $u = 0$
N(2): (2b) $^1/_3\,^2/_3\,u;\ ^2/_3,^1/_3,u+^1/_2$ with $u = {}^3/_4$
N(3): (6c) $xyz;$ $\bar{y},x-y,z;$ $y-x,\bar{x},z;$
  $y,x,z+^1/_2;\ x-y,\bar{y},z+^1/_2;\ \bar{x},y-x,z+^1/_2$
      with $x = {}^1/_3,\ y = z = 0$
N(4): (6c) with $x = {}^1/_3,\ y = {}^1/_3,\ z = {}^1/_4$

Si(1): (6c) with $x = {}^1/_2$, $y = {}^1/_{12}$, $z = {}^1/_4$
Si(2): (6c) with $x = {}^1/_6$, $y = {}^1/_4$, $z = 0$

A basal projection of this arrangement is shown in Figure VD,8.

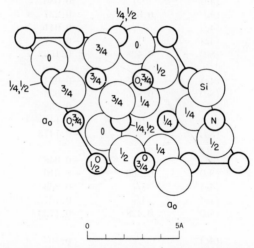

Fig. VD,8. A basal projection of the hexagonal structure described for α-Si$_3$N$_4$.

The α modification of Ge$_3$N$_4$ has the cell dimensions:

$$a_0 = 8.202 \text{ A.}, \qquad c_0 = 5.941 \text{ A.}$$

The β form of the two nitrides has the atomic arrangement that was determined many years ago for phenacite, Be$_2$SiO$_4$ (Chapter VIII). For β-Si$_3$N$_4$, its rhombohedral six-molecule cell has the dimensions:

$$a_0 = 8.15 \text{ A.}, \qquad \alpha = 108°0'$$

The corresponding hexagonal cell containing 18 molecules has the edges:

$$a_0' = 13.16 \text{ A.}, \qquad c_0' = 8.727 \text{ A.}$$

When the two kinds of metallic atom of the phenacite arrangement are alike, as is the case with these nitrides, the rhombohedral unit reduces to one that is truly hexagonal with one-third the volume of the rhombohedron. Such a reduced hexagonal unit containing two molecules per cell has the edges:

$$a_0 = 7.606 \text{ A.}, \qquad c_0 = 2.909 \text{ A.}$$

The relation between the two sets of hexagonal axes is expressed by: $a_0 = a_0'\sqrt{3}$ and $c_0 = {}^1/_3 c_0'$.

This reduced structure for the β modification, based on $C_{6h}{}^2$ ($P6_3/m$) has atoms in the positions:

N(1):   (2c) $\pm(^1/_3\,^2/_3\,^1/_4)$

N(2):   (6h) $\pm(u\,v\,^1/_4;\,\bar{v},u-v,^1/_4;\,v-u,\bar{u},^1/_4)$
       with $u = 0.321$, $v = 0.025$.   In another determination the
       parameters were found to be $u = 0.333$, $v = 0.033$.

Si: (6h)   with $u = 0.174$, $v = -0.234$
       (in the other study $u = 0.172$, $v = -0.231$)

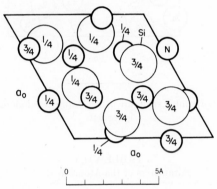

Fig. VD,9. A basal projection of the simpler hexagonal phenacite-like structure found for
β-Si$_3$N$_4$.

In this arrangement (Fig. VD,9) each silicon is surrounded by a tetra-hedron of nitrogen atoms with Si–N = 1.71–1.76 A. The silicon atoms have the positions of the combined metal atoms and the nitrogen those of the oxygen atoms in phenacite.

For the germanium compound, β-Ge$_3$N$_4$ has the cell dimensions:

$$a_0 = 8.62\ \text{A.,} \qquad \alpha = 108°0'$$
$$a_0' = 13.92\ \text{A.,} \qquad c_0' = 9.222\ \text{A.}$$
$$a_0 = 8.038\ \text{A.,} \qquad c_0 = 3.074\ \text{A.}$$

V,d8.   The *tetrarhodium triphosphide*, Rh$_4$P$_3$, is orthorhombic with a tet-ramolecular unit of the dimensions:

$$a_0 = 11.662\ \text{A.;} \quad b_0 = 3.317\ \text{A.;} \quad c_0 = 9.994\ \text{A.}$$

Its space group is $V_h{}^{16}$ ($Pnma$), with all atoms in the special positions:

$$(4c)\quad \pm(u\,^1/_4\,v;\,u+^1/_2,^1/_4,^1/_2-v)$$

The determined parameters are those of Table VD,5.

160 CRYSTAL STRUCTURES

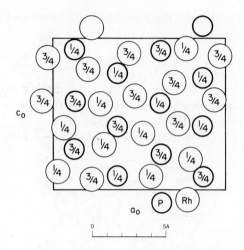

Fig. VD,10a. The orthorhombic structure of Rh₄P₃ projected along its b_0 axis. Origin in lower left.

TABLE VD,5
Parameters of the Atoms in Rh₄P₃

Atom	u	v
Rh(1)	0.0270	0.1172
Rh(2)	0.2717	0.5696
Rh(3)	0.0647	0.4059
Rh(4)	0.2945	0.2911
P(1)	0.3763	0.7616
P(2)	0.1273	0.9212
P(3)	0.3704	0.0792

In this arrangement (Fig. VD,10) each phosphorus atom is at the center of a trigonal prism with P–Rh between 2.27 and 2.61 A. and with Rh–Rh separations ranging upwards from 2.80 A. to ca. 3.70 A. Each rhodium atom has five phosphorus neighbors. The phosphorus atoms are not in contact, the nearest P–P being 3.18 A.

V,d9. *Thorium phosphide*, Th₃P₄, is cubic, with a tetramolecular unit having

$$a_0 = 8.600 \text{ A.}$$

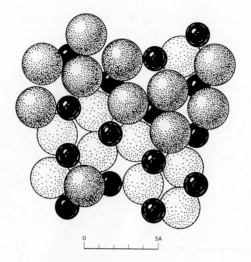

Fig. VD,10b. A packing drawing of the Rh_4P_3 arrangement viewed along its b_0 axis. The phosphorus atoms are black.

Atoms have been placed in the following special positions of T_d^6 ($I\bar{4}3d$):

Th: (12a) $^3/_8\ 0\ ^1/_4;\ ^1/_8\ 0\ ^3/_4;\ ^3/_4\ ^1/_8\ 0;$
              $^1/_4\ ^3/_8\ 0;\ 0\ ^1/_4\ ^3/_8;\ 0\ ^3/_4\ ^1/_8;$ B.C.

P: (16c) $uuu;$ $u,\bar{u},^1/_2-u;$
              $^1/_2-u,u,\bar{u};$ $\bar{u},^1/_2-u,u;$
              $u+^1/_4,u+^1/_4,u+^1/_4;\ ^1/_4-u,u+^1/_4,^3/_4-u;$
              $u+^1/_4,^3/_4-u,^1/_4-u;\ ^3/_4-u,^1/_4-u,u+^1/_4;$ B.C.

with $u = 0.083$

In this structure (Fig. VD,11), each thorium atom has about it eight phosphorus atoms at distances of 2.98 A. The nearest approach of phosphorus atoms to one another is 3.20 A.

Other compounds with this atomic arrangement are listed in Table VD,6.

Though their composition is not R_3X_4, the sesquioxides of rubidium and cesium, Rb_4O_6 and Cs_4O_6, have been described as having the "anti" form of this structure. Alkali atoms have been placed in the positions occupied by phosphorus (16c) and pairs of oxygen atoms have been centered at the points of (12a). The cubic edge lengths for these compounds are

$$Rb_4O_6:\quad a_0 = 9.30\ \text{A}.$$
$$Cs_4O_6:\quad a_0 = 9.86\ \text{A}.$$

Exact positions have not been given the oxygen atoms, but u for both rubidium and cesium has been stated as 0.054.

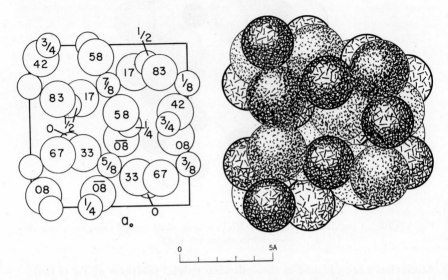

Fig. VD,11a (left). A projection along a cube axis of the Th_3P_4 structure. The small circles are the thorium atoms.

Fig. VD,11b (right). A packing drawing of the Th_3P_4 structure viewed along a cubic axis. The thorium atoms are line shaded.

TABLE VD,6
Compounds with the Cubic Th_3P_4 Structure

Compound	a_0, A.	u
Ce_3S_4	8.6248	—
La_3S_4	8.730	—
Nd_3S_4	8.524	—
Pr_3S_4	8.594	—
Sm_3S_4	8.556	—
Th_3As_4	8.843	$1/12$
Th_3Bi_4	9.559	$1/12$
Th_3Sb_4	9.372	$1/12$
U_3As_4	8.507	$1/12$
U_3Bi_4	9.350	—
U_3P_4	8.197	—
U_3Sb_4	9.095	—
U_3Te_4	9.397	—

Sulfides have been prepared with compositions that vary continuously between Ce_3S_4 and Ce_2S_3. All give x-ray patterns indicative of the Th_3P_4 arrangement, and their cell dimensions are uninfluenced by composition. Thus for Ce_2S_3 the $a_0 = 8.6345$ A. is practically identical with a_0 for Ce_3S_4 (Table VD,6). On this basis, Ce_2S_3 has been interpreted as a deficit structure in which its $5^1/_3$ molecules per cell have the 16 sulfur atoms distributed in $(16c)$ and the $10^2/_3$ metallic atoms in most of the positions of $(12a)$.

A number of other rare-earth and other sulfide preparations have similar deficit Th_3P_4 structures. Cell dimensions for the R_2S_3 end of the series they form are listed in Table VD,7.

TABLE VD,7
Cubic Crystals with the Deficit Th_3P_4 Structure

Crystal	a_0, A.
Ac_2S_3	8.99
Am_2S_3	8.445
La_2S_3	8.731
Nd_2S_3	8.527
Pr_2S_3	8.594
Pu_2S_3	8.4542
Sm_2S_3	8.448

Carbides and Borides

V,d10. *Aluminum carbide*, Al_4C_3, is rhombohedral, with a unimolecular cell having the dimensions:

$$a_0 = 8.53 \text{ A.}, \qquad \alpha = 22°28'$$

The space group is D_{3d}^5 $(R\bar{3}m)$. One carbon atom is in the origin $(1a)$ 000; all other atoms are in $(2c)$ $\pm(uuu)$. Parameters of the four aluminum atoms are $u(Al,1) = 0.293$, $u(Al,2) = 0.128$. The carbon atoms in $(2c)$ have $u = 0.217$.

The trimolecular hexagonal cell corresponding to this rhombohedral unit has

$$a_0' = 3.325 \text{ A.}, \qquad c_0' = 24.94 \text{ A.}$$

In it, atoms with the coordinates $(6c)$ $\pm(00u)$; rh have the parameters stated above.

In this structure (Fig. VD,12) the two sets of carbon atoms have different environments. Each C(1) atom has five aluminum neighbors at dis-

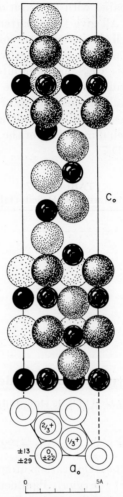

Fig. VD,12. Two projections of the elongated hexagonal Al_4C_3 structure. In the upper projection the carbon atoms are black.

tances between 1.90 and 2.22 A., while a C(2) atom is surrounded by six aluminum atoms 2.17 A. away. The aluminum atoms are at the centers of deformed carbon tetrahedra. The closest approach of aluminum atoms to one another is 3.16 A.; Al–Al separations are from 2.71 A. upward.

This arrangement is closely related to that of the carbonitride Al_5C_3N (**V,e21**) which is a structure consisting of successive layers of the composition of Al_4C_3 and AlN.

V,d11. The *tetranickel triboride*, Ni_4B_3, is dimorphous with an orthorhombic and a monoclinic modification.

The orthorhombic form has a tetramolecular cell of the dimensions:

$$a_0 = 11.953 \text{ A.}; \quad b_0 = 2.981 \text{ A.}; \quad c_0 = 6.569 \text{ A.}$$

on the metal-rich side and

$$a_0 = 11.973 \text{ A.}; \quad b_0 = 2.985 \text{ A.}; \quad c_0 = 6.584 \text{ A.}$$

on the boron-rich side.

The space group is V_h^{16} (*Pnma*) with all atoms in the special positions:

$$(4c) \quad \pm (u \; {}^1/_4 \; v; \; u + {}^1/_2, {}^1/_4, {}^1/_2 - v)$$

The chosen parameters, of limited accuracy for the boron atoms, are stated in Table VD,8.

TABLE VD,8
Parameters of the Atoms in Orthorhombic Ni_4B_3

Atom	u	v
Ni(1)	0.1486	−0.0097
Ni(2)	0.4498	0.7511
Ni(3)	0.1997	0.3780
Ni(4)	0.3763	0.1675
B(1)	0.473	0.430
B(2)	0.037	0.481
B(3)	0.265	0.673

In this arrangement (Fig. VD,13) the boron atoms are at centers of interconnected trigonal prisms having nickel atoms at the corners (Ni–B = 2.0–2.3 A.). The B(1) and B(2) atoms form two separate chains zigzagging in the b_0 direction with B(1)–B(1) = 1.86 A. and B(2)–B(2) = 1.75 A. The B(3) atoms have no close B neighbors.

V,d12. Monoclinic *tetranickel triboride*, Ni_4B_3, has a tetramolecular unit of the dimensions:

$$a_0 = 6.430 \text{ A.}; \quad b_0 = 4.882 \text{ A.}; \quad c_0 = 7.818 \text{ A.}; \quad \beta = 103°18'$$

The space group is C_{2h}^6 (*C2/c*) with atoms in the positions:

$$(4e) \quad \pm (0 \; u \; {}^1/_4; \; {}^1/_2, u + {}^1/_2, {}^1/_4)$$
$$(8f) \quad \pm (xyz; \; \bar{x}, y, {}^1/_2 - z; \; x + {}^1/_2, y + {}^1/_2, z; \; {}^1/_2 - x, y + {}^1/_2, {}^1/_2 - z)$$

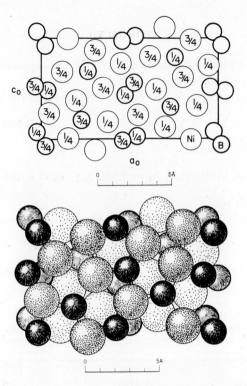

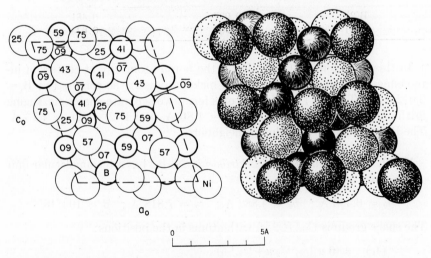

Fig. VD,13a (top). The orthorhombic Ni_4B_3 structure projected along its b_0 axis.
Fig. VD,13b (bottom). A packing drawing of the orthorhombic Ni_4B_3 structure viewed along its b_0 axis. The nickel atoms are dotted.

Fig. VD,14a (left). The monoclinic form of Ni_4B_3 projected along its b_0 axis.
Fig. VD,14b (right). A packing drawing of the monoclinic Ni_4B_3 arrangement seen along its b_0 axis. The nickel atoms are dotted.

Parameters and positions are as follows:

Ni(1):   (8f)   with $x = 0.0435$, $y = 0.250$, $z = 0.4839$
Ni(2):   (8f)   with $x = 0.2029$, $y = 0.568$, $z = 0.2877$
B(1):   (8f)   with $x = 0.234$, $y = 0.091$, $z = 0.443$
B(2):   (4e)   with $u = 0.088$

This is a very different arrangement (Fig. VD,14) from that of the orthorhombic modification (**V,d11**). All the boron atoms lie in chains with B–B between 1.83 and 1.87 A. The B(1) atoms have seven close nickel neighbors, with Ni–B = 2.04–2.18 A; the B(2) have four nickels at 2.11 or 2.14 A. and others more distant. The closest Ni–Ni = 2.52 A.

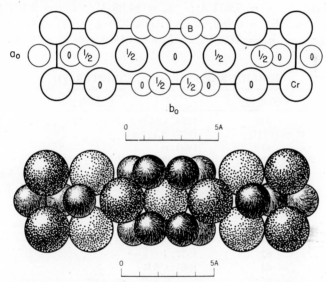

Fig. VD,15a (top). The orthorhombic Cr₃B₄ structure projected along its c_0 axis. Origin in lower left.
Fig. VD,15b (bottom). A packing drawing of the Cr₃B₄ structure seen along its c_0 axis. The chromium atoms are dotted.

V,d13.   Crystals of *trichromium tetraboride*, Cr_3B_4, are orthorhombic with the bimolecular unit:

$$a_0 = 2.986 \text{ A.}; \quad b_0 = 13.020 \text{ A.}; \quad c_0 = 2.952 \text{ A.}$$

The space group is V_h^{25} (*Immm*) with atoms in the positions:

Cr(1):   (2c) $^1/_2 \, ^1/_2 \, 0; \, 0 \, 0 \, ^1/_2$
Cr(2):   (4g) $\pm (0u0; \, ^1/_2, u+^1/_2, ^1/_2)$ with $u = 0.1861$
B(1):   (4g)   with $u' = 0.3607$
B(2):   (4h) $\pm (0 \, u \, ^1/_2; \, ^1/_2, u+^1/_2, 0)$ with $u'' = 0.4351$

In this structure (Fig. VD,15), each Cr(1) has 12 boron neighbors at B–Cr = 2.26 or 2.35 A. There are seven borons around each Cr(2) with two at 2.17 A., four at 2.19 A., and one at 2.27 A. The boron atoms are in chains with B–B = 1.69 and 1.77 A. There are no closer B–B contacts, as was originally proposed.

Other compounds with this structure are as follows:

Mn_3B_4: $a_0 = 3.032$ A.; $b_c = 12.86$ A.; $c_0 = 2.960$ A.
          $u(Mn) = 0.186; u'(B,1) = 0.369; u''(B,2) = 0.442$

Ta_3B_4: $a_0 = 3.29$ A.; $b_0 = 14.0$ A.; $c_c = 3.13$ A.
          $u(Ta) = 0.180; u'(B,1) = 0.375; u''(B,2) = 0.444$

Nb_3B_4: $a_0 = 3.305$ A.; $b_0 = 14.08$ A.; $c_0 = 3.137$ A.
          $u(Nb) = 0.180; u'(B,1) = 0.376; u''(B,2) = 0.444$

V_3B_4: $a_0 = 3.030$ A.; $b_0 = 13.18$ A.; $c_0 = 2.986$ A.

BIBLIOGRAPHY TABLE, CHAPTER VD

Compound	Paragraph	Literature
Al_4C_3	d10	1934: vS&S; 1935: S,S,P&S
Al_3O_4	[VIII]	1957: F,L,A&P
Bi_3Se_4	a19	1954: S
Ce_3S_4	d9	1948: Z; 1949: Z; 1956: P&F
Co_3O_4	[VIII]	1922: H; 1926: N&S; 1928: H&A; N&S; 1929: H; H&K; N&P; 1934: K&T; 1958: W&C
Co_3S_4	[VIII]	1927: dJ&W
Co_3Se_4	d3	1955: R; 1962: C&B
Cr_3B_4	d13	1950: A&K; 1961: E
Cr_3S_4	d3	1957: J
Cr_3Se_4	d3	1961: C&B
Cs_4O_6	d9	1939: H&K
Fe_3O_4	[VIII]	1915: B; N; 1922: H; 1925: W&C; 1926: C; 1928: G; 1929: H; 1931: C,A&B; 1932: O; 1934: E&T; K&T; 1938: M; 1953: A&C; R&W; 1954: Y,K&L; 1957: B; 1958: D&Y; H; vO&R; 1960: D&P
Fe_3S_4	[VIII]	1927: dJ&W; 1960: Y&K; 1962: Y
Fe_3Se_4	d3	1962: C&B
Ge_3N_4	d7	1939: J&H; 1957: P&R; 1958: R&P
La_3S_4	d9	1956: P&F
Mn_3B_4	d13	1950: K

(continued)

BIBLIOGRAPHY TABLE, CHAPTER VD (*continued*)

Compound	Paragraph	Literature
Mn_3O_4	[VIII]	1926: A; 1957: S&S
Nb_3B_4	d13	1950: K
Nd_3S	d9	1956: P&F
Ni_4B_3	d11, d12	1959: R
Ni_3S_4	[VIII]	1926: M; 1927: dJ&W; 1947: L; 1953: G,S&A
Ni_3Se_4	d3	1960: H&W; 1962: C&B
P_4S_3	d5	1955: vH,V&W; L,W&R; 1957: L,W,vH,V,W&W
P_4Se_3	d6	1959: K&V
Pb_3O_4	d1	1941: G; 1942: S; 1943: B&W; G; 1945: B; 1948: K,S&F
Pr_3S_4	d9	1956: P&F
Pt_3O_4	d2	1941: G&R; 1953: A,M,M&R
Rb_4O_6	d9	1939: H&K
Rh_4P_3	d8	1960: R&H
Si_3N_4	d7	1957: G,B&N; H&J; P&R; 1958: R&P; 1959: N&M; 1961: B&S
Sm_3S_4	d9	1956: P&F
Ta_3B_4	d13	1949: K
Th_3As_4	d9	1955: F
Th_3Bi_4	d9	1957: F
Th_3P_4	d9	1938: S,B&M; 1939: M
Th_3Sb_4	d9	1956: F
Ti_3S_4	d4	1956: H&H; 1957: W
Ti_3Se_4	d3	1962: C&B
Ti_3Te_4	d3	1962: C&B
U_3As_4	d9	1952: I
U_3Bi_4	d9	1952: F
U_3P_4	d9	1941: Z
U_3Sb_4	d9	1952: F
U_3Te_4	d9	1954: F
V_3B_4	d13	1956: M
V_3Se_4	d3	1962: C&B
V_3Te_4	d3	1962: C&B

E. COMPOUNDS $R_n X_5$

Compounds of the Type RX_5

V,e1. *Phosphorus pentachloride*, PCl_5, in the solid state is not molecular but an association of PCl_4 and PCl_6 groups. Its symmetry is tetragonal, with a tetramolecular cell of the dimensions:

$$a_0 = 9.22 \text{ A.}, \qquad c_0 = 7.44 \text{ A.}$$

Atoms are in the following positions of C_{4h}^3 $(P4/n)$:

(2a) $^1/_4\, ^1/_4\, 0;\ ^3/_4\, ^3/_4\, 0$
(2c) $^1/_4\, ^3/_4\, u;\ ^3/_4\, ^1/_4\, \bar{u}$
(8g) $\pm\,(xyz;\ ^1/_2-x, ^1/_2-y, z;\ \bar{y}, x+^1/_2, z;\ y+^1/_2, \bar{x}, z)$

with the parameters of Table VE,1.

This structure (Fig. VE,1) is most simply pictured as a CsCl arrangement of its PCl_4 and PCl_6 ions distorted by the packing requirements of these tetrahedral and octahedral groups. In PCl_4^+ the P–Cl distance is 1.97 A., in PCl_6^- it is from 2.04 to 2.08 A.

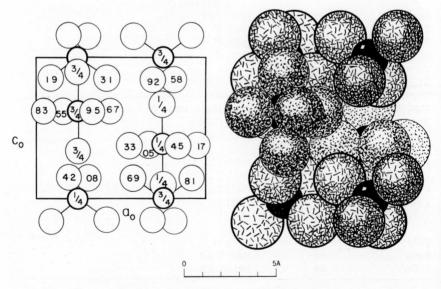

Fig. VE,1a (left). A projection along an a_0 axis of the tetragonal unit of PCl_5. The heavy circles are the phosphorus atoms. Origin in lower left.

Fig. VE,1b (right). A packing drawing of the tetragonal structure of PCl_5 viewed along an a_0 axis. Chlorines of the PCl_6 ions are dotted, those of the PCl_4 cations are line shaded.

TABLE VE,1
Parameters of the Atoms in PCl_5

Atom	Position	x	y	z
P(1)	(2a)	$1/4$	$1/4$	0
P(2)	(2c)	$1/4$	$3/4$	-0.38
Cl(1)	(2c)	$1/4$	$3/4$	-0.10
Cl(2)	(2c)	$1/4$	$3/4$	0.34
Cl(3)	(8g)	0.31	0.084	0.15
Cl(4)	(8g)	0.335	-0.046	-0.38

V,e2. *Phosphorus pentabromide*, PBr_5, like the chloride, forms ionic crystals, but in this case the ions are different. Its symmetry is orthorhombic, with four molecules in the unit:

$$a_0 = 5.62 \text{ A.}; \quad b_0 = 16.91 \text{ A.}; \quad c_0 = 8.29 \text{ A.}$$

All atoms are in one or another of the following positions of V_h^{11} (*Pbcm*):

(4d) $\pm (u \, v \, ^1/_4; \, \bar{u}, v + ^1/_2, ^1/_4)$
(8e) $\pm (xyz; \, \bar{x}, \bar{y}, z + ^1/_2; \, x, ^1/_2 - y, \bar{z}; \, \bar{x}, y + ^1/_2, ^1/_2 - z)$

with the parameters listed below:

Atom	Position	x	y	z
P	(4d)	0.035	0.125	$1/4$
Br(1)	(4d)	0.211	0.038	$1/4$
Br(2)	(4d)	0.154	0.237	$1/4$
Br(3)	(8e)	0.258	0.123	0.040
Br(4)	(4d)	0.605	0.400	$1/4$

The resulting structure is shown in Figure VE,2. In this crystal the PBr_4^+ tetrahedron is less regular than the PCl_4^+ ion in PCl_5. In it the P–Br separation ranges between 2.02 and 2.18 A. The compensatory anion is a simple Br^-; the separations between this bromide Br(4) and the bromine atoms of the PBr_4^+ cation lie between 3.07 and 3.21 A.

V,e3. *Niobium pentachloride*, $NbCl_5$, forms monoclinic crystals whose units with

$$a_0 = 18.30 \text{ A.}; \quad b_0 = 17.96 \text{ A.}; \quad c_0 = 5.888 \text{ A.}; \quad \beta = 90°36'$$

contain 12 molecules.

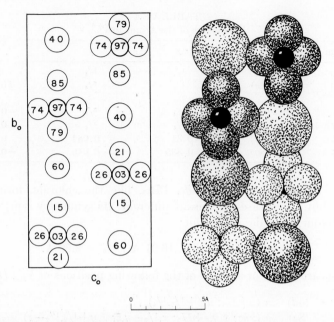

Fig. VE,2a (left). A projection along the a_0 axis of the orthorhombic unit of PBr$_5$. The ringed circles are the phosphorus atoms; the bromide ion is a somewhat larger circle than the other four bromine atoms. Origin in lower left.

Fig. VE,2b (right). A packing drawing of the orthorhombic structure of PBr$_5$ viewed along its a_0 axis. The bromine atoms are dotted, the bromide ions being shown larger than the others.

TABLE VE,2

Positions and Parameters of the Atoms in NbCl$_5$ and MoCl$_5$[a]

Atom	Position	x	y	z
Nb(1) (or Mo)	(4g)	0 (0)	0.1106 (0.1078)	0 (0)
Nb(2)	(8j)	0.3333 (0.3333)	0.1108 (0.1085)	0.525 (0.5652)
Cl(1)	(4i)	0.053 (0.079)	0 (0)	0.225 (0.133)
Cl(2)	(8j)	0.056 (0.077)	0.191 (0.192)	0.240 (0.145)
Cl(3)	(8j)	0.103 (0.075)	0.097 (0.094)	0.782 (0.677)
Cl(4)	(4i)	0.280 (0.260)	0 (0)	0.744 (0.757)
Cl(5)	(8j)	0.279 (0.259)	0.189 (0.193)	0.770 (0.762)
Cl(6)	(8j)	0.232 (0.253)	0.098 (0.093)	0.293 (0.302)
Cl(7)	(4i)	0.381 (0.407)	0 (0)	0.298 (0.372)
Cl(8)	(8j)	0.389 (0.407)	0.190 (0.193)	0.285 (0.371)
Cl(9)	(8j)	0.434 (0.414)	0.098 (0.094)	0.760 (0.833)

[a] Values in parentheses are those for MoCl$_5$.

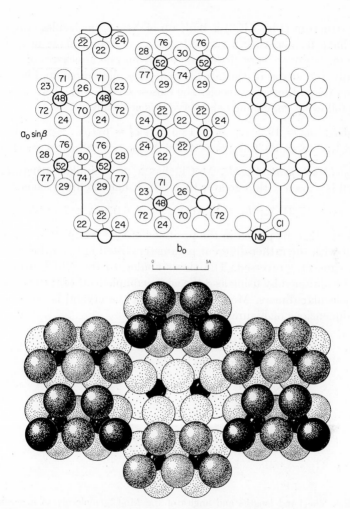

Fig. VE,3a (top). The monoclinic structure of NbCl₅ projected along its c_0 axis. The atoms forming a Nb₂Cl₁₀ molecule are connected by light lines. Origin in lower left.

Fig. VE,3b (bottom). A packing drawing of the NbCl₅ structure seen along its c_0 axis. The niobium atoms are black.

Atoms have been found to be in the following positions of C_{2h}^3 $(C2/m)$:

$(4g)$ $\pm (0u0;\ {}^1/_2,u+{}^1/_2,0)$

$(4i)$ $\pm (u0v;\ u+{}^1/_2,{}^1/_2,v)$

$(8j)$ $\pm (xyz;\ x\bar{y}z;\ x+{}^1/_2,y+{}^1/_2,z;\ x+{}^1/_2,{}^1/_2-y,z)$

The chosen parameters are listed in Table VE,2.

The structure (Fig. VE,3) is built up of Nb_2Cl_{10} molecules only some of which have by necessity a twofold axis as well as a plane of symmetry. They are, however, very little different from one another and have substantially the shape of the $MoCl_5$ molecules illustrated in Figure VE,4. Between these dimeric molecules the closest Cl–Cl = 3.62 A.

The following compounds also have this structure:

$NbBr_5$: $a_0 = 19.2$ A.; $b_0 = 18.6$ A.; $c_0 = 6.0$ A.; $\beta = $ ca. 90°
$TaCl_5$: $a_0 = 18.30$ A.; $b_0 = 17.96$ A.; $c_0 = 5.888$ A.; $\beta = 90°36'$

Molybdenum pentachloride, $MoCl_5$, has a very similar structure. If its unit cell is chosen with β greater than 90°, it has the dimensions:

$$a_0 = 17.31 \text{ A.}; \quad b_0 = 17.81 \text{ A.}; \quad c_0 = 6.079 \text{ A.}; \quad \beta = 95°42'$$

Atoms can then be placed in the positions of C_{2h}^3 occupied in the niobium salt but with the rather different parameters listed in parentheses in Table VE,2, signs of z reversed. The closer analog to the $NbCl_5$ arrangement would be obtained by using as β the acute complement of the chosen angle. The molecular dimers, Mo_2Cl_{10}, from which the crystal is built have the mean dimensions of Figure VE,4.

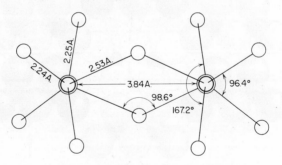

Fig. VE,4. The bond lengths and angles of the Mo_2Cl_{10} molecule as it occurs in its crystals.

V,e4. *Antimony pentachloride*, $SbCl_5$, is hexagonal with a bimolecular unit of the dimensions:

$$a_0 = 7.49 \text{ A.}, \qquad c_0 = 8.01 \text{ A.}$$

The space group is D_{6h}^4 ($P6_3/mmc$), with atoms in the positions:

Sb: (2c) $\pm(1/3 \ 2/3 \ 1/4)$
Cl(1): (4f) $\pm(1/3 \ 2/3 \ v; \ 2/3,1/3,v+1/2)$ with $v = 0.542$
Cl(2): (6h) $\pm(u \ 2u \ 1/4; \ 2\bar{u} \ \bar{u} \ 1/4; \ u \ \bar{u} \ 1/4)$ with $u = 0.157$

The arrangement that results (Fig. VE,5) is essentially molecular in character. Each antimony atom has five chlorine neighbors at the corners of a trigonal bipyramid with Sb–3Cl = 2.29 A. and Sb–2Cl = 2.34 A. Between molecules, chlorine atoms at the apices of the bipyramids are 3.33 A. apart; the next larger Cl–Cl separations are a rather long 4.09 A.

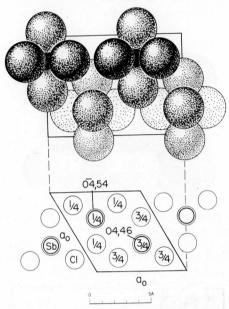

Fig. VE,5. Two projections of the hexagonal SbCl₅ structure. In the upper projection the large chlorine atoms are dotted.

V,e5. *Uranium pentafluoride,* UF₅, occurs in two tetragonal modifications. The *alpha*-form has a bimolecular unit of the dimensions:

$$a_0 = 6.525 \text{ A.}, \qquad c_0 = 4.472 \text{ A.}$$

Atoms have been placed in the following special positions of C_{4h}^5 ($I4/m$):

U: (2a) 000; $\frac{1}{2}\frac{1}{2}\frac{1}{2}$
F(1): (2b) 0 0 $\frac{1}{2}$; $\frac{1}{2}\frac{1}{2}$ 0
F(2): (8h) ±(uv0; v$\bar{u}$0); B.C.   with $u = 0.315, v = 0.113$

In this arrangement (Fig. VE,6), each uranium atom has six fluorine neighbors with U–F(1) = 2.23 A. and U–F(2) = 2.18 A. These octahedra are in strings along the c_0 axis with each F(1) atom belonging to two octahedra. Adjacent strings are in contact through fluorine atoms, with F(1)–F(2) = 2.78 A. and F(2)–F(2) = 2.82 A.

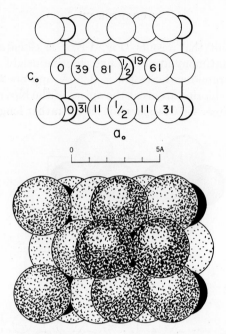

Fig. VE,6a (top). A projection along an a_0 axis of the tetragonal unit of α-UF$_5$. The ringed circles are uranium atoms. Origin in lower left.

Fig. VE,6b (bottom). A packing drawing of the tetragonal structure of α-UF$_5$ viewed along an a_0 axis. The fluorine atoms are dotted.

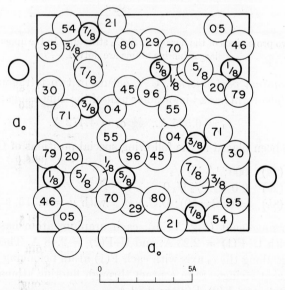

Fig. VE,7a. A projection along the c_0 axis of the tetragonal unit of β-UF$_5$. The ringed circles are uranium atoms. Origin in lower left.

V,e6. The *beta* form of *uranium pentafluoride*, UF_5, has a large tetragonal unit containing eight molecules. Its edges are

$$a_0 = 11.473 \text{ A.}, \qquad c_0 = 5.208 \text{ A.}$$

The uranium atoms have been found to be in the following positions of $V_d{}^{12}$ ($I\bar{4}2d$):

(8d) $u\ ^1/_4\ ^1/_8$; $\bar{u}\ ^3/_4\ ^1/_8$; $^3/_4\ u\ ^7/_8$; $^1/_4\ \bar{u}\ ^7/_8$; B.C.   with $u = 0.083$

Positions for the fluorine atoms could not be established from x-ray observations, but packing considerations have led to the following:

F(1):   (8d)   with $u = 0.273$
F(2):   (16e) xyz; $\bar{x},y+^1/_2,^1/_4-z$; $\bar{x}\bar{y}z$; $x,^1/_2-y,^1/_4-z$;
              $\bar{y}x\bar{z}$; $y,x+^1/_2,z+^1/_4$; $y\bar{x}\bar{z}$; $\bar{y},^1/_2-x,z+^1/_4$; B.C.
                               with $x = 0.15$, $y = 0.07$, $z = 0.05$
F(3):   (16e)   with $x = 0.05$, $y = 0.14$, $z = 0.46$

This results in a sevenfold coordination of the uranium atoms, with U–F lying between 2.18 and 2.29 A. and with four of the fluorine atoms shared by neighboring polyhedra (Fig. VE,7).

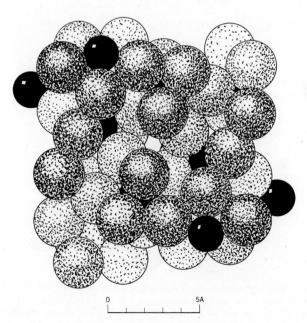

0   5A

Fig. VE,7b. A packing drawing of the tetragonal β-UF_5 structure viewed along the c_0 axis. The fluorine atoms are dotted.

V,e7.   *Iron pentacarbonyl*, $Fe(CO)_5$, is monoclinic with the tetramolecular cell:

$$a_0 = 11.71 \text{ A.;} \quad b_0 = 6.80 \text{ A.;} \quad c_0 = 9.28 \text{ A.;} \quad \beta = 107°36'$$

All atoms have been placed in the positions

$$(4a) \quad xyz; \; x,\bar{y},z+^1/_2; \; x+^1/_2,y+^1/_2,z; \; x+^1/_2,^1/_2-y,z+^1/_2$$

of $C_s^4(Cc)$ with the parameters of Table VE,3.

In the resulting structure (Fig. VE,8), the molecules are trigonal bipyramids of essentially the same dimensions found by electron diffraction

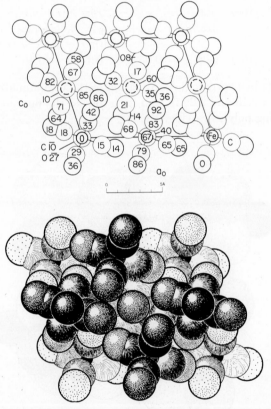

Fig. VE,8a (top). The monoclinic structure of $Fe(CO)_5$ projected along its b_0 axis.

Fig. VE,8b (bottom). A packing drawing of the $Fe(CO)_5$ arrangement viewed along its b_0 axis. The black iron atoms scarcely show; the oxygen atoms are heavily outlined and dotted.

TABLE VE,3
Parameters of the Atoms in Fe(CO)$_5$

Atom	x	y	z
Fe	0.0000	0.1663	0.0000
C(1)	0.0800	0.3271	0.1572
C(2)	0.1293	0.1530	−0.0675
C(3)	−0.0899	0.2927	−0.1704
C(4)	−0.1276	0.1769	0.0671
C(5)	0.0037	−0.1020	0.0135
O(1)	0.1283	0.4204	0.2607
O(2)	0.2106	0.1367	−0.1082
O(3)	−0.1462	0.3633	−0.2763
O(4)	−0.2086	0.1832	0.1120
O(5)	0.0055	−0.2705	0.0173

of the vapor. The Fe–C distances are 1.79–1.84 A. and C–O = 1.11–1.15 A. The shortest O–O between molecules is 3.03 A.

This is presumably a different substance from the Fe$_2$(CO)$_9$ described in **V,f18**. The ascribed symmetry is not the same, but in view of the difficulties encountered in establishing the correct chemical composition of many metallic carbonyls, additional work may be needed to be certain that the two crystalline solids actually exist.

V,e8. *Bromine pentafluoride*, BrF$_5$, at −120°C. is orthorhombic, with the tetramolecular cell:

$$a_0 = 6.422 \text{ A.}; \quad b_0 = 7.245 \text{ A.}; \quad c_0 = 7.846 \text{ A.}$$

As with the trifluoride (**V,b18**) the chosen space group is C_{2v}^{12} ($Cmc2_1$), with most atoms in the same special positions:

(4a) $0uv; \ 0,\bar{u},v+{}^1/_2; \ {}^1/_2,u+{}^1/_2,v; \ {}^1/_2,{}^1/_2-u,v+{}^1/_2$

For the bromine atoms $u = 0.1707$, $v = 0.250$. There are three sets of fluorine atoms in these special positions and a fourth in the general positions:

(8b) $xyz; \qquad \bar{x}yz; \qquad \bar{x},\bar{y},z+{}^1/_2; \qquad x,\bar{y},z+{}^1/_2;$
$x+{}^1/_2,y+{}^1/_2,z; \ {}^1/_2-x,y+{}^1/_2,z; \ {}^1/_2-x,{}^1/_2-y,z+{}^1/_2;$
$x+{}^1/_2,{}^1/_2-y,z+{}^1/_2$

Their parameters are

Atom	Position	x	y	z
F(1)	(4a)	0	0.353	0.382
F(2)	(4a)	0	0.355	0.095
F(3)	(4a)	0	0.028	0.441
F(4)	(8b)	0.271	0.186	0.261

In this structure (Fig. VE,9), the molecules can be described as tetragonal pyramids of fluorine atoms with the bromine atoms lying below the base of the pyramids. In the molecule, Br–F(1) = 1.68 A., Br–F(4) = 1.75 A., Br–F(2) = 1.81 A., and Br–F(3) = 1.82 A. Bond angles are F(1)–Br–F(2) = 80°30′, F(1)–Br–F(3) = 86°30′, and F(1)–Br–F(4) = 85°24′.

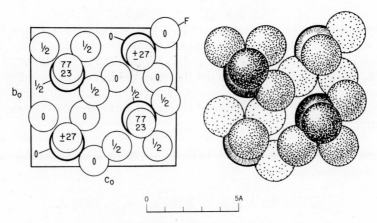

Fig. VE,9a (left). The orthorhombic structure of BrF₅ projected along its a_0 axis. Origin in lower left.

Fig. VE,9b (right). A packing drawing of the BrF₅ arrangement seen along its a_0 axis The fluorine atoms are dotted.

Compounds of the Type R₂X₅

V,e9. The crystal structure of *nitrogen pentoxide* (nitric acid anhydride), N_2O_5, has been determined at ca. −60°C. and at room temperature. Two molecules are contained in a hexagonal unit of the dimensions:

$$a_0 = 5.41 \text{ A.,} \quad c_0 = 6.57 \text{ A. at ca. } -60°\text{C.}$$
$$a_0 = 5.45 \text{ A.,} \quad c_0 = 6.66 \text{ A. at ca. } 20°\text{C.}$$

Atoms are in the following special positions of D_{6h}^4 ($C6/mmc$):

N(1):   (2b) $0\ 0\ ^1/_4;\ 0\ 0\ ^3/_4$
N(2):   (2d) $^2/_3\ ^1/_3\ ^1/_4;\ ^1/_3\ ^2/_3\ ^3/_4$
O(1):   (6h) $\pm(u\ 2u\ ^1/_4;\ 2\bar{u}\ \bar{u}\ ^1/_4;\ u\ \bar{u}\ ^1/_4)$ with $u = 0.133$
O(2):   (4f) $\pm(^1/_3\ ^2/_3\ v;\ ^2/_3,^1/_3,v+^1/_2)$ with $v = -0.074$

This structure is an arrangement of NO_3^- and NO_2^+ ions (Fig. VE,10) in which the NO_3^- anion has its usual trigonal planar shape with N(1)–O(1) = 1.243 A. The NO_2^+ ion is linear with N(2)–O(2) = 1.154 A. The shortest interionic distances are N(2)–N(1) = 2.73 A., O(1)–O(2) = 2.84 and 2.96 A. Between ions of the same sign, N(1)–N(1) = 3.28 A., O(1)–O(1) = 3.26 and 3.34 A., O(2)–O(2) = 3.27 A.

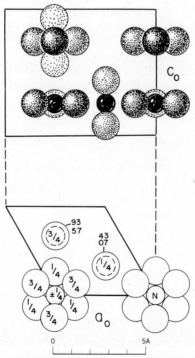

Fig. VE,10. Two projections of the hexagonal unit found for solid N_2O_5. In the upper projection the nitrogen atoms are black.

V,e10. *Metastable phosphorus pentoxide*, P_2O_5, as condensed from the vapor phase is rhombohedral, with

$$a_0 = 7.43\ \text{A.,} \qquad \alpha = 87°$$

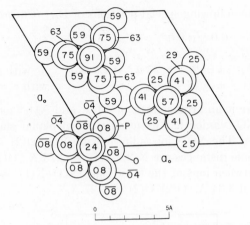

Fig. VE,11. A projection along c_0 of half the contents of the hexagonal cell of the metastable rhombohedral modification of P_2O_5. The smaller circles are phosphorus.

It is composed of the same P_4O_{10} molecules that exist in the vapor. Two of these in each cell have their atoms in the positions of C_{3v}^6 $(R3c)$:

(2a) $uuu;\ u+\frac{1}{2},u+\frac{1}{2},u+\frac{1}{2}$

(6b) $xyz;$ $zxy;$ $yzx;$
$y+\frac{1}{2},x+\frac{1}{2},z+\frac{1}{2};\ x+\frac{1}{2},z+\frac{1}{2},y+\frac{1}{2};\ z+\frac{1}{2},y+\frac{1}{2},x+\frac{1}{2}$

For the hexagonal cell containing six molecules:

$$a_0' = 10.31 \text{ A.}, \qquad c_0' = 13.3 \text{ A.}$$

The hexagonal coordinates are

(6a) $00u;\ 0,0,u+\frac{1}{2};$ rh
(18b) $xyz;$ $\bar{y},x-y,z;$ $y-x,\bar{x},z;$
$\bar{y},\bar{x},z+\frac{1}{2};\ x,x-y,z+\frac{1}{2};\ y-x,y,z+\frac{1}{2};$ rh

The parameters for these atoms in the two types of cell are listed in Table VE,4.

As can be seen from Figure VE,11, this crystal is a body-centered grouping of its P_4O_{10} molecules. The shape of these molecules, as indicated in Figure VE,12, is substantially the same in the solid and in the vapor from which it is grown.

V,e11. When the foregoing metastable P_2O_5 is heated in a closed tube, orthorhombic crystals of *stable phosphorus pentoxide* are produced. There are eight P_2O_5 molecules in its unit which has the edge lengths:

$$a_0 = 16.3 \text{ A.}; \quad b_0 = 8.12 \text{ A.}; \quad c_0 = 5.25 \text{ A.}$$

TABLE VE,4
Positions and Parameters of the Atoms in Metastable P_2O_5

Atom	Position	x	y	z
		Rhombohedral cell		
P(1)	(2a)	0.130	0.130	0.130
P(2)	(6b)	0.122	−0.192	−0.061
O(1)	(2a)	0.236	0.236	0.236
O(2)	(6b)	−0.055	0.061	0.228
O(3)	(6b)	−0.055	0.061	−0.228
O(4)	(6b)	0.225	−0.347	−0.114
		Hexagonal cell		
P(1)	(6a)	0	0	0.130
P(2)	(18b)	−0.166	0.149	−0.043
O(1)	(6a)	0	0	0.236
O(2)	(18b)	−0.134	0.017	0.077
O(3)	(18b)	0.134	−0.017	−0.077
O(4)	(18b)	0.134	0.269	−0.079

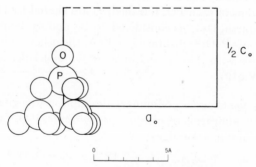

Fig. VE,12. One of the P_4O_{10} molecules in metastable hexagonal (rhombohedral) phosphorus pentoxide viewed along the normal to the a_0c_0 plane.

Atoms are in the following positions of C_{2v}^{19} (*Fdd*):

(8a) $00u; \ ^1/_4,^1/_4,u+^1/_4;$ F.C.

(16b) $xyz; \ \bar{x}\bar{y}z; \ ^1/_4-x,y+^1/_4,z+^1/_4; \ x+^1/_4,^1/_4-y,z+^1/_4;$ F.C.

with the parameters listed below:

Atom	Position	x	y	z
P	(16b)	0.075	0.083	0.153
O(1)	(8a)	0	0	0
O(2)	(16b)	0.114	0.178	0.089
O(3)	(16b)	0.308	0.411	0.133

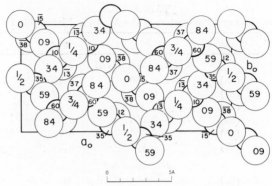

Fig. VE,13a. A projection along the c_0 axis of the stable orthorhombic modification of
P_2O_5. The larger circles are oxygen.

In this structure (Fig. VE,13), atoms are not associated into discrete molecules; instead they go to build up PO_4 tetrahedra which are linked together throughout the solid by sharing oxygen atoms with one another. Three of the four oxygen atoms of each PO_4 tetrahedron are shared by neighboring tetrahedra; the fourth, $O(3)$, is considered to be doubly bound to its phosphorus atom. This $P–O(3)$ distance is 1.40 A, the others being 1.65 A. Such separations agree with those found in the P_4O_{10} molecules of the less stable modification (**V,e10**).

V,e12. Some years ago a preliminary description was given of the structure of a *third* modification of *phosphorus pentoxide*, P_2O_5. Its orthorhombic tetramolecular unit has the edges:

$$a_0 = 9.23 \text{ A.}; \quad b_0 = 7.18 \text{ A.}; \quad c_0 = 4.94 \text{ A.}$$

Atoms are in the following positions of V_h^{16} (*Pnam*):

(4c) $\pm (u \ v \ ^1/_4; \ ^1/_2 - u, v + ^1/_2, ^3/_4)$
(8d) $\pm (xyz; \ x + ^1/_2, ^1/_2 - y, ^1/_2 - z; \ \bar{x}, \bar{y}, z + ^1/_2; \ ^1/_2 - x, y + ^1/_2, \bar{z})$

with the parameters given in Table VE,5.

This structure is like that of the stable modification (**V,e11**), containing separate molecules. Each phosphorus atom here is surrounded by a tetrahedron of oxygen atoms, three of which are shared with neighboring phosphorus atoms to produce (Fig. VE,14) sheets parallel to the a_0 face.

V,e13. Two redeterminations have been made in recent years of the structure of orthorhombic *vanadium pentoxide*, V_2O_5. The bimolecular unit has the edge lengths:

$$a_0 = 11.519 \text{ A.}; \quad b_0 = 3.564 \text{ A.}; \quad c_0 = 4.373 \text{ A.}$$

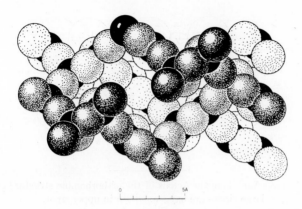

Fig. VE,13b. A packing drawing of the stable P_2O_5 structure seen along its c_0 axis. The phosphorus atoms are black.

TABLE VE,5
Positions and Parameters of the Atoms in $P_2O_5(III)$

Atom	Position	x	y	z
P(1)	(4c)	0.244	0.288	$1/4$
P(2)	(4c)	−0.098	−0.156	$1/4$
O(1)	(4c)	−0.219	−0.011	$1/4$
O(2)	(4c)	−0.142	−0.346	$1/4$
O(3)	(4c)	0.055	−0.089	$1/4$
O(4)	(8d)	0.136	0.282	0.000

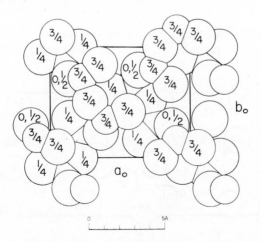

Fig. VE,14. A projection along the c_0 axis of the structure of the third, orthorhombic modification of P_2O_5. The larger circles are oxygen. Origin in lower right.

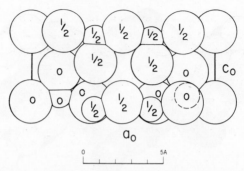

Fig. VE,15. A projection along the b_0 axis of the orthorhombic structure of V_2O_5. The large circles are oxygen. Origin in upper right.

The space group now chosen, V_h^{13} (*Pmmn*), is of higher symmetry than the one originally used. Atoms are placed in its special positions:

$$(2a) \quad 00u; \; ^1/_2 \, ^1/_2 \, \bar{u}$$
$$(4f) \quad u0v; \bar{u}0v; u+^1/_2, ^1/_2, \bar{v}; \; ^1/_2-u, ^1/_2, \bar{v}$$

with the parameters listed below:

Atom	Position	x	y	z
V	(4f)	0.1486 (0.1487)	0	0.105 (0.1086)
O(1)	(4f)	0.149 (0.1460)	0	0.458 (0.4713)
O(2)	(4f)	0.320 (0.3191)	0	0.000 (−0.0026)
O(3)	(2a)	0	0	0.000 (−0.0031)

A recent three-dimensional study based on counter data has led to the parameters given in parentheses. They are more accurate but close to the

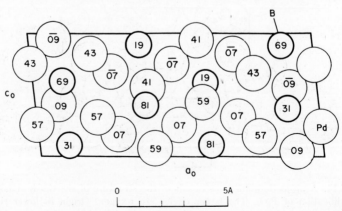

Fig. VE,16a. The monoclinic structure of Pd_5B_2 projected along its b_0 axis. Origin in lower left.

others. In this investigation a detailed consideration was given the possibility that the correct space group might be the C_{2v}^7 ($Pn2_1m$) selected in the original investigation; no experimental evidence was, however, found for this lower symmetry.

In this structure (Fig. VE,15) the positions of the vanadium atoms do not differ too greatly from those originally found, but the positions of the oxygen atoms give more reasonable O–O distances (2.63 A. minimum). Each vanadium atom here has five closest oxygen neighbors, with V–O = 1.54–2.02 A.

V,e14. Crystals of *pentamanganese dicarbide*, Mn_5C_2, and of *pentapalladium diboride*, Pd_5B_2, have the same structure. They are monoclinic with tetramolecular cells of the dimensions:

$$Mn_5C_2: \quad a_0 = 11.66 \text{ A.}; \quad b_0 = 4.573 \text{ A.}; c_0 = 5.086 \text{ A.}; \beta = 97°45'$$
$$Pd_5B_2: \quad a_0 = 12.786 \text{ A.}; b_0 = 4.955 \text{ A.}; c_0 = 5.472 \text{ A.}; \beta = 97°2'$$

The space group is C_{2h}^6 ($C2/c$) with atoms in the positions:

(4e) $\pm(0\ u\ ^1/_4;\ ^1/_2,u+^1/_2,^1/_4)$
(8f) $\pm(xyz;\ \bar{x},y,^1/_2-z;\ x+^1/_2,y+^1/_2,z;\ ^1/_2-x,y+^1/_2,^1/_2-z)$

A determination of parameters for the boride led to the values:

Atom	Position	x	y	z
Pd(1)	(8f)	0.0958	0.0952	0.4213
Pd(2)	(8f)	0.2127	0.5726	0.3138
Pd(3)	(4e)	0	0.5727	$^1/_4$
B	(8f)	0.106	0.311	0.077

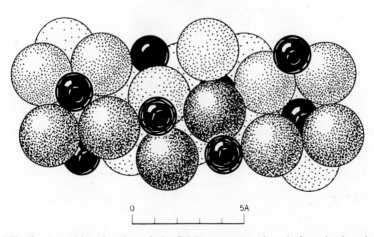

Fig. VE,16b. A packing drawing of the Pd_5B_2 structure viewed along its b_0 axis. The boron atoms are black

Powder data indicate that similar parameters must apply to the carbide. In this structure for Pd_5B_2 (Fig. VE,16), each Pd(1) has two close boron neighbors at a distance of 2.18 A. For the other two kinds of palladium atoms, Pd(2) has a boron neighbor at 2.18 A. and another at 2.19 A. and Pd(3) has four neighbors, two at 2.18 A. and two more at 2.19 A. The close Pd–Pd separations lie between 2.70 and ca. 3.00 A.

V,e15. Crystals of *tungsten pentaboride*, W_2B_5, are hexagonal, with a bimolecular unit of the dimensions:

$$a_0 = 2.982 \text{ A.}, \qquad c_0 = 13.87 \text{ A.}$$

Fig. VE,17. Two projections of the hexagonal W_2B_5 arrangement. In the upper projection the boron atoms are small and dotted.

The space group has been chosen as D_{6h}^4 ($P6/mmc$), with the tungsten atoms in

$$(4f) \quad \pm(^1/_3\,{}^2/_3\,u;\; ^2/_3,^1/_3,u+^1/_2) \qquad \text{with } u = 0.139$$

Boron positions have not been established experimentally, but a reasonable structure is obtained by distributing these light atoms as follows:

$$
\begin{aligned}
&\text{B(1):} \quad (2b) \quad \pm(0\ 0\ ^1/_4) \\
&\text{B(2):} \quad (2d) \quad \pm(^2/_3\ ^1/_3\ ^1/_4) \\
&\text{B(3):} \quad (4f) \quad \text{with } u = -0.028 \\
&\text{B(4):} \quad (2a) \quad 000;\ 0\ 0\ ^1/_2
\end{aligned}
$$

In this arrangement (Fig. VE,17) the W–B distances are 2.31 and 2.32 A.; the B–B separations are 1.72–1.92 A.

Other pentaborides to which this structure has been ascribed are

$$
\begin{aligned}
&\text{Os}_2\text{B}_5\text{:} \quad a_0 = 2.91 \text{ A.}, \quad c_0 = 12.91 \text{ A.} \\
&\text{Ru}_2\text{B}_5\text{:} \quad a_0 = 2.89 \text{ A.}, \quad c_0 = 12.81 \text{ A.} \\
&\text{Ti}_2\text{B}_5\text{:} \quad a_0 = 2.98 \text{ A.}, \quad c_0 = 13.98 \text{ A.}
\end{aligned}
$$

V,e16. The epsilon phase in the molybdenum–boron system, *molybdenum pentaboride*, Mo_2B_5, has a structure related to that of W_2B_5 (**V,e15**). It is rhombohedral, rather than truly hexagonal, and its unimolecular rhombohedron has the dimensions:

$$a_0 = 7.190 \text{ A.}, \qquad \alpha = 24°10'$$

The corresponding trimolecular hexagonal cell has the edge lengths:

$$a_0' = 3.011 \text{ A.}, \qquad c_0' = 20.93 \text{ A.}$$

It thus has about the same base and is about 50% higher than the cell for W_2B_5.

A structure based on D_{3d}^5 ($R\bar{3}m$) places the atoms in the following positions:

$$
\begin{aligned}
&\text{Mo:} \quad (6c) \quad \pm(00u);\,\text{rh} \qquad \text{with } u = 0.075 \\
&\text{B(1):} \quad (6c) \quad \text{with } u = {}^1/_3 \\
&\text{B(2):} \quad (6c) \quad \text{with } u = 0.186 \\
&\text{B(3):} \quad (3b) \quad 0\ 0\ ^1/_2;\,\text{rh}
\end{aligned}
$$

In this arrangement (Fig. VE,18), which evidently is a rhombohedral instead of a hexagonal stacking of the same kinds of atomic layers chosen for W_2B_5, the atomic separations are similar. The Mo–B distances are 2.32 and 2.34 A.; the B–B separations lie between 1.74 and 1.92 A.

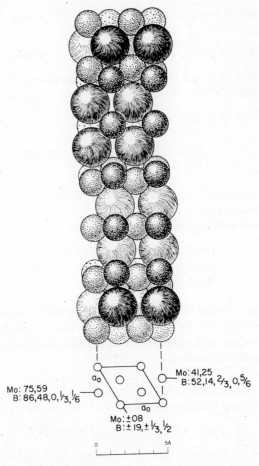

Fig. VE,18. Two projections of the hexagonal Mo_2B_5 arrangement. In the upper projection the boron atoms are small and dotted.

Compounds of the Type R_nX_5 ($n > 2$)

V,e17. *Trititanium pentoxide*, Ti_3O_5, when pure, has an inversion around 120°C. whereby one high-temperature crystal inverts on cooling to a single crystal, and vice versa. When small amounts of iron are present, the high form is stabilized down to room temperature. As would be expected from the fact that there is no crystal fragmentation on inversion, the two forms have cells very similar in dimensions and have been found to have structures not very different one from the other. Both are monoclinic.

TABLE VE,6
Parameters of the Atoms in Low-Ti_3O_5

Atom	u	v
Ti(1)	0.1280	0.0440
Ti(2)	0.7786	0.2669
Ti(3)	0.0538	0.3659
O(1)	0.676	0.060
O(2)	0.241	0.245
O(3)	0.588	0.345
O(4)	0.953	0.158
O(5)	0.866	0.441

For low-temperature Ti_3O_5, the unit containing 12 molecules has the edges:

$$a_0 = 9.752 \text{ A.}; \quad b_0 = 3.8020 \text{ A.}; \quad c_0 = 9.442 \text{ A.}; \quad \beta = 91°33'$$

All atoms have been placed in the following special positions of C_{2h}^3 $(C2/m)$:

$$(4i) \quad \pm (u0v; \, u + {}^1/_2, {}^1/_2, v)$$

with the parameters of Table VE,6.

In this structure (Fig. VE,19) each titanium atom has about it an octahedron of oxygen atoms at distances ranging between 1.94 and 2.18 A.

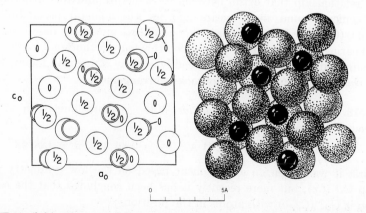

Fig. VE,19a (left). The monoclinic low-temperature Ti_3O_5 structure projected along its b_0 axis. The doubly ringed circles are titanium. Origin in lower left.

Fig. VE,19b (right). A packing drawing of the low-temperature Ti_3O_5 structure viewed along its b_0 axis. The titanium atoms are black.

TABLE VE,7
Parameters of the Atoms in High-Ti_3O_5

Atom	u	v
Ti(1)	0.1308	0.0612
Ti(2)	0.8061	0.2500
Ti(3)	0.1308	0.4388
O(1)	0.690	0.064
O(2)	0.250	0.250
O(3)	0.690	0.436
O(4)	0.963	0.128
O(5)	0.963	0.372

These TiO_6 octahedra can be thought of as building up the crystal by sharing sometimes corners and sometimes edges with one another.

High-temperature Ti_3O_5, photographed at somewhat above 120°C. has the monoclinic cell:

$$a_0 = 9.82 \text{ A.}; \quad b_0 = 3.78 \text{ A.}; \quad c_0 = 9.97 \text{ A.}; \quad \beta = 91°0'$$

Atoms have been found to be in the same special positions (4i) of C_{2h}^3 ($C2/m$) as for the low-temperature modification with parameters (Table VE,7) that are for the most part like those listed in Table VE,6.

When stabilized with iron, high Ti_3O_5 has at room temperature a value of β which is nearer 90° the richer its content of iron. This emphasizes the close relationship that exists between its structure and the orthorhombic arrangement found many years ago for the mineral pseudobrookite, Fe_2TiO_5 (Chapter IX). There is in fact a mineral, anosovite, described as an iron-bearing Ti_3O_5 which has been given this truly orthorhombic structure. The similarity between the two is readily seen by comparing Figure VE,19 with that illustrating the pseudobrookite arrangement.

V,e18. The *trivanadium pentoxide*, V_3O_5, is monoclinic with a tetramolecular unit of the dimensions:

$$a_0 = 10.01 \text{ A.}; \quad b_0 = 5.04 \text{ A.}; \quad c_0 = 9.86 \text{ A.}; \quad \beta = 138°48'$$

At first it was given an arrangement based on the low symmetry space group C_s^4 (Cc), but more recently it has been concluded that the correct group is C_{2h}^6 ($C2/c$) with atoms in the positions:

(4d) $\pm(^1/_4\,^1/_4\,^1/_2;\ ^3/_4\,^1/_4\,0)$
(4e) $\pm(0\ u\ ^1/_4;\ ^1/_2,u+^1/_2,^1/_4)$
(8f) $\pm(xyz;\ x,\bar{y},z+^1/_2;\ x+^1/_2,y+^1/_2,z;\ x+^1/_2,^1/_2-y,z+^1/_2)$

The chosen parameters are

Atom	Position	x	y	z
V(1)	(8f)	0.1601	0.7500	0.2794
V(2)	(4d)	$1/4$	$1/4$	$1/2$
O(1)	(8f)	0.219	0.595	0.168
O(2)	(8f)	0.111	0.901	0.439
O(3)	(4e)	0	0.438	$1/4$

In the structure that results (Fig. VE,20), V–O = 1.77 A., 1.95 A., and larger values. The oxygen atoms are hexagonally close-packed with vanadium atoms in some of the holes. In this respect the arrangement is like Al_2O_3 (**V,a3**) and other oxides, the structures differing in the holes that are occupied by metal atoms and in the distortions required to accommodate them.

Another compound with this structure appears to be

$TiO_2 \cdot V_2O_3$: a_0 = 10.08 A.; b_0 = 5.07 A.; c_0 = 9.95 A.; β = 139°0′

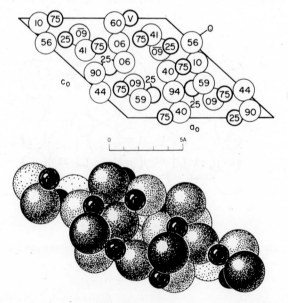

Fig. VE,20a (top). The monoclinic V_3O_5 arrangement projected along its b_0 axis.
Fig. VE,20b (bottom). A packing drawing of the V_3O_5 structure seen along its b_0 axis. The vanadium atoms are black.

V,e19. *Pentachromium triboride*, Cr_5B_3, is tetragonal with the tetramolecular cell:

$$a_0 = 5.46 \text{ A.}, \qquad c_0 = 10.64 \text{ A.}$$

The space group is D_{4h}^{18} ($I4/mcm$), with atoms in the following positions:

Cr(1):   (4c)   000; 0 0 $^1/_2$; B.C.
Cr(2):   (16l) $\pm(u,u+^1/_2,v;\ u,u+^1/_2,\bar{v};\ u+^1/_2,\bar{u},v;\ u+^1/_2,\bar{u},\bar{v})$;   B.C.
with $u = {}^1/_6$, $v = 0.15$

Though the boron positions were not established from x-ray data, they are thought to be in

B(1):   (4a)   0 0 $^1/_4$; 0 0 $^3/_4$; B.C.
B(2):   (8h) $\pm(u,u+^1/_2,0;\ u+^1/_2,\bar{u},0)$; B.C.   with $u = 0.375$

The resulting structure, which seems to be possessed by a number of other borides, is shown in Figure VE,21. These other compounds are

$$Mo_5SiB_2:\ a_0 = 5.998\text{--}6.028 \text{ A.}, \quad c_0 = 11.027\text{--}11.070 \text{ A.}$$

The molybdenum atoms are in (4c) and (16l) with $u = 0.1653$, $v = 0.1388$. The silicon atoms probably are in (4a) and boron in (8h) with $u = 0.375$.

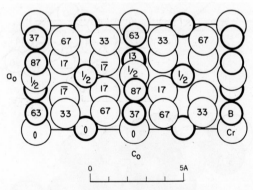

Fig. VE,21a. The tetragonal Cr_5B_3 structure projected along an a_0 axis. Origin in lower left.

Fe_5PB_2: $a_0 = 5.482$ A., $c_0 = 10.332$ A.

The iron atoms are in $(4c)$ and $(16l)$ with $u = 0.1690$, $v = 0.1400$. Phosphorus is in $(4a)$ and the boron atoms are considered to be in $(8h)$ with $u = 0.375$.

Still other compounds for which the parameters have not been established are

$$Fe_5SiB_2:   a_0 = 5.54 \text{ A.}, c_0 = 10.32 \text{ A.}$$
$$Mn_5SiB_2:   a_0 = 5.61 \text{ A.}, c_0 = 10.44 \text{ A.}$$

V,e20. The *titanium telluride*, Ti_5Te_4, is tetragonal with a bimolecular cell of the dimensions:

$$a_0 = 10.164 \text{ A.}, \qquad c_0 = 3.7720 \text{ A.}$$

Atoms are in the following special positions of C_{4h}^5 $(I4/m)$:

Ti(1): $(2a)$ $000; \frac{1}{2} \frac{1}{2} \frac{1}{2}$
Ti(2): $(8h)$ $\pm(uv0; \bar{v}u0)$; B.C.   with $u = 0.3144$, $v = 0.3752$
  Te: $(8h)$ with $u = 0.0589$, $v = 0.2797$

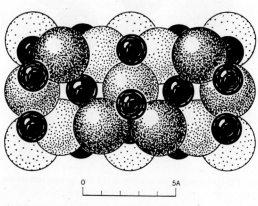

0   5A

Fig. VE,21b. A packing drawing of the Cr_5B_3 arrangement seen along an a_0 axis. The boron atoms are black.

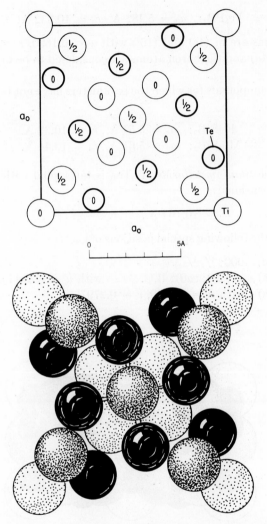

Fig. VE,22a (top). The tetragonal structure of Ti_5Te_4 projected along its c_0 axis. Origin in lower left.

Fig. VE,22b (bottom). A packing drawing of the Ti_5Te_4 arrangement seen along its c_0 axis. The tellurium atoms are black.

In this essentially alloy-type arrangement (Fig. VE,22) each Ti(1) has four quadrilaterally arranged tellurium atoms at Ti–Te = 2.905 A. and eight cubically distributed titanium neighbors with Ti–8Ti(2) = 2.954 A. Each Ti(2) has three tellurium atoms at a distance of 2.773 A. and two more at 2.821 A.; it has two Ti(1) at 2.954 A. and others from 3.215 A. upwards. There is a remote relation to the NiAs arrangement (III,d1).

V,e21. The structure described many years ago for *aluminum carbonitride*, Al_5C_3N, is a composite of the Al_4C_3 structure (**V,d10**) and that of AlN (**III,c2**). Its bimolecular hexagonal unit has the dimensions:

$$a_0 = 3.287 \text{ A.}, \qquad c_0 = 21.59 \text{ A.}$$

Atoms have been placed in the following positions of C_{6v}^4 ($C6mc$):

$$(2a) \quad 00u; \ 0,0,u+{}^1/_2$$
$$(2b) \quad {}^1/_3\,{}^2/_3\,u; \ {}^2/_3,{}^1/_3,u+{}^1/_2$$

with the parameters of Table VE,8. In making this assignment, the carbon and nitrogen atoms could not, of course, be distinguished experimentally.

The structure is shown in Figure VE,23.

TABLE VE,8
Positions and Parameters of the Atoms in Al_5C_3N

Atom	Position	x	y	z
Al(1)	(2a)	0	0	0.150
Al(2)	(2b)	${}^1/_3$	${}^2/_3$	0.044
Al(3)	(2b)	${}^1/_3$	${}^2/_3$	0.456
Al(4)	(2a)	0	0	0.345
Al(5)	(2b)	${}^1/_3$	${}^2/_3$	0.240
C(1)	(2b)	${}^1/_3$	${}^2/_3$	0.133
C(2)	(2a)	0	0	0.001
C(3)	(2b)	${}^1/_8$	${}^2/_3$	0.369
N	(2a)	0	0	0.250

V,e22. The *tetraphosphorus pentasulfide*, P_4S_5, forms monoclinic crystals that have a bimolecular cell of the dimensions:

$$a_0 = 6.41 \text{ A.}; \quad b_0 = 10.94 \text{ A.}; \quad c_0 = 6.69 \text{ A.}; \quad \beta = 111°42'$$

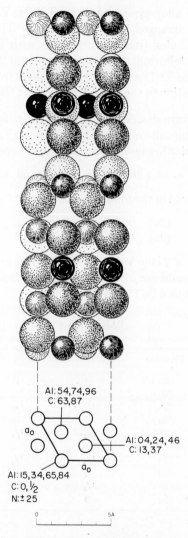

Fig VE,23. Two projections of the hexagonal Al₅C₃N structure. In the upper projection the aluminum atoms are large and dotted, the nitrogen atoms are black.

The space group is the low symmetry C_2^2 ($P2_1$), with all atoms in its general positions:

$$(2a) \quad xyz; \; \bar{x},y+\frac{1}{2},\bar{z}$$

The atomic parameters are those of Table VE,9.

The asymmetric molecules that arise from this determination have the dimensions given in Figure VE,24. The way they pack within the crystal is indicated in Figure VE,25. Between molecules the shortest atomic separation is 3.68 A.

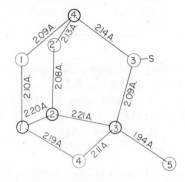

Fig. VE,24. The bond lengths of the P_4S_5 molecule as found in its crystals.

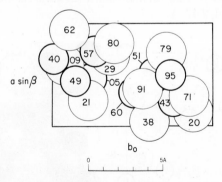

Fig. VE,25. The monoclinic P_4S_5 structure projected along its c_0 axis. The phosphorus atoms are smaller and more heavily outlined. Origin in lower left.

V,e23. Crystals of the complex *nickel phosphide*, $Ni_{12}P_5$, are tetragonal, with the bimolecular unit:

$$a_0 = 8.646 \text{ A.}, \qquad c_0 = 5.070 \text{ A.}$$

TABLE VE,9
Parameters of the Atoms in P_4S_5

Atom	x	y	z
P(1)	0.6845	0.004	0.401
P(2)	0.470	0.132	0.495
P(3)	0.758	0.259	0.575
P(4)	0.480	0.2375	0.052
S(1)	0.627	0.065	0.088
S(2)	0.251	0.209	0.212
S(3)	0.7225	0.349	0.288
S(4)	0.970	0.106	0.6225
S(5)	0.856	0.378	0.805

Atoms are in the following positions of the space group C_{4h}^5 ($I4/m$):

Ni(1): (16i) $\pm(xyz;\ xy\bar{z};\ \bar{y}xz;\ y\bar{x}z)$; B.C.

                            with $x = 0.1160$, $y = 0.1822$, $z = 0.248$

Ni(2): (8h) $\pm(uv0;\ v\bar{u}0)$; B.C.   with $u = 0.368$, $v = 0.060$

P(1):   (8h)   with $u' = 0.195$, $v' = 0.415$

P(2):   (2a) $000;\ ^1/_2\,^1/_2\,^1/_2$

This is an arrangement (Fig. VE,26) closely related to that of Ni_3P (**V,b35**). The shortest Ni–Ni $= 2.45$ A. Each P(1) atom has eight nickel atoms at a distance of 2.25 A., and each P(2) atom six nickel atoms at Ni–P $= 2.24$, 2.41, and 2.47 A.

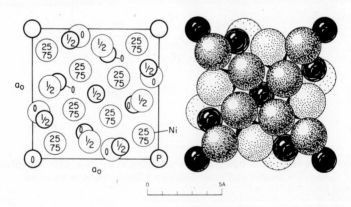

Fig. VE,26a (left). The tetragonal structure of $Ni_{12}P_5$ projected along its c_0 axis. Origin in lower left.

Fig. VE,26b (right). A packing drawing of the $Ni_{12}P_5$ arrangement seen along its c_0 axis. The phosphorus atoms are black.

BIBLIOGRAPHY TABLE, CHAPTER VE

Compound	Paragraph	Literature
Compounds RX_5		
BrF_5	e8	1957: B&B
$Fe(CO)_5$	e7	1962: H
$MoCl_5$	e3	1959: S&Z
$NbBr_5$	e3	1958: R; Z&S
$NbCl_5$	e3	1958: Z&S
PBr_5	e2	1943: D&MG
PCl_5	e1	1940: F,W&M; P,C&W; 1942: C,P&W
$SbCl_5$	e4	1959: O
$TaCl_5$	e3	1958: Z&S
UF_5	e5, e6	1948: Z; 1949: Z
Compounds R_2X_5		
Mn_5C_2	e14	1944: O; 1954: K&P
Mo_2B_5	e16	1947: K
N_2O_5	e9	1950: G,E&dV
Os_2B_5	e15	1961: K&F
P_2O_5	e10, e11, e12	1941: D; D&MG; 1949: MG,D&N
Pd_5B_2	e14	1961: S
Ru_2B_5	e15	1961: K&F
Ti_2B_5	e15	1952: P&G
V_2O_5	e13	1936: K; 1950: B,W&B; 1953: A; 1959: S; 1961: B,A&B
W_2B_5	e15	1947: K
Compounds R_nX_5 ($n > 2$)		
Al_5C_3N	e21	1935: vS,S,P&S
Cr_5B_3	e19	1953: B&B
Fe_5PB_2	e19	1962: R
Fe_5SiB_2	e19	1959: A&L
Mn_5SiB_2	e19	1959: A&L
Mo_5SiB_2	e19	1958: A
$Ni_{12}P_5$	e23	1959: R&L
P_4S_5	e22	1957: vH&W
Ti_3O_5	e17	1957: A&M; 1959: A&M; M,A,K,A,W, H&N
Ti_5Te_4	e20	1961: G,K&R
V_3O_5	e18	1954: A; 1959: A,F,M&A; M,A,K,A,W, H&N

F. COMPOUNDS OF THE TYPE $R_n X_m$ ($m \geqq 6$)

Compounds $R_n X_6$

V,f1. *Calcium hexaboride*, CaB_6, is typical of a large group of hexaborides of divalent metals that have a simple cubic structure. Its unimolecular cell has the edge:

$$a_0 = 4.1450 \text{ A.}$$

The atomic arrangement can be described in terms of the space group O_h^1 (*Pm3m*) using its special positions:

Ca: (1*a*)   000

B: (6*f*) $\pm (^1/_2\,^1/_2\,u;\; ^1/_2\,u\,^1/_2;\; u\,^1/_2\,^1/_2)$ with $u = 0.207$

As is apparent from Figure VF,1, this structure can be considered as a CsCl grouping (**III,b1**) of metal atoms and B_6 octahedra. With the chosen parameter the B–B separation is the same within an octahedron and between atoms of adjacent octahedra: 1.720 A. There is thus in this crystal, as in so many other compounds rich in boron, a skeleton of these atoms reaching throughout the solid with metallic atoms lying in its large holes. It is in

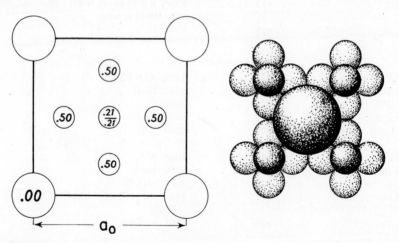

Fig. VF,1a (left). Atoms in the unit of CaB_6 projected on a cube face. The small circles are the boron atoms.

Fig. VF,1b (right). A packing drawing of the CaB_6 structure if the atoms are given their neutral radii. In this figure the calcium atom at the origin of Figure VF,1a is at the center of the drawing with the height 1.00.

TABLE VF,1
Crystals with the Cubic CaB_6 Structure

Crystal	a_0, A.
BaB_6	4.2680
CeB_6	4.1410
DyB_6	4.0976
ErB_6	4.102
EuB_6	4.178[a]
GdB_6	4.1123
HoB_6	4.096
LaB_6	4.1566
LuB_6	4.11
NdB_6	4.1284
PrB_6	4.121
PuB_6	4.115–4.140[b]
SiB_6	4.150
SmB_6	4.1333
SrB_6	4.1984
TbB_6	4.1020
ThB_6	4.1132
TmB_6	4.110
YB_6	4.1132
YbB_6	4.1444

[a] β-deficient have a smaller and β-rich samples have a larger a_0 than this.
[b] This variation in a_0 is considered indicative of excess boron in the structure.

accord with this skeletal picture that the shortest Ca–B distance, 3.0 A., exceeds the sum of either the elemental radii, or of these atoms as ions.

Other compounds with this structure are listed in Table VF,1.

V,f2. *Tungsten hexachloride*, WCl_6, crystallizes with rhombohedral symmetry, its unimolecular cell having the dimensions:

$$a_0 = 6.58 \text{ A.}, \qquad \alpha = 55°0'$$

The trimolecular hexagonal cell corresponding to this has

$$a_0' = 6.088 \text{ A.}, \qquad c_0' = 16.68 \text{ A.}$$

Atoms have been put in the following positions of C_{3i}^2 ($R\overline{3}$):

W: (1a) 000
Cl: (6f) $\pm(xyz; zxy; yzx)$

with $x = 0.37$, $y = 0.29$, $z = 0.21$. In the hexagonal cell these atoms have the following coordinates:

W: (3a) 000; rh
Cl: (18f) $\pm (xyz; \bar{y},x-y,z; y-x,\bar{x},z)$; rh

with $x = y = 0.295$ and $z = 0.080$

This structure (Fig. VF,2) is a distorted hexagonal close-packing of chlorine atoms with tungsten atoms lying in interstices. The forces between

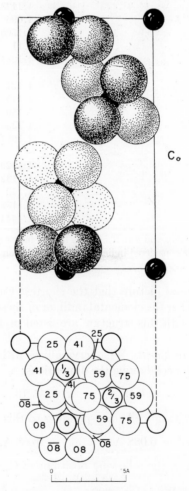

Fig. VF,2. Two projections of the hexagonal structure of WCl_6. In the upper part the tungsten atoms are small and black.

each of these tungsten atoms and the octahedron of chlorine atoms surrounding it are so great that the halogen atoms are drawn much closer to one another (3.11 or 3.21 A.) than to the other chlorines which do not share with them a metallic atom (3.42 or 3.72 A.). The separation between a tungsten atom and its six surrounding chlorine atoms is 2.24 A.

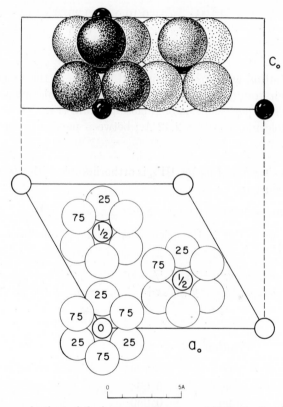

Fig. VF,3. Two projections of the hexagonal structure of UCl$_6$. In the upper projection the uranium atoms are small and black.

V,f3.   Crystals of *uranium hexachloride*, UCl$_6$, have hexagonal symmetry with a trimolecular unit of the dimensions:

$$a_0 = 10.97 \text{ A.}, \qquad c_0 = 6.04 \text{ A.}$$

The space group is D$_{3d}^3$ ($P\bar{3}m1$) with uranium atoms in the positions:

U(1): (1a)   000
U(2): (2d) $\pm(^1/_3\ ^2/_3\ u)$ with $u = {}^1/_2$

The chlorine atoms have been assigned to three sets of the following special positions:

$$(6i) \quad \pm(u\bar{u}v;\ u\ 2u\ v;\ 2\bar{u}\ \bar{u}\ v)$$

with the parameters:

$$\text{Cl}(1): \quad u = 0.10,\ v = 0.25$$
$$\text{Cl}(2): \quad u = 0.43,\ v = 0.25$$
$$\text{Cl}(3): \quad u = 0.77,\ v = 0.25$$

This distribution explains the principal features of the x-ray data, and surrounds each uranium atom by an octahedron of chlorine atoms, themselves in a nearly perfect hexagonal close-packing (Fig. VF,3). Within an octahedral molecule, U–Cl = 2.42 A.; between molecules the closest Cl–Cl = 3.85 A.

V,f4. *Uranium hexafluoride*, UF_6, is orthorhombic with a tetramolecular cell of the dimensions:

$$a_0 = 9.900 \text{ A.}; \quad b_0 = 8.962 \text{ A.}; \quad c_0 = 5.207 \text{ A.}$$

The space group is V_h^{16} in the orientation *Pnma*. Atoms have been put in the positions:

$$(4c) \quad \pm(u\ {}^1/_4\ v;\ u+{}^1/_2, {}^1/_4, {}^1/_2-v)$$
$$(8d) \quad \pm(xyz;\ {}^1/_2-x, y+{}^1/_2, z+{}^1/_2;\ x, {}^1/_2-y, z;\ x+{}^1/_2, y, {}^1/_2-z)$$

with the parameters:

Atom	Position	x	y	z
U	(4c)	0.1295	$^1/_4$	0.081
F(1)	(4c)	0.003	$^1/_4$	−0.250
F(2)	(4c)	0.250	$^1/_4$	0.417
F(3)	(8d)	0.014	0.093	0.250
F(4)	(8d)	0.246	0.093	0.083

This is a structure (Fig. VF,4) which can be described as a double hexagonal close-packing of the fluorine atoms with the uranium atoms in octahedral holes. The close U–F separations lie between 2.01 and 2.13 A. Fluorine-to-fluorine distances range upwards from ca. 2.95 A.

The compound $OsOF_5$ has this arrangement, with

$$a_0 = 9.540 \text{ A.}; \quad b_0 = 8.669 \text{ A.}; \quad c_0 = 5.019 \text{ A.}$$

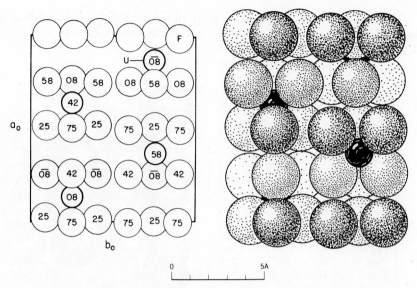

0   5A

Fig. VF,4a (left). The orthorhombic UF_6 arrangement projected along its c_0 axis.
Fig. VF,4b (right). A packing drawing of the UF_6 structure seen along its c_0 axis. The few uranium atoms that show are small and black.

V,f5. *Ruthenium tetracarbonyl diiodide*, $Ru(CO)_4I_2$, is monoclinic with the tetramolecular cell:

$$a_0 = 7.15 \text{ A.}; \quad b_0 = 11.04 \text{ A.}; \quad c_0 = 12.70 \text{ A.}; \beta = 91°18'$$

Atoms have been found to be in the following positions of C_{2h}^6 $(C2/c)$:

(4e) $\pm (0\ u\ ^1/_4;\ ^1/_2, u+^1/_2, ^1/_4)$
(8f) $\pm (xyz;\ x, \bar{y}, z+^1/_2;\ x+^1/_2, y+^1/_2, z;\ x+^1/_2, ^1/_2-y, z+^1/_2)$

with the parameters of Table VF,2.

TABLE VF,2
Positions and Parameters of the Atoms in $Ru(CO)_4I_2$

Atom	Position	x	y	z
Ru	(4e)	0.0000	−0.0110	$^1/_4$
I	(8f)	−0.1564	0.1586	0.3757
C(1)	(8f)	0.230	−0.001	0.342
C(2)	(8f)	0.119	−0.137	0.157
O(1)	(8f)	0.355	−0.010	0.384
O(2)	(8f)	0.180	−0.192	0.099

The structure (Fig. VF,5) is an array of octahedrally shaped molecules each of which has the configuration and dimensions indicated in Figure VF,6.

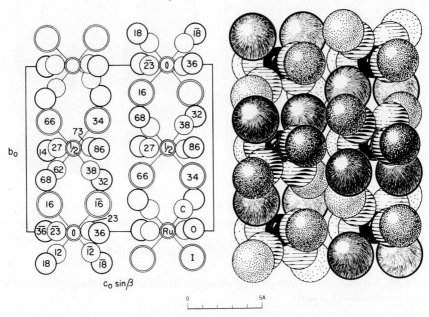

Fig. VF,5a (left). The monoclinic structure of $Ru(CO)_4I_2$ projected along its a_0 axis. Origin in lower left.

Fig. VF,5b (right). A packing drawing of the monoclinic structure of $Ru(CO)_4I_2$ seen along its a_0 axis. The ruthenium atoms are black, the oxygens dotted, and the iodine atoms heavily outlined and fine-line shaded.

V,f6. The phase Cr_5S_6 in the chromium–sulfur system is said to have a bimolecular hexagonal cell, with

$$a_0 = 5.982 \text{ A.}, \qquad c_0 = 11.509 \text{ A.}$$

A majority of the lines in its diffraction photographs can, however, be interpreted in terms of a smaller NiAs-like cell which is half as high and has an $a_0' = a_0\sqrt{3}$. With this in mind it is considered that the atoms of Cr_5S_6 are in the following positions of D_{3d}^2 ($P\bar{3}1c$):

Cr(1):   (2a) $\pm(0\ 0\ 1/4)$
Cr(2):   (2c) $\pm(1/3\ 2/3\ 1/4)$
Cr(3):   (2b) $000;\ 0\ 0\ 1/2$

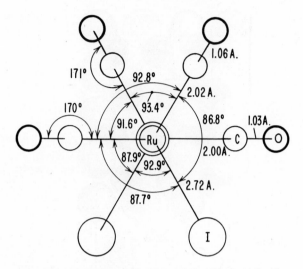

Fig. VF,6. Bond dimensions of the Ru(CO)₄I₂ molecule.

$$Cr(4): \quad (4f) \quad \pm (^1/_3\,^2/_3\,u;\; ^2/_3,^1/_3,u+^1/_2) \qquad \text{with } u = 0$$
$$S \quad (12i) \quad \pm (xyz;\quad \bar{y},x-y,z;\quad y-x,\bar{x},z;$$
$$y,x,z+^1/_2;\; \bar{x},y-x,z+^1/_2;\; x-y,\bar{y},z+^1/_2)$$
$$\text{with } x = {}^1/_3,\; y = 0,\; z = {}^3/_8$$

The resulting arrangement is illustrated in Figure VF,7.

Another phase, which has been given the composition $Cr_{0.88}S$–$Cr_{0.85}S$ and is sometimes designated as Cr_7S_8, has also been given a structure related to that of NiAs. It is hexagonal and for $Cr_{0.88}S$ has the cell edges:

$$a_0 = 3.464 \text{ A.}, \qquad c_0 = 5.763 \text{ A.}$$

Atomic positions based on D_{3d}^3 ($P\bar{3}m1$) have been assigned as follows:

$$Cr(1): \quad (1a) \quad 000 \qquad \text{ca. } {}^3/_4 \; Cr(2): \quad (1b) \quad 0\,0\,{}^1/_2$$
$$S: \quad (2d) \quad \pm ({}^1/_3\,{}^2/_3\,u) \qquad \text{with } u = {}^1/_4$$

Two more phases, both given the composition Cr_2S_3, have been assigned structures related to that of Cr_5S_6.
One of these has the same sized cell, with

$$a_0 = 5.939\text{–}5.943 \text{ A.}, \qquad c_0 = 11.171 \text{ A.}$$

It has the same space group as Cr_5S_6 and atoms in the same positions except that two atoms of chromium are missing from $(2a)$ $\pm (0\,0\,{}^1/_4)$.

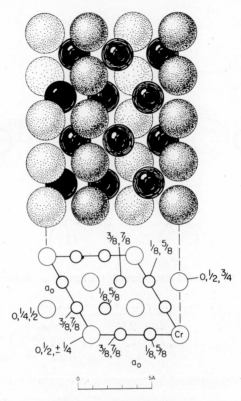

Fig. VF,7. Two projections of the hexagonal Cr_5S_6 arrangement. In the upper projection the sulfur atoms are black.

The other form of Cr_2S_3 is rhombohedral with a hexagonal unit of the same base and a height 50% greater:

$$a_0' = 5.937\text{--}5.939 \text{ A.}, \qquad c_0' = 16.698\text{--}16.65 \text{ A.}$$
$$a_0 = 6.524 \text{ A.}, \qquad \alpha = 54°8'$$

The chosen space group is $C_{3i}{}^2$ $(R\bar{3})$, with atoms in the following positions referred to the hexagonal cell:

Cr(1):   (3b) $0\ 0\ ^1/_2$; $^1/_3\ ^2/_3\ ^1/_6$; $^2/_3\ ^1/_3\ ^5/_6$
Cr(2):   (3a)   000; $^1/_3\ ^2/_3\ ^2/_3$; $^2/_3\ ^1/_3\ ^1/_3$
Cr(3):   (6c) $\pm (00u)$; rh   with $u = ^1/_3$
   S: (18f) $\pm (xyz;\ \bar{y},x-y,z;\ y-x,\bar{x},z)$; rh
                                with $x = ^1/_3,\ y = 0,\ z = ^1/_4$

This structure is illustrated in Figure VF,8.

Still another phase of this system has the composition Cr_3S_4. The structure proposed for it is stated in paragraph **V,d3.**

V,f7. The complex *chromium carbide*, $Cr_{23}C_6$, is cubic, with

$$a_0 = 10.638 \text{ A.}$$

Atoms of the four molecules in this cell were put many years ago in the following special positions of O_h^5 ($Fm3m$):

Cr(1): (4a) 000; 0 $^1/_2$ $^1/_2$; $^1/_2$ 0 $^1/_2$; $^1/_2$ $^1/_2$ 0
Cr(2): (8c) $^1/_4$ $^1/_4$ $^1/_4$; $^3/_4$ $^3/_4$ $^3/_4$; F.C.

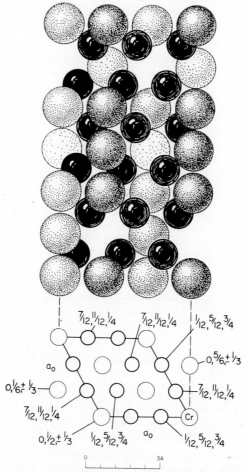

Fig. VF,8. Two projections of the hexagonal Cr_2S_3 structure. In the upper projection the sulfur atoms are black.

Cr(3):　　(32f)　　$\pm(uuu;\ \bar{u}\bar{u}u;\ \bar{u}u\bar{u};\ u\bar{u}\bar{u})$; F.C.　　　　with $u = 0.385$
Cr(4):　　(48h)　　$\pm(0uu;\ u0u;\ uu0;\ 0u\bar{u};\ \bar{u}0u;\ u\bar{u}0)$; F.C.

　　　　　　　　　　　　　　　　　　　　　　　　　　　　with $u = 0.165$

　　C:　(24e)　$\pm(u00;\ 0u0;\ 00u)$; F.C.　　　with $u = 0.275$

Probably such compounds as $Mn_{23}C_6$, $Fe_{21}W_2C_6$, $Fe_{21}Mo_2C_6$, and $Cr_{21}W_2C_6$ are isostructural.

Compounds R_nX_7

V,f8.　*Iodine heptafluoride*, IF_7, freezes near 0°C. to a solid which is cubic with a bimolecular cell of the edge length:

$$a_0 = 6.28\ \text{A.}\ (-110°C.)$$

The diffraction data are indicative of a high degree of disorder.

Below $-120°$C. the structure is orthorhombic with a tetramolecular unit which at $-145°$C. has the dimensions:

$$a_0 = 8.74\ \text{A.};\quad b_0 = 8.87\ \text{A.};\quad c_0 = 6.14\ \text{A.}$$

There has been considerable controversy over the atomic arrangement in this modification. Until recently it seemed probable that the space group was $C_{2v}{}^{17}$ (Aba) with atoms in the positions:

(4a)　　$00u;\ {}^1/_2\ {}^1/_2\ u;\ 0,{}^1/_2,u+{}^1/_2;\ {}^1/_2,0,u+{}^1/_2$
(8b)　　$xyz;\ \bar{x}\bar{y}z;\ {}^1/_2-x,y+{}^1/_2,z;\ x+{}^1/_2,{}^1/_2-y,z;$

　　　　　　　　　　　　and similar points about $0\ {}^1/_2\ {}^1/_2$

Positions and parameters were selected as follows:

Atom	Position	x	y	z
I	(4a)	0	0	0.0000
F(1)	(4a)	0	0	0.2777
F(2)	(8b)	0.0862	0.0974	-0.2361
F(3)	(8b)	0.1235	0.1566	0.0585
F(4)	(8b)	0.1695	-0.1189	0.0310

This led to a molecular structure (Fig. VF,9) in which I–F = 1.71–1.85 A.

A very recent reconsideration of the original data suggests that the arrangement differs somewhat from the foregoing and is partially disordered. This can be described by saying that the correct space group is in fact $V_h{}^{18}$ ($Abam$). The higher symmetry group differs from Aba by the addition of a center of symmetry (and a doubling of the number of equivalent positions per cell). The disorder in the orientation of the IF_7 molecules is

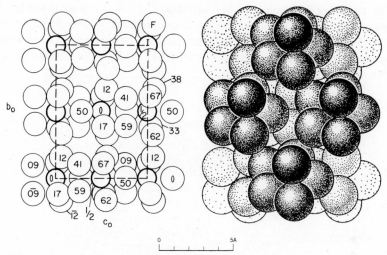

Fig. VF,9a (left). The low-temperature orthorhombic modification of IF₇ projected along its a_0 axis.

Fig. VF,9b (right). A packing drawing of the low-temperature IF₇ seen along its a_0 axis. Only the dotted fluorine atoms show.

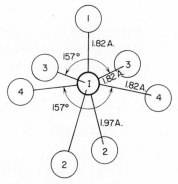

Fig. VF,10. The bond dimensions of the IF₇ molecule in its crystals.

expressed by placing half-atoms in the corresponding positions (8d) and (16g) of *Abam;* and a revision of the parameters leads to the values:

Atom	Position	x	y	z
I	(8d)	0	0	0
F(1)	(8d)	0	0	0.2942
F(2)	(16g)	0.0717	0.0869	−0.2763
F(3)	(16g)	0.1163	0.1632	0.0678
F(4)	(16g)	−0.1673	0.1234	−0.0495

The resulting molecules, whose haphazard orientations give the crystal its seeming center of symmetry, have the appearance of Figure VF,10 with I–F between 1.81 and 1.97 A. They very nearly (or exactly) have two planes of symmetry at right angles to one another and the bond lengths and angles of Figure VF,10.

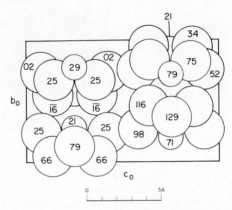

Fig. VF,11. The orthorhombic structure of $(NH)S_7$ seen along its a_0 axis. The small circle is the NH radical. The same molecular distribution prevails for $(NH)_2S_6$ and $(NH)_4S_4$.

V,f9. The sulfurimides, NHS_7, $(NH)_2S_6$, and $(NH)_4S_4$ have molecules like that of sulfur itself, S_8, and the solids they form have similar structures.

The *heptasulfurimide*, NHS_7, is orthorhombic like rhombic sulfur (**II,n1**) but with a smaller unit containing four molecules:

$$a_0 = 7.61 \text{ A.}; \quad b_0 = 8.04 \text{ A.}; \quad c_0 = 13.03 \text{ A.}$$

The space group is V_h^{16} (*Pbnm*) with all atoms in the positions:

(4c) $\pm (u \, v \, {}^1/_4; \; {}^1/_2 - u, v + {}^1/_2, {}^1/_4)$
(8d) $\pm (xyz; \; x + {}^1/_2, {}^1/_2 - y, z + {}^1/_2; \; x, y, {}^1/_2 - z; \; {}^1/_2 - x, y + {}^1/_2, z)$

For the *hexasulfurdiimide*, $(NH)_2S_6$, the orthorhombic cell has the dimensions:

$$a_0 = 7.386 \text{ A.}; \quad b_0 = 7.864 \text{ A.}; \quad c_0 = 12.828 \text{ A.}$$

For the *tetrasulfur tetrimide*, $(NH)_4S_4$, the dimensions are

$$a_0 = 6.727 \text{ A.}; \quad b_0 = 8.010 \text{ A.}; \quad c_0 = 12.20 \text{ A.}$$

The similar parameters that apply to the atoms in these three compounds are listed in Table VF,3.

TABLE VF,3
Parameters of the Atoms in NHS_7, $(NH)_2S_6$ and $(NH)_4S_4$

Atom	Position	x	y	z
		For NHS_7		
S(1)	(8d)	0.247	0.161	0.142
S(2)	(8d)	0.485	0.228	0.080
S(3)	(8d)	0.660	0.045	0.136
S(4)	(4c)	0.790	0.127	1/4
N	(4c)	0.214	0.272	1/4
		For $(NH)_2S_6$		
S(1)	(8d)	0.235	0.129	0.141
S(2)	(8d)	0.487	0.204	0.095
S(3)	(8d)	0.673	0.025	0.138
N(1)	(4c)	0.174	0.215	1/4
N(2)	(4c)	0.769	0.093	1/4
		For $(NH)_4S_4$[a]		
S(1)	(8d)	0.680 (0.1810)	0.043 (0.0415)	0.131 (0.1300)
S(2)	(8d)	0.259 (0.7575)	0.137 (0.1372)	0.131 (0.1304)
N(1)	(8d)	0.494 (0.9940)	0.171 (0.1677)	0.094 (0.0925)
N(2)	(4c)	0.221 (0.7192)	0.225 (0.2325)	1/4
N(3)	(4c)	0.774 (0.2717)	0.108 (0.1115)	1/4

[a] The parameters in parentheses are from 1958: S&D. In this paper the chosen origin is displaced by 1/2 0 0 from that used to describe the other structures.

In all these structures (Fig. VF,11) the molecules are the same kind of puckered eight-membered ring that exists for rhombic sulfur. Their configurations are exemplified by the drawing of Figure VF,12, made for $(NH_2)_2S_6$. The bond lengths and angles for the molecules of the three different compounds are given in Figures VF,13, VF,14, and VF,15. All distances between the atoms of neighboring molecules exceed 3.0 A.

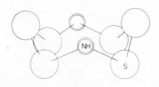

Fig. VF,12. A drawing to indicate the sulfur-like shape of the $(NH)_2S_6$ molecule. Molecules of $(NH)S_7$ and $(NH)_4S_4$ have the same configuration with appropriate exchanges of S and (NH).

CRYSTAL STRUCTURES

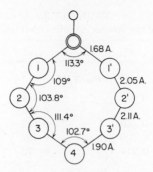

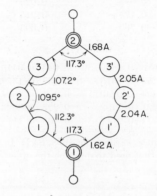

Fig. VF,13. The bond dimensions of the (NH)S$_7$ molecule.

Fig. VF,14. The bond dimensions of the (NH)$_2$S$_6$ molecule.

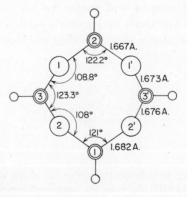

Fig. VF,15. The bond dimensions of the (NH)$_4$S$_4$ molecule.

V,f10. Crystals of the *ruthenium boride*, Ru_7B_3, are hexagonal with a bimolecular cell of the dimensions:

$$a_0 = 7.467 \text{ A.}, \qquad c_0 = 4.713 \text{ A.}$$

Atoms are in the following positions of C_{6v}^4 ($P6_3mc$):

Ru(1): (6c) $u\bar{u}v;$ $\quad u\,2u\,v;$ $\quad 2\bar{u}\,\bar{u}\,v;$
$\bar{u},u,v+^1/_2;$ $\bar{u},2\bar{u},v+^1/_2;$ $2u,u,v+^1/_2$
with $u = 0.4563,\ v = 0.318$

Ru(2): (6c) with $u = 0.1219,\ v = 0$ (arbitrary)

Ru(3): (2b) $^1/_3\,^2/_3\,u;\ ^2/_3,^1/_3,u+^1/_2$ with $u = 0.818$

B: (6c) with a probable, approximate $u = 0.19,\ v = 0.58$

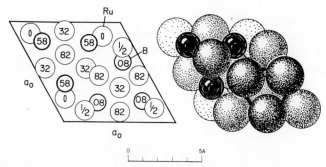

Fig. VF,16a (left). A basal projection of the hexagonal Ru_7B_3 structure.
Fig. VF,16b (right). A packing drawing of the Ru_7B_3 arrangement seen along its c_0 axis. The boron atoms are black.

In this structure (Fig. VF,16) each boron atom is at the center of a trigonal pyramid of ruthenium atoms with Ru–B between 2.15 and 2.20 A. The Ru–Ru separations range upwards from 2.61 A.

The following additional borides have this structure:

$$Re_7B_3: \quad a_0 = 7.504 \text{ A.}, c_0 = 4.772 \text{ A.}$$
$$Rh_7B_3: \quad a_0 = 7.471 \text{ A.}, c_0 = 4.777 \text{ A.}$$

This is the arrangement that has also been found for the intermetallic compound Th_7Fe_3.

This structure is simpler than but very like that given many years ago to the chromium carbide, Cr_7C_3. Its unit was then stated as a four times larger one, with

$$a_0 = 13.90 \text{ A.}, \qquad c_0 = 4.54 \text{ A.}$$

but it was shown later that the positions of the chromium atoms could be described in terms of the space group C_{6v}^4 and a cell with half the value of a_0, i.e., 6.95 A. The data are satisfied if the chromium atoms are placed in the same $(6c)$ and $(2b)$ used above for the ruthenium atoms of Ru_7B_3 with $u(Cr,1) = 0.455$ and $u(Cr,2) = 0.125$ (the z parameters not having been established).

V,f11. The compound *tetraphosphorus heptasulfide*, P_4S_7, crystallizes in the monoclinic system with a tetramolecular cell of the dimensions:

$$a_0 = 8.87 \text{ A.}; \quad b_0 = 17.35 \text{ A.}; \quad c_0 = 6.83 \text{ A.}; \quad \beta = 92°42'$$

All atoms are in general positions of C_{2h}^5 $(P2_1/n)$:

$$(4e) \quad \pm(xyz; \ x+\tfrac{1}{2}, \tfrac{1}{2}-y, z+\tfrac{1}{2})$$

with the parameters of Table VF,4.

The molecule in the structure that results (Fig. VF,17) is a nine-membered cage with two additional sulfur atoms attached to phosphorus. Its general shape and its bond lengths are indicated in Figure VF,18. The bond

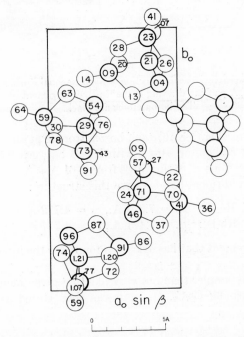

Fig. VF,17a. A projection along the c_0 axis of the monoclinic structure of P_4S_7. The lighter ringed circles are sulfur. Origin in lower right.

TABLE VF,4. Parameters of the Atoms in P_4S_7

Atom	x	y	z
P(1)	0.2821	0.3791	0.7088
P(2)	0.3531	0.2932	0.4637
P(3)	0.271	0.461	0.274
P(4)	0.003	0.335	0.409
S(1)	0.043	0.425	0.220
S(2)	0.307	0.544	0.094
S(3)	−0.212	0.308	0.363
S(4)	0.305	0.487	0.569
S(5)	0.413	0.367	0.239
S(6)	0.150	0.245	0.366
S(7)	0.052	0.366	0.697

angles all lie between 101° and 116°. Between molecules the shortest atomic separation is ca. 3.4 A.

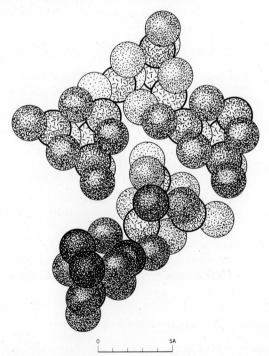

Fig. VF,17b. A packing drawing of the monoclinic P_4S_7 structure viewed along its c_0 axis. The sulfur atoms are dotted.

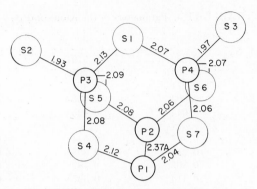

Fig. VF,18. The dimensions of the molecule of P_4S_7.

Compounds R_nX_m ($m \geqq 8$)

V,f12. The *uranium oxide*, U_3O_8, is dimorphous with an orthorhombic low-temperature form. Neutron diffraction measurements have led to a structure different from that earlier suggested on the basis of x-ray data. Its bimolecular unit has the edges:

$$a_0 = 6.704 \text{ A.}; \quad b_0 = 11.95 \text{ A.}; \quad c_0 = 4.142 \text{ A.}$$

The space group has been found to be V^6 ($C222$), with atoms in its positions:

U(1): (2a)   000; $^1/_2 \, ^1/_2 \, ^1/_2$
U(2): (4g) $\pm (0u0)$; B.C.   with $u = 0.315$
O(1): (2b)   0 $^1/_2$ 0; $^1/_2$ 0 $^1/_2$
O(2): (2d)   0 0 $^1/_2$; $^1/_2 \, ^1/_2$ 0
O(3): (4h) $\pm (0 \, u \, ^1/_2)$; B.C.   with $u = 0.315$
O(4): (8l) $xyz; \, \bar{x}\bar{y}z; \, \bar{x}y\bar{z}; \, x\bar{y}\bar{z};$ B.C.
                  with $x = 0.19$, $y = 0.145$, $z = 0.08$

The resulting arrangement is shown in Figure VF,19. The U(1) atoms are surrounded by six oxygens at corners of a distorted octahedron with U–O = 2.07 or 2.18 A. The U(2) atoms have seven oxygen neighbors with U–6O lying between 2.07 and 2.21 A. and with the seventh 2.42 A. away.

The mixed oxide UTa_2O_8 is isostructural, with

$$a_0 = 6.41 \text{ A.}; \quad b_0 = 11.10 \text{ A.}; \quad c_0 = 3.95 \text{ A.}$$

There is still much disagreement concerning both the composition and the atomic arrangement in the several uranium oxides that seem to exist.

Thus, a compound of the composition U_3O_7 has been described as cubic with $a_0 = 5.42$ A. It would contain 3.15 molecules per cell and it has been said on the basis of neutron diffraction data that the atoms are fractionally distributed over various special positions of T_h^6 ($Pa3$).

Another oxide with the composition U_4O_9 has also been said to be cubic with a large unit having $a_0 = 21.77$ A. (about four times the cell edge of UO_2). For it, a very complicated arrangement with many partly filled positions has been proposed.

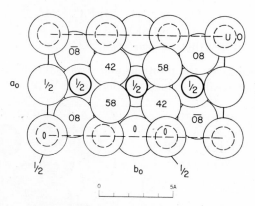

Fig. VF,19. The orthorhombic U_3O_8 arrangement projected along its c_0 axis. Origin in lower left.

V,f13. When the orthorhombic U_3O_8 is heated between 445 and 600°C. its diffraction pattern alters somewhat to correspond to that of a hexagonal unit, with

$$a_0 = 6.815 \text{ A.,} \qquad c_0 = 4.136 \text{ A.}$$

A structure has been proposed, but not proved, for this high-U_3O_8. It places atoms in the following positions of C_{3i}^1 ($P\bar{3}$):

$$
\begin{aligned}
&U(1){:} \quad (1a) \quad 000 \\
&U(2){:} \quad (2d) \quad \pm(^1/_3\ ^2/_3\ u) \qquad \text{with } u = 0 \\
&O(1){:} \quad (2d) \quad \text{with } u = ^1/_2 \\
&O(2){:} \quad (6g) \quad \pm(xyz;\ y,y-x,\bar{z};\ x-y,x,\bar{z}) \\
&\qquad\qquad\qquad\qquad \text{with } x = ^1/_3,\ y = 0,\ z = 0.01
\end{aligned}
$$

In this suggested arrangement the uranium atoms are in about the same positions as they are in the low-temperature form (**V,f12**).

V,f14. Recently, a monoclinic structure has been ascribed to the *chromium selenide*, Cr_7Se_8, which is unlike that described above (**V,f6**) for the analogous sulfide. Its large tetramolecular face-centered cell has

$$a_0 = 12.67 \text{ A.}; \quad b_0 = 7.37 \text{ A.}; \quad c_0 = 11.98 \text{ A.}; \quad \beta = 90°57'$$

The space group is C_{2h}^3, which when face-centered has the symbol $F2/m$. The special positions employed are

$$(4b) \quad 0 \, ^1/_2 \, 0; \text{ F.C.}$$
$$(8e) \quad ^1/_4 \, ^1/_4 \, 0; \, ^1/_4 \, ^3/_4 \, 0; \text{ F.C.}$$
$$(8h) \quad \pm(^1/_4 \, u \, ^1/_4); \text{ F.C.}$$
$$(8i) \quad \pm(u0v); \text{ F.C.}$$
$$(16j) \quad \pm(xyz; \, x\bar{y}z); \text{ F.C.}$$

Atoms have been given the positions and parameters of Table VF,5. The arrangement that results is illustrated in Figure VF,20.

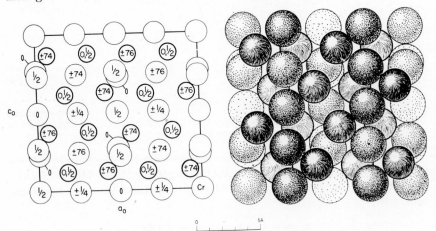

Fig. VF,20a (left). The monoclinic structure given Cr_7Se_8 as projected along its b_0 axis.

Fig. VF,20b (right). A packing drawing of the Cr_7Se_8 arrangement viewed along its b_0 axis. The chromium atoms are dotted.

The bimolecular cell corresponding to the more usual c_0-centered orientation has the dimensions:

$$a_0' = a_0; \quad b_0' = -b_0; \quad c_0' = 8.65 \text{ A.}; \quad \beta' = 136°9'$$

V,f15. An atomic arrangement has been described for the cubic *cobalt sulfide*, Co_9S_8. The four molecules in its unit with

$$a_0 = 9.9284 \text{ A. (Co-rich)}–9.9256 \text{ A. (Co-poor)}$$

TABLE VF,5
Positions and Parameters of the Atoms in Cr_7Se_8

Atom	Position	x	y	z
Cr(1)	(4b)	0	$1/2$	0
Cr(2)	(8e)	$1/4$	$1/4$	0
Cr(3)	(8h)	$1/4$	0.244	$1/4$
Cr(4)	(8i)	0.494	0	0.268
Se(1)	(8i)	0.173	0	0.122
Se(2)	(8i)	0.669	0	0.126
Se(3)	(16j)	0.422	0.240	0.131

have atoms in special positions of O_h^5 $(Fm3m)$:

Co(1): (4b) $1/2\ 1/2\ 1/2$; F.C.

Co(2): (32f) $\pm(uuu;\ \bar{u}\bar{u}u;\ \bar{u}u\bar{u};\ u\bar{u}\bar{u})$; F.C.   with $u = 0.1260$

S(1): (8c) $1/4\ 1/4\ 1/4;\ 3/4\ 3/4\ 3/4$; F.C.

S(2): (24e) $\pm(u00;\ 0u0;\ 00u)$; F.C.   with $u = 0.2591$

This arrangement can be considered as a cubic close-packing of the sulfur atoms with cobalt atoms in the interstices (Fig. VF,21). In it, Co(1) atoms are surrounded by octahedra of sulfur atoms at a distance of 2.39 A. and Co(2) atoms by four much nearer sulfurs (2.13 A.). The sulfur atoms are 3.38 A. from one another.

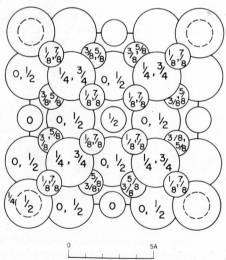

Fig. VF,21. A projection along a cubic axis of the contents of the unit cube of Co_9S_8. The small circles are the cobalt atoms.

The mineral *pentlandite*, $(Ni,Fe)_9S_8$, is isomorphous, with

$$a_0 = 10.02 \text{ A.}$$

Another isostructural compound is the selenide, Co_9Se_8, with

$$a_0 = 10.431 \text{ A.}$$

V,f16.   The complex *ruthenium boride*, $Ru_{11}B_8$, is orthorhombic, with a bimolecular unit of the dimensions:

$$a_0 = 11.609 \text{ A.}; \quad b_0 = 11.342 \text{ A.}; \quad c_0 = 2.836 \text{ A.}$$

Atoms are in the following special positions of V_h^9 (*Pbam*):

$$(2a) \quad 000; \ {}^1/_2 \, {}^1/_2 \, 0$$
$$(4g) \quad \pm(uv0; \ {}^1/_2-u,v+{}^1/_2,0)$$
$$(4h) \quad \pm(u \ v \ {}^1/_2; \ {}^1/_2-u,v+{}^1/_2,{}^1/_2)$$

with the parameters of Table VF,6.

In this arrangement (Fig. VF,22), each of the four kinds of boron atoms has nine neighbors. Six of these are ruthenium atoms at the corners of an enveloping trigonal prism while the other three atoms are at the corners of a triangle whose plane is normal to the axis of the prism. For B(1) the three

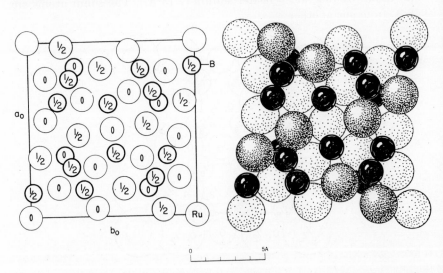

Fig. VF,22a (left). The orthorhombic $Ru_{11}B_8$ arrangement projected along its c_0 axis.
Fig. VF,22b (right). A packing drawing of the $Ru_{11}B_8$ structure viewed along its c_0 axis. The boron atoms are black.

TABLE VF,6
Positions and Parameters of the Atoms in $Ru_{11}B_8$

Atom	Position	u	v
Ru(1)	(2a)	—	—
Ru(2)	(4g)	0.2844	0.3913
Ru(3)	(4g)	0.0429	0.3952
Ru(4)	(4g)	0.1686	0.1740
Ru(5)	(4h)	0.4636	0.2962
Ru(6)	(4h)	0.3404	0.0616
B(1)	(4h)	0.140	0.018
B(2)	(4g)	0.348	0.216
B(3)	(4h)	0.152	0.326
B(4)	(4h)	0.280	0.253

atoms in this plane are ruthenium, for B(4) they are boron, for B(2) they are two boron and one ruthenium, and for B(3) they are two ruthenium and one boron. In all cases the atoms at the corners of the trigonal prisms are ruthenium. The distances Ru–B in the prisms lie between 2.06 and 2.29 A.; in the normal planes they are greater (2.38–2.94 A.). The boron atoms are in contact with one another (1.68–1.70 A.) to form chains of the type shown in Figure VF,23.

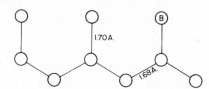

Fig. VF,23. The chain-like arrangement of the boron atoms in $Ru_{11}B_8$.

V,f17. There are four molecules of *diuranium nonafluoride*, U_2F_9, in a cubic unit of the edge length:

$$a_0 = 8.4714 \text{ A.}$$

Uranium atoms are in the following special positions of $T_d{}^3$ $(I\overline{4}3m)$:

(8c) $uuu; u\bar{u}\bar{u}; \bar{u}u\bar{u}; \bar{u}\bar{u}u;$ B.C.   with $u = 0.187$

Though the fluorine positions could not be determined from the x-ray data, it was found that the following distribution yields acceptable interatomic distances:

F(1): (12e) $\pm(u00; 0u0; 00u)$; B.C.   with $u = 0.225$
F(2): (24g) $uuv; u\bar{u}\bar{v}; \bar{u}u\bar{v}; \bar{u}\bar{u}v;$ B.C.; tr   with $u = 0.20, v = 0.46$

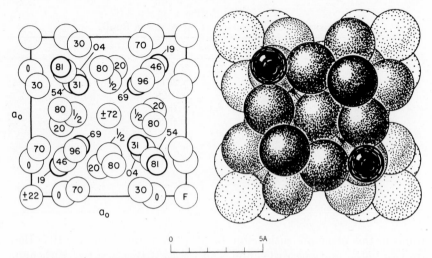

Fig. VF,24a (left). The cubic U_2F_9 structure projected along an a_0 axis. The uranium atoms are small and more heavily ringed.

Fig. VF,24b (right). A packing drawing of the U_2F_9 arrangement seen along a cube axis. The uranium atoms are black.

This is an arrangement (Fig. VF,24) that surrounds each uranium with nine fluorine atoms, three at a distance of 2.26 A., three at 2.31 A., and three more at 2.34 A. The closest approach of fluorine atoms to one another is $F(1)-F(2) = 2.57$ A.

There is known a substance to which the formula $NaTh_2F_9$ has been given. It is cubic with a unit only a little larger than that of U_2F_9: $a_0 = 8.705$ A. It has been proposed that its thorium and fluorine atoms are in nearly the same positions as the uranium and fluorine atoms in U_2F_9.

V,f18. An iron carbonyl different from that described in **V,e7** has been said to form hexagonal crystals. This *iron nonacarbonyl*, $Fe_2(CO)_9$, has two molecules in a cell with the edges:

$$a_0 = 6.45 \text{ A.}, \qquad c_0 = 15.98 \text{ A.}$$

Its iron atoms are in the special positions of C_{6h}^2 $(P6_3/m)$:

$$(4f) \quad \pm (^1/_3 \, ^2/_3 \, u; \, ^2/_3, ^1/_3, u + ^1/_2) \qquad \text{with } u = 0.173$$

The carbon and oxygen atoms of the carbonyl groups have been distributed as follows:

$$C(1): \quad (6h) \quad \pm (u \, v \, ^1/_4; \, \bar{v}, u - v, ^1/_4; \, v - u, \bar{u}, ^1/_4)$$
$$\text{with } u = 0.32, \quad v = 0.86$$

C(2):   (12i) $\pm (xyz;$ $\bar{y},x-y,z;$ $y-x,\bar{x},z;$
                $\bar{x},\bar{y},z+{}^1/_2; y,y-x,z+{}^1/_2; x-y,x,z+{}^1/_2)$
                        with $x = 0.35$, $y = 0.43$, $z = 0.11$

O(1):   (6h)   with $u = 0.315$, $v = 0.055$

O(2):   (12i)   with $x = 0.36$, $y = 0.285$, $z = 0.07$

The $Fe_2(CO)_9$ molecules in such an arrangement have the shape and interatomic distances shown in Figure VF,25. They are distributed with their centers at the points of a hexagonal close-packing.

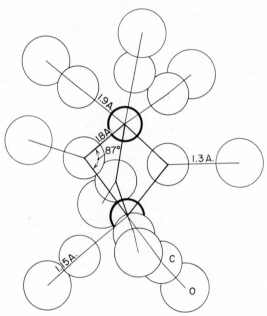

Fig. VF,25. The shape of the $Fe_2(CO)_9$ molecule in the structure ascribed to its crystals.

V,f19.   Crystals of Ti_5O_9 are triclinic with the bimolecular cell:

$$a_0 = 5.569 \text{ A.}; \ b_0 = 7.120 \text{ A.}; \ c_0 = 8.865 \text{ A.}$$
$$\alpha = 97°33' \quad \beta = 112°20' \quad \gamma = 108°30'$$

The atoms have been put in general positions of C_1^1 ($P1$): xyz with the parameters of Table VF,7. If this is correct it appears to be one of the very rare examples of a crystal devoid of symmetry—practically all triclinic crystals have belonged to the point group C_i and its only space group C_i^1 ($P\bar{1}$). In this crystal all atomic positions have in fact the center characteristic of this higher symmetry except Ti(5) and Ti(6); other sets of coordinates in Table VF,7 therefore apply to two atoms in $\pm (xyz)$.

TABLE VF,7
Parameters of the Atoms in Ti_5O_9

Atom	x	y	z
Ti(1)	0.276	0.170	0.107
Ti(2)	0.276	0.670	0.107
Ti(3)	0.852	0.008	0.316
Ti(4)	0.852	0.502	0.316
Ti(5)	0.575	0.170	0.470
Ti(6)	0.575	0.670	0.470
O(1)	0.63	0.32	0.11
O(2)	0.11	0.38	0.03
O(3)	0.92	0.01	0.12
O(4)	0.47	0.97	0.20
O(5)	0.67	0.70	0.23
O(6)	0.20	0.65	0.30
O(7)	0.51	0.37	0.36
O(8)	0.99	0.30	0.43
O(9)	0.24	0.04	0.46

Note: All triplets except Ti(5) and Ti(6) designate pairs of atoms in $\pm(xyz)$.

In this arrangement the titanium atoms are at centers of oxygen octahedra with Ti–O = 1.75–2.35 A. The shortest Ti–Ti = 2.81 A. and the shortest O–O = 2.4 A. The TiO_6 octahedra build up the crystal by sharing faces, edges, and corners to give a deformed hexagonal close-packing of the oxygens. The arrangement resembles that of rutile (**IV,b1**) but differs in having rutile-like blocks that are limited in a third direction.

V,f20. Crystals of P_4S_{10} are triclinic with a bimolecular cell of the dimensions:

$$a_0 = 9.07 \text{ A.}; \quad b_0 = 9.18 \text{ A.}; \quad c_0 = 9.19 \text{ A.},$$
$$\alpha = 92°24'; \quad \beta = 101°12'; \quad \gamma = 110°30'$$

Atoms are in general positions $\pm(xyz)$ of C_i^1 ($P\bar{1}$) with the parameters of Table VF,8.

The molecule that arises from this determination is an eleven-membered P_4S_7 cage with three additional sulfur atoms attached to phosphorus. Its shape and the bond lengths that prevail are indicated in Figure VF,26. All bond angles are essentially tetrahedral, lying between 113° and 107°.

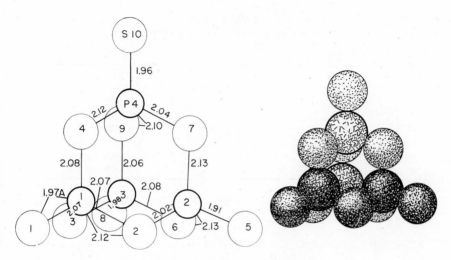

Fig. VF,26a (left). The bond lengths in the molecule of P_4S_{10}.
Fig. VF,26b (right). A packing drawing to show the general shape of the P_4S_{10} molecule.
The sulfur atoms are dotted.

TABLE VF,8
Parameters of the Atoms in P_4S_{10}

Atom	x	y	z^a
P(1)	0.259	0.087	−0.188
P(2)	−0.090	0.153	−0.289
P(3)	0.267	0.435	−0.297
P(4)	0.178	0.345	0.035
S(1)	0.357	−0.074	−0.196
S(2)	0.008	−0.014	−0.285
S(3)	0.367	0.265	−0.308
S(4)	0.288	0.176	0.031
S(5)	−0.316	0.065	−0.373
S(6)	0.024	0.343	−0.404
S(7)	−0.062	0.245	−0.064
S(8)	0.375	0.609	−0.407
S(9)	0.298	0.526	−0.079
S(10)	0.199	0.430	0.241

[a] Simple averages of values given in original. These differ by ca. 0.002 except for S(10).

TABLE VF,9
Compounds with the Cubic UB_{12} Structure

Compound	a_0, A.
DyB_{12}	7.501
ErB_{12}	7.484
HoB_{12}	7.492
LuB_{12}	7.464
TmB_{12}	7.476
YB_{12}	7.500
ZrB_{12}	7.408

V,f21. *Uranium dodecaboride*, UB_{12}, is typical of a series of higher borides with cubic symmetry. For it,

$$a_0 = 7.473 \text{ A.}$$

There are four molecules in this unit, and atoms have been reported to be in the following positions of O_h^5 (*Fm3m*):

U: (4a)   000; F.C.
B: (48i) $\pm(1/2\,u\,u;\ u\,1/2\,u;\ u\,u\,1/2;\ 1/2\,u\,\bar{u};\ \bar{u}\,1/2\,u;\ u\,\bar{u}\,1/2)$; F.C.
with $u = 1/6$

This is a structure (Fig. VF,27) that places each uranium atom at the center of a cube-octahedron of 24 boron atoms shared by adjacent poly-

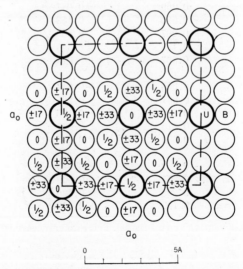

Fig. VF,27a. The cubic structure of UB_{12} projected along an a_0 axis.

hedra. This atomic distribution can, however, also be described as an NaCl arrangement of metal atoms and B_{12} groups. In the latter, B–B = 1.76 A. The shortest B–U = 2.79 A.

A number of other dodecaborides have now been shown to have this structure. Their unit cubes are listed in Table VF,9.

V,f22. A partially disordered structure has been proposed for the Th_7S_{12} phase of the Th–S system. Its symmetry is hexagonal and it has one molecule in a cell of the dimensions:

$$a_0 = 11.063 \text{ A.}, \qquad c_0 = 3.991 \text{ A.}$$

The following arrangement based on C_{6h}^2 $(P6_3/m)$ has been proposed:

One thorium atom would be in half the positions of $(2a)$ $0\ 0\ ^1/_4$; $0\ 0\ ^3/_4$, the rest of the thorium atoms and the sulfur atoms in three sets of

$$(6h) \quad \pm (u\,v\,\,^1/_4;\ \bar{v},u-v,^1/_4;\ v-u,\bar{u},^1/_4)$$

with the parameters:

$$\text{Th}(2): \quad u = 0.153, v = -0.283$$
$$\text{S}(1): \quad u = 0.514, v = 0.375$$
$$\text{S}(2): \quad u = 0.235, v = 0.000$$

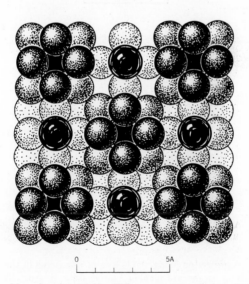

Fig. VF,27b. A packing drawing of the UB_{12} arrangement viewed along a cube axis. The uranium atoms are black.

Such an arrangement would have improbably short Th(1)–S(2) distances. To avoid this it was suggested that the value of u for S(2) is an average one with the result that where $0\ 0\ ^1/_4$ is occupied by thorium, u actually is 0.255 and where $0\ 0\ ^3/_4$ is occupied, u for S(2) is 0.215.

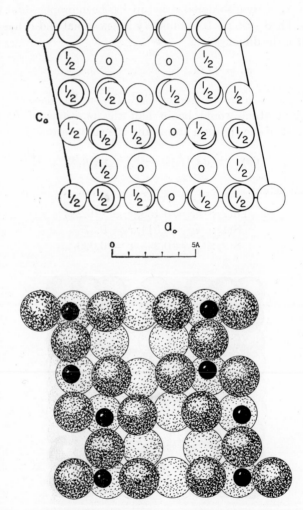

Fig. VF,28a (top). A projection along the b_0 axis of the atoms in the monoclinic unit of V_6O_{13}. The vanadium atoms are smaller and heavily ringed. Origin in lower left.

Fig. VF,28b (bottom). A packing drawing of the monoclinic V_6O_{13} structure viewed along its b_0 axis. The oxygen atoms are dotted.

The corresponding selenium compound, Th_7Se_{12}, has this arrangement, with

$$a_0 = 11.569 \text{ A.}, \qquad c_0 = 4.23 \text{ A.}$$

The chosen atomic parameters are

$$\begin{aligned}
&\text{Th(2):} \quad (6h) \quad u = 0.15, v = -0.28 \\
&\text{Se(1):} \quad (6h) \quad u = 0.51, v = 0.38 \\
&\text{Se(2):} \quad (6h) \quad u = 0.24, v = 0.00
\end{aligned}$$

V,f23. The *vanadium oxide*, V_6O_{13}, is monoclinic, with a bimolecular cell of the dimensions:

$$a_0 = 11.90 \text{ A.}; \quad b_0 = 3.67 \text{ A.}; c_0 = 10.12 \text{ A.}; \quad \beta = 100°52'$$

Atoms have been placed in the following positions of C_{2h}^3 $(C2/m)$:

$$(2b) \quad 0 \; {}^1\!/_2 \, 0; \; {}^1\!/_2 \, 0 \, 0 \qquad (4i) \quad \pm(u0v; \; u+{}^1\!/_2, {}^1\!/_2, v)$$

Chosen atomic positions and parameters are those of Table VF,10.

TABLE VF,10
Positions and Parameters of the Atoms in V_6O_{13}

Atom	Position	x	y	z
V(1)	$(4i)$	0.349	0	0.000
V(2)	$(4i)$	0.409	0	0.369
V(3)	$(4i)$	0.719	0	0.369
O(1)	$(2b)$	0	${}^1\!/_2$	0
O(2)	$(4i)$	0.17	${}^1\!/_2$	0.00
O(3)	$(4i)$	0.38	${}^1\!/_2$	0.18
O(4)	$(4i)$	0.68	${}^1\!/_2$	0.18
O(5)	$(4i)$	0.21	${}^1\!/_2$	0.38
O(6)	$(4i)$	0.89	${}^1\!/_2$	0.38
O(7)	$(4i)$	0.57	${}^1\!/_2$	0.40

In this arrangement (Fig. VF,28), each vanadium atom is tetrahedrally surrounded by oxygen atoms and these tetrahedra are bound together in some cases by sharing corners, in others by sharing edges.

V,f24. Thorium has a complex hydride, Th_4H_{15}, which is cubic in symmetry. Its tetramolecular unit has the edge:

$$a_0 = 9.11 \text{ A.}$$

The thorium atoms are in positions

$$(16c) \quad uuu; \qquad\qquad u+{}^1/_2, {}^1/_2-u, \bar{u};$$
$$\bar{u}, u+{}^1/_2, {}^1/_2-u; \qquad {}^1/_2-u, \bar{u}, u+{}^1/_2;$$
$$u+{}^1/_4, u+{}^1/_4, u+{}^1/_4; \ u+{}^3/_4, {}^1/_4-u, {}^3/_4-u;$$
$${}^3/_4-u, u+{}^3/_4, {}^1/_4-u; \ {}^1/_4-u, {}^3/_4-u, u+{}^3/_4; \ \text{B.C.}$$

of $T_d{}^6$ ($I\bar{4}3d$), with u redetermined as 0.208. It has been suggested, though not demonstrated, that the 60 hydrogen atoms are in positions:

$$(12a) \quad {}^3/_8 \, 0 \, {}^1/_4; {}^1/_8 \, 0 \, {}^3/_4; \text{tr}; \text{B.C.}$$
$$(48e) \quad xyz; \qquad\qquad x+{}^1/_2, {}^1/_2-y, \bar{z};$$
$$\bar{x}, y+{}^1/_2, {}^1/_2-z; \qquad {}^1/_2-x, \bar{y}, z+{}^1/_2;$$
$$x+{}^1/_4, z+{}^1/_4, y+{}^1/_4; \ x+{}^3/_4, {}^1/_4-z, {}^3/_4-y;$$
$${}^3/_4-x, z+{}^3/_4, {}^1/_4-y; \ {}^1/_4-x, {}^3/_4-z, y+{}^3/_4; \ \text{tr}; \ \text{B.C.}$$
$$\text{with } x = 0.400, \ y = 0.230, \ z = 0.372$$

These positions would lead to the following shortest interatomic distances: Th–Th = 3.87–4.10 A., Th–H = 2.29–2.46 A., H–H = 2.02–2.38 A.

V,f25. The *selenide of palladium* which formerly would have been considered to have the composition Pd_9Se_8 has in fact the formula $Pd_{17}Se_{15}$. It is cubic, with a bimolecular cell of the edge length:

$$a_0 = 10.606 \text{ A.}$$

For this crystal, similar structures based on different space groups gave approximately equally good agreement with the experimental data and therefore no final choice between them could be made.

In one of them, based on $O_h{}^1$ ($Pm3m$), atoms would be in the positions:

Pd(1):   (1b) ${}^1/_2 \, {}^1/_2 \, {}^1/_2$
Pd(2):   (3d) ${}^1/_2 \, 0 \, 0; \, 0 \, {}^1/_2 \, 0; \, 0 \, 0 \, {}^1/_2$
Pd(3):   (6e) $\pm (u00; 0u0; 00u)$ with $u = 0.2378$
Pd(4):   (24m) $\pm (uuv; u\bar{u}\bar{v}; \bar{u}u\bar{v}; \bar{u}\bar{u}v)$; tr

                                  with $u = 0.3521, v = 0.1501$

Se(1):   (6f) $\pm (u \, {}^1/_2 \, {}^1/_2; {}^1/_2 \, u \, {}^1/_2; {}^1/_2 \, {}^1/_2 \, u)$ with $u = 0.2571$
Se(2):   (12i) $\pm (0uu; 0u\bar{u})$; tr   with $u = 0.2297$
Se(3):   (12j) $\pm ({}^1/_2 \, u \, u; {}^1/_2 \, u \, \bar{u})$; tr   with $u = 0.1684$

In a second, based on $T_d{}^1$ ($P\bar{4}3m$), the atoms would have the similar distribution:

Pd(1), Pd(2), Pd(3), Se(1) as before
Pd(4):   (12i) $uuv; \bar{u}u\bar{v}; \bar{u}\bar{u}v; u\bar{u}\bar{v}$; tr   with $u = 0.3541, v = 0.1439$

Pd(5): (12i) with $u = -0.3500, v = -0.1565$
Se(2): (12i) with $u = 0.2301, v = -0.0010$
Se(3): (12i) with $u = 0.1680, v = 0.5064$

A third structure based on O^1 ($P432$) is not very different. Refinement of the existing data did not lead to a satisfactory convergence, however, and accordingly it is considered less probable.

Of the other two, that developed from $O_h{}^1$ appears somewhat preferable. With it, the close Pd–Se separations lie between 2.430 and 2.576 A. and the close Pd–Pd range upwards from 2.781 A.

The compound $Rh_{17}S_{15}$, which formerly was described as Rh_9S_8, is isostructural, with

$$a_0 = 9.911 \text{ A.}$$

The atomic positions and parameters in terms of the arrangement based on $O_h{}^1$ ($Pm3m$) are (see above):

Rh in (1b), (3d), (6e), with $u = 0.2388$, (24m) with $u = 0.3564, v = 0.1435$
S in (6f) with $u = 0.2643$, (12i) with $u = 0.2310$, (12j) with $u = 0.1696$

The compound Co_9S_8 (**V,f15**), though it has a very similar cell, does not have this composition and structure.

V,f26. Crystals of the ordered form of *tetraborane*, B_4H_{10} (m.p. $-120°$C.), are monoclinic, with a tetramolecular unit which at $-150°$C. has the dimensions:

$$a_0 = 8.68 \text{ A.}; \quad b_0 = 10.14 \text{ A.}; \quad c_0 = 5.78 \text{ A.}; \quad \beta = 105°54'$$

Atoms are in general positions of $C_{2h}{}^5$ ($P2_1/n$):

$$(4e) \quad \pm(xyz; x+\tfrac{1}{2}, \tfrac{1}{2}-y, z+\tfrac{1}{2})$$

with the parameters, including those of hydrogen, listed in Table VF,11.

The resulting structure is illustrated in Figure VF,29. Its molecules have the general shape indicated in Figure VF,30. Though it is not required by space group considerations, they have (at least approximately) $C_2{}^v$ symmetry. The B(1)–B(3) separation across this plane is 1.750 A.; on one side of the plane, B(1)–B(2) = 1.848 A. and B(3)–B(4) = 1.842 A. Where hydrogen atoms are close to one boron only, B–H = 1.03–1.17 A.; for hydrogen atoms that are close to two borons, B–H = 1.16–1.49 A.

TABLE VF,11
Parameters of the Atoms in Tetraborane

Atom	x	y	z
B(1)	0.277	0.414	0.172
B(2)	0.116	0.296	0.176
B(3)	0.267	0.271	0.012
B(4)	0.244	0.427	0.844
H(1)	0.393	0.433	0.324
H(2)	0.169	0.427	0.243
H(3)	0.142	0.241	0.354
H(4)	0.999	0.323	0.061
H(5)	0.148	0.211	0.001
H(6)	0.366	0.206	0.072
H(7)	0.227	0.296	0.812
H(8)	0.356	0.457	0.798
H(9)	0.130	0.450	0.746
H(10)	0.247	0.500	0.031

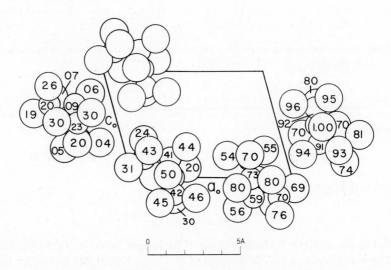

Fig. VF,29. A projection along the b_0 axis of molecules in the monoclinic unit of B_4H_{10}. The smaller circles are the boron atoms.

There has also been observed a second, hexagonal, modification which has two molecules in a cell of the dimensions:

$$a_0 = 5.79 \text{ A.}, \qquad c_0 = 9.36 \text{ A.}$$

The rapid uniform decline of intensities with reflection angle leads to the conclusion that the crystal is a close-packing of disordered molecules with centers in the positions:

$$\tfrac{1}{3}\,\tfrac{2}{3}\,\tfrac{1}{4};\ \tfrac{2}{3}\,\tfrac{1}{3}\,\tfrac{3}{4}$$

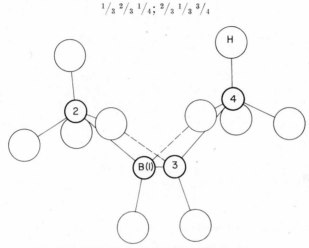

Fig. VF,30. The general shape of the B_4H_{10} molecule.

V,f27. Crystals of *pentaborane*, B_5H_9, which melt at $-47°$C., have at ca. $-115°$C. a bimolecular tetragonal cell of the dimensions:

$$a_0 = 7.16\ \text{A.}, \qquad c_0 = 5.38\ \text{A.}$$

Atoms, including the hydrogens, have been determined to be in the following positions of C_{4v}^9 (*I4mm*):

B(1): (2a) 00u; B.C.   with $u = 0.202$
B(2): (8d) u0v; ū0v; 0uv; 0ūv; B.C.   with $u = 0.175, v = 0.000$
H(1): (2a) with $u = 0.427$
H(2): (8d) with $u = 0.328, v = 0.092$
H(3): (8c) uuv; uūv; ūuv; ūūv; B.C.   with $u = 0.136, v = -0.165$

The structure thus derived is illustrated in Figure VF,31. In the square pyramid of boron atoms that forms the core of the molecule, B(1)–B(2) = 1.66 A. and B(2)–B(2) = 1.77 A. The B(1)–H(1) distance is 1.21 A.; B(2)–H(2) = 1.20 A. and B(2)–H(3) = 1.35 A. The closest intermolecular distance is an H(2)–H(2) = 2.46 A.

V,f28. Crystals of B_5H_{11} (m.p. $-129°$C.) photographed at $-140°$C. are monoclinic, with a tetramolecular unit of the dimensions:

$$a_0 = 6.76\ \text{A.};\quad b_0 = 8.51\ \text{A.};\quad c_0 = 10.14\ \text{A.};\quad \beta = 94°18'$$

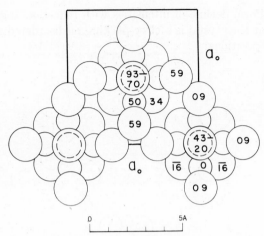

Fig. VF,31. A projection along the c_0 axis of atoms in the tetragonal unit of B_5H_9. The smaller circles are the boron atoms.

All atoms are in general positions of C_{2h}^5 $(P2_1/n)$:

$$(4e) \quad \pm (xyz; x+\tfrac{1}{2}, \tfrac{1}{2}-y, z+\tfrac{1}{2})$$

with the parameters, including those determined for hydrogen, of Table VF,12.

TABLE VF,12
Parameters of the Atoms in B_5H_{11}

Atom	x	y	z
B(1)	0.290	0.107	0.320
B(2)	0.132	0.106	0.175
B(3)	0.024	0.047	0.316
B(4)	0.889	0.165	0.207
B(5)	0.878	0.216	0.375
H(1)	0.400	0.006	0.322
H(2)	0.325	0.218	0.380
H(3)	0.154	0.026	0.089
H(4)	0.974	0.935	0.327
H(5)	0.767	0.111	0.160
H(6)	0.965	0.300	0.428
H(7)	0.722	0.168	0.406
H(8)	0.255	0.173	0.220
H(9)	0.038	0.195	0.147
H(10)	0.841	0.308	0.257
H(11)	0.076	0.099	0.410

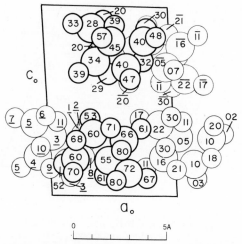

Fig. VF,32. A projection along the b_0 axis of the monoclinic structure of B_5H_{11}. The smaller circles are hydrogen. Atoms of the two molecules farther above the plane of the paper are heavily outlined. The underlined numbers of the lower molecule at lower left correspond to atomic designations of the table and to Figure VF,33. Origin at lower left.

The resulting structure is shown in Figure VF,32. The B_5H_{11} molecule to which it gives rise is indicated in Figure VF,33. Though space group considerations do not require it to have symmetry, it has in effect a plane passing through B(3) and H(4), H(9), and H(11). In this molecule, B–B = either ca. 1.70, 1.77, or 1.87 A. [between B(3) and B(1) or B(5)]. Atoms H(1)–H(7) are joined to one boron with B–H = ca. 1.07 A.; atoms H(8)–H(11) are bound to more than one boron.

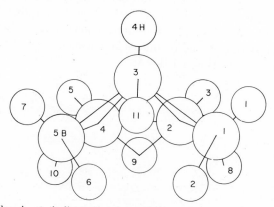

Fig. VF,33. A drawing to indicate the shape of the molecule of B_5H_{11}. The larger circles are boron.

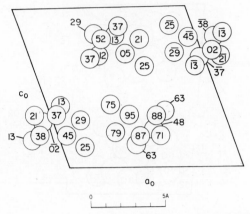

Fig. VF,34. The monoclinic structure of B_9H_{15} projected along its b_0 axis. Only the boron positions are shown.

V,f29. Crystals of the *nonaboron hydride*, B_9H_{15}, are monoclinic, with a tetramolecular cell of the dimensions:

$$a_0 = 11.80 \text{ A.}; \quad b_0 = 6.94 \text{ A.}; \quad c_0 = 11.25 \text{ A.}; \quad \beta = 109°9'$$

All atoms are in general positions of C_{2h}^5 $(P2_1/n)$:

$$(4e) \quad \pm (xyz; \; x+^1/_2, ^1/_2-y, z+^1/_2)$$

with the parameters listed in Table VF,13.

The structure is shown in Figure VF,34. The boron framework of the molecule at the bottom right of this figure is reproduced in Figure VF,35a to indicate how these atoms are interconnected. Some of the hydrogen atoms are attached to a single boron; others serve as bridges between two such atoms (Fig. VF,35b). The B–B separations range between 1.74 and 1.95 A. The B–H single bond distances lie between 0.9 and 1.3 A.; for the bridging hydrogens, B–H = 1.1–1.6 A.

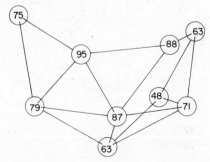

Fig. VF,35a. The boron framework of one of the molecules of B_9H_{15} shown in Figure VF,34.

TABLE VF,13. Parameters of the Atoms in B_9H_{15}

Atom	x	y	z
B(1)	0.370	0.755	0.384
B(2)	0.337	0.790	0.213
B(3)	0.446	0.946	0.319
B(4)	0.469	0.872	0.182
B(5)	0.430	0.633	0.134
B(6)	0.603	0.880	0.308
B(7)	0.586	0.714	0.180
B(8)	0.555	0.481	0.222
B(9)	0.659	0.627	0.337
H(1)	0.693	0.002	0.321
H(2)	0.465	0.991	0.097
H(3)	0.435	0.102	0.338
H(4)	0.752	0.595	0.396
H(5)	0.619	0.752	0.094
H(6)	0.250	0.866	0.145
H(7)	0.558	0.340	0.187
H(8)	0.375	0.584	0.049
H(9)	0.446	0.499	0.229
H(10)	0.568	0.483	0.344
H(11)	0.634	0.750	0.392
H(12)	0.470	0.870	0.407
H(13)	0.283	0.636	0.279
H(14)	0.304	0.859	0.434
H(15)	0.411	0.640	0.416

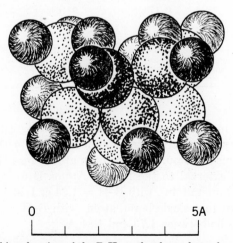

0   5A

Fig. VF,35b. A packing drawing of the B_9H_{15} molecule made to show the distribution of the fine-line shaded hydrogen atoms.

V,f30. Crystals of *decaborane*, $B_{10}H_{14}$, are monoclinic pseudo-orthorhombic, with a unit containing eight molecules. In the published description, the chosen space group C_{2h}^4 ($P112/a$) is so oriented that γ rather than is the monoclinic angle. On this basis the unit cell has the dimensions:

$$a_0 = 14.37 \text{ A.}; \quad b_0 = 20.98 \text{ A.}; \quad c_0 = 5.69 \text{ A.}; \quad \gamma = 90°0'$$

and atoms are given the positions:

$$(8g) \quad \pm (xyz; \; x,y+^1/_2,\bar{z}; \; x+^1/_2,y+^1/_2,z; \; x+^1/_2,y,\bar{z})$$

The parameters are those of Table VF,14.

TABLE VF,14
Parameters of the Atoms in Decaborane

Atom	x	y	z
B(1)	0.034	0.330	0.000
B(2)	0.100	0.273	0.168
B(3)	0.116	0.278	−0.131
B(4)	0.097	0.202	0.000
B(5)	0.018	0.211	0.239
B'(1)–(5)	$^1/_4-x$(B)	y(B)$+^3/_4$	z(B)$+^1/_2$
H(1)	0.041	0.388	0.000
H(2)	0.164	0.276	0.320
H(3)	0.195	0.292	−0.220
H(4)	0.173	0.164	0.000
H(5)	0.031	0.174	0.388
H(6)	0.049	0.319	−0.239
H(7)	0.085	0.213	−0.214
H'(1)–(7)	$^1/_4-x$(H)	y(H)$+^3/_4$	z(H)$+^1/_2$

The resulting molecule is a basket-like framework of boron atoms (Fig. VF,36) which has the general form of an icosahedron with two vertices un occupied by atoms of boron. The hydrogen atoms, resolved in Fourier projections, are all external to the basket. Atoms H(6) and H(7) are each attached to two boron atoms, with B–H = 1.34 or 1.40 A.; all other hydrogens are attached to a single boron atom, with B–H = 1.25–1.29 A. The B–B separations lie between 1.73 and 1.81 A.

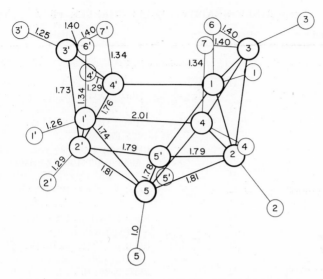

Fig. VF,36a. Dimensions of the molecules of $B_{10}H_{14}$. The heavy circles are the boron atoms. The numbers within circles are the atomic identifications used in Table VF,13.

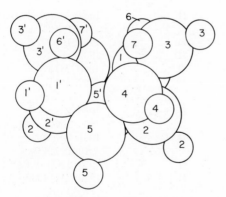

Fig. VF,36b. A packing of the atoms in the molecule of $B_{10}H_{14}$. Here the large circles are the boron atoms.

These crystals all produce diffuse $h+k$ odd reflections considered indicative of a very intimate polysynthetic twinning of individuals with small dimensions in the b_0 direction. This can be interpreted as the equivalent of partial disorder in the molecular arrangement, and has been described in terms of the orthorhombic V_h^{12} (*Pnnm*).

BIBLIOGRAPHY TABLE, CHAPTER VF

Compound	Paragraph	Literature

Compounds R_nX_6

BaB_6	f1	1931: vS; 1932: A; vS&N; 1954: B&B; 1956: S&G; 1961: Z,S,P&S
CaB_6	f1	1931: vS; 1932: A; vS&N; 1934: P&W; 1954: B&B; K; 1956: S&G; 1961: Z,S,P&S
CeB_6	f1	1931: vS; 1932: A; vS&N; 1954: B&B; 1956: S&G; 1961: Z,S,P&S
DyB_6	f1	1957: N&S; 1958: N&S; 1959: E&G
ErB_6	f1	1932: A; vS&N; 1958: N&S
EuB_6	f1	1958: F,B&P; S,D&S; 1959: S,D&S; 1961: Z,S,P&S
GdB_6	f1	1932: A; 1954: B&B; 1958: N&S; 1959: E&G; 1961: Z,S,P&S
HoB_6	f1	1957: N&S; 1958: N&S; 1959: E&G
LaB_6	f1	1931: vS; 1932: A; vS&N; 1954: B&B; 1956: S&G; 1961: Z,S,P&S
LuB_6	f1	1957: N&S; 1958: N&S
NdB_6	f1	1932: A; vS&N; 1954: B&B; 1959: E&G; 1961: Z,S,P&S
$OsOF_5$	f4	1962: B,J&T
PrB_6	f1	1932: vS&N
PuB_6	f1	1960: MD&S
$Ru(CO)_4I_2$	f5	1962: D&W
SiB_6	f1	1956: Z
SmB_6	f1	1959: E&G; 1961: Z,S,P&S
SrB_6	f1	1931: vS; 1932: A; vS&N; 1954: B&B; 1961: Z,S,P&S
TbB_6	f1	1959: E&G; S,P&S
ThB_6	f1	1929: A; 1932: A; 1954: B&B; 1961: Z,S,P&S
TmB_6	f1	1961: P&S
UCl_6	f3	1948: Z
UF_6	f4	1958: H&S
WCl_6	f2	1943: K&O
YB_6	f1	1932: A
YbB_6	f1	1932: A; 1954: B&B; 1958: S&Z; 1959: E&G; 1961: Z,S,P&S

(continued)

BIBLIOGRAPHY TABLE, CHAPTER VF *(continued)*

Compound	Paragraph	Literature
$(NH)_2S_6$	f9	1960: W
Cr_5S_6	f6	1955: J; 1957: J
$Cr_{23}C_6$	f7	1933: W
$Fe_{21}Mo_2C_6$	f7	1933: W
$Fe_{21}W_2C_6$	f7	1933: W
$Mn_{23}C_6$	f7	1933: W

Compounds R_nX_7

IF_7	f8	1957: B&B; 1959: B; D; 1962: B; L&L
NHS_7	f9	1960: W
$(NH)_4S_4$	f9	1957: L&S; 1958: S&D
Cr_7C_3	f10	1926: W&P; 1931: W
Mn_7C_3	f10	1935: W
Re_7B_3	f10	1960: A,S&A
Rh_7B_3	f10	1960: A,S&A
Ru_7B_3	f10	1959: A; A,A&S
U_3O_7	f12	1960: M,L,D&M
P_4S_7	f11	1954: V&W; 1955: V&W; 1956: V&W

Compounds R_nX_m $(m \geqq 8)$

U_3O_8	f12, f13	1948: G; 1951: M; 1952: H&P; 1955: S; 1958: A
$(U,Ta_2)O_8$	f12	1957: G
Cr_7S_8	f6	1957: J
Cr_7Se_8	f14	1961: C&B
Co_9S_8	f15	1932: C&R; 1936: H&W; L,L&W; 1959: K,H&V; 1961: D,A&B; K&I; 1962: G; P,B&C
Co_9Se_8	f15	1955: B,G,H&P
$(Ni,Fe)_9S_8$ (pentlandite)	f15	1936: L,L&W; 1947: L; 1955: E; 1956: P&B; 1961: K&I
$Ru_{11}B_8$	f16	1960: A
$Fe_2(CO)_9$	f18	1926: B; 1939: P&E
U_2F_9	f17	1948: Z; 1949: Z
U_4O_9	f12	1961: B,P&P
B_5H_9	f27	1951: D&L; 1952: D&L
Ti_5O_9	f19	1960: A
B_4H_{10}	f26	1953: N&L; 1957: M,D&L

(continued)

BIBLIOGRAPHY TABLE, CHAPTER VF (*continued*)

Compound	Paragraph	Literature
P_4S_{10}	f20	1954: V&W; 1955: V&W
Mo_4O_{11}	b25	1948: M; 1959: M,A,K,A,W,H&N
W_4O_{11}	b25	1934: E&F
B_5H_{11}	f28	1953: L&L; 1954: L&L; 1957: M,D&L
BeB_{12}	[II,f1]	1960: B
DyB_{12}	f21	1960: B,LP&P
ErB_{12}	f21	1960: B,LP&P
HoB_{12}	f21	1960: B,LP&P
LuB_{12}	f21	1960: B,LP&P
TmB_{12}	f21	1960: B,LP&P
UB_{12}	f21	1949: B&B; 1954: B&B
YB_{12}	f21	1960: B,LP&P
ZrB_{12}	f21	1953: G&P
Th_7S_{12}	f22	1949: Z
Th_7Se_{12}	f22	1952: DE,S&M; 1953: DE
As_2B_{13}	c11	1961: LP&P
P_2B_{13}	c11	1961: M
V_6O_{13}	f23	1948: A
$B_{10}H_{14}$	f30	1931: M; 1948: K,L&H; 1950: K,L&H; 1957: M,D&L
Th_4H_{15}	f24	1953: Z
B_9H_{15}	f29	1956: D,W,H,L&S; 1957: D,W,H&L; 1961: S&L
$Pd_{17}Se_{15}$	f25	1962: G
$Rh_{17}S_{15}$	f25	1962: G
Mo_8O_{23}	b25	1948: M; 1953: B,K&M
Mo_9O_{26}	b25	1944: H&M; 1948: M; 1953: B,K&M
$(Mo,W)_{10}O_{29}$	b25	1953: B,K&M
$W_{10}O_{29}$	b25	1950: M; 1951: M; 1953: B,K&M; M
$(Mo,W)_{11}O_{32}$	b25	1953: B,K&M
$Mo_{17}O_{47}$	b25	1959: M,A,K,A,W,H&N; 1960: K
$W_{18}O_{49}$	b25	1944: H&M; 1949: M
$W_{20}O_{58}$	b25	1950: M

BIBLIOGRAPHY, CHAPTER V

1915

Bragg, W. H., "The Structure of the Spinel Group of Crystals," *Nature*, **95**, 561; *Phil. Mag.*, **30**, 305.

Nishikawa, S., "The Structure of Some Crystals of the Spinel Group," *Proc. Math. Phys. Soc. Tokyo*, **8**, 199.

1922

Backhurst, I., "Variations of the Intensity of Reflected X-radiation with the Temperature of the Crystal," *Proc. Roy. Soc. (London)*, **102A**, 340.

Hedvall, J. A., "Changes in Properties Produced by Different Methods of Preparation of Some Ignited Oxides, as Studied by X-ray Interference," *Arkiv Kemi Mineral. Geol.*, **8**, No. 11; *Z. Anorg. Allgem. Chem.*, **120**, 327.

Westgren, A., and Phragmén, G., "X-ray Studies on the Crystal Structure of Steel," *J. Iron Steel Inst.*, **105**, 241; *Engineering*, **113**, 630.

Wever, F., "Iron Carbide," *Mitt. Kaiser Wilhelm-Inst. Eisenforsch.*, **4**, 67.

1923

Bozorth, R. M., "The Crystal Structures of the Cubic Forms of Arsenious and Antimonous Oxides," *J. Am. Chem. Soc.*, **45**, 1621.

Davey, W. P., "Precision Measurements of the Crystal Structures of Al_2O_3, Fe_2O_3, and Cr_2O_3," *Phys. Rev.*, **21**, 716.

Dickinson, R. G., "The Crystal Structure of Tin Tetraiodide," *J. Am. Chem. Soc.*, **45**, 958.

Goldschmidt, V. M., and Thomassen, L., "The Crystal Structure of Natural and Synthetic Oxides of Uranium, Thorium and Cerium," *Skrifter Norske Videnskaps-Akad. Oslo I. Mat.-Naturv. Klasse*, **1923**, No. 2.

Mark, H., and Weissenberg, K., "X-Ray Determination of the Structure of Urea and of Tin Tetraiodide," *Z. Physik*, **16**, 1.

1924

Mark, H., "The Application of X-Ray Crystal Analysis to the Problem of the Structure of Organic Compounds," *Ber.*, **57B**, 1820.

Mauguin, C., "The Crystal Structure of Corundum and Hematite," *Compt. Rend.*, **178**, 785.

Westgren, A., and Phragmén, G., "X-Ray Studies on the Crystal Structure of Steel," *J. Iron Steel Inst.*, **109**, 159; *Nature*, **114**, 94.

1925

Alsén, N., "X-Ray Investigation of the Crystal Structures of Pyrrhotite, Breithauptite, Pentlandite, Millerite and Related Compounds," *Geol. För. Förh.*, **47**, 19.

Bragg, W. H., and Bragg, W. L., *X-Rays and Crystal Structure*, 5th ed., G. Bell & Sons, Ltd., London.

Goldschmidt, V. M., Barth, T., and Lunde, G., "Isomorphy and Polymorphy of the Sesquioxides. The Lanthanide Contraction and Its Consequences," *Skrifter Norske Videnskaps-Akad. Oslo 1. Mat.-Naturv. Klasse*, 1925, No. 7.

Goldschmidt, V. M., Ulrich, F., and Barth, T., "The Crystal Structure of the Oxides of the Rare Earth Metals," *Skrifter Norske Videnskaps-Akad. Oslo I. Mat.-Naturv. Klasse*, 1925, No. 5.

Jaeger, F. M., Terpstra, P., and Westenbrink, H. G. K., "The Crystal Structure of Germanium Tetraiodide," *Verslag. Akad. Wetens. Amsterdam*, **34**, 721; *Proc. Acad. Sci. Amsterdam*, **28**, 747.

Mark, H., Basche, W., and Pohland, E., "Determination of the Structure of Some Simple Inorganic Substances," *Z. Elektrochem.*, **31**, 523.

Mark, H., and Pohland, E., "The Structure of Ammonia," *Z. Krist.*, **61**, 532.

Mark, H., and Pohland, E., "The Structures of Ethane and Diborane," *Z. Krist.*, **62**, 103.

Oftedal, I., "The Crystal Structure of Skutterudite and Related Minerals," *Norsk Geol. Tidsskr.*, **8**, 250.

Pauling, L., and Hendricks, S. B., "The Crystal Structures of Hematite and Corundum," *J. Am. Chem. Soc.*, **47**, 781.

Smedt, J. de, "The Structure of Solid Ammonia as Determined by X-Rays," *Bull. Classe Sci. Acad. Roy. Belg.*, **11**, 655.

Ulrich, F., "Note on the Crystal Structure of the Corundum-Hematite Group," *Norsk Geol. Tidsskr.*, **8**, 115.

Welo, A., and Baudisch, O., "The Two-stage Transformation of Magnetite and Hematite," *Phil. Mag.*, **50**, 399.

Wyckoff, R. W. G., and Crittenden, E. D., "An X-Ray Examination of Some Ammonia Catalysts," *J. Am. Chem. Soc.*, **47**, 2866.

1926

Aminoff, G., "The Crystal Structure of Hausmannite (MnMn$_2$O$_4$)," *Z. Krist.*, **64**, 475.

Brill, R., "X-ray Investigation of the Iron Carbonyl Fe$_2$(CO)$_9$," *Z. Krist.*, **65**, 85.

Brill, R., "The Crystal Structure of Lithium Nitride, Li$_3$N," *Z. Krist.*, **65**, 94.

Claassen, A., "The Scattering Power of Oxygen and Iron for X-Rays," *Proc. Phys. Soc. London*, **38**, 482.

Dickinson, R. G., "The Symmetry and the Unit Cell of Stannic Iodide," *Z. Krist.*, **64**, 400.

Frankenberger, W., "The Chemical Union of Nitrogen and Lithium and the Mechanism of This Reaction," *Z. Elektrochem.*, **32**, 481.

Kirkpatrick, L. M., and Pauling, L., "The Crystal Structure of Cubic Telluric Acid," *Z. Krist.*, **63**, 502.

Menzer, G., "The Crystal Structure of Linnaeite, Polydymite, and Sychnodymite," *Z. Krist.*, **64**, 506; *Mineralog. Petrog. Mitt.*, **37**, 247; *Fortschr. Min. Kryst. Petrog.*, **11**, 315 (1927).

Natta, G., and Schmid, F., "The Crystalline Structure of Co$_3$O$_4$," *Rend. Accad. Lincei*, **4**, 145.

Ooe, Z., "The Crystal Structure of Stibnite," *J. Geol. Soc. Tokyo*, **33**, 187; *Neues Jahrb. Mineral.*, 1928, I, 61.

Ott, H., "The Structures of MnO, MnS, AgF, NiS, SnI$_4$, SrCl$_2$, BaF$_2$; Precision Measurements upon Various Alkali Halides," *Z. Krist.*, **63**, 222.

Westgren, A., and Phragmén, G., "X-Ray Analysis of Chromium–Carbon Systems," *Kgl. Svenska Vetenskapsakad. Handl.*, **2**, No. 5.

Zachariasen, W. H., "The Crystal Structure of the A-Modifications of the Sesquioxides of the Rare Earths (La$_2$O$_3$, Ce$_2$O$_3$, Pr$_2$O$_3$, Nd$_2$O$_3$)," *Z. Physik. Chem.*, **123**, 134.

1927

Dehlinger, U., "The Crystal Structure of the Antimony Oxides," *Z. Krist.*, **66**, 108.

Dehlinger, U., and Glocker, R., "The Atomic Structure of the Antimony Oxides," *Z. Anorg. Allgem. Chem.*, **165**, 41.

Gottfried, C., "The Structure of Stibnite," *Z. Krist.*, **65**, 428.

Harrington, E. A., "X-Ray Diffraction Measurements on Some of the Pure Compounds Concerned in the Study of Portland Cement," *Am. J. Sci.*, **13**, 467.

Jong, W. F. de, and Willems, H. W. V., "The Compounds Fe_3S_4, Co_3S_4, Ni_3S_4 and Their Structure," *Z. Anorg. Allgem. Chem.*, **161**, 311.

Lunde, G., "The Existence and Preparation of Certain Oxides of the Platinum Metals (with a Supplement Regarding Amorphous Oxides)," *Z. Anorg. Allgem. Chem.*, **163**, 345.

Lunde, G., "Titanium Sesquioxide," *Z. Anorg. Allgem. Chem.*, **164**, 341.

Natta, G., "On the Crystal Structure of the Chlorides of Trivalent Metals. I. Chromic Chloride," *Rend. Accad. Lincei*, **5**, 592.

Simon, A., "The Density, Crystal Structure and Nature of Oxygen Linkage in Oxides of Antimony," *Z. Anorg. Allgem. Chem.*, **165**, 31.

Zachariasen, W. H., "The Crystal Structure of the Modification C of the Sesquioxides of the Rare Earth Metals and of Indium and Thallium," *Norsk Geol. Tidssk.*, **9**, 310.

1928

Brill, R., "The Structure of Iron Nitride, Fe_4N," *Naturwiss.*, **16**, 593; *Z. Krist.*, **68**, 379.

Brill, R., and Mark, H., "X-Ray Studies on the Structure of Complex Iron Cyanides," *Z. Physik. Chem.*, **133**, 443.

Groebler, H., "X-Ray Studies of the Structure of the Oxides of Iron," *Z. Physik*, **48**, 567.

Hägg, G., "X-Ray Studies of the Binary Systems of Iron with Phosphorus, Arsenic, Antimony and Bismuth," *Z. Krist.*, **68**, 470.

Hägg, G., "X-Ray Studies on the 'Nitrides' of Iron," *Nature*, **121**, 826; **122**, 314, 962.

Hendricks, S. B., and Albrecht, W. H., "X-Ray and Chemical Investigations of Various Oxides of Iron and Cobalt," *Ber.*, **61B**, 2153.

Natta, G., and Passerini, L., "Arsenides of Magnesium and Zinc," *Gazz. Chim. Ital.*, **58**, 541.

Natta, G., and Strada, M., "Spinels of Trivalent Cobalt: Cobaltous and Zinc Cobaltite," *Rend. Accad. Lincei*, **7**, 1024.

Natta, G., and Strada, M., "Oxides and Hydroxides of Cobalt," *Gazz. Chim. Ital.*, **58**, 419.

Oftedal, I., "The Crystal Structure of Skutterudite and Speiscobaltchloanthite," *Z. Krist.*, **66**, 517.

Ôsawa, A., and Iwaizumi, S., "X-Ray Investigation of Iron–Nitrogen Alloys," *Z. Krist.*, **69**, 26; *Sci. Repts. Tohoku Imp. Univ.*, **18**, 79 (1929).

Passerini, L., "Crystalline Structure of Some Phosphides of Bivalent and Trivalent Metals," *Gazz. Chim. Ital.*, **58**, 655.

Passerini, L., "X-Ray Analyses of Cadmium Arsenide and of Arsenious Anhydride," *Gazz. Chim. Ital.*, **58**, 775.

Zachariasen, W. H., "The Crystal Structure of the Sesquioxides and Compounds of the Type ABO_3," *Skrifter Norske Videnskaps-Akad. Oslo I. Mat.-Naturv. Klasse*, **1928**, No. 4.

Zachariasen, W. H., "The Crystal Structure of Bixbyite and Artificial Mn_2O_3," *Z. Krist.*, **67**, 455.

1929

Allard, G., "On the Crystalline Structure of Thorium Boride," *Compt. Rend.*, **189**, 108.

Dehlinger, U., "Addition of Gas Atoms to Crystal Lattices," *Z. Physik. Chem.*, **6B**, 127.

Ferrari, A., and Scherillo, A., "The Crystal Structure of Aluminum Fluoride," *Gazz. Chim. Ital.*, **59**, 927.

Gottfried, C., and Lubberger, E., "The Space Group of Stibnite Sb_2S_3," *Z. Krist.*, **71**, 257.

Hägg, G., "X-Ray Investigation of the Nitrides of Manganese," *Z. Physik. Chem.*, **4B**, 346.

Hägg, G., "X-Ray Studies on the Binary Systems of Iron with Nitrogen, Phosphorus, Arsenic, Antimony and Bismuth," *Nova Acta Reg. Soc. Sci. Upsaliensis IV*, **7**, No. 1.

Halla, F., "Reaction of Titanium Sesquioxide with the Oxides of Iron," *Z. Anorg. Allgem. Chem.*, **184**, 421.

Hassel, O., and Nilssen, S., "The Crystal Structure of BiF_3," *Z. Anorg. Allgem. Chem.*, **181**, 172.

Holgersson, S., "X-Ray Examination of the Minerals of the Spinel Group and of Synthesized Substances of the Spinel Type," *Lunds Univ. Arsskr.*, *II*, **23**, 9; *Kgl. Fysiograf. Sällskap. Handl.*, **38**, 1.

Holgersson, S., and Karlsson, A., "A Few New Cobaltites of the Spinel Type," *Z. Anorg. Allgem. Chem.*, **183**, 384.

Natta, G., and Passerini, L., "Spinels of the Type $Me_2^{++}Me^{++++}O_4$," *Rend. Accad. Lincei*, **9**, 557.

Natta, G., and Passerini, L., "Spinels of Bivalent Cobalt, Cobaltous Aluminate, Chromite, Ferrite, and Cobaltite," *Gazz. Chim. Ital.*, **59**, 280.

Natta, G., and Passerini, L., "The Constitution of Rinman Green, Thénard Blue, and Other Colored Solid Derivatives of Cobalt Oxide," *Gazz. Chim. Ital.*, **59**, 620.

Oftedal, I., "The Crystal Structure of Tysonite and of Some Artificial Rare Earth Fluorides," *Z. Physik. Chem.*, **5B**, 272.

Pauling, L., "The Crystal Structure of the A-Modification of the Rare Earth Sesquioxides," *Z. Krist.*, **69**, 415; *cf.* Zachariasen, *ibid.*, **70**, 187.

Ramsdell, L. S., "An X-Ray Study of the Domeykite Group," *Am. Mineralogist*, **14**, 188.

1930

Bräkken, H., "The Crystal Structure of the Triiodides of Arsenic, Antimony and Bismuth," *Z. Krist.*, **74**, 67.

Hägg. G., "The Crystal Structure of Magnesium Nitride Mg_3N_2," *Z. Krist.*, **74**, 95.

Hägg, G., "X-Ray Investigation of Iron Nitrides," *Z. Physik. Chem.*, **8B**, 455.

Hansen, H., "The Coordination Properties of Halides Near the Point of Volatilization with Determinations of Structure for the Halides AX_4," *Z. Physik. Chem.*, **8B**, 1.

Hendricks, S. B., "The Crystal Structure of Cementite," *Z. Krist.*, **74**, 534.

Heyworth, D., "Note on the Space Group of AsI_3," *Z. Krist.*, **75**, 574; see also Bräkken, H., *ibid.*, **75**, 574.

Lashkarev, V. E., "The Structure of Aluminum Chloride," *Z. Anorg. Allgem. Chem.*, **193**, 270.

Machatschki, F., "Cu_3As, Domeykite," *Zentral. Mineral. Geol.*, **1930A**, 19.

Natta, G., "Structure of Silicon Tetrafluoride," *Gazz. Chim. Ital.*, **60**, 911.

Natta, G., and Casazza, E., "The Structure of Hydrogen Phosphide and of Hydrogen Arsenide," *Gazz. Chim. Ital.*, **60**, 851.

Passerini, L., "Solid Solutions, Isomorphism and Symmorphism of the Oxides of Trivalent Metals. The Systems: Al_2O_3–Cr_2O_3; Al_2O_3–Fe_2O_3; Cr_2O_3–Fe_2O_3," *Gazz. Chim. Ital.*, 60, 544.

Pauling, L., and Shappell, M. D., "The Crystal Structure of Bixbyite and the C-Modification of the Sesquioxides," *Z. Krist.*, 75, 128.

Shimura, S., "A Study on the Structure of Cementite," *Proc. Imp. Acad. (Tokyo)*, 6, 269.

Wooster, N., "The Structure of Chromium Trichloride $CrCl_3$," *Z. Krist.*, 74, 363.

Wretblad, P. E., "X-Ray Investigation of the Systems Fe_2O_3–Cr_2O_3 and Fe_2O_3–Mn_2O_3," *Z. Anorg. Allgem. Chem.*, 189, 329.

1931

Biltz, W., Lehrer, G. A., and Meisel, K., "Rhenium Trioxide I, II," *Nachr. Ges. Wiss. Göttingen, Math.-Phys. Kl.*, 1931, 191; *Z. Anorg. Allgem. Chem.*, 207, 113 (1932).

Bräkken, H., "The Crystal Structures of the Trioxides of Chromium, Molybdenum and Tungsten," *Z. Krist.*, 78, 484.

Brill, R., "X-Ray Investigation of Iron Tetracarbonyl," *Z. Krist.*, 77, 36.

Clark, G. L., Ally, A., and Badger, A. E., "The Lattice Dimensions of Spinels," *Am. J. Sci.*, 22, 539.

Ebert, F., "The Crystal Structure of Some Fluorides of the Eighth Group of the Periodic System," *Z. Anorg. Allgem. Chem.*, 196, 395.

Forestier, H., and Galand, M., "Beryllium Ferrite and Ferric Oxide from Its Decomposition," *Compt. Rend.*, 193, 733.

Hassel, O., and Kringstad, H., "The Crystal Structure of the Tetrahalogenides of Light Elements. Determination of the Structure of Silicon Tetraiodide," *Z. Physik. Chem.*, 13B, 1.

Hassel, O., and Kringstad, H., "Crystalline Carbon Tetraiodide," *Tek. Ukeblad*, 78, 230.

Heyworth, D., "Crystal Structure of Arsenic Triiodide," *Phys. Rev.*, 38, 351, 1792.

Ketelaar, J. A. A., "Structure of the Trifluorides of Aluminum, Iron, Cobalt, Rhodium and Palladium," *Nature*, 128, 303.

Möller, H., "The Crystal Structure of $B_{10}H_{14}$," *Z. Krist.*, 76, 500.

Oftedal, I., "The Crystal Structure of Tysonite (Ce, La, ...)F_3," *Z. Physik. Chem.*, 13B, 190.

Ôsawa, A., and Takeda, S., "An X-Ray Investigation of Alloys of the Iron–Tungsten System and Their Carbides," *Kinzoku-no-Kenkyu*, 8, 181; *J. Inst. Metals*, 47, 534.

Stackelberg, M. v., "The Crystal Structures of Several Carbides and Borides," *Z. Elektrochem.*, 37, 542.

Thewlis, J., "The Structure of Ferromagnetic Ferric Oxide," *Phil. Mag.*, 12, 1089.

Westgren, A., "High-Chromium Steels," *Metal Progr.*, 20, No. 5, 57.

Wooster, N., "Crystal Structure of Molybdenum Trioxide," *Nature*, 127, 93; *Z. Krist.*, 80, 504.

Wooster, W. A., and Wooster, N., "Crystal Structure of Chromium Trioxide," *Nature*, 127, 782.

1932

Allard, G., "X-Ray Study of Some Borides," *Bull. Soc. Chim. France*, 51, 1213.

Barlett, H. B., "Occurrence and Properties of Crystalline Alumina in Silicate Melts," *J. Am. Ceram. Soc.*, 15, 361.

Bräkken, H., "The Crystal Structure of Chromium Tribromide," *Kgl. Norske Videnskab. Selskab Forh.*, 5, No. 11.

Brill, R., "The Lattice Constants of α-Fe_2O_3 and γ-Al_2O_3," *Z. Krist.*, 83, 323.

Caglioti, V., and Roberti, G., "X-Ray Investigation of a Subsulfide of Cobalt Employed as a Catalyst in the Hydrogenation of Phenol," *Gazz. Chim. Ital.*, **62, 19.**

Hägg, G., "The Density and Crystal Structure of Magnesium Nitride, Mg_3N_2," *Z. Krist.*, **82, 471.**

Hassel, O., and Kringstad, H., "Crystal Structure of the Tetrahalides of the Lighter Elements II," *Z. Physik. Chem.*, **15B, 274.**

Heide, F., Herschkowitsch, E., and Preuss, E., "X-Ray Investigation of Schreibersite, $(Fe,Ni,Co)_3P$," *Chem. Erde*, **7, 483.**

Lihl, F., "Precision Measurements of the Lattice Constant of Arsenic Trioxide," *Z. Krist.*, **81, 142.**

Meisel, K., "Rhenium Trioxide. III. The Crystal Structure of Rhenium Trioxide," *Z. Anorg. Allgem. Chem.*, **207, 121.**

Okamura, T., "On the Transformation of Magnetite at Low Temperature," *Sci. Rept. Tôhoku Imp. Univ.*, **21, 231.**

Schleede, A., and Wellmann, M., "Structure of Products of Action of Alkali Metals on Graphite," *Z. Physik. Chem.*, **18B, 1.**

Stackelberg, M. v., and Neumann, F., "Crystal Structure of Borides of Composition MeB_6," *Z. Physik. Chem.*, **19B, 314.**

Westgren, A., "Crystal Structure of Cementite," *Jernkontorets Ann.*, **116, 457.**

Wooster, N., "The Crystal Structure of Ferric Chloride," *Z. Krist.*, **83, 35.**

1933

Biltz, W., "Rhenium Trioxide and Rhenium Dioxide," *Z. Anorg. Allgem. Chem.*, **214, 225.**

Bräkken, H., "Crystal Structure of Phosphorus Triiodide," *Kgl. Norske Videnskab. Selskab Forh.*, **5, No. 52, 202.**

Franck, H. H., Bredig, M. A., and Hoffmann, G., "The Crystal Structure of Calcium–Nitrogen Compounds," *Naturwiss.*, **21, 330.**

Halla, F., Nowotny, H., and Tompa, H., "X-Ray Investigations in the System (Zinc, Cadmium)–Antimony II," *Z. Anorg. Allgem. Chem.*, **214, 196.**

Hofmann, W., "The Structure of the Minerals of the Stibnite Group," *Z. Krist.*, **86, 225.**

Katzoff, S., and Ott, E., "On the Lattice Constants of Ferric Oxide," *Z. Krist.*, **86, 311.**

Ketelaar, J. A. A., "The Crystal Structure of the Aluminum Halides. I. The Structure of Aluminum Trifluoride," *Z. Krist.*, **85, 119.**

Stackelberg, M. v., and Paulus, R., "Research on the Crystal Structure of Nitrides and Phosphides of Bivalent Metals," *Z. Physik. Chem.*, **22B, 305.**

Westgren, A., "Crystal Structure and Composition of Cubic Chromium Carbide," *Jernkontorets Ann.*, **117, 501.**

Wooster, N., "Note on the Structure of the Trifluorides of the Transition Metals," *Z. Krist.*, **84, 320.**

Zintl, E., and Husemann, E., "Metals and Alloys. XII. Binding Type and Lattice Structure of Magnesium Compounds," *Z. Physik. Chem.*, **21B, 138.**

1934

Barth, T. F. W., and Posnjak, E., "Notes on Some Structures of the Ilmenite Type," *Z. Krist.*, **88, 271.**

Bräkken, H., "The Crystal Structure of Ag_2O_3," *Norske Videnskab. Selskab. Forh.*, **7, 143.**

Brill, R., "The Lattice Constants of α-Fe$_2$O$_3$," *Z. Krist.*, **88**, 177.
Ebert, F., and Flasch, H., "X-Ray Determination of New Forms of Combination. I. The Tungsten Oxides W$_4$O$_{11}$ and W$_8$O$_{23}$," *Z. Anorg. Allgem. Chem.*, **217**, 95.
Ellefson, B. S., and Taylor, N. W., "Crystal Structures and Expansion Anomalies of MnO, MnS, FeO, Fe$_3$O$_4$ between 100°K and 200°K," *J. Chem. Phys.*, **2**, 58.
Gossner, B., and Kraus, O., "The Crystal Lattice of Telluric Acid," *Z. Krist.*, **88**, 298.
Harker, D., "The Crystal Structure of the Mineral Tetradymite," *Z. Krist.*, **89**, 175.
Hartmann, H., and Fröhlich, H. J., "The Nitriding of Calcium," *Z. Anorg. Allgem. Chem.*, **218**, 190.
Krause, O., and Theil, W., "On Ceramic Coloring Matters," *Ber. Keram. Ges.*, **15**, 101.
Megaw, H. D., "The Crystal Structure of Hydrargillite," *Z. Krist.*, **87**, 185.
Pauling, L., and Weinbaum, S., "The Structure of Calcium Boride," *Z. Krist.*, **87**, 181.
Ridgway, R. R., "On the Structure of Boron Carbide," *Trans. Am. Electrochem. Soc.*, **66**, 293.
Schulze, G. E. R., "Crystal Form and Space Lattice of ZrF$_4$ and HfF$_4$," *Z. Krist.*, **89**, 477.
Stackelberg, M. v., and Schnorrenberg, E., "The Structure of Aluminum Carbide," *Z. Physik. Chem.*, **27B**, 37.

1935

Cole, S. S., and Taylor, N. W., "Crystalline B$_2$O$_3$," *J. Am. Ceram. Soc.*, **18**, 55.
Hägg, G., "The Crystal Structure of Magnetic Ferric Oxide," *Z. Physik. Chem.*, **29B**, 95.
Ketelaar, J. A. A., "The Crystal Structure of AlCl$_3$," *Z. Krist.*, **90A**, 237.
Kordes, E., "Crystal Chemical Investigations of Aluminum Compounds with a Spinel-like Structure and of gamma Fe$_2$O$_3$," *Z. Krist.*, **91A**, 193.
Löhberg, K., "Mixed Crystals of Mg$_3$Sb$_2$ and Zn$_3$Sb$_2$," *Z. Physik. Chem.*, **27B**, 381.
Löhberg, K., "The C-form of Nd$_2$O$_3$," *Z. Physik. Chem.*, **28B**, 402.
Passerini, L., and Rollier, M. A., "The Constitution of Telluric Acid," *Rend. Accad. Lincei*, **21**, 364.
Pauling, L., "The Unit Structure of Telluric Acid, Te(OH)$_6$," *Z. Krist.*, **91A**, 367.
Rüdorff, W., and Hofmann, U., "Crystal Structure of Chromium, Molybdenum and Tungsten Hexacarbonyl," *Z. Physik. Chem.*, **28B**, 351.
Stackelberg, M. v., and Paulus, R., "Investigations on the Phosphides and Arsenides of Zinc and Cadmium. The Zn$_3$P$_2$ Lattice," *Z. Physik. Chem.*, **28B**, 427.
Stackelberg, M. v., Schnorrenberg, E., Paulus, R., and Spiess, K. F., "The Structures of Al$_4$C$_3$ and Al$_5$C$_3$N," *Z. Physik. Chem.*, **175A**, 127, 140.
Verwey, E. J. W., "The Structure of Electrolytic Oxide Layer on Aluminum," *Z. Krist.*, **91A**, 317.
Verwey, E. J. W., and Bruggen, M. G. van, "The Structure of Solid Solutions of Fe$_2$O$_3$ in Mn$_3$O$_4$," *Z. Krist.*, **92A**, 136.
Westgren, A., "The Structure and Composition of the Trigonal Chromium and Manganese Carbides," *Jernkontorets Ann.*, **1935**, 231.
Zintl, E., and Brauer, G., "On the Constitution of Li$_3$N," *Z. Elektrochem.*, **41**, 102.
Zintl, E., and Brauer, G., "The Constitution of Lithium–Bismuth Alloys," *Z. Elektrochem.*, **41**, 297.

1936

Buerger, M. J., "On the Structure of Valentinite, Sb_2O_3," *Am. Mineralogist*, **21**, 206.

Caglioti, V., and D'Agostino, O., "On the Structure of Aerogels. I. The Structure of Metallic Oxides," *Gazz. Chim. Ital.*, **66**, 543.

Hartmann, H., and Orban, F., "A New Tungsten Phosphide W_4P," *Z. Anorg. Allgem. Chem.*, **226**, 257.

Hülsemann, O., and Weibke, F., "On the Lower Sulfides of Co. The Composition-diagram of the System Co–CoS," *Z. Anorg. Allgem. Chem.*, **227**, 113.

Ketelaar, J. A. A., "The Crystal Structure of V_2O_5," *Z. Krist.*, **95A**, 9; *Chem. Weekbl.*, **36**, 51; *Nature*, **137**, 316.

Lundqvist, M., Lundqvist, D., and Westgren, A., "The Crystal Structure of Co_9S_8 and of Pentlandite, $(Ni,Fe)_9S_8$," *Svensk. Kem. Tidsk.*, **48**, 156.

Miller, W. S., and King, A. J., "The Structure of BaS_3," *Z. Krist.*, **94A**, 439.

Naráy-Szabó, S. v., "X-Ray Investigation of AlB_{12}," *Z. Krist.*, **94A**, 367; *Naturwiss.*, **24**, 77.

Verwey, E. J. W., and Boer, J. H. de, "Cation Arrangement in a Few Oxides with Crystal Structures of the Spinel Type," *Rec. Trav. Chim.*, **55**, 531.

1937

Årstad, O., and Nowotny, H., "X-Ray Investigations in the System Mn–P," *Z. Physik. Chem.*, **38B**, 356.

Brauer, G., and Zintl, E., "The Constitution of the Phosphides, Arsenides, Antimonides, and Bismuthides of Lithium, Sodium, and Potassium," *Z. Physik. Chem.*, **37B**, 323.

Clark, G. L., Schieltz, N. C., and Quirke, T. T., "Studies on Lead Oxides. I. A New Study of the Preparations of the Higher Oxides of Lead," *J. Am. Chem. Soc.*, **59**, 2305.

Finbak, C., and Hassel, O., "Crystal and Molecular Structure of CI_4 and CBr_4," *Z. Physik. Chem.*, **36B**, 301.

Mark, H., "On the Crystal and Molecular Structure of CI_4 and CBr_4," *Z. Physik. Chem.*, **38B**, 209.

Maxwell, L. R., Hendricks, S. B., and Deming, L. S., "The Molecular Structure of P_4O_6, P_4O_8, P_4O_{10}, and As_4O_6 by Electron Diffraction," *J. Chem. Phys.*, **5**, 626.

Meyer, W. F., "Investigations on Cobalt and in the System Co–C," *Z. Krist.*, **97A**, 145.

Sillén, L. G., "X-Ray Studies on Bi_2O_3," *Arkiv Kemi Mineral. Geol.*, **12A**, No. 18.

1938

Garrido, J., and Feo, R., "The Sulfotellurides of Bismuth," *Bull. Soc. Franc. Mineral.*, **61**, 196.

Gorbunova, O. E., and Vaganova, L. I., "X-Ray Investigation of the Transformation of gamma-Alumina into alpha-Alumina," *Tr. Tsentr. Nauk Issled. Lab. Kamnei Samotsv. Tr. Russ. Samotsv.*, No. 4, 66.

Hampson, P. C., and Stosick, A. J., "The Molecular Structure of As_4O_6, P_4O_6, P_4O_{10}, and $(CH_2)_6N_4$ by Electron Diffraction," *J. Am. Chem. Soc.*, **60**, 1814.

Juza, R., and Hahn, H., "On the Crystal Structures of Cu_3N, GaN, and InN; Metal Amides and Nitrides, V," *Z. Anorg. Allgem. Chem.*, **239**, 282.

Montoro, V., "Miscibility of Fe_3O_4 and Mn_3O_4," *Gazz. Chim. Ital.*, **68**, 728.

Nowacki, W., "The Crystal Structure of YF_3," *Naturwiss.*, **26**, 431; Nowacki, W., and Beck, G., *Z. Krist.*, **100A**, 242.

Nowotny, H., and Henglein, E., "Investigations in the System Cr–P," Z. Anorg. Allgem. Chem., **239**, 14.

Steenberg, B., "The Crystal Structure of Cu₃As and Cu₃P," Arkiv Kemi Mineral. Geol., **12A**, No. 26.

Strotzer, E. F., Biltz, W., and Meisel, K. "Affinity. LXXXI. Thorium Phosphide," Z. Anorg. Allgem. Chem., **238**, 69.

Westgren, A., "The Crystal Structure of Ni₃S₂," Z. Anorg. Allgem. Chem., **239**, 82.

1939

Akiyama, K., "Crystalline Modifications of Alumina. I. The Inversion of the Crystalline Forms of Al(OH)₃ by Heating," J. Soc. Chem. Ind. Japan, **42**, Suppl. Binding, 394.

Bommer, H., "Lattice Constants of the C-form of the Rare Earth Oxides," Z. Anorg. Allgem. Chem., **241**, 273.

Caglioti, V., "AlF and Double Salts Derived Therefrom," Chim. e Ind., **20**, 274; Neues Jahrb. Mineral. Geol., Ref. I, **1939**, 628.

Frondel, C., "Redefinition of Tellurobismuthite, Bi₂Te₃." Am. Mineralogist, **24**, #12, Pt. 2, 7.

Glemser, O., and Gwinner, E., "A New Ferromagnetic Modification of Fe₂O₃," Naturwiss., **26**, 739 (1938); Z. Anorg. Allgem. Chem., **240**, 161.

Halla, F., and Weil, R., "X-Ray Investigation of 'Crystalline Boron,'" Z. Krist., **101A**, 435; Naturwiss., **27**, 96.

Haraldsen, H., "The Phosphides of Copper," Z. Anorg. Allgem. Chem., **240**, 337.

Haul, R., and Schoon, T., "The Structure of Ferromagnetic Ferric Oxides, γ-Fe₂O₃," Z. Physik. Chem., **44B**, 216.

Helms, A., and Klemm, W, "The Crystal Structure of Rubidium and Cesium Sesquioxides," Z. Anorg. Allgem. Chem., **242**, 201.

Juza, R., and Hahn, H., "Crystal Structure of Ge₃N₄," Naturwiss., **27**, 32.

Lange, P. W., "A Comparison of Bi₂Te₃ with Bi₂Te₂S," Naturwiss., **27**, 133.

Meisel, K., "The Crystal Structure of Thorium Phosphides," Z. Anorg. Allgem. Chem., **240**, 300.

Nowacki, W., "The Crystal Structure of ScF₃," Naturwiss., **26**, 801 (1938); Z. Krist., **101A**, 273.

Powell, H. M., and Ewens, R. V. G., "The Crystal Structure of Iron Enneacarbonyl," J. Chem. Soc. (London), **1939**, 286.

Sevast'yanov, N. G., "X-Ray Analysis of Boron Carbide," Zavodskaya Lab., **8**, #12, 1317.

Wallbaum, H. J., "Vanadium Silicide, V₃Si," Z. Metallk., **31**, 362.

1940

Ficquelmont, A. M. de, Wetroff, G., and Moureu, H., "The X-Ray Spectra of Crystallized PCl₅," Compt. Rend., **211**, 566.

Juza, R., and Hahn, H., "Metal Amides and Metal Nitrides. IX. The Crystal Structure of Zn₃N₂, Cd₃N₂ and Ge₃N₄," Z. Anorg. Allgem. Chem., **244**, 111.

Lipson, H., and Petch, N. J., "The Crystal Structure of Cementite, Fe₃C," J. Iron Steel Inst. (London), **142**, 95.

Montoro, V., "Miscibility of the Sesquioxides of Iron and Manganese," Gazz. Chim. Ital., **70**, 145.

256 CRYSTAL STRUCTURES

Powell, H. M., Clark, D., and Wells, A. F., "The Crystal Structure of PCl₅," *Nature*, **145**, 149.
Sillén, L. G., "The Crystal Structure of alpha-Bi₂O₃," *Naturwiss.*, **28**, 206.

1941

Decker, H. C. J. de, "Crystal Structure of the Stable P₂O₅," *Rec. Trav. Chim.*, **60**, 413.
Decker, H. C. J. de, and MacGillavry, C. H., "The Crystal Structure of Volatile Metastable P₂O₅," *Rec. Trav. Chim.*, **60**, 153.
Galloni, E. E., and Roffo, A. E., Jr., "The Crystalline Structure of Pt₃O₄," *J. Chem. Phys.*, **9**, 875; *Bol. Inst. Med. Exptl. Estud. Cancer*, **18**, #56, 177.
Gross, S. T., "Unit Cell Measurements of Pb₃O₄, Pb₂O₃ and Tl₂SO₄," *J. Am. Chem. Soc.*, **63**, 1168.
Hofmann, W., "The Superstructure of Cu₃Sb," *Z. Metallk.*, **33**, 373.
Juza, R., "Metal Amides and Metal Nitrides, XI. The Crystal Structure of Cu₃N," *Z. Anorg. Allgem. Chem.*, **248**, 118.
Sillén, L. G., "The Crystal Structure of Monoclinic alpha-Bi₂O₃," *Z. Krist.*, **103A**, 274.
Westrink, R., and MacGillavry, C. H., "The Crystal Structure of the Ice-like Form of SO₃(γ)," *Rec. Trav. Chim.*, **60**, 794.
Zhdanov, G. S., and Sevast'yanov, N. G., "The Crystal Structure of B₄C," *Compt. Rend. Acad. Sci. URSS*, **32**, 432.
Zumbusch, M., "Structural Analogy of Uranium and Thorium Phosphides," *Z. Anorg. Allgem. Chem.*, **245**, 402.

1942

Almin, K. E., and Westgren, A., "Lattice Parameters of Cubic As₄O₆ and Sb₄O₆," *Arkiv Kemi Mineral. Geol.*, **15B**, No. 22.
Brody, S. B., "An X-Ray Investigation of the Structure of Lead Chromate," *J. Chem. Phys.*, **10**, 650.
Clark, D., Powell, H. M., and Wells, A. F., "The Crystal Structure of PCl₅," *J. Chem. Soc. (London)*, **1942**, 642.
Dyatkina, M. E., and Syrkin, Y. K., "The Structure of Boron Hydrides," *Compt. Rend. Acad. Sci. URSS*, **35**, 180.
Hume-Rothery, W., Rayner, G. V., and Little, A. T., "The Lattice Spacing of Cementite," *J. Iron Steel Inst. (London)*, **145**, 143.
Montoro, V., "Crystalline Structure of Bayerite," *Ric. Sci.*, **13**, 565.
Seifert, F. H., "The Crystallography of Cesium Fluorosulfonate, CsSO₃F," *Z. Krist.*, **104**, 385.
Straumanis, M., "Lattice Constants of Minium," *Z. Physik. Chem.*, **52B**, 127.
Vegard, L., and Hillesund, S., "Structure of a Few D Compounds and Comparison with that of the Corresponding H Compounds," *Avh. Norske Videnskaps-Akad. Oslo I. Mat.-Naturv. Klasse*, **1942**, No. 8, 24 pp.

1943

Byström, A., and Westgren, A., "The Crystal Structure of Pb₃O₄ and SnPb₂O₄," *Arkiv Kemi Mineral. Geol.*, **16B**, No. 14, 7 pp.
Byström, A., and Westgren, A., "X-Ray Analysis of Antimony Trifluoride," *Arkiv Kemi Mineral. Geol.*, **17B**, No. 2.
Clark, H. K., and Hoard, J. L., "The Crystal Structure of B₄C," *J. Am. Chem. Soc.*, **65**, 2115.

Driel, M. van, and MacGillavry, C. H., "The Crystal Structure of PBr₅," *Rec. Trav. Chim.*, **62**, 167.

Gross, S. T., "The Crystal Structure of Pb₃O₄," *J. Am. Chem. Soc.*, **65**, 1107.

Ketelaar, J. A. A., and Oosterhout, G. W. van, "The Crystal Structure of WCl₆," *Rec. Trav. Chim.*, **62**, 197.

Zhdanov, G. S., and Sevast'yanov, N. G., "X-Ray Investigation of the Structure of B₄C," *J. Phys. Chem.* (*USSR*), **17**, 326.

1944

Byström, A., "X-Ray Studies on Lead Sesquioxide, Pb₂O₃," *Arkiv Kemi Mineral. Geol.*, **18A**, No. 23.

Hägg, G., and Magnéli, A., "X-Ray Studies on Molybdenum and Tungsten Oxides," *Arkiv Kemi Mineral. Geol.*, **19A**, No. 2.

Öhrman, E., "Carbides in the System Fe–Mn–C," *Jernkontorets Ann.*, **128**, 13.

Petch, N. J., "The Interpretation of the Crystal Structure of Cementite," *J. Iron Steel. Inst.* (*London*), **149**, 143.

1945

Aurivillius, B., and Sillén, L. G., "Polymorphy of Bi₂O₃," *Nature*, **155**, 305.

Byström, A., "X-Ray Studies on Lead Sesquioxide," *Arkiv Kemi Mineral. Geol.*, **A18** No. 23, 8 pp.

Byström, A., "The Decomposition Products of Lead Peroxide and Oxidation Products of Lead Oxide," *Arkiv Kemi Mineral. Geol.*, **A20**, No. 11, 31 pp.

Juza, R., and Sachsze, W., "Metal Amides and Metal Nitrides. XIII. The System Cobalt–Nitrogen," *Z. Anorg. Allgem. Chem.*, **253**, 95.

Renes, P. A., and MacGillavry, C. H., "Crystal Structure of Aluminum Bromide," *Rec. Trav. Chim.*, **64**, 275.

Warren, H. V., and Peacock, M. A., "Hedleyite, a Bismuth Telluride from British Columbia, with Notes on Wehrite and Some Bi–Te Alloys," *Univ. Toronto Studies Geol. Ser.*, **49**, 55.

1946

Guillaud, C., and Wyart, J., "Relations between the Nitrogen Content, Magnetic Moments, and Curie Points, and the Distances between Adjacent Atoms of Manganese in the Ferromagnetic Solid Solutions of Nitrogen in Manganese," *Compt. Rend.*, **219**, 203.

Hägg, G., and Jerslev, B., "Unit Dimensions and Space Group of Tl₂Cl₃," *Experientia*, **2**, 495.

Peacock, M. A., "Heazlewoodite and the Artificial Compound Ni₃S₂," *Univ. Toronto Studies, Geol. Ser.*, **51**, 59.

Schubert, K., and Seitz, A., "The Crystal Structure of Y(OH)₃," *Z. Naturforsch.*, **1**, 321.

Seitz, A., "The Existence and Stability of Crystalline Hydroxides of the Rare Earths," *Z. Naturforsch.*, **1**, 321.

Thompson, R. M., and Peacock, M. A., "Montbrayite, a New Gold Telluride," *Am. Mineralogist*, **31**, 515.

1947

Brauer, G., Nowotny, H., and Rudolph, R., "X-Ray Investigations in the System Magnesium–Mercury," *Metallforsch.*, **2**, 81.

Fricke, R., and Seitz, A., "Crystalline Hydroxides of the Rare Earths," *Z. Anorg. Allgem. Chem.*, **254**, 107.

Fricke, R., and Seitz, A., "Crystalline Hydroxides of Indium and Scandium," *Z. Anorg. Allgem. Chem.*, **255**, 13.

Iandelli, A., "Modifications of Sesquioxides of the Rare Earths," *Gazz. Chim. Ital.*, **77**, 312.

Ketelaar, J. A. A., MacGillavry, C. H., and Renes, P. A., "The Crystal Structure of Aluminum Chloride," *Rec. Trav. Chim.*, **66**, 501.

Kiessling, R., "Crystal Structures of Molybdenum and Tungsten Borides," *Acta Chem. Scand.*, **1**, 893.

Klemm, W., and Krose, E., "The Crystal Structure of $ScCl_3$, $TiCl_3$ and VCl_3," *Z. Anorg. Allgem. Chem.*, **253**, 218.

Lundqvist, D., "X-Ray Studies of the Binary System Ni–S," *Arkiv Kemi Mineral. Geol.*, **24A**, No. 21, 12 pp.

Rollier, M. A., and Riva, A., "The X-Ray Determination of the Crystal Structure of Boron Chloride and of Boron Bromide in the Solid State," *Gazz. Chim. Ital.*, **77**, 361.

Rundle, R. E., "The Structure of Uranium Hydride and Deuteride," *J. Am. Chem. Soc.*, **69**, 1719.

Schubert, K., and Seitz, A., "Crystal Structure of $Y(OH)_3$," *Z. Anorg. Allgem. Chem.*, **254**, 116.

Vegard, L., "Investigation into the Structure and Properties of Solid Matter with the Help of X-Rays," *Skrifter Norske Videnskaps-Akad. Oslo I. Mat.-Naturv. Klasse*, **1947**, No. 2, 83 pp.

1948

Aebi, F., ' Phase Studies in the System Vanadium–Oxygen, and the Crystal Structure of $V_{12}O_{26}$," *Helv. Chim. Acta*, **31**, 8.

Grønvold, F., "Crystal Structure of Uranium Oxide (U_3O_8)," *Nature*, **162**, 70.

Kasper, J. S., Lucht, C. M., and Harker, D., "The Crystal Structure of Decaborane," *Am. Mineralogist*, **33**, 768.

Katz, T., Siramy, M., and Faivre, R., "Preparation and Study by X-Ray Diffraction of a Pseudocubic Lead Oxide of Variable Composition," *Compt. Rend.*, **227**, 282.

König, H., "The Lattice Constants of γ-Alumina," *Naturwiss.*, **35**, 92.

Magnéli, A., "The Crystal Structures of Mo_9O_{26} (β'-Molybdenum Oxide) and Mo_8O_{23} (β-Molybdenum Oxide)," *Acta Chem. Scand.*, **2**, 501.

Magnéli, A., "The Crystal Structure of Mo_4O_{11} (γ-Molybdenum Oxide)," *Acta Chem. Scand.*, **2**, 861.

Palm, A., "X-Ray Diffraction of Indium Hydroxide and Indium Chromate," *J. Phys. Coll. Chem.*, **52**, 959.

Pauling, L., and Ewing, F. J., "The Structure of Uranium Hydride," *J. Am. Chem. Soc.*, **70**, 1660.

Ramdohr, P., "The Mineral Species Guanajuatite and Paraguanajuatite," *Com. Direct. Invest. Recursos Mineral. Mex., Bol.*, **20**, 1.

Rundle, R. E., Baenziger, N. C., Wilson, A. S., and McDonald, R. A., "The Structures of Carbides, Nitrides and Oxides of Uranium," *J. Am. Chem. Soc.*, **70**, 99.

Schubert, K., and Seitz, A., "Crystal Structure of Scandium and Indium Hydroxides," *Z. Anorg. Allgem. Chem.*, **256**, 226.

Wilsdorf, H., "The Crystal Structure of Univalent Copper Azide, CuN_3," *Acta Cryst.*, **1**, 115.

Zachariasen, W. H., "Crystal Chemical Studies of the 5f-Series of Elements. I. New Structure Types," *Acta Cryst.*, 1, 265.
Zachariasen, W. H., "Crystal Chemical Studies of the 5f-Series of Elements. III. The Disorder in the Crystal Structure of Anhydrous Uranyl Fluoride," *Acta Cryst.*, 1, 277.
Zachariasen, W. H., "Crystal Chemical Studies of the 5f-Series of Elements. V. The Crystal Structure of Uranium Hexa-Chloride," *Acta Cryst.*, 1, 285.

1949

Bertaut, F., and Blum, P., "The Structure of the Uranium Borides," *Compt. Rend.*, 229, 666.
Fricke, R., and Dürrwächter, W., "Further Investigations on the Crystalline Hydroxides of the Rare Earths," *Z. Anorg. Allgem. Chem.*, 259, 305.
Geiersberger, K., "Note on the Lattice Constants of FeCl₃," *Z. Anorg. Allgem. Chem,.* 258, 361.
Hahn, H., and Gilbert, E., "Metal Amides and Metal Nitrides. XIX. Silver Nitride," *Z. Anorg. Allgem. Chem.*, 258, 77.
Hahn, H., and Klingler, W., "The Crystal Structure of Gallium Sulfide, Selenide and Telluride," *Z. Anorg. Allgem. Chem.*, 250, 135.
Hund, F., and Fricke, R., "The Crystal Structure of α-Bismuth Fluoride," *Z. Anorg. Allgem. Chem.*, 258, 198.
Kiessling, R., "The Borides of Tantalum," *Acta Chem. Scand.*, 3, 603.
Kürbs, E., Plieth, K., and Stranski, I. N., "X-Ray Investigation of Condensed Arsenic," *Z. Anorg. Allgem. Chem.*, 258, 238.
MacGillavry, C. H., Decker, H. C. J. de, and Nijland, L. M., "Crystal Structure of the Third Form of Phosphorus Pentoxide," *Nature*, 164, 448.
Magnéli, A., "Crystal Structure Studies on γ-Tungsten Oxide," *Arkiv Kemi*, 1, 223.
Mikheev, V. I., "α-, β- and γ-Domeykite," *Zap. Vses. Mineralog. Obshchestva*, 78, No. 1, 3.
Mooney, R. C. L., "The Crystal Structure of Thorium Chloride and Uranium Chloride," *Acta Cryst.*, 2, 189.
Morimoto, N., "The Crystal Structure of Orpiment," *X-Sen (X-Rays)*, 5, 115.
Morozov, I. S., and Kuznetsov, V. G., "The γ-Modification of Manganese Dioxide," *Izvest. Akad. Nauk SSSR, Otdel. Khim. Nauk*, 1949, 343.
Zachariasen, W. H., "Crystal Chemical Studies of the 5f-Series of Elements. VI. The Ce₂S₃–Ce₃S₄ Type of Structure," *Acta Cryst.*, 2, 57.
Zachariasen, W. H., "Crystal Chemical Studies of the 5f-Series of Elements. VII. The Crystal Structure of Ce₂O₂S, La₂O₂S, and Pu₂O₂S," *Acta Cryst.*, 2, 60.
Zachariasen, W. H., "Crystal Chemical Studies of the 5f-Series of Elements. IX. The Crystal Structure of Th₇S₁₂," *Acta Cryst.*, 2, 288.
Zachariasen, W. H., "Crystal Chemical Studies of the 5f-Series of Elements. X. Sulfides and Oxysulfides," *Acta Cryst.*, 2, 291.
Zachariasen, W. H., "Crystal Chemical Studies of the 5f-Series of Elements. XI. The Crystal Structure of α-UF₅ and of β-UF₅," *Acta Cryst.*, 2, 296.
Zachariasen, W. H., "Crystal Chemical Studies of the 5f-Series of Elements. XII. New Compounds Representing Known Structure Types," *Acta Cryst.*, 2, 388.
Zachariasen, W. H., "Crystal Chemical Studies of the 5f-Series of Elements. XIII. The Crystal Structures of U₂F₉ and NaTh₂F₉," *Acta Cryst.*, 2, 390.

1950

Andersson, L.-H., and Kiessling, R., "Investigations on the Binary Systems of Boron with Chromium, Columbium, Nickel, and Thorium Including a Discussion of the Phase 'TiB' in the Titanium Boron System," *Acta Chem. Scand.*, **4**, 146.

Andersson, G., and Magnéli, A., "Crystal Structure of Molybdenum Trioxide," *Acta Chem. Scand.*, **4**, 793.

Arbuzov, M. P., "Crystal Structure and Particle Size of the Carbide Phase in Tempered Steel," *Dokl. Akad. Nauk SSSR*, **73**, 83.

Byström, A., and Wilhelmi, K. A., "The Crystal Structure of Chromium Trioxide," *Acta Chem. Scand.*, **4**, 1131.

Byström, A., Wilhelmi, K. A., and Brotzen, O., "Vanadium Pentoxide, a Compound with Five-Coordinated Vanadium Atoms," *Acta Chem. Scand.*, **4**, 1119.

D'Eye, R. W. M., "The Crystal Structure of Thorium Tetrabromide," *J. Chem. Soc. (London)*, 1950, 2764.

Dönges, E., "On the Selenohalogenides of Trivalent Antimony and Bismuth and on Antimony(III) Selenide," *Z. Anorg. Allgem. Chem.*, **263**, 280.

Early, J. W., "Description and Synthesis of the Selenide Minerals," *Am. Mineralogist*, **35**, 337.

Elson, R., Fried, S., Sellers, P., and Zachariasen, W. H., "Quadrivalent and Quinquevalent States of Protoactinium," *J. Am. Chem. Soc.*, **72**, 5791.

Grison, E., Eriks, K., and Vries, J. L. de, "The Crystal Structure of Nitric Acid Anhydride, N_2O_5," *Acta Cryst.*, **3**, 290.

Kasper, J. S., Lucht, C. M., and Harker, D., "The Crystal Structure of Decaborane," *Acta Cryst.*, **3**, 436.

Kiessling, R., "Borides of Manganese," *Acta Chem. Scand.*, **4**, 146.

Magnéli, A., "Structure of β-Tungsten Oxide," *Nature*, **165**, 356; *Arkiv Kemi*, **1**, 513.

Shinoda, G., and Amano, Y., "X-Ray Investigation of Artificially Prepared Jewels," *X-Sen (X-Rays)*, **6**, 7.

Tertian, R., "The Constitution and Crystal Structure of Activated Alumina," *Compt. Rend.*, **230**, 1677.

Ueda, R., and Ichinokawa, T., "The Domain Structure of Tungsten Trioxide," *Phys. Rev.*, **80**, 1106.

Zalkin, A., and Templeton, D. H., "The Crystal Structures of CeB_4, ThB_4, and UB_4," *J. Chem. Phys.*, **18**, 391.

Zemann, J., "The Crystal Chemistry of Bismuth," *Tschermaks Mineral. Petrog. Mitt.*, **1**, 361.

1951

Becker, K. A., Plieth, K., and Stranski, I. N., "Structure Investigation of the Monoclinic Arsenic Trioxide Modification Claudetite," *Z. Anorg. Allgem. Chem.*, **266**, 293.

Bogatskii, D. P., "Diagram of the State of the System Nickel–Oxygen," *Zh. Obshchei Khim.*, **21**, 3.

Dönges, E., "Chalcogenides of Trivalent Antimony and Bismuth. III. The Tellurohalides of Trivalent Antimony and Bismuth; Telluride of Antimony and Bismuth; Selenide of Bismuth," *Z. Anorg. Allgem. Chem.*, **265**, 56.

Dulmage, W. J., and Lipscomb, W. N., "The Molecular Structure of Pentaborane," *J. Am. Chem. Soc.*, **73**, 3539.

Flahaut, J., "A Variety of Aluminum Sulfide Stable at High Temperature," *Compt. Rend.*, **232**, 2100.

Frueh, A. J., Jr., "The Crystal Structure of Claudetite (Monoclinic As_2O_3)," *Am. Mineralogist*, **36**, 833.

Gregory, N. W., "The Crystal Structure of Ferric Bromide," *J. Am. Chem. Soc.*, **73**, 472.

Gutmann, V., and Jack, K. H., "The Crystal Structures of Molybdenum Trifluoride, MoF_3, and Tantalum Trifluoride, TaF_3," *Acta Cryst.*, **4**, 244.

Jack, K. H., and Gutmann, V., "The Crystal Structure of Vanadium Trifluoride," *Acta Cryst.*, **4**, 246.

Kojima, T., Inoue, T., and Ishiyama, T., "Metallurgical Research on Cerium Metal. III. Studies on Cerium Compounds by X-Ray Analysis of Crystal Structure," *J. Electrochem. Soc. Japan*, **19**, 285.

Magnéli, A., "Diffraction Effects in X-Ray Fourier Syntheses Due to Nonobserved 'Weak Reflections,'" *Acta Cryst.*, **4**, 447.

Mallett, M. W., Gerds, A. F., and Vaughan, D. A., "Uranium Sesquicarbide," *J. Electrochem. Soc.*, **98**, 505.

Matthias, B. T., and Wood, E. A., "Low Temperature Polymorphic Transformation in WO_3," *Phys. Rev.*, **84**, 1255.

Milne, I. H., "Radioactive Compounds. III. Uranouranic Oxide (U_3O_8)," *Am. Mineralogist*, **36**, 415.

Nowotny, H., Funk, R., and Pesl, J., "Crystal Chemical Studies in the Systems Mn–As, V–Sb and Ti–Sb," *Monatsh.*, **82**, 513.

Rundle, R. E., "Hydrogen Positions in Uranium Hydride by Neutron Diffraction," *J. Am. Chem. Soc.*, **73**, 4172.

Ueda, R., and Ichinokawa, T., "Phase Transition of Tungsten Oxide. II," *Busseiron Kenkyu*, No. **36**, 64; *Phys. Rev.*, **82**, 563.

Wyart, J., and Foex, M., "Polymorphism of Tungstic Anhydride Studies at High Temperatures by Means of X-Rays," *Compt. Rend.*, **232**, 2459.

Zhdanov, G. S., and Zvonkova, Z. V., "Crystal Structure of KO_3," *Zh. Fiz. Khim.*, **25**, 100.

1952

Abrahams, S. C., Grison, E., and Kalnajs, J., "The Crystal Chemistry of Cesium Penta- and Hexasulfide," *J. Am. Chem. Soc.*, **74**, 3761.

Becker, K. A., Plieth, K., and Stranski, I. N., "Correction to the Work: 'Structure Investigation of the Monoclinic Arsenic Trioxide Modification Claudetite,'" *Z. Anorg. Allgem. Chem.*, **269**, 92.

Berger, S. V., "The Crystal Structure of B_2O_3," *Acta Cryst.*, **5**, 389.

Bertaut, F., and Blum, P., "Hexaborides and Alkaline Substitutions," *Compt. Rend.*, **234**, 2621.

D'Eye, R. W. M., Sellman, P. G., and Murray, J. R., "The Thorium–Selenium System," *J. Chem. Soc. (London)*, **1952**, 2555.

Dulmage, W. J., and Lipscomb, W. N., "The Crystal and Molecular Structure of Pentaborane," *Acta Cryst.*, **5**, 260.

Ferro, R., "Compounds of Uranium with Antimony," *Atti Accad. Lincei, Rend., Cl. Sci. Fis. Mat. Nat.*, **13**, 53.

Ferro, R., "Alloys of Uranium with Bismuth," *Atti Accad. Lincei, Rend., Cl. Sci. Fis. Mat. Nat.*, **13**, 401.

Flahaut, J., "Contribution to the Study of Aluminum Sulfide," *Ann. Chim. (Paris)*, **7**, 632.

Hahn, H., "Structure of Gallium Chalcogenides," *Angew. Chem.*, **64**, 203.

Handy, L. L., and Gregory, N. W., "Structural Properties of Chromium(III) Iodide and some Chromium(III) Mixed Halides," *J. Am. Chem. Soc.*, **74**, 891.

Hund, F., and Peetz, U., "Investigations of the Systems La_2O_3, Nd_2O_3, Sm_2O_3, Yb_2O_3, Sc_2O_3 with U_3O_8," *Z. Anorg. Allgem. Chem.*, **271**, 6.

Iandelli, A., "Uranium Arsenides," *Atti Accad. Lincei, Rend., Cl. Sci. Fis. Mat. Nat.*, **13**, 138.

Kehl, W. L., Hay, R. G., and Wahl, D., "The Structure of Tetragonal Tungsten Trioxide," *J. Appl. Phys.*, **23**, 212.

Ladell, J., Post, B., and Fankuchen, I., "The Crystal Structure of Nickel Carbonyl," *Acta Cryst.*, **5**, 795.

Leslie, W. C., Carroll, K. G., and Fisher, R. M., "Diffraction Patterns and Crystal Structure of Silicon Nitride and Germanium Nitride," *J. Metals*, **4**, *Trans.*, 204.

Pinsker, Z. G., "Electron Diffraction Investigation of Bismuth Triiodide; Ideas on the Structure of Layer Lattices," *Tr. Inst. Krist., Akad. Nauk SSSR*, **7**, 35.

Post, B., and Glaser, F. W., "Borides of Some Transition Metals," *J. Chem. Phys.*, **20**, 1050.

Post, B., and Glaser, F. W., "Crystal Structure of ZrB_{12}," *J. Metals*, **4**, *Trans.*, 631.

Schlyter, K., "Crystal Structure of Fluorides of the Tysonite, or LaF_3, Type," *Arkiv Kemi*, **5**, 73.

Templeton, D. H., and Dauben, C. H., "The Crystal Structures of NbC and Pu_2O_3," *Univ. California Rept. UCRL-1886*, July 14, 1952.

Zachariasen, W. H., "Crystal Chemical Studies of the 5f-Series of Elements. XV. The Crystal Structure of Plutonium Sesquicarbide," *Acta Cryst.*, **5**, 17.

1953

Abrahams, S. C., and Calhoun, B. A., "The Low-Temperature Transition in Magnetite," *Acta Cryst.*, **6**, 105.

Abrahams, S. C., and Grison, E., "The Crystal Structure of Cesium Hexasulfide," *Acta Cryst.*, **6**, 206.

Andersson, G., "X-Ray Studies on Vanadium Oxides," *Research*, **6**, 45S.

Andersson, G., "The Crystal Structure of Tungsten Trioxide," *Acta Chem. Scand.*, **7**, 154.

Ariya, S. M., Morozova, M. P., Markevich, G. S., and Reikhardt, A. A., "The System Platinum–Oxygen," *Sb. Stat. Obshch. Khim. Akad. Nauk SSSR*, **1**, 76.

Berger, S. V., "The Crystal Structure of Boron Oxide," *Acta Chem. Scand.*, **7**, 611.

Bertaut, F., and Blum, P., "Borides of Chromium," *Compt. Rend.*, **236**, 1055.

Blomberg, B., Kihlborg, L., and Magnéli, A., "Crystal Structures of $(Mo,W)_{10}O_{29}$ and $(Mo,W)_{11}O_{32}$," *Arkiv Kemi*, **6**, 133.

Burbank, R. D., and Bensey, F. N., "The Structure of the Interhalogen Compounds. I. Chlorine Trifluoride at $-120°$," *J. Chem. Phys.*, **21**, 602.

Caillat, R., Coriou, H., and Pério, P., "A New Variety of Uranium Hydride," *Compt. Rend.*, **237**, 812.

Cowley, J. M., "Stacking Faults in γ-Alumina," *Acta Cryst.*, **6**, 53.

Cowley, J. M., "Structure Analysis of Single Crystals by Electron Diffraction. III. Modifications of Alumina," *Acta Cryst.*, **6**, 846.

Curlook, W., and Pidgeon, L. M., "The Cobalt–Iron–Sulfur System," *Trans. Can. Inst. Mining Met. Engrs.*, **56** (in *Can. Mining Met. Bull.*, **493**, 297).

D'Eye, R. W. M., "The Crystal Structures of ThSe₂ and Th₇Se₁₂," *J. Chem. Soc. (London)*, **1953**, 1670.

Glaser, F. W., Moskowitz, D., and Post, B., "An Investigation of Boron Carbide," *J. Appl. Phys.*, **24**, 731.

Glaser, F. W., and Post, B., "The System: Zirconium–Boron," *J. Metals*, **5**, *AIME Trans.*, **197**, 1117.

Gritsaenko, G. S., Sludskaya, N. N., and Aidinyan, N. K., "Synthesis of Vaesite and Polydymite," *Zap. Vses. Mineral. Obshchestva*, **82**, 42.

Koehler, W. C., and Wollan, E. O., "Neutron Diffraction Study of the Structure of the A-Form of the Rare Earth Sesquioxides," *Acta Cryst.*, **6**, 741.

Lavine, L. R., and Lipscomb, W. N., "The Molecular Structure of B₅H₁₁," *J. Chem. Phys.*, **21**, 2087.

Magnéli, A., "Structure of the ReO₃ Type with Recurrent Dislocations of Atoms; 'Homologous Series' of Molybdenum and Tungsten Oxides," *Acta Cryst.*, **6**, 495.

Natta, G., and Corradini, P., "Structure of Some Cobalt Carbonyl Compounds," *Atti Accad. Lincei, Rend., Cl. Sci. Fis. Mat. Nat.*, **15**, 248.

Nordman, C. E., "The Crystal and Molecular Structure of Tetraborane," *Univ. Microfilms (Ann Arbor)* Publ. No. 5550, p. 86.

Nordman, C. E., and Lipscomb, W. N., "The Molecular Structure of B₄H₁₀," *J. Am. Chem. Soc.*, **75**, 4116.

Nordman, C. E., and Lipscomb, W. N., "The Crystal and Molecular Structure of Tetraborane," *J. Chem. Phys.*, **21**, 1856.

Pério, P., "Crystalline Varieties of UO₃," *Bull. Soc. Chim. France*, **1953**, 776.

Reed, T. B., and Lipscomb, W. N., "The Crystallography of Solid Dinitrogen Trioxide at −115°," *Acta Cryst.*, **6**, 781.

Rooksby, H. P., and Willis, B. T. M., "The Low Temperature Crystal Structure of Magnetite," *Acta Cryst.*, **6**, 565.

Roy, R., and McKinstry, H. A., "The So-called Y(OH)₃ Type Structure, and the Structure of La(OH)₃," *Acta Cryst.*, **6**, 365.

Scatturin, V., and Tornati, M., "Lattice Imperfection of Thallium Oxides," *Ric. Sci.*, **23**, 1805.

Schubert, K., and Fricke, H., "Crystallochemistry of the B-Subgroup Metals. II. Trigonal Distorted NaCl Structures," *Z. Metallk.*, **44**, 457.

Templetion, D. H., and Dauben, C. H., "Crystal Structures of Americium Compounds," *J. Am. Chem. Soc.*, **75**, 4560.

Ueda, R., and Kobayashi, J., "Antiparallel Dipole Arrangement in Tungsten Trioxide," *Phys. Rev.*, **91**, 1565.

Zachariasen, W. H., "Crystal Chemical Studies of the 5f-Series of Elements. XIX. The Crystal Structure of the Higher Thorium Hydride, Th₄H₁₅," *Acta Cryst.*, **6**, 393.

Zalkin, A., and Templeton, D. H., "The Crystal Structures of Yttrium Fluoride and Related Compounds," *J. Am. Chem. Soc.*, **75**, 2453.

Zalkin, A., and Templeton, D. H., "The Crystal Structure of Osmium Tetroxide," *Acta Cryst.*, **6**, 106.

Zalkin, A., and Templeton, D. H., "The Crystal Structures of CeB₄, ThB₄ and UB₄," *Acta Cryst.*, **6**, 269.

Zhdanov, G. S., Zhuravlev, N. N., and Zevin, L. S., "X-Ray Establishment of Formation of Solid Solutions in Boron Carbide," *Dokl. Akad. Nauk SSSR*, **92**, 767.

1954

Andersson, G., "Vanadium Oxides," *Acta Chem. Scand.*, **8**, 1599.

Asprey, L. B., "New Compounds of Quadrivalent Americium, AmF_4, $KAmF_5$," *J. Am. Chem. Soc.*, **76**, 2019.

Atoji, M., and Lipscomb, W. N., "The Structure of SiF_4," *Acta Cryst.*, **7**, 597.

Batuecas, T., "Determination of Atomic Masses by the Pykno-X-ray Method," *Nature*, **173**, 345.

Becker, K. A., Plieth, K., and Stranski, I. N., "Structure Investigation of the Monoclinic Arsenic Oxide Modification Claudetite. II," *Z. Anorg. Allgem. Chem.*, **275**, 297.

Blum, P., and Bertaut, F., "Borides with Higher Content of Boron," *Acta Cryst.*, **7**, 81.

Boswijk, K. H., and Wiebenga, E. H., "The Crystal Structure of I_2Cl_6," *Acta Cryst.*, **7**, 417.

Brauer, I. G., and Gradinger, H., "Mixed Phases of Rare Earth Oxides," *Z. Anorg. Allgem. Chem.*, **276**, 209.

Brimm, E. O., Lynch, M. A., Jr., and Sesny, W. J., "Preparation and Properties of Manganese Carbonyl," *J. Am. Chem. Soc.*, **76**, 3831.

Cunningham, B. B., Feay, D. C., and Rollier, M. A., "Terbium Tetrafluoride; Preparation and Properties," *J. Am. Chem. Soc.*, **76**, 3361.

Ehrlich, P., and Pietzka, G., "Titanium Trifluoride," *Z. Anorg. Allgem. Chem.*, **275**, 121.

Ferro, R., "Several Selenium and Tellurium Compounds of Uranium," *Z. Anorg. Allgem. Chem.*, **275**, 320.

Geth, E. D., Holden, J. R., Baenziger, N. C., and Eyring, L., "Praseodymium Oxides, II. X-ray and Differential Thermal Analyses," *J. Am. Chem. Soc.*, **76**, 5239.

Hägg, G., and Schönberg, N., "β-Tungsten as a Tungsten Oxide," *Acta Cryst.*, **7**, 351.

Havinga, E. E., Boswijk, K. H., and Wiebenga, E. H., "The Crystal Structure of Cs_2I_8-(CsI_4)," *Acta Cryst.*, **7**, 487.

Hirakawa, K., "Tungsten Trioxide Single Crystals," *Mem. Fac. Sci. Kyushu Univ.*, Ser. B, **1**, No. 4, 112.

Inuzuka, H., and Sugaike, S., "Indium Telluride, Preparation and Lattice Constant," *Proc. Japan Acad.*, **30**, 383.

Kondrashev, Y. D., "The Parameter of Boron in the Structure of CaB_6," *Dokl. Akad. Nauk SSSR*, **94**, 471.

Kuo, K., and Persson, L. E., "A Contribution to the Constitution of the Ternary System Fe–Mn–C," *J. Iron Steel Inst.*, **178**, 39.

Lapitskii, A. V., Simanov, Y. P., Semenenko, K. N., and Yarembash, E. I., "The Properties of Tantalum Pentoxide," *Vestn. Mosk. Univ.*, **9**, No. 3, Ser. Fiz.-Mat. Estesven. Nauk, No. 2, 85.

Lavine, L. R., and Lipscomb, W. N., "The Crystal and Molecular Structure of B_5H_{11}," *J. Chem. Phys.*, **22**, 614.

Morimoto, N., "The Crystal Structure of Orpiment (As_2S_3) Redefined," *Mineralog. J.* (*Japan*), **1**, 160.

Mulford, R. N. R., Ellinger, F. H., and Zachariasen, W. H., "A New Form of Uranium Hydride," *J. Am. Chem. Soc.*, **76**, 297.

Schönberg, N., "X-Ray Studies on Vanadium and Chromium Oxides with Low Oxygen Content," *Acta Chem. Scand.*, **8**, 221.

Schönberg, N., "An X-Ray Investigation of Transition Metal Phosphides," *Acta Chem. Scand.*, **8**, 226.

Schönberg, N., "The Existence of a Metallic Molybdenum Oxide," *Acta Chem. Scand.* **8**, 617.

Schönberg, ..., "Existence of Metallic Ternary Oxides M'M"O with the Metal Atoms in Hexagonal Close Packing," *Acta Chem. Scand.*, **8**, 630.

Schönberg, N., "On the Existence of Ternary Transition Metal Oxides," *Acta Chem. Scand.*, **8**, 932.

Schönberg, N., "Ternary Transition Metal Oxide Phases of the Fluorite Structure," *Acta Chem. Scand.*, **8**, 1347.

Sellers, P. A., Fried, S., Elson, R. E., and Zachariasen, W. H., "The Preparation of Some Protoactinium Compounds and the Metal," *J. Am. Chem. Soc.*, **76**, 5935.

Semiletov, S. A., "Electron Diffraction Investigation of the Structure of Sublimed Films of the Composition Bi–Se and Bi–Te," *Tr. Inst. Krist., Akad. Nauk SSSR*, **10**, 76.

Simanov, Y. P., Lapitskii, A. V., and Artamonova, E. P., "The Properties of Tantalum Pentoxide. II," *Vestn. Mosk. Univ.*, **9**, No. 9, *Ser. Fiz.-Mat. Estestven. Nauk*, **6**, 109.

Templeton, D. H., and Carter, G. F., "The Crystal Structure of Yttrium Chloride and Similar Compounds," *J. Phys. Chem.*, **58**, 940.

Templeton, D. H., and Dauben, C. H., "Lattice Parameters of Some Rare Earth Compounds and a Set of Crystal Radii," *J. Am. Chem. Soc.*, **76**, 5237.

Vos, V. A., and Wiebenga, E. H., "Crystal Structures of P_4S_{10} and P_4S_7," *Koninkl. Ned. Akad. Wetenschap., Proc.*, **57B**, 497.

Westrik, R., and MacGillavry, C. H., "The Crystal Structure of the Asbestos-like Form of Sulfur Trioxide," *Acta Cryst.*, **7**, 764.

Wilhelmi, K. A., "X-Ray Investigation of Re_2O_7," *Acta Chem. Scand.*, **8**, 693.

Yearian, H. J., Kortright, J. M., and Langenheim, R. H., "Lattice Parameters of the $FeFe_{(2-x)}Cr_xO_4$ Spinel System," *J. Chem. Phys.*, **22**, 1196.

Zhdanov, G. S., Meerson, G. A., Zhuravlev, N. N., and Samsonov, G. V., "Solubility of Boron and Carbon in Boron Carbide $B_{12}C_3$ (B_4C)," *Zh. Fiz. Khim.*, **28**, 1076.

1955

Aronsson, B., "The Crystal Structure of Ni_3P (Fe_3P-type)," *Acta Chem. Scand.*, **9**, 137.

Asprey, L. B., Ellinger, F. H., Fried, S., and Zachariasen, W. H., "Evidence for Quadrivalent Curium; X-Ray Data on Curium Oxides," *J. Am. Chem. Soc.*, **77**, 1707.

Aurivillius, B., "X-Ray Studies on the System BiF_3–Bi_2O_3. I. Preliminary Phase Analysis and a Note on the Structure of BiF_3," *Acta Chem. Scand.*, **9**, 1206.

Aurivillius, B., and Lundqvist, T., "X-Ray Studies of the System BiF_3–Bi_2O_3. II. A Bismuth Oxide Fluoride of Defective Tysonite Type," *Acta Chem. Scand.*, **9**, 1209.

Bevan, D. J. M., "Ordered Intermediate Phases in the System CeO_2–Ce_2O_3," *J. Inorg. Nucl. Chem.*, **1**, 49.

Bøhm, F., Grønvold, F., Haraldsen, H., and Prydz, H., "X-Ray and Magnetic Study of the System Cobalt–Selenium," *Acta Chem. Scand.*, **9**, 1510.

Buerger, M. J., and Robinson, D. W., "Crystal Structure and Twinning of Co_2S_3," *Proc. Natl. Acad. Sci. U.S.*, **41**, 199.

Burdese, A., "The System Iron–Nitrogen," *Met. Ital.*, **47**, 357.

Ehrlich, P., Plöger, F., and Pietzka, G., "Niobium Trifluoride," *Z. Anorg. Allgem. Chem.*, **282**, 19.

Eliseev, E. N., "Chemical Composition and Crystal Structure of Pentlandite," *Zap. Vses. Mineralog. Obshchestva*, **84**, 53.

Ferro, R., "The Crystal Structure of Thorium Arsenides," *Acta Cryst.*, **8**, 360.

Goryunova, N. A., Kotovich, V. A., and Frank-Kamenetskii, V. A., "X-Ray Investigation of the Isomorphism of Gallium and Zinc Compounds," *Dokl. Akad. Nauk SSSR*, **103**, 659.

Hahn, H., and Frank, G., "Structure of Ga₂S₃," Z. Anorg. Allgem. Chem., 278, 333.

Houten, S., van, Vos, A., and Wiegers, G. A., "The Crystal Structure of P₄S₃," Rec. Trav. Chim., 74, 1167.

Jellinek, F., "The Crystal Structures of Chromium Sulphides," Koninkl. Ned. Akad. Wetenschap., Proc., 58B, 213.

Leung, Y. C., Waser, J., and Roberts, L. R., "The Crystal Structure of P₄S₃," Chem. Ind. (London), 1955, 948.

Meller, F., and Fankuchen, I., "The Crystal Structure of Tin Tetraiodide," Acta Cryst., 8, 343.

Ramdohr, P., "Four New Natural Cobalt Selenides from the Trogtal Quarry near Lautental (Harz Mountains)," Neues Jahrb. Mineral., Monatsh., 6, 133.

Siegel, S., "The Crystal Structure of Trigonal U₃O₈," Acta Cryst., 8, 617.

Singer, J., and Spencer, C. W., "X-Ray Crystallographic Data on As₂Te₃," J. Metals, 7, AIME Trans., 203, 144.

Tasman, H. A., and Boswijk, K. H., "Reinvestigation of the Crystal Structure of CsI₃," Acta Cryst., 8, 59.

Vorres, K., and Donohue, J., "The Structure of Titanium Oxydifluoride," Acta Cryst., 8, 25.

Vos, A., and Wiebenga, E. H., "The Crystal Structures of P₄S₁₀ and P₄S₇," Acta Cryst., 8, 217.

Wait, E., "A Cubic Form of Uranium Trioxide," J. Inorg. Nucl. Chem., 1, 309.

1956

Burbank. R. D., and Bensey, F. N., Jr., "The Crystal Structure of Zirconium Tetrafluoride," U.S. At. Energy Comm. Rept. K-1280, 19 pp.

Dachs, H., "Crystal Structure of Bixbyite (Fe,Mn)₂O₃," Z. Krist., 107, 370.

Dickerson, R. E., Wheatley, P. J., Howell, P. A., Lipscomb, W. N., and Schaeffer, R., "Boron Arrangement in a B₉ Hydride," J. Chem. Phys., 25, 606.

Ferro, R., "The Crystal Structures of Thorium Antimonides," Acta Cryst., 9, 817.

Frevel, L. K., and Rinn, H. W., "The Crystal Structure of NbO₂F and TaO₂F," Acta Cryst., 9, 626.

Hahn, H., and Harder, B., "The Crystal Structures of the Titanium Sulfides," Z. Anorg. Allgem. Chem., 288, 241.

Lindqvist, I., and Niggli, A., "The Crystal Structure of Antimony Trichloride," J. Inorg. Nucl. Chem., 2, 345.

Moskowitz, D., "New Vanadium Boride of the Composition V₃B₄," J. Metals, 8, AIME Trans., 206, 1325.

Pearson, A. D., and Buerger, M. J., "Confirmation of the Crystal Structure of Pentlandite," Am. Mineralogist, 41, 804.

Picon, M., and Flahaut, J., "S₄Me₃ Sulfides of the Ceric Rare Earths," Compt. Rend., 243, 2074.

Picon, M., and Patrie, M., "Oxysulfides of the Rare Earth Elements of the Cerium Groups," Compt. Rend., 242, 516.

Samsonov, G., and Grodshtein, A. E., "Some of the Properties of the Hexaborides of the Alkaline Earth and of the Rare Earth Metals," Zh. Fiz. Khim., 30, 379.

Sasvári, K., "The Crystal Structure of α-Bayerite, Al(OH)₃," Acta Geol. Acad. Sci. Hung., 4, 123.

Scatturin, V., Zannetti, R., and Censolo, G., "Nonstoichiometric Compounds of Trivalent Thallium Oxide," Ric. Sci., 26, 3108.

Semiletov, S. A., "Electron Diffraction Determination of the Antimony Telluride Structure," *Kristallografiya*, 1, 403.

Shal'nikova, N. A., and Yakovlev, I. A., "X-Ray Determination of the Crystal Lattice Constants and the Coefficients of Thermal Expansion of Leucosapphire and Ruby," *Kristallografiya*, 1, 531.

Siegel, S., "The Structure of Titanium Fluoride," *Acta Cryst.*, 9, 684.

Tsai, K.-R., Harris, P. M., and Lassettre, E. N., "The Crystal Structure of Tricesium Monoxide," *J. Phys. Chem.*, 60, 345.

Vos, A., and Wiebenga, E. H., "Refinement of the P–P Bond Length in P_4S_7," *Acta Cryst.*, 9, 92.

Zhuravlev, N. N., "X-Ray Determination of the Structure of SiB_6," *Kristallografiya*, 1, 666.

Zhuravlev, N. N., and Zhdanov, G. S., "X-Ray Determination of the Structures of $CoSb_3$, $RhSb_3$, and $IrSb_3$," *Kristallografiya*, 1, 509.

1957

Agarwala, R. P., and Sinha, A. P. B., "Crystal Structure of Nickel Selenide, Ni_3Se_2," *Z. Anorg. Allgem. Chem.*, 289, 203.

Åsbrink, A., and Magnéli, A., "Note on the Crystal Structure of Trititanium Pentoxide," *Acta Chem. Scand.*, 11, 1606.

Asprey, L. B., Ellinger, F. H., Fried, S., and Zachariasen, W. H., "Evidence for Quadrivalent Curium. II. Curium Tetrafluoride," *J. Am. Chem. Soc.*, 79, 5825.

Atoji, M., and Lipscomb, W. N., "B—Cl Distance in Boron Trichloride," *J. Chem. Phys.*, 27, 195.

Basta, E. Z., "Accurate Determination of the Cell Dimensions of Magnetite," *Mineralog. Mag.*, 31, 431.

Burbank, R. D., and Bensey, F. N., Jr., "Structure of the Interhalogen Compounds. II. Iodine Heptafluoride at −110°C. and at −145°C.," *J. Chem. Phys.*, 27, 981.

Burbank, R. D., and Bensey, F. N., Jr., "Structure of the Interhalogen Compounds. III. Concluding Note on Bromine Trifluoride, Bromine Pentafluoride, and Iodine Pentafluoride," *J. Chem. Phys.*, 27, 982.

Cromer, D. T., "The Crystal Structure of Monoclinic Sm_2O_3," *J. Phys. Chem.*, 61, 753.

Dahl, L. F., and Rundle, R. E., "Polynuclear Metal Carbonyls. II. Structure of Iron Tetracarbonyl by X-Ray Diffraction," *J. Chem. Phys.*, 26, 1751.

Dickerson, R. E., Wheatley, P. J., Howell, P. A., and Lipscomb, W. N., "Crystal and Molecular Structure of B_9H_{15}," *J. Chem. Phys.*, 27, 200.

Ferro, R., "The Crystal Structures of Thorium Bismuthides," *Acta Cryst.*, 10, 476.

Filonenko, N. E., Lavrov, I. V., Andreeva, O. V., and Pevzner, R. L., "The Aluminous Spinel $AlO \cdot Al_2O_3$," *Dokl. Akad. Nauk SSSR*, 115, 583.

Forsberg, H. E., "The Crystal Structure of $InOHF_2$," *Acta Chem. Scand.*, 11, 676.

Fruchart, R., and Michel, A., "A New Boride of Nickel, Ni_3B, of the Same Structure as Cementite," *Compt. Rend.*, 245, 171.

Gasperin, M., "Crystallographic Study of a New Double Oxide of Tantalum and Uranium," *Compt. Rend.*, 244, 1225.

Glemser, O., Beltz, K., and Naumann, P., "The Silicon–Nitrogen System," *Z. Anorg. Allgem. Chem.*, 291, 51.

Hahn, H., and Frank, G., "On the Crystal Structure of In_2Se_3," *Naturwiss.*, 44, 533.

Hardie, D., and Jack, K. H., "Crystal Structures of Silicon Nitride," *Nature*, 180, 332.

Hepworth, M. A., and Jack K. H., "The Crystal Structure of Manganese Trifluoride, MnF₃," *Acta Cryst.*, **10**, 345.

Hepworth, M. A., Jack, K. H., Peacock, R. D., and Westland, G. J., "The Crystal Structures of the Trifluorides of Iron, Cobalt, Ruthenium, Rhodium, Palladium, and Iridium," *Acta Cryst.*, **10**, 63.

Houten, S., van, and Wiebenga, E. H., "The Crystal Structure of P_4S_5," *Acta Cryst.*, **10**, 156.

Jack, K. H., and Maitland, R., "The Crystal Structures and Interatomic Bonding of Chromous and Chromic Fluorides," *Proc. Chem. Soc.*, **1957**, 232.

Jack, K. H., and Wachtel, M. M., "The Characterization and Crystal Structure of Cesium Antimonide, a Photoelectric Surface Material," *Proc. Roy. Soc. (London)*, **A239**, 46.

Jagodzinski, H., "The Determination of a Structure Occurring During Precipitation of Mg–Al Spinels Supersaturated with Al_2O_3," *Z. Krist.*, **109**, 388.

Jellinek, F., "The Structures of the Chromium Sulphides," *Acta Cryst.*, **10**, 620.

Khodadad, P., "Uranium Oxyselenide, OSeU," *Compt. Rend.*, **245**, 2286.

Khodadad, P., and Flahaut, J., "On the Polyselenide of Uranium Se_3U," *Compt. Rend.*, **244**, 462.

Leung, Y. C., Waser, J., Houten, S. van, Vos, A., Wiegers, G. A., and Wiebenga, E. H., "The Crystal Structure of P_4S_3," *Acta Cryst.*, **10**, 574.

Lund, E. W., and Svendson, S. R., "The Crystal Structure of $N_4S_4H_4$," *Acta Chem. Scand.*, **11**, 940.

Mills, O. S., "Structure of Iron Dodecacarbonyl," *Chem. Ind. (London)*, **1957**, 73.

Moore, E. B., Jr., Dickerson, R. E., and Lipscomb, W. N., "Least Squares Refinements of $B_{10}H_{14}$, B_4H_{10}, and B_5H_{11}," *J. Chem. Phys.*, **27**, 209.

Neshpor, V. S., and Samsonov, G. V., "New Borides of the Rare Earth Elements," *Dopovidi Akad. Nauk Ukr. RSR*, **1957**, 478.

Nicholson, M. E., "Solubility of Boron in Fe_3C and Variation of Saturation Magnetization, Curie Temperature, and Lattice Parameter of $Fe_3(C,B)$ with Composition," *J. Metals*, **9**, *AIME Trans.*, **209**, 1.

Picon, M., and Flahaut, J., "Physical Characterization of Manganese Carbides," *Compt. Rend.*, **245**, 534.

Popper, P., and Ruddlesden, S. N., "Structure of the Nitrides of Silicon and Germanium," *Nature*, **179**, 1129.

Sinha, K. P., and Sinha, A. P. B., "Vacancy Distribution and Bonding in Some Oxides of Spinel Structure," *J. Phys. Chem.*, **61**, 758.

Sinha, K. P., and Sinha, A. P. B., "A Vacancy Superstructure Model for γ-Fe_2O_3," *Z. Anorg. Allgem. Chem.*, **293**, 228.

Stroganov, E. V., and Ovchinnikov, K. V., "Crystal Structure of Ruthenium Trichloride," *Vestn. Leningr. Univ.*, **12**, No. 22, *Ser. Fiz. Khim.*, No. 4, 152.

Tideswell, N. W., Kruse, F. H., and McCullough, J. D., "The Crystal Structure of Antimony Selenide, Sb_2Se_3," *Acta Cryst.*, **10**, 99.

Ventriglia, U., "Structural Investigations on Cobalt Arsenides," *Periodico Mineral. (Rome)*, **26**, 345.

Wadsley, A. D., "Partial Disorder in the Non-Stoichiometric Phase $Ti_{2+x}S_4(0.2<x<1)$," *Acta Cryst.*, **10**, 715.

Zubenko, V. V., and Umanskii, M. M., "X-Ray Determination of the Thermal Expansion of Single Crystals," *Kristallografiya*, **2**, 508.

1958

Andresen, A. F., "The Structure of U_3O_8 Determined by Neutron Diffraction," *Acta Cryst.*, **11**, 612.

Aronsson, B., "The Crystal Structure of Mo_5SiB_2," *Acta Chem. Scand.*, **12**, 31; *Congr. Intern. Chim. Pure Appl. 16^e, Paris 1957, Mem. Sect. Chim. Minerale*, 211.

Atoji, M., Gschneider, K., Jr., Daane, A. H., Rundle, R. E., and Spedding, F. H., "The Structures of Lanthanum Dicarbide and Sesquicarbide by X-Ray and Neutron Diffraction," *J. Am. Chem. Soc.*, **80**, 1804.

Beattie, H. J., Jr., "The Crystal Structure of a M_3B_2-Type Double Boride," *Acta Cryst.*, **11**, 607.

Bonnevie-Svendsen, M., "β-Fe_2O_3—A New Iron(III) Oxide Form," *Naturwiss.*, **45**, 542.

Clark, E. S., Templeton, D. H., and MacGillavry, C. H., "The Crystal Structure of Gold(III) Chloride," *Acta Cryst.*, **11**, 284.

Derbyshire, W. D., and Yearian, H. J., "X-Ray Diffraction and Magnetic Measurements of the Iron–Chromium Spinels," *Phys. Rev.*, **112**, 1603.

Dudkin, L. D., "The Chemical Bond in Semiconducting $CoSb_3$," *Soviet Phys.-Tech. Phys.*, **3**, 216; *Zh. Tekh. Fiz.*, **28**, 240.

Dvoryankina, G. G., and Pinsker, Z. G., "An Investigation of the Structure of Fe_4N," *Kristallografiya*, **3**, 438.

Eick, H. A., "The Preparation, Lattice Parameters and Some Chemical Properties of the Rare Earth Mono-Thio Oxides," *J. Am. Chem. Soc.*, **80**, 43.

Falqui, M. T., "Crystal Structure of Some Halides of Elements of the Eighth Group. II. The Crystal Structure of $PtCl_4$," *Ann. Chim. (Rome)*, 48, 1160.

Felten, E. J., Binder, I., and Post, B., "Europium Hexaboride and Lanthanum Tetraboride," *J. Am. Chem. Soc.*, **80**, 3479.

Ferguson, G. A., Jr., and Hass, M., "Magnetic Structure and Vacancy Distribution in γ-Fe_2O_3 by Neutron Diffraction," *Phys. Rev.*, **112**, 1130.

Francombe, M. H., "Structure-cell Data and Expansion Coefficients of Bismuth Telluride," *Brit. J. Appl. Phys.*, **9**, 415.

Guentert, O. J., and Mozzi, R. L., "The Monoclinic Modification of Gadolinium Sesquioxide, Gd_2O_3," *Acta Cryst.*, **11**, 746.

Hamilton, W. C., "Neutron Diffraction Investigation of the 119°K. Transition in Magnetite," *Phys. Rev.*, **110**, 1050.

Hoard, J. L., and Stroupe, J. D., "The Structure of Crystalline Uranium Hexafluoride," U.S. At. Energy Comm. Rept. TID-5290, Bk. 1, 325.

Jeannin, Y., and Bénard, J., "Structure and Stability of Titanium Trisulfide," *Compt. Rend.*, **246**, 614.

Kouvo, O., and Vuorelainen, Y., "Eskolaite, a New Chromium Mineral," *Am. Mineralogist*, **43**, 1098.

Krönert, W., and Plieth, K., "Crystal Structure of $ZrSe_3$," *Naturwiss.*, **45**, 416.

Kudintseva, G. A., Polyakova, M. D., Samsonov, G. V., and Tsarev, B. M., "Preparation and Some Properties of the Hexaboride of Yttrium," *Fiz. Metal. Metalloved.*, **6**, 272.

Lindqvist, I., and Weiss, J., "Crystal Structure of Platinum Tetrathionitrosyl," *J. Inorg. Nucl. Chem.*, **6**, 184.

Nagakura, S., "Study of Metallic Carbides by Electron Diffraction. II. Crystal Structure Analysis of Nickel Carbide," *J. Phys. Soc. Japan*, **13**, 1005.

Natta, G., Corradini, P., Bassi, I. W., and Porri, L., "Polymorphism of Crystalline Titanium Trichloride," *Atti Accad. Lincei, Rend., Classe Sci. Fis. Mat. Nat.*, **24**, 121.

Neshpor, V. S., and Samsonov, G. V., "New Borides of Rare Earth Metals," *Zh. Fiz. Khim.*, **32**, 1328.

Nowotny, H., and Wittmann, A., "The Structure of Metal-Rich Boride Phases of Vanadium, Niobium and Tantalum," *Monatsh.*, **89**, 220.

Oosterhout, G. W. van, and Rooijmans, C. J. M., "A New Superstructure in γ-Ferric Oxide," *Nature*, **181**, 44.

Rolsten, R. F., "An X-Ray Study of Niobium Pentabromide," *J. Phys. Chem.*, **62**, 126.

Rolsten, R. F., and Sisler, H. H., "An X-Ray Study of Titanium Tribromide," *J. Phys. Chem.*, **62**, 1024.

Ruddlesden, S. N., and Popper, P., "On the Crystal Structures of the Nitrides of Silicon and Germanium," *Acta Cryst.*, **11**, 465.

Rundqvist, S., "Crystal Structure of Ni_3B and Co_3B," *Acta Chem. Scand.*, **12**, 658.

Samsonov, G. V., Dzeganovskii, V. P., and Semashko, I. A., "Europium Hexaboride," *Dokl. Akad. Nauk SSSR*, **119**, 506.

Sass, R. L., and Donohue, J., "The Crystal Structure of $S_4N_4H_4$," *Acta Cryst.*, **11**, 497.

Spedding, F. H., Gschneider, K., Jr., and Daane, A. H., "The Crystal Structures of Some of the Rare Earth Carbides," *J. Am. Chem. Soc.*, **80**, 4499.

Stehlík, B., Weidenthaler, P., and Vlach, J., "Crystal Structure of Silver(III) Oxide," *Chem. Listy*, **52**, 2230.

Stepanova, A. A., and Zhuravlev, N. N., "X-Ray Diffraction Study of YbB_6, LuB_4, HoB_4, and GdB_4," *Kristallografiya*, **3**, 94.

Swoboda, T. J., Toole, R. C., and Vaughan, J. D., "New Magnetic Compounds of the Ilmenite-type Structure," *Phys. Chem. Solids*, **5**, 293.

Wickham, D. G., and Croft, W. J., "Crystallographic and Magnetic Properties of Several Spinels Containing Trivalent JA-1044 Manganese," *Phys. Chem. Solids*, **7**, 351.

Yamaguchi, G., and Sakamoto, K., "Crystal Structure of Bayerite," *Bull. Chem. Soc. Japan*, **31**, 140.

Zalkin, A., and Sands, D. E., "The Crystal Structure of $NbCl_5$," *Acta Cryst.*, **11**, 615.

1959

Aronsson, B., "The Crystal Structure of Ru_7B_3," *Acta Chem. Scand.*, **13**, 109.

Aronsson, B., and Lundgren, G., "X-Ray Investigations on Me–Si–B Systems (Me=Mn, Fe,Co). I. Some Features of the Co–Si–B System at 1000°C. Intermediate Phases in the Co–Si–B and Fe–Si–B Systems," *Acta Chem. Scand.*, **13**, 433.

Aronsson, B., Aselius, J., and Stenberg, E., "Borides and Silicides of the Platinum Metals," *Nature*, **183**, 1318.

Åsbrink, S., and Magnéli, A., "Crystal Structure Studies on Trititanium Pentoxide, Ti_3O_5," *Acta Cryst.*, **12**, 575.

Åsbrink, S., Friberg, S., Magnéli, A., and Andersson, G., "Note on the Crystal Structure of Trivanadium Pentoxide," *Acta Chem. Scand.*, **13**, 603.

Atoji, M., and Williams, D. E., "Deuterium Positions in Lanthanum Deuteroxide by Neutron Diffraction," *J. Chem. Phys.*, **31**, 329.

Austin, A. E., "Carbon Positions in Uranium Carbides," *Acta Cryst.*, **12**, 159.

Brewer, F. M., Garton, G., and Goodgame, D. M. L., "The Preparation and Crystal Structure of Gallium Trifluoride," *J. Inorg. Nucl. Chem.*, **9**, 56.

Brindley, G. W., and Nakahira, M., "Evidence for Variable Long-Range Order in Nearly Anhydrous Gamma Alumina," *Nature*, **183**, 1620.

Burbank, R. D., "Comments on J. Donohue's 'Molecular Symmetry of Iodine Heptafluoride'," *J. Chem. Phys.*, **30**, 1619.

Domange, L., Flahaut, J., and Guittard, M., "The Sulfides and Oxysulfide of Europium," *Compt. Rend.*, **249**, 697.

Donohue, J., "Molecular Symmetry of Iodine Heptafluoride," *J. Chem. Phys.*, **30**, 1618.

Eick, H. A., and Gilles, P. W., "Precise Lattice Parameters of Selected Rare Earth Tetra- and Hexaborides," *J. Am. Chem. Soc.*, **81**, 5030.

Keulen, E., and Vos, A., "The Crystal Structure of P_4Se_3," *Acta Cryst.*, **12**, 323.

Kouvo, O., Huhma, M., and Vuorelainen, Y., "A Natural Cobalt Analog of Pentlandite," *Am. Mineralogist*, **44**, 897.

Magnéli, A., Andersson, S., Kihlborg, L., Åsbrink, S., Westman, S., Holmberg, B., and Nordmark, C., "The Crystal Chemistry of Titanium, Vanadium and Molybdenum Oxides at Elevated Temperatures," *U.S. At. Energy Comm. Rept.* NP-8054, 141 pp.

Narita, K., and Mori, K., "Crystal Structures of Silicon Nitride," *Bull. Chem. Soc. Japan*, **32**, 417.

Natta, G., Corradini, P., and Allegra, G., "Crystal Structure of the γ-Form of Titanium Trichloride," *Atti Accad. Lincei, Rend., Classe Sci. Fis. Mat. Nat.*, **26**, 155.

Natta, G., Ercoli, R., Calderazzo, F., Alberola, A., Corradini, P., and Allegra, G., "Properties and Structure of a New Metal Carbonyl: $V(CO)_6$," *Atti Accad. Lincei, Rend., Classe Sci. Fis. Mat. Nat.*, **27**, 107.

Ohlberg, S. M., "The Crystal Structure of Antimony Pentachloride at $-30°$," *J. Am. Chem. Soc.*, **81**, 811.

Olovsson, I., and Templeton, D. H., "X-Ray Study of Solid Ammonia," *Acta Cryst.*, **12**, 832.

Paderno, Y. B., Serebryakova, T. I., and Samsonov, G. V., "Compounds of Terbium with Boron and the Electronic Configuration of the Terbium Atom," *Dokl. Akad. Nauk SSSR*, **125**, 317.

Pedersen, B., and Grønvold, F., "The Crystal Structures of α-V_3S and β-V_3S," *Acta Cryst.*, **12**, 1022.

Rooymans, C. J. M., "New Type of Cation-Vacancy Ordering in the Spinel Lattice of Indium Trisulfide," *J. Inorg. Nucl. Chem.*, **11**, 78.

Rundqvist, S., "An X-Ray Investigation of the Nickel–Boron System. The Crystal Structures of Orthorhombic and Monoclinic Ni_4B_3," *Acta Chem. Scand.*, **13**, 1193.

Rundqvist, S., and Larsson, E., "The Crystal Structure of $Ni_{12}P_5$," *Acta Chem. Scand.*, **13**, 551.

Samsonov, G. V., Dzeganovskii, V. P., and Semashko, I. A., "Europium Hexaboride," *Kristallografiya*, **4**, 119.

Samsonov, G. V., Paderno, Y. B., and Serebryakova, T. I., "Borides of Praseodymium, Erbium and Terbium," *Kristallografiya*, **4**, 542.

Sands, D. E., and Zalkin, A., "The Crystal Structure of $MoCl_5$," *Acta Cryst.*, **12**, 723.

Scheer, J. J., and Zalm, P., "Crystal Structure of Sodium Potassium Antimonide (Na_2KSb)," *Philips Res. Rept.*, **14**, 143.

Sgarlata, F., "Physical Properties of Crystals Containing Atoms with d Valence Electrons (Pyrite, Hauerite, Hematite, Rutile, Vanadium Pentoxide)," *Rend. Ist. Super. Sanità*, **22**, 851.

Shirane, G., Pickart, S. J., and Ishikawa, Y., "Neutron Diffraction Study of Antiferromagnetic $MnTiO_3$ and $NiTiO_3$," *J. Phys. Soc. Japan*, **14**, 1352.

Shirane, G., Pickart, S. J., Nathans, R., and Ishikawa, Y., "Neutron Diffraction Study of Antiferromagnetic $FeTiO_3$ and Its Solid Solutions with $\alpha\text{-}Fe_2O_3$," *Phys. Chem. Solids*, **10**, 35.

Terao, N., "Existence of a New Form of Nickel Nitride: Ni_4N," *Naturwiss.*, **46**, 204.

Zhuravlev, N. N., and Smirnov, V. A., "X-Ray Determination of the Structure of Cs_3Bi," *Kristallografiya*, **4**, 534.

1960

Andersson, S., "The Crystal Structure of Ti_5O_9," *Acta Chem. Scand.*, **14**, 1161.

Arnott, R. J., and Wold, A., "The Preparation and Crystallography of FeNiN and the Series $Fe_{4-x}Ni_xN$," *Phys. Chem. Solids*, **15**, 152.

Aronsson, B., Bäckman, M., and Rundqvist, S., "The Crystal Structure of Re_3B," *Acta Chem. Scand.*, **14**, 1001; U.S. Dept. Comm., Off. Tech. Serv., PB Rept. 145,859.

Aronsson, B., Stenberg, E., and Aselius, J., "Borides of Rhenium and the Platinum Metals. The Crystal Structures of Re_7B_3, ReB_3, Rh_7B_3, $RhB_{\sim 1.1}$, $IrB_{\sim 1.1}$ and PtB," *Acta Chem. Scand.*, **14**, 733; U.S. Dept. Comm., Off. Tech. Serv., PB Rept. 145,866.

Aselius, J., "The Crystal Structure of $Ru_{11}B_8$," *Acta Chem. Scand.*, **14**, 2169.

Atoji, M., and Rundle, R. E., "Neutron Diffraction Study on Sodium Tungsten Bronzes Na_xWO_3 ($x = 0.9 \sim 0.6$)," *J. Chem. Phys.*, **32**, 627.

Becher, H. J., "Beryllium Boride, BeB_{12}, with β-Boron Structure," *Z. Anorg. Allgem. Chem.*, **306**, 266.

Bertaut, F., Corliss, L., and Forrat, F., "Crystallographic and Magnetic Structure of Niobates and Tantalates of Bivalent Transition Metals," *Compt. Rend.*, **251**, 1733.

Binder, I., LaPlaca, S., and Post, B., "Some New Rare Earth Borides," *Boron Synthesis, Structure, Properties, Proc. Conf. Asbury Park, N.J., 1959*, p. 86.

Brosset, C., and Magnusson, B., "The Silicon–Boron System," *Nature*, **187**, 54.

Dvoryankina, G. G., and Pinsker, Z. J., "Electron Diffraction Study of Ferro-Ferric Oxide," *Dokl. Akad. Nauk SSSR*, **132**, 110.

Eckerlin, P., and Rabenau, A., "The System Be_3N_2–Si_3N_4. The Structure of a New Modification of Be_3N_2," *Z. Anorg. Allgem. Chem.*, **304**, 218.

Eick, H. A., "The Crystal Structure and Lattice Parameters of Some Rare Earth Monoseleno Oxides," *Acta Cryst.*, **13**, 161.

Geller, S., "Crystal Structure of $\beta\text{-}Ga_2O_3$," *J. Chem. Phys.*, **33**, 676.

Hanic, F., and Štempelová, D., "Structure of Chromium Trioxide," *Chem. Zvesti*, **14**, 165.

Hiller, J. E., and Wegener, W., "The System Nickel–Selenium," *Neues Jahrb. Mineral., Abh.*, **94**, 1147.

Kihlborg, L., "The Crystal Structure of $Mo_{17}O_{47}$," *Acta Chem. Scand.*, **14**, 1612.

Knox, K., "Structure of Chromium(III) Fluoride," *Acta Cryst.*, **13**, 507.

Kudryavtsev, V. I., and Sofronov, G. V., "Precision Lattice Parameter Determination of Boron Carbide of Compositions $B_{2.75}C$ to $B_{6.75}C$ by X-Ray Exposures Obtained in the Region of Large Angle Scattering ($0 \rightarrow 90°$)," *Tr. Seminara po Zharostoikim Materialam, Akad. Nauk Ukr. SSR, Inst. Metallokeram. i Spets. Splavov, Kiev*, 1958 (Pub. 1960), No. 5, p. 52.

LaValle, D. E., Steele, R. M., Wilkinson, M. K., and Yakel, H. L., Jr., "The Preparation and Crystal Structure of Molybdenum(III) Fluoride," *J. Am. Chem. Soc.*, **82**, 2433.

Makarov, E. S., Lipova, I. M., Dolmanova, I. F., and Melik'yan, A. A., "Crystal Structure of Uraninites and Pitchblendes," *Geokhimya*, **1960**, 193; *Geochemistry*, **1960**, 229.

Matkovich, V. I., "A New Form of Boron Silicide, B_4Si," *Acta Cryst.*, **13**, 679.

McDonald, B. J., and Stuart, W. I., "The Crystal Structures of Some Plutonium Borides," *Acta Cryst.*, **13**, 447.

Meinhardt, D., and Krisement, O., "Structure Investigation of the Chromium Carbide Cr_3C_2 by Thermal Neutrons," *Z. Naturforsch.*, **15a**, 880.

Paoletti, A., and Pickart, S. J., "Study of Rhombohedral V_2O_3 by Neutron Diffraction," *J. Chem. Phys.*, **32**, 308.

Rundqvist, S., "Phosphides of the Platinum Metals," *Nature*, **185**, 31.

Rundqvist, S., and Gullman, L.-O., "The Crystal Structure of Pd_3P," *Acta Chem. Scand.*, **14**, 2246.

Rundqvist, S., and Hede, A., "X-Ray Investigation on Rhodium Phosphides. The Crystal Structure of Rh_4P_3," *Acta Chem. Scand.*, **14**, 893; U.S. Dept. Comm., Off. Tech. Serv., PB Rept. 145,860 (1961).

Saalfeld, H., "The Structures of Gibbsite and of the Intermediate Products of Its Dehydration," *Neues Jahrb. Mineral., Abh.*, **95**, 1.

Ščavničar, S., "The Crystal Structure of Stibnite. A Redetermination of Atomic Positions," *Z. Krist.*, **114**, 85.

Semiletov, S. A., "Crystal Structure of the High Temperature Modification in In_2Se_3," *Kristallografiya*, **5**, 704.

Tanisaki, S., "Crystal Structure of Monoclinic Tungsten Trioxide at Room Temperature," *J. Phys. Soc. Japan*, **15**, 573.

Weiss, J., "The Crystal and Molecular Structure of Hexasulfurdiimide, $S_6(NH)_2$, and Heptasulfurimide, S_7NH," *Z. Anorg. Allgem. Chem.*, **305**, 190.

Wiese, J. R., and Muldawer, L., "Lattice Constants of Bi_2Te_3–Bi_2Se_3 Solid Solution Alloys," *Phys. Chem. Solids*, **15**, 13.

Yamaguchi, S., and Katsurai, T., "Formation of Ferromagnetic Fe_3S_4," *Kolloid-Z.*, **170**, 147.

Zhuravlev, N. N., Smirnov, V. A., and Mingazin, T. A., "X-Ray Study of Rb_3Bi and Rb_3Sb," *Kristallografiya*, **5**, 134.

1961

Atoji, M., and Williams, D. E., "Neutron Diffraction Studies of La_2C_3, Ce_2C_3, Pr_2C_3, and Tb_2C_3," *J. Chem. Phys.*, **35**, 1960.

Bachmann, H. G., Ahmed, F. R., and Barnes, W. H., "The Crystal Structure of Vanadium Pentoxide," *Z. Krist.*, **115**, 110.

Baenziger, N. C., Eick, H. A., Schuldt, H. S., and Eyring, L., "Terbium Oxides. III. X-Ray Diffraction Studies of Several Stable Phases," *J. Am. Chem. Soc.*, **83**, 2219.

Belbéoch, B., Piekarski, C., and Pério, P., "Structure of U_4O_9," *Acta Cryst.*, **14**, 837.

Bertaut, E. F., Corliss, L., Forrat, F., Aléonard, R., and Pauthenet, R., "Niobates and Tantalates of Bivalent Transition Metals," *Phys. Chem. Solids*, **21**, 234.

Bland, J. A., and Basinski, S. J., "The Crystal Structure of Bi_2Te_2Se," *Can. J. Phys.*, **39**, 1040.

Borgen, O., and Seip, H. M., "Crystal Structure of β-Si_3N_4," *Acta Chem. Scand.*, **15**, 1789.

Chevreton, M., and Bertaut, F., "Chromium Selenides," *Compt. Rend.*, **253**, 145.

274   CRYSTAL STRUCTURES

Chikawa, J., Imamura, S., Tanaka, K., and Shiojiri, M., "Crystal Structures and Electrical Properties of Alkali Antimonides," *J. Phys. Soc. Japan*, 16, 1175.

Corey, E. R., and Dahl, L. F., "Trinuclear Osmium and Ruthenium Carbonyls and Their Identities with Previously Reported $Os_2(CO)_9$ and $Ru_2(CO)_9$," *J. Am. Chem. Soc.*, 83, 2203.

Delafosse, D., Abon, M., and Barret, P., "Action of the Mixture H_2–H_2S on Anhydrous Cobalt Sulfate, Preparation of Co_9S_8," *Bull. Soc. Chim. France*, 1961, 1110.

Elfstrom, M., "The Crystal Structure of Cr_3B_4," *Acta Chem. Scand.*, 15, 1178.

Gnutzmann, G., Dorn, F. W., and Klemm, W., "The Behaviour of Alkali Metals towards Metalloids. VII. Some A_3B and AB_2 Compounds of the Heavy Alkali Metals with Elements of the Fifth Group," *Z. Anorg. Allgem. Chem.*, 309, 210.

Goodyear, J., Duffin, W. J., and Steigmann, G. A., "The Unit Cell of α-Ga_2S_3," *Acta Cryst.*, 14, 1168.

Grønvold, F., Kjekshus, A., and Raaum, F., "The Crystal Structure of Ti_5Te_4," *Acta Cryst.*, 14, 930.

Heyding, R. D., and Calvert, L. D., "Arsenides of the Transition Metals. IV. A Note on the Platinum Metal Arsenides," *Can. J. Chem.*, 39, 955.

Jellinek, F., "Structure of Molybdenum Sesquisulphide," *Nature*, 192, 1065.

Kempter, C. P., and Fries, R. J., "Crystallography of the Ru–B and Os–B Systems," *J. Chem. Phys.*, 34, 1994.

Kjekshus, A., and Pedersen, G., "The Crystal Structures of $IrAs_3$ and $IrSb_3$," *Acta Cryst.*, 14, 1065.

Knop, O., and Ibrahim, M. A., "Chalkogenides of the Transition Elements. II. Existence of the π Phase in the M_9S_8 Section of the System Fe–Co–Ni–S," *Can. J. Chem.*, 39, 297.

LaPlaca, S., and Post, B., "The Boron Carbide Structure Type," *Planseeber. Pulvermet.*, 9, 109.

Löhberg, K., "Cementite, Fe_3C, as an Ordered Substitution Crystal," *Naturwiss.*, 48, 46.

Matkovich, V. I., "Unit Cell, Space Group and Composition of a Lower Boron Phosphide," *Acta Cryst.*, 14, 93.

Men'kov, A. A., Komissarova, L. N., Simanov, Y. P., and Spitsyn, V. I., "Scandium Chalcogenides," *Dokl. Akad. Nauk SSSR*, 141, 364.

Muetterties, E. L., and Castle, J. E., "Reaction of Hydrogen Fluoride with Metals and Metalloids," *J. Inorg. Nucl. Chem.*, 18, 148.

Nagakura, S., "Study of Metallic Carbides by Electron Diffraction. IV. Cobalt Carbides," *J. Phys. Soc. Japan*, 16, 1213.

Natta, G., Corradini, P., and Allegra, G., "The Different Crystalline Modifications of $TiCl_3$, a Catalyst Component for the Polymerization of α-Olefins. I. α-, β-, γ-$TiCl_3$. II. δ-$TiCl_3$," *J. Polymer Sci.*, 51, 399.

Paderno, Y. B., and Samsonov, G. V., "Thulium Borides," *Zh. Strukt. Khim.*, 2, 213.

Perez y Jorba, M., Queyroux, F., and Collongues, R., "Existence of a Continuous Transition between the Fluorite Structure and the Tl_2O_3 Typical Structure in the Rare Earth–Zirconium Oxide Systems," *Compt. Rend.*, 253, 670.

Reed, J. W., and Harris, P. M., "Neutron Diffraction Study of Solid Deuteroammonia," *J. Chem. Phys.*, 35, 1730.

Saalfeld, H., "Corundum Structure," *Naturwiss.*, 48, 24.

Semiletov, S. A., "Electron Diffraction Investigation of the Structure of Indium Selenide Thin Films," *Fiz. Tverd. Tela*, 3, 746.

Semiletov, S. A., "Crystal Structure of the Low-Temperature Modification of In_2Se_3," *Kristallografiya*, **6**, 200.

Simpson, P. G., and Lipscomb, W. N., "Refinement of the B_9H_{15} Structure and Test for Solid Solutions," *J. Chem. Phys.*, **35**, 1340.

Stenberg, E., "The Crystal Structures of Pd_5B_2, (Mn_5C_2), and Pd_3B," *Acta Chem. Scand.*, **15**, 861.

Stuckens, W., and Michel, A., "Variations in the Stoichiometry of Pure Cementite," *Compt. Rend.*, **253**, 2358.

Zhuravlev, N. N., Stepanova, A. A., Paderno, Y. B., and Samsonov, G. V., "X-Ray Determination of the Expansion Coefficients of Hexaborides," *Kristallografiya*, **6**, 791.

1962

Aronsson, B., and Rundqvist, S., "Structural Features of Some Phases Related to Cementite," *Acta Cryst.*, **15**, 878.

Bartlett, N., Jha, N. K., and Trotter, J., "Osmium Oxide Pentafluoride $OsOF_5$," *Proc. Chem. Soc.*, **1962**, 277.

Berdonosov, S. S., Lapitskii, A. V., Vlasov, L. G., and Berdonosova, D. G., "X-Ray Studies of Zirconium Tetrabromide," *Zh. Neorg. Khim.*, **7**, 1465.

Burbank, R. D., "A Redetermination of the Orthorhombic IF_7 Structure," *Acta Cryst.*, **15**, 1207.

Chevreton, M., and Bertaut, F., "Titanium and Vanadium Selenide and Titanium Telluride," *Compt. Rend.*, **255**, 1275.

Corey, E. R., and Dahl, L. F., "Molecular and Crystal Structure of $Os_3(CO)_{12}$," *Inorg. Chem.*, **1**, 521.

Cras, J. A., "Crystal Structure of β-$TiCl_3$," *Nature*, **194**, 678.

Cushen, D. W., and Hulme, R., "The Crystal and Molecular Structure of Antimony Tribromide: β-Antimony Tribromide," *J. Chem. Soc.*, **1962**, 2218.

Dahl, L. F., and Wampler, D. L., "The Crystal Structure of α-Niobium Tetraiodide," *Acta Cryst.*, **15**, 903.

Dahl, L. F., and Wampler, D. L., "The Crystal Structure of Ruthenium Tetracarbonyl," *Acta Cryst.*, **15**, 946.

Eyring, L., and Baenziger, N. C., "On the Structure and Related Properties of the Oxides of Praseodymium," *J. Appl. Phys.*, **33**, 428.

Fert, A., "Structure of Some Rare Earth Oxides," *Bull. Soc. Franc. Mineral. Crist.*, **85**, 267.

Geller, S., "The Crystal Structure of $Pd_{17}Se_{15}$," *Acta Cryst.*, **15**, 713.

Geller, S., "Refinement of the Crystal Structure of Co_9S_8," *Acta Cryst.*, **15**, 1195.

Geller, S., "The Crystal Structure of the Superconductor $Rh_{17}S_{15}$," *Acta Cryst.*, **15**, 1198.

Grønvold, F., and Røst, E., "The Crystal Structure of Pd_4Se and Pd_4S," *Acta Cryst.*, **15**, 11.

Hanson, A. W., "The Crystal Structure of Iron Pentacarbonyl," *Acta Cryst.*, **15**, 930.

Hoppe, R., and Dähne, W., "The Crystal Structure of SnF_4 and PbF_4," *Naturwiss.*, **49**, 254.

King, G. S. D., "The Space Group of β-In_2S_3," *Acta Cryst.*, **15**, 512.

Kjekshus, A., Grønvold, F., and Thorbjörnsen, J., "On the Phase Relations in the Titanium–Antimony System. The Crystal Structures of Ti_3Sb," *Acta Chem. Scand.*, **16**, 1493.

Lohr, L. L., Jr., and Lipscomb, W. N., "Molecular Symmetry of IF_7," *J. Chem. Phys.*, **36**, 2225.

Magnusson, B., and Brosset, C., "The Crystal Structure of $B_{2.89}Si$," *Acta Chem. Scand.*, **16**, 449.

Newnham, R. E., and Haan, Y. M. de, "Refinement of the α-Al_2O_3, Ti_2O_3, V_2O_3 and Cr_2O_3 Structures," *Z. Krist.*, **117**, 235.

Pannetier, G., Bugli, G., and Courtine, P., "Radiocrystallographic Study of Co_9S_8," *Bull. Soc. Chim. France*, **1962**, 107.

Pebler, A., and Wallace, W. E., "Crystal Structures of Some Lanthanide Hydrides," *J. Phys. Chem.*, **66**, 148.

Ring, M. A., Donnay, J. D. H., and Koski, W. S., "The Crystal Structure of Boron Triiodide," *Inorg. Chem.*, **1**, 109.

Rundqvist, S., "X-Ray Investigations of the Ternary System Fe–P–B. Some Features of the Systems Cr–P–B, Mn–P–B, Co–P–B and Ni–P–B," *Acta Chem. Scand.*, **16**, 1.

Rundqvist, S., "X-Ray Investigations of Mn_3P, Mn_2P, and Ni_2P," *Acta Chem. Scand.*, **16**, 992.

Rundqvist, S., Hassler, E., and Lundvik, L., "Refinement of the Ni_3P Structure," *Acta Chem. Scand.*, **16**, 242.

Straumanis, M. E., and Ejima, T., "Imperfections within the Phase Ti_2O_3 and Its Structure Found by the Lattice Parameter and Density Method," *Acta Cryst.*, **15**, 404.

Terao, N., "Transformation of Metallic Lattices by Insertion of Nitrogen Atoms. I. Structure of Nickel Nitrides," *J. Phys. Soc. Japan*, **17**, *Suppl.* B-II, 238.

Yamaguchi, S., "Analysis of Berthollides by Electron Diffraction," *Z. Anal. Chem.*, **185**, 121.

Chapter VI

STRUCTURES OF THE COMPOUNDS R(MX$_2$)$_n$

Many of the compounds of this and the immediately succeeding chapters contain complex ions which, in the way they associate together, are distortions of some of the simple arrangements described in previous chapters. It is tempting to try to classify them according to these affinities. But when this is done, too many fall outside such a classification either because they are complex ionic associations that bear little relation to simple structures or because their bondings are predominantly non-ionic. With this in mind it has seemed better to group the crystals of the type R(MX$_2$)$_n$ together in terms of the nature of the X atoms they contain, that is into halides, oxides, etc. It will be noted that most of the halogen-containing crystals are built up of clear-cut complex ions, that complex ionic groupings are less evident in the oxides and quite lost when sulfides and related substances are reached.

Halides and Pseudo-Halides

VI,1. The acid fluorides and trinitrides of potassium and other large cations are clearly ionic with arrangements that are evidently like that of CsCl (**III,b1**). Their symmetry is tetragonal, and in the typical case of *potassium acid fluoride*, KHF$_2$, the tetramolecular cell has the dimensions:

$$a_0 = 5.67 \text{ A.}, \qquad c_0 = 6.81 \text{ A.}$$

Its atoms are in the following special positions of D$_{4h}^{18}$ (*I4/mcm*):

K: (4a) $\pm(0\ 0\ ^1/_4)$; B.C.
H: (4d) $0\ ^1/_2\ 0;\ ^1/_2\ 0\ 0$; B.C.
F: (8h) $\pm(u,u+^1/_2,0;\ ^1/_2-u,u,0)$; B.C.   with $u = 0.1408$

This structure, as illustrated by KN$_3$, is shown in Figure VI,1. The CsCl-like distribution of its K and HF$_2$ ions is apparent if one considers the pseudo-cell which has a c_0' axis (3.40 A.) that is half the length of the c_0 axis, and a_0' axes (4.00 A.) half the face diagonals of the square cell-base having a_0 as edge. In the HF$_2$ ions, fluorine atoms are 2.26 A. apart, this

277

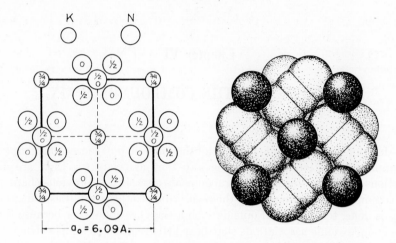

Fig. VI,1a (left). A projection on a basal plane of the tetragonal structure of KN_3. The smaller circles represent potassium atoms.
Fig. VI,1b (right). A packing drawing showing the distribution of the atoms of KN_3.

short distance being attributed to bondings to the hydrogen atom that lies between. There has been much discussion as to whether or not the hydrogen atom is symmetrically placed in this ion. Neutron diffraction studies of KHF_2 show that in this crystal the hydrogen cannot depart by more than 0.1 A. from a central position.

Other acid fluorides with this structure have the cell dimensions:

$$\alpha\text{-}RbHF_2: \quad a_0 = 5.90 \text{ A.}, \quad c_0 = 7.26 \text{ A. } (25^\circ C.)$$
$$\alpha\text{-}CsHF_2: \quad a_0 = 6.14 \text{ A.}, \quad c_0 = 7.84 \text{ A. } (40^\circ C.)$$

The potassium salt involving the halide-like trinitride ion, *potassium trinitride*, KN_3, has this atomic arrangement, with a unit of the dimensions:

$$a_0 = 6.094 \text{ A.}, \quad c_0 = 7.056 \text{ A.}$$

One nitrogen atom replaces the hydrogen atom of KHF_2 in (4*d*), and the parameter of the nitrogen atoms in (8*h*) has been determined as 0.133. As Figure VI,1 indicates, this results in linear N_3 ions in which the two equal N–N separations have the value 1.15 A.

Rubidium trinitride, RbN_3, also has this structure, with

$$a_0 = 6.36 \text{ A.}, \quad c_0 = 7.41 \text{ A.}$$

Several alkali cyanates have this type of arrangement, but a precise study has not been made of any one of them. Undoubtedly the CNO ions in these compounds replace the trinitride ions in KN_3, but since C–N is not necessarily equal to N–O, the symmetry is less than holohedral and the space group is other than $D_{4h}{}^{18}$ unless there is disorder in the way the CNO ions point in the crystal. Cell dimensions for these substances are

$$\begin{aligned}
\text{KCNO:} \quad & a_0 = 6.070 \text{ A.}, \quad && c_0 = 7.030 \text{ A.} \\
\text{RbCNO:} \quad & a_0 = 6.35 \text{ A.}, \quad && c_0 = 7.38 \text{ A.} \\
\text{CsCNO:} \quad & a_0 = 6.71 \text{ A.}, \quad && c_0 = 8.04 \text{ A.} \\
\text{TlCNO:} \quad & a_0 = 6.232 \text{ A.}, \quad && c_0 = 7.320 \text{ A.}
\end{aligned}$$

VI,2. The orthorhombic structure found for *ammonium acid fluoride*, NH_4HF_2, is a different distortion of the CsCl arrangement. Its tetramolecular unit has the dimensions:

$$a_0 = 8.408 \text{ A.}; \quad b_0 = 8.163 \text{ A.}; \quad c_0 = 3.670 \text{ A.}$$

Atoms are in the following positions of $V_h{}^7$ (*Pman*):

N:	(4g)	$\pm(\frac{1}{4}\,\frac{1}{4}\,u;\ \frac{3}{4}\,\frac{1}{4}\,u)$	with $u = 0.4498$
H(1):	(2a)	$000;\ \frac{1}{2}\,\frac{1}{2}\,0$	
H(2):	(2b)	$\frac{1}{2}\,0\,0;\ 0\,\frac{1}{2}\,0$	
F(1):	(4e)	$\pm(u00;\ u+\frac{1}{2},\frac{1}{2},0)$	with $u = 0.1353$
F(2):	(4h)	$\pm(0uv;\ \frac{1}{2},\frac{1}{2}-u,v)$	with $u = 0.3706,\ v = 0.8872$

The parameters given here are those of a recent detailed redetermination. They are close to but more precise than those originally found many years ago (direction of the c_0 axis and of the z parameters reversed).

In this recent study, positions have been assigned to the hydrogen atoms of the ammonium ions. Put in two sets of general positions:

$$(8i) \quad \pm(xyz;\ x\bar{y}\bar{z};\ \tfrac{1}{2}-x,\tfrac{1}{2}-y,z;\ \tfrac{1}{2}-x,y+\tfrac{1}{2},\bar{z})$$

they have been given the parameters:

$$\begin{aligned}
\text{H(3):} \quad & x = 0.212; \quad y = 0.173; \quad z = 0.307 \\
\text{H(4):} \quad & x = 0.327; \quad y = 0.212; \quad z = 0.586
\end{aligned}$$

The resulting ion is an almost regular tetrahedron with bond angles between 106°48′ and 110°54′. In it, N–H(3) = 0.878 A. and N–H(4) = 0.875 A. It is bound to four adjacent fluorine atoms through hydrogen bonds which have the lengths:

$$\text{N–H(3)–F(1)} = 2.797 \text{ A. and N–H(4)–F(2)} = 2.822 \text{ A.}$$

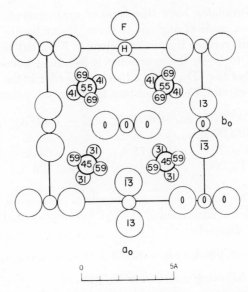

Fig. VI,2a. A projection of the orthorhombic structure of NH_4HF_2 along its c_0 axis. Origin in lower right.

The crystallographically unlike linear HF_2 ions have almost identical dimensions, with

$$F(1)-H-F(1) = 2.275 \text{ A. and } F(2)-H-F(2) = 2.269 \text{ A.}$$

As is evident from Figure VI,2, the CsCl-like pseudocell of this arrangement has a_0' and b_0' axes that are half those of the true unit.

Ammonium trinitride, NH_4N_3, also has this structure, with

$$a_0 = 8.930 \text{ A.}; \quad b_0 = 8.642 \text{ A.}; \quad c_0 = 3.800 \text{ A.}$$

Parameters found for it are $u(NH_4) = 0.467$; $u(F1) = 0.131$; $u(F2) = 0.377$; $v(F2) = 0.880$. The central nitrogen atoms of N_3 ions are in $(2a)$ and $(2b)$ listed above. In the crystallographically different N_3 ions, $N(1)-N(3) = 1.16$ A. and $N(2)-N(3) = 1.17$ A. Hydrogen positions were not determined, but, as in NH_4HF_2, each ammonium ion has four close neighbors with 2.94 A. as the shortest NH_4-N separation.

VI,3. *Silver trinitride*, AgN_3, though orthorhombic, has a unit which closely resembles in size that of KN_3 (**VI,1**). Its tetramolecular cell has the dimensions:

$$a_0 = 5.6170 \text{ A.}; \quad b_0 = 5.9146 \text{ A.}; \quad c_0 = 6.0057 \text{ A.}$$

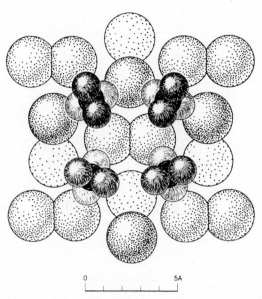

0   5A

Fig. VI,2b. A packing drawing of the NH_4HF_2 structure. The nitrogen atoms at the centers of the NH_4 ions are black. The hydrogen atoms of the HF_2 ions are not shown.

A structure has been described based on the following special positions of $V_h{}^{26}$ (*Ibam*):

Ag: (4b) $^1/_2\,0\,^1/_4$; $^1/_2\,0\,^3/_4$; B.C.
N(1):   (4c) 000; $0\,0\,^1/_2$; B.C.
N(2):   (8j) $\pm(uv0;\ \bar{u}\,v\,^1/_2)$; B.C.   with $u = v = 0.145$

The resulting arrangement is shown in Figure VI,3. With these parameters, N–N = 1.14 A., and in this structure Ag–N = 3.33 A.

VI,4. *Strontium trinitride*, $Sr(N_3)_2$, has a structure which does not closely resemble simpler ionic types. It forms orthorhombic crystals whose unit cell containing eight molecules has the edge lengths:

$$a_0 = 11.82 \text{ A.};\quad b_0 = 11.47 \text{ A.};\quad c_0 = 6.08 \text{ A.}$$

Atoms have been placed in the following positions of $V_h{}^{24}$ (*Fddd*):

Sr: (8a) 000; $^1/_4\,^1/_4\,^1/_4$; F.C.
N(1): (16e) $u00$; $\bar{u}00$; $u+^1/_4,^1/_4,^1/_4$; $^1/_4-u,^1/_4,^1/_4$; F.C.
                        with $u = 0.383$
N(2): (32h) xyz; $x\bar{y}\bar{z}$; $^1/_4-x,^1/_4-y,^1/_4-z$; $^1/_4-x,y+^1/_4,z+^1/_4$;
                        $\bar{x}y\bar{z}$; $\bar{x}\bar{y}z$; $x+^1/_4,^1/_4-y,z+^1/_4$; $x+^1/_4,y+^1/_4,^1/_4-z$; F.C.
                        with $x = 0.383$, $y = 0.058$, $z = 0.148$

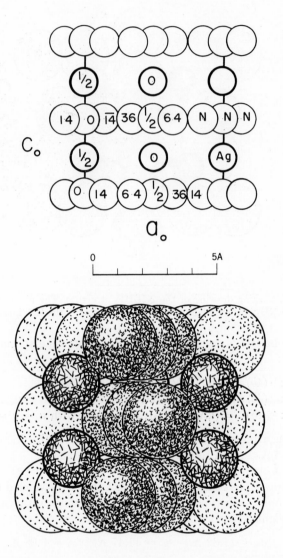

Fig. VI,3a (top). A projection along the b_0 axis of the orthorhombic structure of AgN_3. Origin in lower left.

Fig. VI,3b (bottom). A packing drawing of the orthorhombic structure of AgN_3 viewed along its b_0 axis. The nitrogen atoms are dotted.

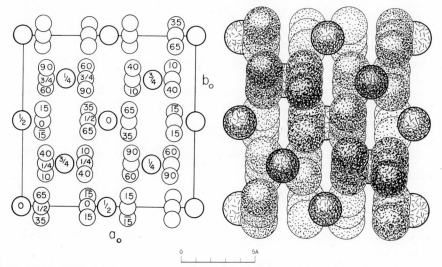

Fig. VI,4a (left). A projection along the c_0 axis of the orthorhombic structure of $Sr(N_3)_2$. The strontium atoms are more heavily ringed. Origin in lower right.

Fig. VI,4b (right). A packing drawing of the orthorhombic structure of $Sr(N_3)_2$ viewed along its c_0 axis. The nitrogen atoms are dotted.

In this arrangement (Fig. VI,4), the central atom of the N_3 group is separated from its two neighbors by 1.12 A. Each strontium atom has eight nitrogen neighbors, four at 2.63 A. and four more at 2.77 A.

VI,5. *Potassium isothiocyanate,* KSCN, is orthorhombic with a cell that is similar in shape to that of AgN_3 **(VI,3)**. Its tetramolecular unit has the edge lengths:

$$a_0 = 6.66 \text{ A.}; \quad b_0 = 6.635 \text{ A.}; \quad c_0 = 7.58 \text{ A.}$$

The space group is V_h^{11} (*Pbcm*), and potassium atoms are in the special positions:

$$(4c) \quad \pm (u \; ^1/_4 \; 0; \; u \; ^1/_4 \; ^1/_2) \qquad \text{with } u = -0.212$$

All other atoms are in

$$(4d) \quad \pm (u \; v \; ^1/_4; \; \bar{u}, v + ^1/_2, ^1/_4)$$

with the parameters $u(C) = 0.205$; $v(C) = 0.280$; $u(N) = 0.080$; $v(N) = 0.400$; $u(S) = 0.400$; $v(S) = 0.095$. Another determination, also made many years ago, led to substantially the same arrangement (except for somewhat different positions for carbon).

This structure, as illustrated in Figure VI,5, is much like that of AgN_3 (**VI,3**). Comparison with Figure VI,3 makes it apparent that they differ mainly in a small displacement of the centers of the SCN anions due presumably to their asymmetric shapes compared with the trinitride ion.

Thallium isothiocyanate, TlSCN, undoubtedly has the KSCN structure, with

$$a_0 = 6.78 \text{ A.}; \quad b_0 = 6.80 \text{ A.}; \quad c_0 = 7.52 \text{ A.}$$

Silver isocyanate, AgNCO, apparently is orthorhombic, with a somewhat similar cell:

$$a_0 = 6.37 \text{ A.}; \quad b_0 = 6.82 \text{ A.}; \quad c_0 = 5.48 \text{ A.}$$

It is not apparent from the existing data if it has the KSCN structure or how the axial sequence given here would be related to this arrangement.

VI,6. *Ammonium isothiocyanate,* NH_4SCN, forms monoclinic crystals, with a tetramolecular unit of the dimensions:

$$a_0 = 4.3 \text{ A.}; \quad b_0 = 7.2 \text{ A.}; \quad c_0 = 13.0 \text{ A.}; \quad \beta = 97°40'$$

All atoms are in general positions of C_{2h}^5 ($P2_1/c$):

$$(4e) \quad \pm(xyz; \bar{x},y+^1/_2,^1/_2-z)$$

with the following parameters:

Atom	x	y	z
S	0.992	0	0.195
N	0.165	0.714	0.061
C	0.086	0.840	0.120
NH₄	0.442	0.333	0.111

The structure that results (Fig. VI,6) is unlike others being discussed here. Nevertheless, it gives linear SCN ions similar to those found in other isothiocyanates with S–C = 1.59 A. and C–N = 1.25 A.

VI,7. A preliminary study of the structure of *silver isothiocyanate,* AgSCN, throws some more light on the thiocyanate group. This monoclinic crystal has eight molecules in a unit of the dimensions:

$$a_0 = 8.74 \text{ A.}; \quad b_0 = 7.96 \text{ A.}; \quad c_0 = 12.32 \text{ A.}; \quad \beta = 138°36'$$

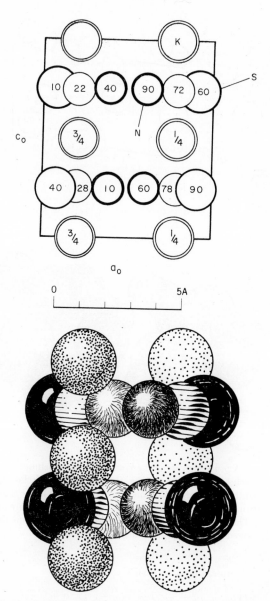

Fig. VI,5a (top). The orthorhombic structure of KSCN viewed along its b_0 axis. Origin in lower left.

Fig. VI,5b (bottom). A packing drawing of the orthorhombic KSCN structure seen along its b_0 axis. The sulfur atoms are large and black, the potassium atoms are dotted. The nitrogen atoms are fine-line shaded.

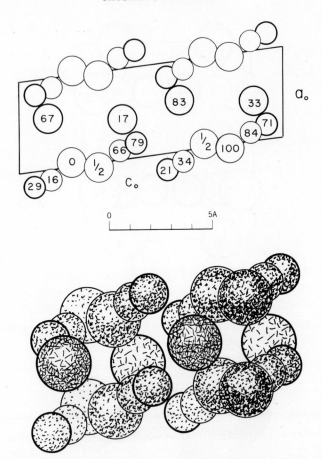

Fig. VI,6a (top). A projection along the b_0 axis of the monoclinic structure of NH$_4$SCN. The large heavy circles are ammonium ions, the small heavy circles are nitrogen atoms. Large and small lighter circles are sulfur and carbon. Origin in lower right.

Fig. VI,6b (bottom). A packing drawing of the monoclinic structure of NH$_4$SCN viewed along its b_0 axis. The ammonium ions are line-shaded. Of the smaller dotted atoms the more heavily ringed are the nitrogen, the others are carbon.

All atoms are in general positions of C_{2h}^6 $(C2/c)$:

$$(8f) \quad \pm(xyz;\ x,\bar{y},z+{}^1\!/_2;\ x+{}^1\!/_2,y+{}^1\!/_2,z;\ x+{}^1\!/_2,{}^1\!/_2-y,z+{}^1\!/_2)$$

For the heavier silver and sulfur, the parameters were found to be

$$\text{Ag:} \quad x = 0.048; \quad y = 0.105; \quad z = 0.404$$
$$\text{S:} \quad x = 0.211; \quad y = 0.217; \quad z = 0.700$$

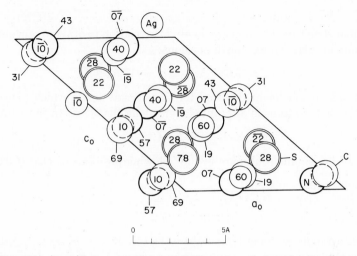

Fig. VI,7. The monoclinic structure of AgSCN projected along its b_0 axis. Origin in lower left.

The positions of the light carbon and nitrogen atoms are considered to be close to

$$N: \quad x = 0.349; \quad y = 0.067; \quad z = 0.047$$
$$C: \quad x = 0.465; \quad y = 0.186; \quad z = 0.089$$

The structure as shown in Figure VI,7 consists of atomic sheets roughly normal to the c_0 axis. Within these sheets, atoms lie in endless chains (Fig. VI,8) bent at the sulfur atoms but with straight SCN portions. The chains are considered to be held together in the crystal by silver–sulfur interactions with Ag–S = 2.997 and 2.886 A.

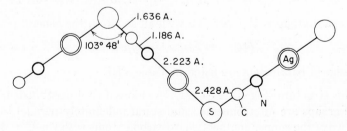

Fig. VI,8. A drawing to show the interatomic distances and bond angle in one of the chains of atoms found in AgSCN.

VI,8. The crystalline salt *potassium silver cyanide*, $KAg(CN)_2$, has a hexagonal structure which bears no obvious relation to other arrangements. Its hexamolecular cell has the dimensions:

$$a_0 = 7.384 \text{ A.}, \qquad c_0 = 17.55 \text{ A.}$$

The atomic arrangement deduced for this crystal is one which places the various kinds of atoms in alternate layers along the c_0 axis in the sequence:

$$\text{Ag-CN-K-CN-Ag-CN-K} \ . \ . \ .$$

Atoms of the six molecules are in the following positions of D_{3d}^2 ($C\bar{3}1c$):

K(1):   (2b)   000; 0 0 $^1/_2$
K(2):   (4f) $\pm(^1/_3\ ^2/_3\ u;\ ^1/_3,^2/_3,^1/_2-u)$ with $u = 0.010$
Ag: (6h) $\pm(u\ \bar{u}\ ^1/_4;\ u\ 2u\ ^1/_4;\ 2\bar{u}\ \bar{u}\ ^1/_4)$ with $u = 0.833$

In most cyanides at room temperature the CN group has been "rotating," that is, separate positions could not be assigned the carbon and nitrogen atoms. In the structure described for $KAg(CN)_2$, however, the carbon and nitrogen atoms have been given the definite positions:

$$(12i) \quad \pm(xyz; \qquad \bar{y},x-y,z; \qquad y-x,\bar{x},z;$$
$$y,x,z+^1/_2; \bar{x},y-x,z+^1/_2; x-y,\bar{y},z+^1/_2)$$

with $x(C) = 0.045$, $y(C) = 0.083$, $z(C) = -0.371$, and $x(N) = 0.115$, $y(N) = 0.083$, and $z(N) = -0.333$. These cannot be considered as firmly established by the existing x-ray data.

Sodium aurous cyanide, $NaAu(CN)_2$, is isomorphous, with

$$a_0 = 7.2 \text{ A.}, \qquad c_0 = 17.8 \text{ A.}$$

VI,9. Crystals of *potassium cuprous cyanide*, $KCu(CN)_2$, are monoclinic, with a tetramolecular cell of the dimensions:

$$a_0 = 7.57 \text{ A.}; \quad b_0 = 7.82 \text{ A.}; \quad c_0 = 7.45 \text{ A.}; \quad \beta = 102°12'$$

The space group is C_{2h}^5 ($P2_1/c$), with all atoms in the general positions:

$$(4e) \quad \pm(xyz; x,^1/_2-y,z+^1/_2)$$

The assigned parameters are listed in Table VI,1.

In this structure (Fig. VI,9), $Cu(CN)_2$ anions do not exist; instead, these atomic groups are interconnected and spiral indefinitely parallel to the b_0 axis. Two of the copper contacts are to carbon atoms with Cu–C = 1.92 A., the third is to nitrogen with Cu–N = 2.05 A. In the cyanide radicals, C–N = 1.15 or 1.13 A. The three atoms closest to a copper atom are almost exactly in a plane with it. The potassium ions lie in the spaces between the $[Cu(CN)_2]_\infty$ spirals and presumably bond them together. Each K^+ is coordinated to seven cyanide radicals with K–C(or N) distances of between ca. 2.80 and ca. 3.25 A.

TABLE VI,1
Parameters of the Atoms in KCu(CN)$_2$

Atom	x	y	z
K	0.3077	0.5700	0.1887
Cu	0.1219	0.1180	0.1074
C(1)	0.3349	0.1989	0.0334
C(2)	0.0649	0.9190	0.2393
N(1)	0.4656	0.2486	0.0003
N(2)	0.0340	0.8154	0.3324

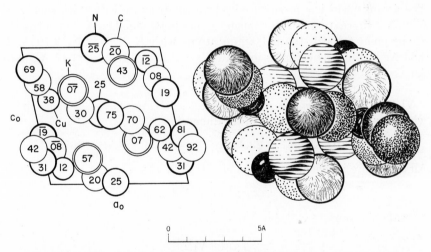

Fig. VI,9a (left). The monoclinic structure of KCu(CN)$_2$ seen along its b_0 axis.
Fig. VI,9b (right). A packing drawing of the KCu(CN)$_2$ arrangement viewed along its b_0 axis. The copper atoms are black, the carbon atoms are dotted, and the nitrogen atoms heavily outlined.

VI,10. *Potassium aurous cyanide*, KAu(CN)$_2$, is hexagonal, like the silver salt **(VI,8)**, but with a larger unit that corresponds to a more complicated stacking along the principal axis.

The symmetry is rhombohedral, rather than purely hexagonal, with the trimolecular unit:

$$a_0 = 9.74 \text{ A.}, \qquad \alpha = 43°54'$$

The corresponding cell in terms of hexagonal axes, containing nine molecules, has the edges:

$$a_0' = 7.28 \text{ A.}, \qquad c_0' = 26.36 \text{ A.}$$

This is a base of about the same size and a height about 50% greater than the unit of $KAg(CN)_2$.

Atoms have been found to be in the following positions of C_{3i}^2 ($R\bar{3}$) expressed in terms of the hexagonal cell:

K(1):   (3a)   000; rh

K(2):   (6b) $\pm(00u)$; rh $u = 0.3441$

Au: (9d) $^1/_2\,0\,^1/_2$; $0\,^1/_2\,^1/_2$; $^1/_2\,^1/_2\,^1/_2$; rh

C: (18f) $\pm(xyz;\ \bar{y},x-y,z;\ y-x,\bar{x},z)$; rh

            with $x = 0.403$, $y = 0.406$, $z = 0.424$

N: (18f)   with $x = 0.337$, $y = 0.348$, $z = 0.383$

The arrangement is a pronouncedly layered structure with thick sheets of $(AuCN)_2$ alternating with the much thinner potassium layers. The nitrogen atoms point towards the potassium atoms with K–N = 2.70, 2.79, and 2.83 A. Within the complex ion, Au–C = 2.12 A., C–N = 1.17 A., and Au–C–N = 172°48'.

VI,11. *Ammonium silver isothiocyanate*, $NH_4Ag(SCN)_2$, is monoclinic, with a long tetramolecular cell of the dimensions:

$$a_0 = 4.02 \text{ A.}; \quad b_0 = 23.86 \text{ A.}; \quad c_0 = 7.23 \text{ A.}; \quad \beta = 96°5'$$

All atoms are in general positions of C_{2h}^5 ($P2_1/n$):

$$(4e) \quad \pm(xyz;\ ^1/_2-x,y+^1/_2,^1/_2-z)$$

with the parameters of Table VI,2.

TABLE VI,2
Parameters of the Atoms in $NH_4Ag(SCN)_2$

Atom	x	y	z
Ag	0.411	0.1987	0.040
S(1)	0.630	0.1022	0.021
S(2)	0.397	0.2473	0.369
C(1)	0.508	0.0650	0.185
C(2)	0.517	0.1915	0.508
N(1)	0.378	0.0383	0.306
N(2)	0.577	0.1505	0.571
NH_4	0.062	0.0615	0.662

The structure, as shown in Figure VI,10, does not contain $Ag(SCN)_2$ ions. Instead, it can be interpreted as a mixture of NH_4^+, $(SCN)^-$ ions and neutral AgSCN molecular units. Within AgSCN, Ag–S = 2.474 A.; from a silver to the sulfurs of three neighboring SCN ions, the separations are 2.654, 2.630, and 2.742 A. In the AgSCN molecular units, S(1)–C(1) = 1.599 A. and C(1)–N(1) = 1.241 A. In the $(SCN)^-$ ions, S(2)–C(2) = 1.707 A. and C(2)–N(2) = 1.095 A.

In the crystal, each NH_4^+ ion has five nitrogen neighbors at distances between 2.911 and 3.268 A., and two sulfurs at 3.412 and 3.414 A.

VI,12. The structure possessed by *cesium dichloroiodide*, $CsCl_2I$, and numerous other compounds listed in Table VI,3, forms a bridge between the simple CsCl and NaCl arrangements of Chapter III. Its symmetry is

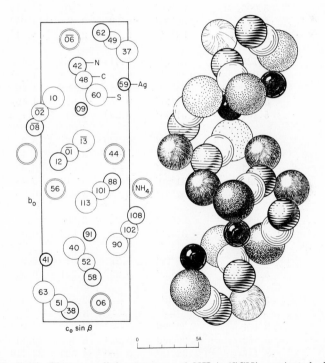

Fig. VI,10a (left). The monoclinic structure of $NH_4Ag(SCN)_2$ projected along its a_0 axis. Origin in lower left.

Fig. VI,10b (right). A packing drawing of the monoclinic $NH_4Ag(SCN)_2$ arrangement viewed along its a_0 axis. The silver atoms are black. In the SCN groups the sulfur atoms are the largest and the nitrogen atoms are the more heavily outlined circles.

TABLE VI,3. Crystals with the CsCl₂I Arrangement

Compound	Rhombohedral			Hexagonal		Element at origin
	a_0, A.	α	u	a_0', A.	c_0', A.	
AgBiSe₂[a]	7.022(240°C.)	34°30′	0.26	4.164	19.791	Ag
AgCrO₂	6.317	27°2′	0.125	2.946	18.25	Ag
AgFeO₂	6.44	27°6′	—	3.018	18.60	—
CaCN₂	5.35	40°28′	0.415	3.67	14.85	Ca
CuAlO₂	5.896	28°6′	0.12	2.870	16.98	Cu
CuCoO₂	5.95	27°42′	—	2.85	17.16	Cu
CuCrO₂	5.951	28°56′	—	2.975	17.096	Cu
CuFeO₂	5.959	29°26′	0.111	3.028	17.094	Cu
CuGaO₂	5.966	29°24′	0.11	3.033	17.12	Cu
CuRhO₂	5.99	29°45′	0.11	3.075	17.165	Cu
KCrS₂	7.36	28°28′	—	3.618	21.16	—
KTlO₂	6.416	30°59′	0.262	3.427	18 31	Tl
LiAlO₂	5.006	32°29′	—	2.801	14.214	—
LiCoO₂	4.96	32°58′	0.26	2.8166	14.052	Co
LiCrO₂	5.10	33°7′	0.25	2.905	14.44	Cr
LiHF₂	4.725	37°3′	0.414	3.003	13.186	Li
LiNO₂	5.013	32°2′	0.25	2.878	14.19	—
LiNiO₂	5.02	33°22′	0.25	2.88	14.20	Ni
LiRhO₂	5.075	34°37′	0.25	3.02	14.30	Rh
LiVO₂	5.167	31°54′	0.239	2.84	14.7	—
NaCrO₂	5.569	30°50′	0.220	2.96	15.9	—
NaCrS₂	6.87	29°48′	0.264	3.534	19.49	Cr
NaCrSe₂	7.094	30°18′	—	3.708	20.29	—
NaFeO₂	5.59	31°20′	0.222	3.019	15.934	Na
NaInO₂	5.761	32°37′	0.257	3.235	16.35	In
NaInS₂	6.984	31°36′	0.26	3.803	19.89	In
NaInSe₂	7.331	31°26′	0.26	3.972	20.89	In
NaNiO₂(high)[b]	5.527	31°4′	—	2.96	15 77	—
NaTiO₂	5 694	30°52′	0.22	3.02	16.2	Na
NaTlO₂	5.837	33°22′	0.25	3.351	16.52	Tl
NaVO₂	5.902	28°3′	—	2.86	17.0	—
RbCrS₂	—	—	—	3.39	?	—
RbCrSe₂	9.171	20°59′	—	3.32	26.9	—
RbTlO₂	6.652	29°59′	0.267	3.458	19.14	Tl
TlBiTe₂	8.137	32°18′	0.250	4.526	23.116	Tl
TlSbTe₂	8.177	31°24′	0.243	4.425	23.302	Tl

[a] See VI,41 for the room-temperature form of AgBiSe₂.
[b] The form of NaNiO₂ stable above 220°C. For the room-temperature modification see VI,16.

rhombohedral, with a unimolecular cell having ionic or other centers in the positions 000 and $1/2\ 1/2\ 1/2$. If the rhombohedral angle is ca. 90°, this is the CsCl grouping; if it is ca. 60°, the arrangement is that of NaCl. For most crystals with this structure, α is less than 60° and thus they appear as NaCl groupings that have been elongated along their body diagonals, often by the linear MX_2 ions they contain. For all these substances the space group is D_{3d}^5 ($R\bar{3}m$) with atoms in the positions:

R: (1a) 000   M: (1b) $\frac{1}{2}\frac{1}{2}\frac{1}{2}$ X: (2c) $\pm(uuu)$

or rhombohedral and

R: (3a) 000;rh   M: (3b) $0\ 0\ \frac{1}{2}$;rh   X: (6c) $\pm(00u)$;rh

for hexagonal axes and cells.

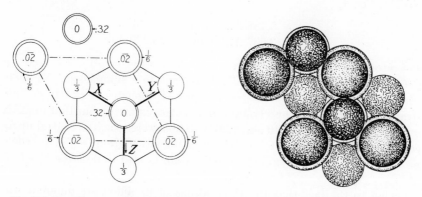

Fig. VI,11a (left). A projection of a portion of the $CsCl_2I$ structure on a plane through the apex of its unit rhombohedron and normal to the threefold axes. Cesium atoms are the smallest and iodine atoms the largest circles.

Fig. VI,11b (right). A packing drawing of the atoms of Figure VI,11a giving them their usual ionic sizes. The line-shaded spheres are the cesium ions.

For $CsCl_2I$, the unimolecular rhombohedron has the dimensions:

$$a_0 = 5.470\ \text{A.,} \qquad \alpha = 70°41'$$

Cesium is in the origin (1a), and $u(\text{Cl}) = 0.315$.
The trimolecular hexagonal cell has

$$a_0' = 6.328\ \text{A.,} \qquad c_0' = 12.216\ \text{A. (26°C.)}$$

In this structure (Fig. VI,11), Cs–Cl $= 3.66$ A. and I–Cl $= 2.25$ A.

Sodium acid fluoride, $NaHF_2$, is representative of the compounds with this structure that, having small values of α, are more like NaCl. For it,

$$a_0 = 5.004 \text{ A.}, \qquad \alpha = 40°33' \text{ for the rhombohedral cell}$$

and

$$a_0' = 3.468 \text{ A.}, \qquad c_0' = 13.76 \text{ A. (25°C.) for the hexagonal cell}$$

Its atoms are in the positions of D_{3d}^5 listed above, with H presumably at $1/2 \, 1/2 \, 1/2$ and u for fluorine around 0.410. Such a value of u yields the short F–H–F separation which is a consequence of the hydrogen bond existing in this ion. The Na–F distance, 2.30 A., is equal to the sum of the radii of these atoms as simple ions.

The unit of *sodium trinitride,* NaN_3, is similar in size and shape to that of $NaHF_2$. For it,

$$a_0 = 5.488 \text{ A.}, \qquad \alpha = 38°43'$$
$$a_0' = 3.637 \text{ A.}, \qquad c_0' = 15.209 \text{ A.}$$

The atomic positions have been expressed by the coordinates used for $CsCl_2I$ and $NaHF_2$, with $u = 0.425$. This leads to a symmetrical N_3 ion in which N–N = 1.17 A. It has been suggested, though not clearly demonstrated, that the N_3 in this crystal is not centrosymmetric but that its N–N separations instead are 1.10 and 1.26 A. If this were true, the central nitrogen atom would not be at $1/2 \, 1/2 \, 1/2$ and the space group of NaN_3 would probably be the lower symmetry space group needed to describe NaOCN.

Owing to the fact that the three atoms of its anion are different, the similar structure of *sodium isocyanate,* NaOCN, is based on C_{3v}^5 (*R3m*) with all atoms in the singly equivalent positions uuu. Parameters chosen for them [$u(\text{Na}) = 0$; $u(\text{N}) = 0.575$; $u(\text{C}) = 0.495$; $u(\text{O}) = 0.420$] result in separations within the OCN ion of C–O = 1.13 A. and C–N = 1.21 A. The unimolecular rhombohedron of this crystal has the dimensions:

$$a_0 = 5.44 \text{ A.}, \qquad \alpha = 38°37'$$

The trimolecular hexagonal cell has

$$a_0' = 3.576 \text{ A.}, \quad c_0' = 15.10 \text{ A.}$$

According to recent work, *sodium cyanate,* NaCNO, has substantially the same unit:

$$a_0 = 3.5851 \text{ A.}, \qquad c_0 = 15.110 \text{ A. (25°C.)}$$

Chromous acid, HCrO₂, and its deuterium analog, DCrO₂, have been shown by both x-ray and neutron analysis to have the CsCl₂I structure. For HCrO₂,

$$a_0 = 4.787 \text{ A.}, \qquad \alpha = 36°18'$$
$$a_0' = 2.984 \text{ A.}, \qquad c_0' = 13.40 \text{ A.}$$

The chromium atom is in (1*a*) 000 and $u(O) = 0.4050$. The neutron data suggest that the hydrogen atoms very possibly are exactly in (1*b*) $^1/_2\,^1/_2\,^1/_2$.

For DCrO₂, on the contrary, such a symmetrical position for the deuterium atoms seems improbable. Instead, they are described as statistically distributed about this position with half-atoms in (2*c*) $\pm(uuu)$ and $u(D) = 0.4834$. The parameter of the oxygens is $u(O) = 0.4082$. These values lead to a D–O = 1.02 A. Though complete cell data have not been published for DCrO₂, its c_0' axis is definitely longer ($c_0' = 13.50$ A.) than that of HCrO₂, and the best evidence points to an O–H–O = 2.51 A. in HCrO₂ shorter than O–D–O (2.57 A.) in DCrO₂.

Some of the other crystals with this atomic arrangement (Table VI,3) do not seem to contain stable ionic complexes and are probably best looked at from a different point of view. Thus when *u* is near $^1/_4$, this arrangement can be thought of as an NaCl grouping of X and undifferentiated R+M metallic atoms elongated along a trigonal axis; or it is a superlattice on NaCl which takes into account the difference in size between the R and M atoms. With small cations, the NaCl grouping itself appears as a close-packed framework of its large anions; from this standpoint such a compound as CuFeO₂ or NaFeO₂ is a distorted close-packing of O″ ions with interleaved metallic cations. This packing is illustrated, for α-NaFeO₂, in Figure VI,12.

Several compounds involving sulfur, or selenium, have unit cells like the others discussed here and apparently the reduced symmetry described for NaOCN. The electronegative atoms are approximately close-packed as in other CsCl₂I-like crystals, but it has been concluded that the metallic atoms are differently distributed. These crystals and the parameters chosen for them are as follows:

CuCrS₂: $a_0 = 6.605$ A., $\alpha = 30°42'$
$a_0' = 3.492$ A., $c_0' = 18.87$ A.
$u(Cu) = 0$; $u(Cr) = 0.86$; $u(S) = 0.13$ and 0.26

AgCrS₂: $a_0 = 7.130$ A., $\alpha = 28°24'$
$a_0' = 3.487$ A., $c_0' = 20.31$ A.
$u(Ag) = 0$; $u(Cr) = 0.858$, $u(S) = 0.142$ and 0.26

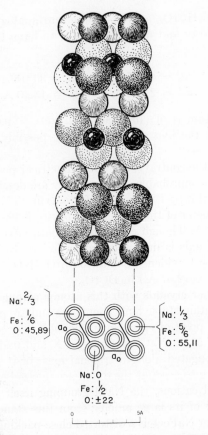

Fig. VI,12. Two projections in terms of the hexagonal cell of the $CsCl_2I$ structure as illustrated by $NaFeO_2$. In the packing drawing the iron atoms are black and the oxygens dotted.

$CuCrSe_2$: $a_0 = 6.804$ A., $\alpha = 31°18'$
$a_0' = 3.669$ A., $c_0' = 19.40$ A.
$u(Cu) = 0.26$; $u(Cr) = 0.74$; $u(Se) = 0$ and 0.13

$AgCrSe_2$: $a_0 = 7.392$ A., $\alpha = 28°54'$
$a_0' = 3.691$ A., $c_0' = 21.22$ A.
$u(Ag) = 0.27$; $u(Cr) = 0.73$; $u(Se) = 0$ and 0.14

VI,13. *Tetrachlorophosphonium dichloroiodide*, PCl_4ICl_2, has a structure which is a different distortion of the CsCl grouping. Two molecules are in a tetragonal unit of the dimensions:

$$a_0 = 9.26 \text{ A.,} \qquad c_0 = 5.68 \text{ A.}$$

The space group is V_d^3 ($P\bar{4}2_1m$), with atoms in the positions:

P: (2b) $0\,0\,{}^1/_2;$ ${}^1/_2\,{}^1/_2\,{}^1/_2$

I: (2c) $0\,{}^1/_2\,u;$ ${}^1/_2\,0\,\bar{u}$ with $u = 0.161$

Cl(1): (4e) $u,u+{}^1/_2,v;\ \bar{u},{}^1/_2-u,v;\ u+{}^1/_2,\bar{u},\bar{v};\ {}^1/_2-u,u,\bar{v}$
                                              with $u = 0.18,\ v = 0.16$

Cl(2): (8f) $xyz;\ \bar{x}\bar{y}z;\ {}^1/_2-x,y+{}^1/_2,\bar{z};\ x+{}^1/_2,{}^1/_2-y,\bar{z};$
             $\bar{y}x\bar{z};\ yx\bar{z};\ y+{}^1/_2,x+{}^1/_2,z;\ {}^1/_2-y,{}^1/_2-x,z$
                                              with $x = 0.15,\ y = 0.08,\ z = 0.29$

This structure was first described for *tetramethyl ammonium dichloro-iodide*, $N(CH_3)_4ICl_2$ (Chapter XIII loose-leaf). For it,

$$a_0 = 9.18 \text{ A.,} \qquad c_0 = 5.80 \text{ A.}$$

Its atoms are in the positions:

N: (2b)
I: (2c) with $u = 0.106$
Cl(1): (4e) with $u = 0.181,\ v = 0.106$
Cl(2): (8f) with $x = 0.114;\ y = 0.053;\ z = 0.361$

The arrangement, pictured for the substituted ammonium compound, is shown in Figure VI,13. As is evident from this drawing, the CsCl-like pseudo-cell with its single molecule has a base with edges that are half the diagonals of the true unit.

In both these dichloroiodides the ICl_2 group is linear, with I–Cl = 2.34 A.

VI,14. Cesium dichloroiodide (**VI,12**) is the only trihalide crystallizing with hexagonal symmetry. Most of the others are orthorhombic; and of these, two ammonium and two cesium salts have been studied with x-rays. They are isomorphous, with tetramolecular units of the dimensions:

CsI_3: $a_0 = 9.98$ A.; $b_0 = 11.09$ A.; $c_0 = 6.86$ A.
$CsIBr_2$: $a_0 = 9.18$ A.; $b_0 = 10.66$ A.; $c_0 = 6.57$ A.
NH_4I_3: $a_0 = 9.66$ A.; $b_0 = 10.82$ A.; $c_0 = 6.64$ A.
NH_4IBrCl: $a_0 = 8.58$ A.; $b_0 = 10.03$ A.; $c_0 = 6.14$ A.

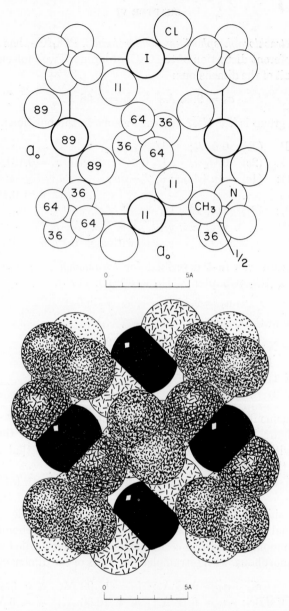

Fig. VI,13a (top). The tetragonal structure of $N(CH_3)_4ICl_2$ projected along its c_0 axis. Origin in lower left.

Fig. VI,13b (bottom). A packing drawing of the tetragonal structure of $N(CH_3)_4ICl_2$ viewed along its c_0 axis. The iodine atoms are black and the chlorine atoms line-shaded. The nitrogen atoms do not show within the enveloping tetrahedra of large dotted methyl groups.

Their space group is $V_h{}^{16}$ (*Pbnm*) and all atoms are in the positions:

$$(4c) \quad \pm(u\,v\,{}^1/_4;\,{}^1/_2-u,v+{}^1/_2,{}^1/_4)$$

An accurate determination has most recently been made of *cesium triiodide*, CsI_3. For it, the parameters are

Atom	u	v
Cs	0.4640	0.8259
I(1)	0.3598	0.1641
I(2)	0.5500	0.3778
I(3)	0.7367	0.5697

The structure that results yields triiodide ions that are almost linear, with I(1)–I(2)–I(3) = 176°18′ and I(1)–I(2) = 2.83 A., I(2)–I(3) = 3.03 A.

For *ammonium triiodide*, NH_4I_3, and *ammonium chlorobromoiodide*, NH_4ClBrI, the parameters are the following:

	NH_4I_3			NH_4ClBrI	
Atom	u	v	Atom	u	v
NH_4	0.478	0.831	NH_4	0.472	0.864
I(2)	0.544	0.381	I	0.561	0.375
I(1)	0.35	0.153	Cl	0.375	0.181
I(3)	0.736	0.575	Br	0.736	0.564

They lead to a linear BrICl ion but to an I_3 that is bent. The atomic separations within a BrICl group (I–Cl = 2.38 A. and I–Br = 2.50 A.) are, as in the case of the Cl_2I ion, close to the sums of the neutral radii. The shortest NH_4–Cl distance, 3.23 A., is about the sum of the ionic radii, but there is also an equally short NH_4–Br of 3.21 A. This atomic arrangement is illustrated in Figure VI,14. No detailed study has been published for $CsBr_2I$.

Oxides

VI,15. In addition to its rhombohedral $CsCl_2I$ form (**VI,12**) *sodium ferrite*, $NaFeO_2$, has a second, β, modification that is orthorhombic, with a tetramolecular cell of the dimensions:

$$a_0 = 5.672 \text{ A.}; \quad b_0 = 7.316 \text{ A.}; \quad c_0 = 5.377 \text{ A.}$$

All atoms are in general positions of $C_{2v}{}^9$ (*Pna*):

$$(4a) \quad xyz;\ \bar{x},\bar{y},z+{}^1/_2;\ {}^1/_2-x,y+{}^1/_2,z+{}^1/_2;\ x+{}^1/_2,{}^1/_2-y,z$$

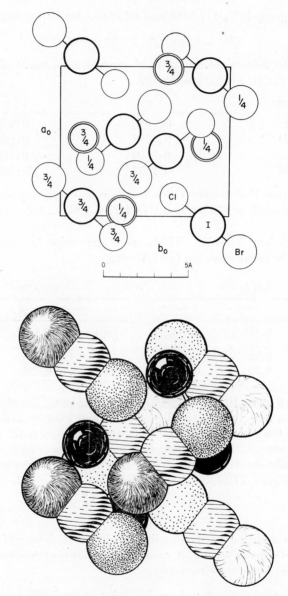

Fig. VI,14a (top). The orthorhombic structure of NH₄ClIBr projected along its c_0 axis. Origin in lower left.

Fig. VI,14b (bottom). A packing drawing of the orthorhombic NH₄ClIBr arrangement seen along its c_0 axis. The NH₄ ions are the black circles. The chlorine atoms are dotted and the bromine atoms are fine-line shaded.

with the following parameters:

Atom	x	y	z
Na	0.40	0.135	0.5
Fe	0.075	0.128	0.0
O(1)	0.05	0.10	0.33
O(2)	0.38	0.125	−0.10

Unlike most other structures of this sort, this involves a tetrahedral co-ordination of oxygen around the metallic atoms (Fig. VI,15).

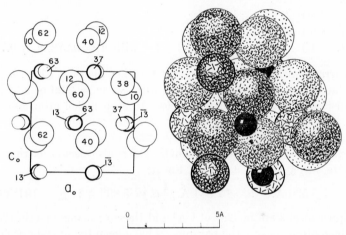

Fig. VI,15a (left). A projection of the orthorhombic, β, form of $NaFeO_2$ along the b_0 axis. The small heavy circles are the iron, the small light circles the sodium atoms. Origin in lower left.

Fig. VI,15b (right). A packing drawing of the orthorhombic, β, form of $NaFeO_2$ viewed along the b_0 axis. The sodium atoms are line-shaded, the oxygen atoms are dotted.

The beta form of $NaAlO_2$ has the same structure. For it,

$$a_0 = 5.376 \text{ A.}; \quad b_0 = 7.075 \text{ A.}; \quad c_0 = 5.216 \text{ A.}$$

A third, high-temperature, form of $NaFeO_2$ has tetragonal symmetry. Its tetramolecular cell with

$$a_0 = 5.56 \text{ A.}, \qquad c_0 = 7.30 \text{ A.}$$

does not differ greatly in size and shape from the orthorhombic, β, unit. Other substances with a similar modification are:

$$\gamma\text{-}NaAlO_2: \quad a_0 = 5.325 \text{ A.}, \quad c_0 = 7.058 \text{ A.}$$
$$\gamma\text{-}LiAlO_2: \quad a_0 = 5.181 \text{ A.}, \quad c_0 = 6.309 \text{ A.}$$

Their structure does not seem to have been determined. Presumably it is not isomorphous with KHF_2 (**VI,1**) and it is unlikely to be like $LiFeO_2$ (**VI,24**) which, though tetragonal, has a cell of different shape.

VI,16. The low-temperature form of *sodium nickel dioxide*, $NaNiO_2$, has a monoclinic structure that is a distortion of its $CsCl_2I$-like high-temperature modification (**VI,12**). The bimolecular unit has the dimensions:

$$a_0 = 5.33 \text{ A.}; \quad b_0 = 2.86 \text{ A.}; \quad c_0 = 5.59 \text{ A.}; \quad \beta = 110°30'$$

Atoms have been placed in the following special positions of C_{2h}^3 ($C2/m$):

Na: (2d) $0 \frac{1}{2} \frac{1}{2}; \frac{1}{2} 0 \frac{1}{2}$
Ni: (2a) $000; \frac{1}{2} \frac{1}{2} 0$
O: (4i) $\pm(u0v; u+\frac{1}{2}, \frac{1}{2}, v)$ with $u = 0.278, v = 0.795$

This arrangement (Fig. VI,16) is NaCl- rather than CsCl-like. In it each sodium atom has six oxygen neighbors with Na–O = 2.29 or 2.34 A.; the Ni–O distances for the six oxygens close to each nickel atom are 1.95 or 2.17 A.

Cupric manganous dioxide, $CuMnO_2$, which also occurs as the mineral crednerite, has a related structure. For it,

$$a_0 = 5.530 \text{ A.}; \quad b_0 = 2.884 \text{ A.}; \quad c_0 = 5.898 \text{ A.}; \quad \beta = 104°36'$$

The copper atoms are in (2d) of C_{2h}^3 and those of manganese in (2a). For the oxygen atoms in (4i) the parameters have been found to be $u = 0.416$, $v = 0.143$, values which are significantly different from those found for $NaNiO_2$. The manganese atoms can be thought of as octahedrally surrounded by oxygens with Mn–4O = 1.92 A., and Mn–2O = 2.28 A. Copper atoms located between these octahedra are 1.80 A. away from two oxygens.

No high-temperature rhombohedral phase analogous to that for $NaNiO_2$ was found.

VI,17. The alkali nitrites have NaCl-like structures with a low symmetry dictated by the spatial requirements of their V-shaped NO_2 ions. *Sodium nitrite*, $NaNO_2$, is orthorhombic, with

$$a_0 = 3.569 \text{ A.}; \quad b_0 = 5.563 \text{ A.}; \quad c_0 = 5.384 \text{ A.}$$

Its two molecules are in the special positions of C_{2v}^{20} ($Im2m$):

Na and N: (2a) $0u0; \frac{1}{2}, u+\frac{1}{2}, \frac{1}{2}$
O: (4c) $0uv; 0u\bar{v};$ B.C.

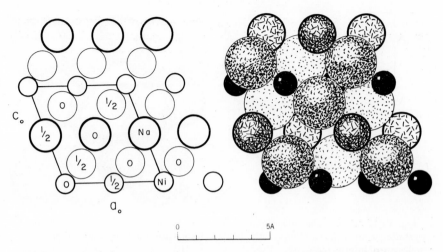

Fig. VI,16a (left). A projection along the b_0 axis of the monoclinic low-temperature form of NaNiO$_2$. The heavy large circles are sodium, the small circles are nickel atoms. Origin in lower left.

Fig. VI,16b (right). A packing drawing of the monoclinic form of NaNiO$_2$ viewed along the b_0 axis. The sodium atoms are line-shaded, the oxygen atoms are dotted.

Three precise determinations have recently been made of these parameters, two with x rays and a third using neutrons. The values, the last in each case being due to the neutron work, are

Na: $u = 0.5852, 0.5862, 0.5853$
N: $u = 0.1217, 0.1188, 0.1200$
O: $u = 0$ (arbitrary); $v = 0.1946, 0.1944, 0.1941$

In the resulting structure (Fig. VI,17), each sodium atom has about it oxygen atoms at a distance close to 2.470 A. and a nitrogen 2.590 A. away. In the NO$_2$ groups, N–O = ca. 1.240 A. and O–N–O = ca. 115°0'.

Silver nitrite, AgNO$_2$, has this structure, with

$a_0 = 3.528$ A.; $b_0 = 5.17$ A.; $c_0 = 6.171$ A.

A different origin along b_0 has been chosen for this crystal [u(Ag) = 0 rather than u(O) = 0] but bearing this in mind, the parameters found in a recent study are much like those established for NaNO$_2$: u(Ag) = 0 (arbitrary); u(N) = 0.478; u(O) = 0.576; v(O) = 0.168. An earlier determination placed the silver atoms closest to nitrogen rather than to oxygen atoms, but the parameters listed above make the distances about equal.

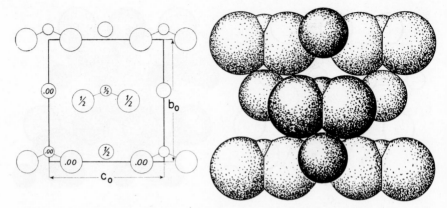

Fig. VI,17a (left). The orthorhombic structure found for NaNO₂ projected on the crystallographic a face. Atoms of the bent NO₂ ions are joined by light lines, the Na⁺ circles are of intermediate size.

Fig. VI,17b (right). A packing drawing of the atoms of NaNO₂. The oxygen atoms are the larger circles and the nitrogen atoms appear as pie-shaped segments.

Thus the short Ag–N = 2.47 A. while each silver atom also has two oxygen neighbors 2.47 and 2.42 A. away, and four more at 2.73 A.

The piezoelectricity shown by NaNO₂ at room temperature is gradually lost in the range between 160° and 166°C., and at higher temperatures the cell dimensions change at rates significantly different from those that prevail below 160°C. This slow transition expresses an increase in symmetry but not a radical alteration in the atomic arrangement. At 205°C.,

$$a_0 = 3.69 \text{ A.}; \quad b_0 = 5.68 \text{ A.}; \quad c_0 = 5.33 \text{ A.}$$

It has been concluded that the change is brought about by a developing oscillation of the ions along the b_0 axis around positions defined by the higher symmetry space group V_h^{25} (*Immm*). The atoms of sodium and nitrogen are to be considered as being in half the following positions:

$$(4g) \quad \pm (0u0); \text{ B.C.}$$

with parameter values which, as determined at 185°C., are

$$u(\text{Na}) = 0.4599, \quad u(\text{N}) = 0.0725$$

The atoms of oxygen are similarly distributed over half the positions:

$$(8l) \quad \pm (0uv; 0u\bar{v}); \text{ B.C.}$$

with $u(\text{O}) = -0.0416$ and $v(\text{O}) = 0.1920$. Such parameters give rise to

NO_2 ions in which the N–O is increased by ca. 5% and O–N–O is reduced compared to the dimensions prevailing in the room-temperature form.

VI,18. Crystals of *potassium nitrite*, KNO_2, are monoclinic with a bimolecular cell that has the following dimensions:

$$a_0 = 4.45 \text{ A.}; \quad b_0 = 4.99 \text{ A.}; \quad c_0 = 7.31 \text{ A.}; \quad \beta = 114°50'$$

Atoms have been put in special positions of the space group C_s^3 (Am):

K: (2a) $u0v; u,^1/_2,v+^1/_2$ with $u = v = 0$
N: (2a) with $u = 0.500$, $v = 0.486$
O: (4b) $xyz; x\bar{y}z; x,y+^1/_2,z+^1/_2; x,^1/_2-y,z+^1/_2$
    with $x = 0.444$, $y = 0.194$, $z = 0.417$

The resulting NO_2 ion has the same shape as in other nitrites. This structure (Fig. VI,18) is like $NaNO_2$ in having the shortest metallic contacts to oxygen. Each potassium atom is surrounded by four oxygen atoms of four different NO_2 groups (K–O = 2.75 A.) and by two more oxygen atoms belonging to a single NO_2 ion (K–O = 3.01 A.).

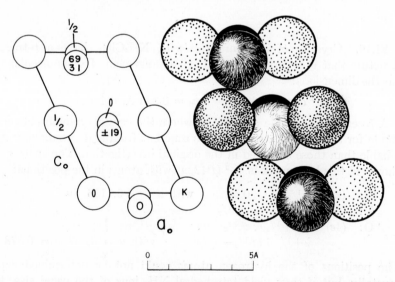

Fig. VI,18a (left). A projection along the b_0 axis of the monoclinic structure of KNO_2. Origin in lower left.

Fig. VI,18b (right). A packing drawing of the monoclinic KNO_2 arrangement viewed along its b_0 axis. The potassium atoms are dotted. Only segments of the black nitrogen atoms show behind the fine-line shaded atoms of oxygen.

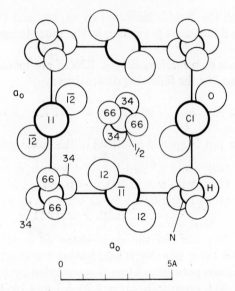

Fig. VI,19a. The tetragonal structure of NH_4ClO_2 projected along its c_0 axis.

VI,19. Crystals of *ammonium chlorite*, NH_4ClO_2, have a tetragonal structure that is a variant of the CsCl arrangement. Its bimolecular unit has the dimensions:

$$a_0 \doteqdot 6.362 \text{ A.}, \qquad c_0 = 3.823 \text{ A. (24°C.)}$$

A recent study, made at $-35°C$. where the crystals remain chemically stable for several days, gives results different from and presumably more reliable than those obtained in the original investigation. The space group chosen in this later work is $V_d{}^3$ ($P\bar{4}2_1m$) with atoms in the positions:

Cl: (2c)　$0 \, ^1/_2 \, u; \, ^1/_2 \, 0 \, \bar{u}$　　with $u = -0.1136$

N: (2b)　$0 \, 0 \, ^1/_2; \, ^1/_2 \, ^1/_2 \, ^1/_2$

O: (4e)　$u,u+^1/_2,v; \, \bar{u},^1/_2-u,v;$

　　　　$u+^1/_2,\bar{u},\bar{v}; \, ^1/_2-u,u,\bar{v}$　　with $u = 0.357, v = 0.878$

The positions of the hydrogen atoms could not be determined experimentally, but if they yield tetrahedral NH_4 ions of the usual size, they would be in (8f):

(8f)　$xyz; \, \bar{x}\bar{y}z; \, ^1/_2-x,y+^1/_2,\bar{z}; \, x+^1/_2,^1/_2-y,\bar{z};$

　　　$\bar{y}x\bar{z}; \, yx\bar{z}; \, y+^1/_2,x+^1/_2,z; \, ^1/_2-y,^1/_2-x,z$

　　　　　　with $x = 0.033, y = 0.100, z = 0.341$

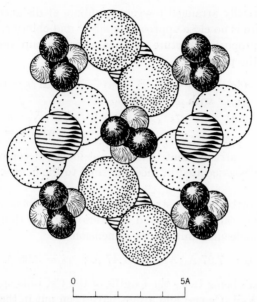

0   5A

Fig. VI,19b. A packing drawing of the tetragonal structure of NH_4ClO_2 seen along its c_0 axis. In the ammonium groups the hydrogen atoms are fine-line shaded and the nitrogens black. The oxygen atoms are dotted.

The resulting structure is shown in Figure VI,19. The ClO_2 ions are V-shaped with Cl–O = 1.57 A. and O–Cl–O = 110°30′. The N–H–O separations are 2.83 A.

The ClO_2 ions as described here are similar in dimensions to those found in the most recent study of $AgClO_2$ (**VI,20**).

VI,20. Two recent but disagreeing structures have been described for *silver chlorite*, $AgClO_2$. The crystal is orthorhombic, with a tetramolecular cell of the dimensions:

$$a_0 = 6.07 \text{ A.}; \quad b_0 = 6.13 \text{ A.}; \quad c_0 = 6.68 \text{ A.}$$

According to the later study, the space group is V_h^8 (*Pcac*), with atoms in the positions:

Ag: (4c) $\pm(0\ ^1/_4\ u;\ ^1/_2\ ^3/_4\ u)$ with $u = 0.112$
Cl: (4c)   with $u = 0.638$
O: (8f) $\pm(xyz;\ ^1/_2-x,y,\bar{z};\ x+^1/_2,^1/_2-y,\bar{z};\ \bar{x},^1/_2-y,z)$
          with $x = 0.36,\ y = 0.095,\ z = 0.23$

This gives a strongly layered structure (Fig. VI,20) composed of silver and ClO_2 ions. In the anion Cl–O = 1.55 A. and O–Cl–O = 111°. Each Ag has six prismatically arranged oxygen neighbors at distances between 2.4 and 2.6 A. There is no close approach of silver and chlorine.

In the other proposed structure the chosen space group was V_h^{21} ($Cmma$), with atoms in the positions:

$$\text{Ag:} \quad (4g) \quad \pm(0\ ^1/_4\ u;\ ^1/_2\ ^3/_4\ u) \qquad \text{with } u = 0.12$$
$$\text{Cl:} \quad (4g) \quad \text{with } u = 0.79$$
$$\text{O:} \quad (8n) \quad \pm(u\ ^1/_4\ v;\ \bar{u}\ ^1/_4\ v;\ u+^1/_2,^3/_4,v;\ ^1/_2-u,^3/_4,v)$$
$$\text{with } u = 0.22,\ v = 0.66$$

This would be an arrangement that was not ionic but composed of planar $AgClO_2$ molecules associated in sheets normal to the b_0 axis.

VI,21. The orthorhombic structure proposed for *ammonium hypophosphite*, $NH_4H_2PO_2$, can be considered as a seriously distorted CsCl arrangement. Its tetramolecular unit has the dimensions:

$$a_0 = 7.57 \text{ A.}; \quad b_0 = 11.47 \text{ A.}; \quad c_0 = 3.98 \text{ A.}$$

the a_0 and b_0 axes being twice the lengths of the CsCl-like pseudocell. The space group is V_h^{21} ($Cmma$) and atoms have been put in the positions:

$$\text{N:} \quad (4a) \quad \pm(^1/_4\ 0\ 0;\ ^3/_4\ ^1/_2\ 0)$$
$$\text{P:} \quad (4g) \quad \pm(0\ ^1/_4\ u;\ ^1/_2\ ^3/_4\ u)$$
$$\text{O:} \quad (8m) \quad \pm(0uv;\ ^1/_2\ u\ \bar{v};\ ^1/_2,u+^1/_2,v;\ 0,u+^1/_2,\bar{v})$$

with $u(P) = 0.542$, $u(O) = 0.136$ and $v(O) = 0.348$.

In the original description of this structure (Fig. VI,21), the hydrogen atoms were put in $(8n)$ $u\ ^1/_4\ v$; etc. to form tetrahedral H_2PO_2 ions; this may well be the case but such hydrogen positions were not established from the x-ray data. The significant interatomic distances are P–O = 1.52 A., O–O = 2.62 A. within the H_2PO_2 ion and N–O = 2.82 A. between NH_4 and H_2PO_2 ions.

VI,22. A structure has been assigned to *calcium hypophosphite*, $Ca(H_2PO_2)_2$, based on x-ray and neutron diffraction data. The symmetry is monoclinic, and there are four molecules in a cell of the dimensions:

$$a_0 = 15.08 \text{ A.}; \quad b_0 = 5.66 \text{ A.}; \quad c_0 = 6.73 \text{ A.}; \quad \beta = 102°8'$$

The space group is C_{2h}^6 ($C2/c$), with atoms in the positions:

$$(4e) \quad \pm(0\ u\ ^1/_4;\ ^1/_2,u+^1/_2,^1/_4)$$
$$(8f) \quad \pm(xyz;\ x,\bar{y},z+^1/_2;\ x+^1/_2,y+^1/_2,z;\ x+^1/_2,^1/_2-y,z+^1/_2)$$

The parameters as established with neutrons are those of Table VI,4.

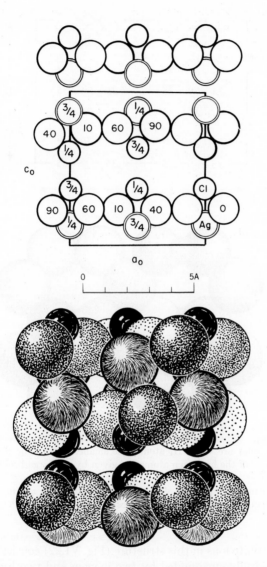

Fig. VI,20a (top). The orthorhombic structure of AgClO$_2$ based on V$_h^8$ projected along its b_0 axis. Origin in lower left.

Fig. VI,20b (bottom). A packing drawing of the orthorhombic AgClO$_2$ arrangement viewed along its b_0 axis. The chlorine atoms are black, the oxygen atoms dotted.

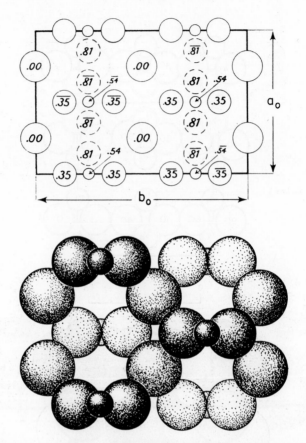

Fig. VI,21a (top). A projection along the c_0 axis of the orthorhombic structure of NH₄-H₂PO₂. Large circles are the NH₄ atoms, the smallest circles the phosphorus atoms. Proposed positions of the hydrogen atoms are given by the large broken circles. Origin at lower left.

Fig. VI,21b (bottom). A packing drawing of the NH₄H₂PO₂ arrangement giving the positions of the NH₄⁺ ions (the largest circles) and of the PO₂ groups.

The hypophosphite ions in this structure (Fig. VI,22) consist of phosphorus atoms tetrahedrally surrounded by two oxygen and two hydrogen atoms, with P–O = 1.49 and 1.52 A. and P–H = 1.38 and 1.40 A. The calcium ions are closest to oxygens, with Ca–O = 2.31, 2.39, and 2.41 A. This is an arrangement which, as the figure illustrates, consists of sheets parallel to the b_0c_0 plane; these sheets make contact with one another only through H–H separations.

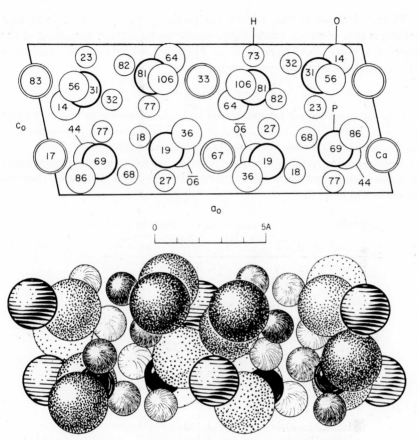

Fig. VI,22a (top). The monoclinic structure of Ca(H₂PO₂)₂ projected along its b_0 axis. Origin in lower left.

Fig. VI,22b (bottom). A packing drawing of the monoclinic Ca(H₂PO₂)₂ arrangement seen along its b_0 axis. The phosphorus atoms are black, the oxygen atoms dotted, and the hydrogens fine-line shaded.

TABLE VI,4

Positions and Parameters of the Atoms in Ca(H₂PO₂)₂

Atom	Position	x	y	z
Ca	(4e)	0	0.166	$^1/_4$
P	(8f)	0.143	0.686	0.218
O(1)	(8f)	0.075	0.861	0.103
O(2)	(8f)	0.112	0.437	0.230
H(1)	(8f)	0.170	0.770	0.417
H(2)	(8f)	0.220	0.685	0.138

VI,23. *Lithium ferrite*, $LiFeO_2$, as prepared by quenching from high temperatures gives an exceedingly simple x-ray pattern consisting of the lines of a cubic NaCl arrangement. There are accordingly two molecules in the unit which has the edge length:

$$a_0 = 4.141 \text{ A.}$$

The oxygen atoms may be considered to be in the positions of the chlorine atoms of NaCl ($1/2 \ 1/2 \ 1/2$; F.C.) and the lithium and iron atoms together distributed haphazardly in the sodium positions (000; F.C.).

A considerable number of oxides, sulfides, selenides, and tellurides, many of them high-temperature forms, have this disordered structure. They are listed in Table VI,5.

TABLE VI,5. Compounds with the Disordered $LiFeO_2$ Structure

Crystal	a_0, A.
$AgBiS_2$	5.648
$AgBiSe_2$[a]	5.887 (300°C.); 5.833 (25°C.)
$AgBiTe_2$	6.155
$AgSbS_2$	5.647 (25°C.)
$AgSbSe_2$	5.786 (25°C.)
$AgSbTe_2$	6.078 (25°C.)
$CuBiSe_2$	5.69
$KBiS_2$	6.04
$KBiSe_2$	5.922
$LiBiS_2$	5.603
$LiCo_{0.5}Mn_{0.5}O_2$	4.150
$LiFe_{0.5}Mn_{0.5}O_2$	4.184
$LiNi_{0.5}Mn_{0.5}O_2$	4.137
$LiCo_{0.5}Ti_{0.5}O_2$	4.166
$LiFe_{0.5}Ti_{0.5}O_2$	4.137
$LiMn_{0.5}Ti_{0.5}O_2$	4.216
$LiNi_{0.5}Ti_{0.5}O_2$	4.144
$LiTiO_2$	4.140
γ-$LiTlO_2$	4.568
$NaBiS_2$	5.77
$NaBiSe_2$	5.852
$TlBiS_2$	6.18
$TlSbS_2$	5.87–5.94

[a] For $AgBiSe_2$, the stable phase between 287° and ca. 120°C. is the rhombohedral $CsCl_2I$ structure described in **VI,12.** Below 120°C. it has the hexagonal form given in paragraph **VI,41.**

At a high temperature, KHF_2 has this cubic structure with $a_0 = 6.36$ A. (200°C.). The β form of $RbHF_2$ has the same structure with $a_0 = 6.71$ A. (180°C.). The corresponding β-$CsHF_2$ is body-centered though cubic, with a unimolecular cell having $a_0 = 4.21$ A. (80°C.). The fluorine atoms in these β forms are not considered to be in positions fixed by symmetry, but their centers are. For the potassium and rubidium salts fluorine atoms are said to be aligned along randomly chosen body diagonals of the cube; in β-$CsHF_2$ the alignment is random along cube edges. Two nitrites of large cations also are body-centered, like β-$CsHF_2$:

For $CsNO_2$: $a_0 = 4.34$ A.   For $TlNO_2$: $a_0 = 4.21$ A.

VI,24. When *lithium ferrite*, $LiFeO_2$, is annealed at 570°C., order develops in the distribution of the metallic atoms. The symmetry of this α form is tetragonal and it has a tetramolecular cell of the dimensions:

$$a_0 = 4.057 \text{ A.}, \qquad c_0 = 8.759 \text{ A.}$$

The space group is D_{4h}^{19} ($I4_1/amd$), with atoms in the special positions:

Fe: (4a) $000; 0\,^1/_2\,^1/_4$; B.C.
Li: (4b) $00\,^1/_2; 0\,^1/_2\,^3/_4$; B.C.
O: (8e) $\pm(00u;\, 0,u,u+^1/_4)$; B.C.   with $u = 0.25$

As Figure VI,23 indicates, this arrangement is the same kind of NaCl-like grouping as the disordered form (**VI,23**), the departure from cubic symmetry being a result of the need to accommodate metallic atoms of unequal sizes.

The following substances have been found to have this structure with $u = 0.25$ and the heavy metal in (4a):

$LiInO_2$: $a_0 = 4.316$ A., $c_0 = 9.347$ A.
β-$LiTlO_2$: $a_0 = 4.547$ A., $c_0 = 9.255$ A.
$LiScO_2$: $a_0 = 4.182$ A., $c_0 = 9.318$ A.

Still other compounds to which this structure has been ascribed are

$LiErO_2$: $a_0 = 4.42$ A., $c_0 = 10.20$ A.
$LiLuO_2$: $a_0 = 4.37$ A., $c_0 = 9.95$ A.
$LiTmO_2$: $a_0 = 4.405$ A., $c_0 = 10.15$ A.
$LiYO_2$: $a_0 = 4.44$ A., $c_0 = 10.35$ A.
$LiYbO_2$: $a_0 = 4.39$ A., $c_0 = 10.06$ A.

The following substance has a unit of the same general shape and size, though it has not yet been shown definitely to have this arrangement:

$LiPN_2$: $a_0 = 4.566$ A., $c_0 = 7.145$ A.

Another ordered form of LiFeO₂ has been reported as existing at temperatures below 400°C. It, too, is tetragonal, with a unit half as high as that of the α modification:

$$a_0 = 4.07 \text{ A.}, \qquad c_0 = 4.28 \text{ A.}$$

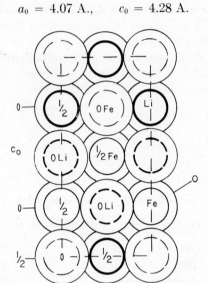

Fig. VI,23. The tetragonal ordered form of LiFeO₂ projected along an a_0 axis. The lithium atoms are heavily outlined, the oxygen atoms are the large, lightly ringed circles.

VI,25. Two determinations have recently been made of the hexagonal structure of *barium zinc dioxide*, $BaZnO_2$. They agree in the general type of arrangement but differ somewhat in the parameters assigned to the various atoms.

There are three molecules in a unit prism which according to the two studies have the dimensions:

$$a_0 = 5.927 \text{ A. } (5.886 \text{ A.}), \qquad c_0 = 6.707 \text{ A. } (6.734 \text{ A.})$$

The space group is D_3^4 ($P3_12$), with atoms in the following positions:

(3a) $u \, 0 \, ^1/_3; \; 0 \, u \, ^2/_3; \; \bar{u}\bar{u}0$

(3b) $u \, 0 \, ^5/_6; \; 0 \, u \, ^1/_6; \; \bar{u} \, \bar{u} \, ^1/_2$

(6c) $xyz; \; \bar{y}, x-y, z+^1/_3; \; y-x, \bar{x}, z+^2/_3;$

$yx\bar{z}; \; \bar{x}, y-x, ^1/_3-z; \; x-y, \bar{y}, ^2/_3-z$

According to one investigation, the barium atoms are in $(3b)$ with $u = 0.65$, the zinc atoms in $(3a)$ with $u = 0.50$, and the oxygens in $(6c)$ with $x = 0.41$, $y = 0.19$, $z = 0.15$.

According to the other (1960: vS,H&Z), bariums are in $(3a)$ with $u = 0.347$, zincs in $(3b)$ with $u = 0.418$, and oxygens in $(6c)$ with $x = 0.698$, $y = 0.118$, $z = 0.040$.

The close similarity between the results of these two determinations can be seen by comparing Figures VI,24a and VI,24b. As these two drawings indicate, origins have been taken at diagonally opposite corners of the base and the origin of one description is displaced by $^1/_2c_0$ with respect to the other. Figure VI,24c is a packing drawing of Figure VI,24a.

This structure is best considered as a distortion of that of low quartz (**IV,e3**) in which half the metallic atoms are different from the others and all atoms are displaced from their positions in quartz to take into account the different sizes of the two kinds of metal atoms. According to both determinations the zinc atoms are at the centers of tetrahedra of oxygen atoms with Zn–O = 1.92 A. in one case and 1.97–2.00 A. in the other. In the determination that placed the barium atoms in $(3a)$, they, too, are tetrahedrally surrounded by oxygens at 2.64 and 2.68 A. but with two more oxygens at 2.97 A. In the other determination the barium coordination is somewhat less definite, but the shortest Ba–O also is 2.64 A.

The compound *barium cobalt dioxide*, $BaCoO_2$, is isomorphous, with

$$a_0 = 5.85 \text{ A.}, \qquad c_0 = 6.73 \text{ A.}$$

Its atomic parameters have not been established.

VI,26. *Barium nickel dioxide*, $BaNiO_2$, is orthorhombic, with a tetramolecular cell of the edge lengths:

$$a_0 = 5.73 \text{ A.}; \quad b_0 = 9.20 \text{ A.}; \quad c_0 = 4.73 \text{ A.}$$

Atoms have been found to be in the following positions of V_h^{17} $(Cmcm)$:

Ni: $(4a)$ $000; \, ^1/_2\,^1/_2\,0; \, 0\,0\,^1/_2; \, ^1/_2\,^1/_2\,^1/_2$

Ba: $(4c)$ $\pm(0 \, u \, ^1/_4; \, ^1/_2,u+^1/_2,^1/_4)$ with $u = \, ^{11}/_{32}$

O: $(8g)$ $\pm(u \, v \, ^1/_4; \, \bar{u} \, v \, ^1/_4; \, u+^1/_2,v+^1/_2,^1/_4; \, ^1/_2-u,v+^1/_2,^1/_4)$

with $u = $ ca. $^1/_4$, $v = $ ca. $^1/_{12}$

In this arrangement (Fig. VI,25), which does not closely resemble those found for other crystals, Ni–O = 2.01 A., Ba–O = 2.80 and 2.84 A., but Ni–Ni = 2.36 A.

316 CRYSTAL STRUCTURES

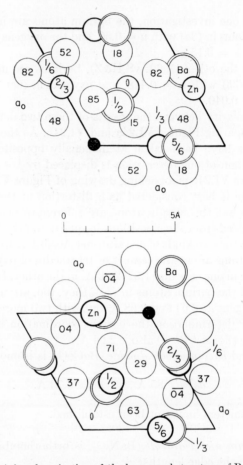

Fig. VI,24a (top). A basal projection of the hexagonal structure of BaZnO₂ according to the determination that places the zinc atoms in (3a). Origin in lower left.

Fig. VI,24b (bottom). A basal projection of the BaZnO₂ structure that placed the zinc atoms in (3b). Right-handed axes and origin in upper right.

VI,27. Crystals of *strontium zinc dioxide*, $SrZnO_2$, are orthorhombic, with a tetramolecular unit:

$$a_0 = 5.84 \text{ A.}; \quad b_0 = 3.35 \text{ A.}; \quad c_0 = 11.37 \text{ A.}$$

All atoms have been placed in the following special positions of V_h^{16} (*Pnma*):

$$(4c) \quad \pm(u\ ^1/_4\ v;\ u+^1/_2, ^1/_4, ^1/_2-v)$$

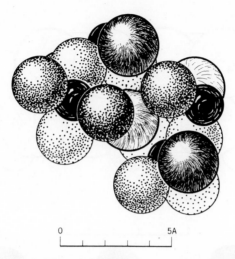

Fig. VI,24c. A packing drawing of the $BaZnO_2$ arrangement as portrayed in Figure VI,24a. The zinc atoms are black, the bariums are fine-line shaded.

with the parameters:

$$\text{Sr:} \quad u = 0.189, \quad v = 0.131$$
$$\text{Zn:} \quad u = 0.796, \quad v = 0.631$$
$$\text{O(1):} \quad u = 0.546, \quad v = 0.755$$
$$\text{O(2):} \quad u = 0.249, \quad v = 0.461$$

These lead to a structure (Fig. VI,26) in which the zinc atoms are tetrahedrally surrounded by four oxygens at distances between 1.97 and 2.01 A. Strontium atoms have seven closest oxygen neighbors with Sr–O = 2.58–2.78 A.

Barium cadmium dioxide, $BaCdO_2$, has a similar structure, with

$$a_0 = 6.168 \quad \text{A.} ; b_0 = 3.660 \text{ A.}; \quad c_0 = 11.950 \text{ A.}$$

The parameters of the atoms have, however, been assigned values which, especially for the metal atoms, are somewhat different from those given above:

$$\text{Ba:} \quad u = 0.1046, \quad v = 0.1424$$
$$\text{Cd:} \quad u = 0.9041, \quad v = 0.6074$$
$$\text{O(1):} \quad u = 0.651, \quad v = 0.748$$
$$\text{O(2):} \quad u = 0.236, \quad v = 0.470$$

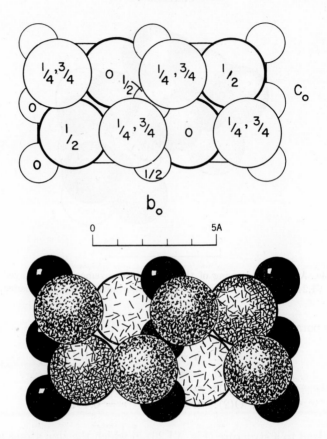

Fig. VI,25a (top). A projection along the a_0 axis of the orthorhombic structure of Ba-NiO₂. The heavy large circles are the barium, the light large circles the oxygen atoms. Origin in lower right.

Fig. VI,25b (bottom). A packing drawing of the orthorhombic structure of BaNiO₂ viewed along its a_0 axis. The barium atoms are line-shaded, the oxygen atoms dotted.

These result in cadmium having a fifth rather close oxygen neighbor, the four tetrahedrally distributed oxygens having Cd–O = 2.23–2.30 A. and the fifth 2.62 A. The seven oxygens nearest to barium are at distances between 2.64 and 2.92 A.

VI,28. *Silver lead dioxide*, Ag_2PbO_2, is monoclinic, with a tetramolecular cell of the dimensions:

$$a_0 = 6.082 \text{ A.}; \quad b_0 = 8.715 \text{ A.}; \quad c_0 = 6.556 \text{ A.}; \quad \beta = 93°42'$$

Atoms have been placed in the following positions of C_{2h}^6 ($I2/c$):

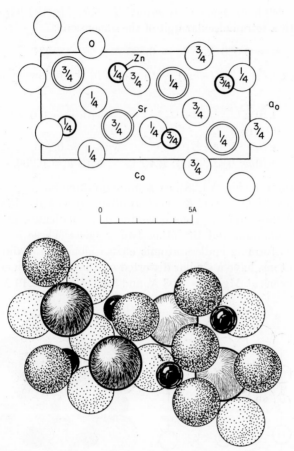

Fig. VI,26a (top). The orthorhombic structure of SrZnO₂ projected along its b_0 axis. Origin in lower right.

Fig. VI,26b (bottom). A packing drawing of the orthorhombic SrZnO₂ arrangement viewed along its b_0 axis. The zinc atoms are small and black, the oxygens dotted.

Ag(1):   (4b) $0 \frac{1}{2} 0; 0 \frac{1}{2} \frac{1}{2}$; B.C.
Ag(2):   (4d) $\frac{1}{4} \frac{1}{4} \frac{3}{4}; \frac{3}{4} \frac{1}{4} \frac{3}{4}$; B.C.
Pb: (4e) $0 u \frac{1}{4}; 0 \bar{u} \frac{3}{4}$; B.C.   with $u = 0.125$
O: (8f) $\pm (xyz; \bar{x},y,\frac{1}{2}-z)$; B.C.
                        with $x = 0.311$, $y = 0.195$, $z = 0.446$

The resulting structure is shown in Figure VI,27. Each atom of lead is surrounded by four oxygens in an approximately square arrangement, with Pb–O = 2.28 and 2.37 A. Oxygen atoms are shared by adjacent lead atoms along the a_0 direction to produce PbO₂ chains. The silver atoms interleave these chains, with Ag–O = 2.08 and 2.10 A.

VI,29. Crystals of *magnesium uranyl dioxide*, $Mg(UO_2)O_2$, are ortho-rhombic, with a tetramolecular unit of the dimensions:

$$a_0 = 6.520 \text{ A.}; \quad b_0 = 6.595 \text{ A.}; \quad c_0 = 6.924 \text{ A.}$$

The space group is V_h^{28} (*Imam*), with atoms in the following special positions:

U: (4e) $\pm (0\ u\ ^1/_4;\ ^1/_2, u + ^1/_2, ^3/_4)$ with $u = 0.0222$
Mg: (4b) $0\ ^1/_2\ 0;\ 0\ ^1/_2\ ^1/_2;\ ^1/_2\ 0\ ^1/_2;\ ^1/_2\ 0\ 0$
O(1):   (8i) $\pm (u\ v\ ^1/_4;\ \bar{u}\ v\ ^1/_4)$; B.C.   with $u = 0.295, v = 0.022$
O(2):   (8h) $\pm (0uv;\ 0, u, ^1/_2 - v)$; B.C.   with $u = 0.200, v = 0.012$

In this structure (Fig. VI,28), each uranium atom has two oxygens of a uranyl group at U–O = 1.92 A. and four other oxygens at 2.16 and 2.20 A. The plane of these four oxygens is, as the figure indicates, normal to the line through uranium and the other two oxygens. These atoms may be thought of as forming endless atomic chains along the c_0 direction. The magnesium atoms have a similar distorted octahedral oxygen environment with two oxygens at Mg–O = 1.98 A. and four others at 2.19 A.

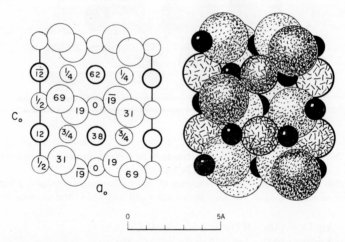

Fig. VI,27a (left). A projection along the b_0 axis of the monoclinic structure of Ag_2-PbO_2. The small heavy circles are the lead, the small light circles the silver atoms. Origin in lower left.
Fig. VI,27b (right). A packing drawing of the monoclinic structure of Ag_2PbO_2 viewed along the b_0 axis. The lead atoms are line-shaded, the oxygens are dotted.

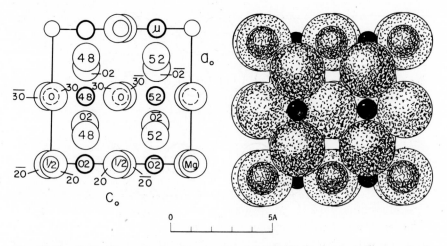

Fig. VI,28a (left). The orthorhombic structure of $Mg(UO_2)O_2$ projected along its b_0 axis. Origin in lower right.

Fig. VI,28b (right). A packing drawing of the $Mg(UO_2)O_2$ arrangement viewed along its b_0 axis. The uranium atoms are black, the magnesium atoms small and line-shaded.

VI,30. *Barium uranyl dioxide*, $Ba(UO_2)O_2$, is orthorhombic like the magnesium compound (**VI,29**) but with a different structure. Its tetramolecular unit has the dimensions:

$$a_0 = 5.751 \text{ A.}; \quad b_0 = 8.135 \text{ A.}; \quad c_0 = 8.236 \text{ A.}$$

The heavy atoms have been found to be in the following special positions of V_h^{11} (*Pbcm*):

U: (4a) $000; 0\,^1/_2\,0; 00\,^1/_2; 0\,^1/_2\,^1/_2$
Ba: (4d) $\pm(u\,v\,^1/_4; u,^1/_2-v,^3/_4)$ with $u = 0.474$, $v = 0.200$

Plausible positions for the oxygen atoms have been chosen as

O(1): (4c) $\pm(u\,^1/_4\,0; u\,^1/_4\,^1/_2)$ with $u = 0.11$
O(2): (4d) with $u = -0.13$, $v = -0.04$
O(3): (8e) $\pm(xyz; x,y,^1/_2-z; \bar{x},y+^1/_2,z; \bar{x},y+^1/_2,^1/_2-z)$
          with $x = 0.29$, $y = -0.06$, $z = 0.09$

This is a structure (Fig. VI,29) that surrounds each uranium atom with six octahedrally distributed oxygen neighbors; of these, four are shared (at

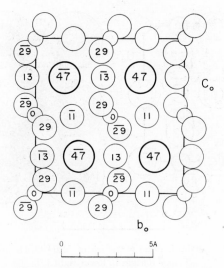

Fig. VI,29a. A projection along the a_0 axis of the orthorhombic structure of $Ba(UO_2)O_2$. The largest, heavy circles are the barium atoms, the smallest circles are uranium. Origin in lower right.

distances of 2.12 and 2.22 A.), the other two are those of UO_2 groups with U–O = 1.90 A. These octahedra form a net or sheet in the b_0c_0 plane. The sheets are held together along the a_0 axis by barium atoms midway between. Though the chemical formula of this compound, $BaUO_4$, like that of the other uranyl dioxides, corresponds to a uranate, the structure clearly contains uranyl cations.

$Pb(UO_2)O_2$, or $PbUO_4$, also has this structure with

$$a_0 = 5.528 \text{ A.}; \quad b_0 = 7.952 \text{ A.}; \quad c_0 = 8.180 \text{ A.}$$

A second form of $PbUO_4$ also exists with the cubic fluorite structure possessed by UO_2 (**IV,a1**) and with $a_0 = 5.600$ A.

VI,31. The structures found for *calcium uranyl dioxide*, $Ca(UO_2)O_2$, and for the isomorphous $Sr(UO_2)O_2$ and $K(AmO_2)F_2$ differ from that of the barium salt, though they, too, indicate formulas of the type $R(MO_2)X_2$. These crystals have hexagonal symmetry with unimolecular rhombohedral cells of the dimensions:

$$
\begin{array}{lll}
Ca(UO_2)O_2: & a_0 = 6.266 \text{ A.}, & \alpha = 36°2' \\
Sr(UO_2)O_2: & a_0 = 6.54 \text{ A.}, & \alpha = 35°32' \\
K(AmO_2)F_2: & a_0 = 6.78 \text{ A.}, & \alpha = 36°15'
\end{array}
$$

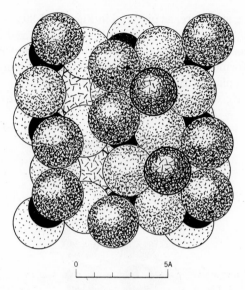

Fig. VI,29b. A packing drawing of the orthorhombic structure of Ba(UO$_2$)O$_2$ viewed along its a_0 axis. The barium atoms are line-shaded, the oxygen atoms are dotted.

Referred to hexagonal axes, their trimolecular cells have the edge lengths:

Ca(UO$_2$)O$_2$: $a_0' = 3.876$ A., $c_0' = 17.558$ A.
Sr(UO$_2$)O$_2$: $a_0' = 3.991$ A., $c_0' = 18.361$ A.
K(AmO$_2$)F$_2$: $a_0' = 4.218$ A., $c_0' = 18.982$ A.

As determined for the calcium compound, atoms are in the following positions of D_{3d}^5 ($R\overline{3}m$):

U: (1a)   000
Ca: (1b) $^1/_2\,^1/_2\,^1/_2$
O(1): (2c) $\pm(uuu)$ with $u = 0.109$
O(2): (2c)   with $u = 0.361$

for the unit rhombohedron and

U: (3a)   000; rh.,
Ca: (3b)   00 $^1/_2$; rh
O(1): (6c) $\pm(00u)$; rh   with $u = 0.109$
O(2): (6c)   with $u = 0.361$

for the hexagonal cell.

In the resulting structure, as pictured in Figure VI,30, U–O(1) = 1.91 A. and U–O(2) = 2.29 A. The Ca–O separations are 2.44 and 2.45 A. This is a structure that contains the usual uranyl groups but that can also be considered as a deformed CaF_2 arrangement with the calcium and uranium atoms together replacing the calcium ions of CaF_2.

For $K(AmO_2)F_2$, the parameters were determined as $u(O)$ = 0.102 and $u(F)$ = 0.356.

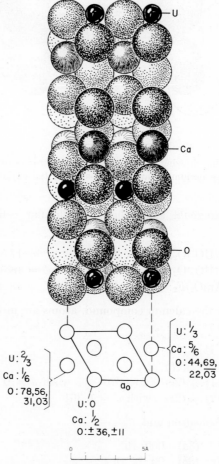

Fig. VI,30. Two projections of the hexagonal $Ca(UO_2)O_2$ structure.

VI,32. An orthorhombic beta modification of *cadmium uranyl dioxide*, $Cd(UO_2)O_2$, has been described. Its bimolecular cell has the dimensions:

$$a_0 = 7.01 \text{ A.}; \quad b_0 = 6.836 \text{ A.}; \quad c_0 = 3.52 \text{ A.}$$

A structure has been assigned based on V_h^{19} ($Cmmm$) with atoms in the positions:

U: (2a) $000; \frac{1}{2}\frac{1}{2}0$

Cd: (2c) $\frac{1}{2}0\frac{1}{2}; 0\frac{1}{2}\frac{1}{2}$

O(1): (4i) $\pm(0u0; \frac{1}{2},u+\frac{1}{2},0)$ with $u = 0.278$

O(2): (4h) $\pm(u\,0\,\frac{1}{2}; u+\frac{1}{2},\frac{1}{2},\frac{1}{2})$ with $u = 0.200$

The suggestion has been made that the structure may be similar to the foregoing but based on V_h^9 ($Pbam$). Evidently, additional work should be carried out on this, as on several of the other uranyl oxides.

VI,33. Several not entirely satisfactory structures have been suggested for alkali uranyl dioxides. Thus the α form of *sodium uranyl dioxide*, $Na_2(UO_2)O_2$, or Na_2UO_4, is described as orthorhombic with a bimolecular cell of the dimensions:

$$a_0 = 9.74 \text{ A.}; \quad b_0 = 5.72 \text{ A.}; \quad c_0 = 3.49 \text{ A.}$$

Its atoms have been placed in the following positions of V_h^{19} ($Cmmm$):

U: (2a) $000; \frac{1}{2}\frac{1}{2}0$

Na: (4f) $\frac{1}{4}\frac{1}{4}\frac{1}{2}; \frac{3}{4}\frac{1}{4}\frac{1}{2}; \frac{3}{4}\frac{3}{4}\frac{1}{2}; \frac{1}{4}\frac{3}{4}\frac{1}{2}$

O(1): (4i) $\pm(0u0; \frac{1}{2},u+\frac{1}{2},0)$ with $u = 0.195$

O(2): (4h) $\pm(u\,0\,\frac{1}{2}; u+\frac{1}{2},\frac{1}{2},\frac{1}{2})$ with $u = 0.245$

This is an arrangement which gives unacceptably short Na–O(2) separations (ca. $\frac{1}{4}b_0 = 1.43$ A.); perhaps these oxygens should be in (4g) with the same value of u.

The beta modification of $Na_2(UO_2)O_2$ is also orthorhombic. Its tetramolecular unit has the dimensions:

$$a_0 = 5.97 \text{ A.}; \quad b_0 = 5.795 \text{ A.}; \quad c_0 = 11.68 \text{ A.}$$

The space group has been chosen as V_h^{23} ($Fmmm$), and atoms have been put in the following positions:

U: (4a) 000; F.C.

Na: (8i) $\pm(00u)$; F.C. with $u = 0.38$

O(1): (8i) with $u = 0.165$

O(2): (8e) $\frac{1}{4}\frac{1}{4}0; \frac{1}{4}\frac{1}{4}\frac{1}{2}$; F.C.

This grouping gives uranyl ions with the usual U–O = ca. 1.90 A. and the other interatomic distances are reasonable.

The alpha form of *lithium uranyl dioxide*, $Li_2(UO_2)O_2$, has an orthorhombic unit of somewhat similar shape:

$$a_0 = 6.06 \text{ A.}; \quad b_0 = 5.13 \text{ A.}; \quad c_0 = 10.52 \text{ A.}$$

It is thought probable that the structure resembles that of $\beta\text{-}Na_2(UO_2)O_2$ but that in view of the differences in axial lengths of the two crystals, the lithium atoms could be in $(8f)$ $^1/_4\,^1/_4\,^1/_4$; $^1/_4\,^1/_4\,^3/_4$; F.C. instead of in $(8i)$.

It has been reported that the corresponding potassium, rubidium, and cesium compounds are not uranyl oxides but have the K_2NiF_4 structure described in Chapter VIII.

Clearly, additional work on all these compounds is needed.

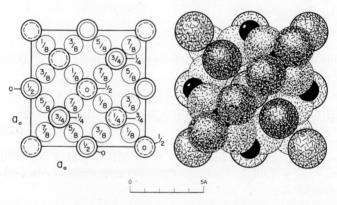

Fig. VI,31a (left). A projection along the cubic axis of the structure of $KAlO_2$. The heavy circles are potassium, the larger, light circles oxygen.

Fig. VI,31b (right). A packing drawing of the $KAlO_2$ structure viewed along a cubic axis. The aluminum atoms are black, the potassium atoms line-shaded.

VI,34. *Potassium ferric dioxide*, $KFeO_2$, and the analogous $KAlO_2$ have a cubic structure related to that of the high-cristobalite modification of silica (**IV,e7**). Their unit cubes, containing eight molecules, have the edge lengths:

$$KFeO_2: \quad a_0 = 7.958 \text{ A.}$$
$$KAlO_2: \quad a_0 = 7.69 \text{ A.}$$

To a first approximation it would appear that the iron, or aluminum, and the oxygen atoms have the positions which were originally given the silicon and oxygen atoms of high-cristobalite. Based on O_h^7 ($Fd3m$), they are:

O: $(16c)$ $\frac{1}{8}\frac{1}{8}\frac{1}{8}$; $\frac{1}{8}\frac{3}{8}\frac{3}{8}$; $\frac{3}{8}\frac{1}{8}\frac{3}{8}$; $\frac{3}{8}\frac{3}{8}\frac{1}{8}$; F.C.
Fe or Al: $(8f)$ 000; $\frac{1}{4}\frac{1}{4}\frac{1}{4}$; F.C.

With potassium atoms in $(8g)$ $\frac{1}{2}\frac{1}{2}\frac{1}{2}$; $\frac{3}{4}\frac{3}{4}\frac{3}{4}$; F.C., this would give a structure (Fig. VI,31) in which the closest atomic approaches are O–O = 2.72 A., Al–O = 1.66 A., Fe–O = 1.73 A., and K–O = 3.19 A. in the aluminum salt and 3.32 A. in the iron compound. Certain intensities are not adequately explained by this arrangement and it has therefore been concluded that, as in the case of cristobalite itself, the true atomic grouping may be a distortion of this involving the less symmetrical space group T^4.

VI,35. Borates show the same multiplicity of combination between boron and oxygen that the silicates do between silicon and oxygen. It is in keeping with this fact that the boron–oxygen association in the alkali metaborates should be totally unlike that found in the $Ca(BO_2)_2$ described in **VI,36.** There are six molecules in the following rhombohedral units of *sodium metaborate*, $NaBO_2$, and the corresponding KBO_2:

$$NaBO_2: \quad a_0 = 7.22 \text{ A.}, \quad \alpha = 111°29'$$
$$KBO_2: \quad a_0 = 7.76 \text{ A.}, \quad \alpha = 110°36'$$

The larger, 18-molecule, cells referred to hexagonal axes have the edges:

$$NaBO_2: \quad a_0' = 11.93 \text{ A.}, \quad c_0' = 6.46 \text{ A.}$$
$$KBO_2: \quad a_0' = 12.77 \text{ A.}, \quad c_0' = 7.35 \text{ A.}$$

All atoms have been put in the same special positions of D_{3d}^6 ($R\bar{3}c$):

$$(6e) \quad \pm(u,\tfrac{1}{2}-u,\tfrac{1}{4}; \ \tfrac{1}{2}-u,\tfrac{1}{4},u; \ \tfrac{1}{4},u,\tfrac{1}{2}-u)$$

for rhombohedral axes and

$$(18e) \quad \pm(u \ 0 \ \tfrac{1}{4}; \ 0 \ u \ \tfrac{1}{4}; \ \bar{u} \ \bar{u} \ \tfrac{1}{4}); \text{rh}$$

for hexagonal axes. The parameters, with those applying to the hexagonal axes in parentheses, are

Atom	u, $NaBO_2$	u, KBO_2
Na or K	0.696 (0.446)	0.689 (0.439)
B	0.362 (0.113)	0.361 (0.112)
O(1)	0.479 (0.229)	0.465 (0.215)
O(2)	0.138 (−0.112)	0.146 (−0.104)

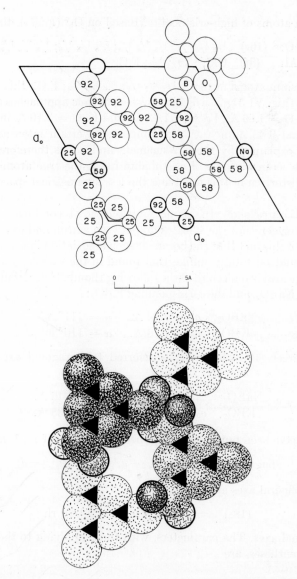

Fig. VI,32a (top). A projection along the c_0 axis of the hexagonal cell of the rhombohedral structure of NaBO₂.

Fig. VI,32b (bottom). A packing drawing of the hexagonal structure of NaBO₂ viewed along the c_0 axis. The sodium atoms are line-shaded, the oxygen atoms dotted.

The structure that results (Fig. VI,32) is composed of R and B_3O_6 ions. The B_3O_6 groups are planar with a trigonal axis normal to the plane. There are two kinds of oxygen atoms. Those with the parameters of O(2) are joined to two boron atoms; the O(1) atoms are connected only with a single boron atom. B–O separations are between 1.33 and 1.38 A. The Na–O distances range between 2.51 and 2.58 A. and the seven oxygens around a K atom are at distances between 2.80 and 2.85 A.

VI,36. The structure given *calcium metaborate*, $Ca(BO_2)_2$, differs radically from that of $NaBO_2$ (**VI,35**) in that discrete borate ions do not exist. Instead, boron and oxygen atoms are linked to produce endless chains running parallel to the c_0 axis of the crystal. The symmetry is orthorhombic, with a tetramolecular cell having

$$a_0 = 6.19 \text{ A.}; \quad b_0 = 11.60 \text{ A.}; \quad c_0 = 4.28 \text{ A.}$$

Atoms are in the following positions of V_h^{14} (*Pnca*):

$(4c)$ $\pm(^1/_4\ 0\ u;\ ^1/_4,^1/_2,u+^1/_2)$
$(8d)$ $\pm(xyz;\ x+^1/_2,^1/_2-y,^1/_2-z;\ \bar{x},y+^1/_2,^1/_2-z;\ ^1/_2-x,\bar{y},z)$

with the parameters:

Atom	Position	x	y	z
Ca	$(4c)$	$^1/_4$	0	0.26
B	$(8d)$	0.12	0.20	0.88
O(1)	$(8d)$	0.125	0.21	0.19
O(2)	$(8d)$	0.11	0.09	0.75

In the borate chains that exist (Fig. VI,33), boron and oxygen atoms have the same environment as in the alkali salts. The three oxygens around a boron atom are at distances between 1.34 and 1.38 A. Eight oxygens are nearest to a calcium atom with Ca–O = 2.38–2.68 A.

The presence of separate ions in KBO_2 and $NaBO_2$ and not in $Ca(BO_2)_2$ is perhaps associated with the fact that the alkali salts were obtained from solution while the calcium compound is crystallized from a melt.

VI,37. In a preliminary notice it has been stated that crystals of the *basic zinc metaborate*, $Zn_4O(BO_2)_6$, are cubic, with a bimolecular unit of the edge length:

$$a_0 = 7.48 \text{ A.}$$

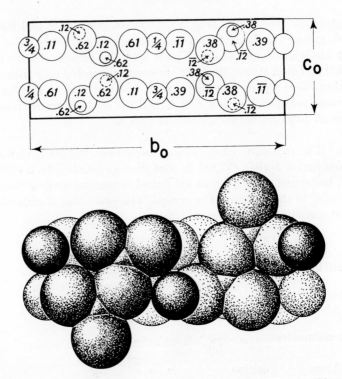

Fig. VI,33a (top). A projection on an *a* face of the atoms in the orthorhombic unit of Ca-B₂O₄. The largest circles are the oxygen atoms, the smallest the atoms of boron.

Fig. VI,33b (bottom). A packing drawing of the atoms of CaB₂O₄. The larger circles are oxygen, the borons are triangular in outline. The sharing of oxygen atoms by adjacent BO₄ tetrahedra is evident.

Atoms have been placed in the following positions of T_d^3 $(I\bar{4}3m)$:

Zn: (8c) $uuu; u\bar{u}\bar{u}; \bar{u}u\bar{u}; \bar{u}\bar{u}u;$ B.C.   with $u = 0.155$

B: (12d) $^1/_4\,^1/_2\,0; ^1/_4\,0\,^1/_2;$ tr; B.C.

O(1): (2a) $000; ^1/_2\,^1/_2\,^1/_2$

O(2): (24g) $uuv; u\bar{u}\bar{v}; v\bar{u}\bar{u}; u\bar{v}\bar{u};$ tr; B.C.   with $u = 0.135,$
$v = 0.400$

This results in a structure (Fig. VI,34) in which boron and oxygen atoms (in 24g) are bound together to form an indefinitely extended, three-dimensional network. Each boron is at the center of four tetrahedrally distributed oxygens, with B–O = 1.52 A. and B–B = 1.87 A.

It is pointed out that there is a close relationship between this structure and that of sodalite, Na₄(Si₃Al₃O₁₂)Cl (Chapter XII).

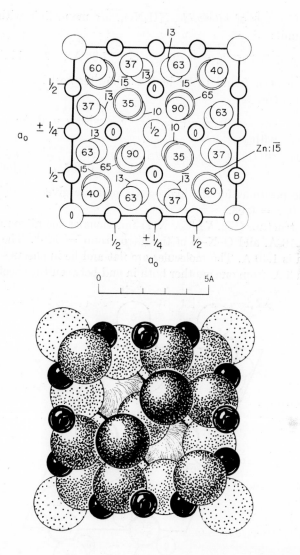

Fig. VI,34a (top). A projection along a cube axis of the structure found for Zn₄O(BO₂)₆.
Fig. VI,34b (bottom). A packing drawing of the Zn₄O(BO₂)₆ structure seen down a
cube axis. The boron atoms are black, the zinc atoms fine-line shaded.

VI,38.   Crystals of *nitramide*, NH_2NO_2, are monoclinic with the tetra-molecular unit:

$$a_0 = 6.65 \text{ A.;} \quad b_0 = 4.79 \text{ A.;} \quad c_0 = 7.86 \text{ A.;} \quad \beta = 112°24'$$

Atoms have been found to be in the following positions of C_{2h}^6 in the orientation $A2/a$.

N(1):   (4e) $\pm (^1/_4\,u\,0;\ ^1/_4,u+^1/_2,^1/_2)$ with $u = 0.575$
N(2):   (4e)   with $u = 0.866$
  O: (8f) $\pm (xyz;\ x+^1/_2,\bar{y},z;\ x,y+^1/_2,z+^1/_2;\ x+^1/_2,^1/_2-y,z+^1/_2)$
                       with $x = 0.367$, $y = 0.466$, $z = 0.137$

Probable positions for the hydrogen atoms are given as (8f) with $x = 0.42$, $y = 0$, $z = 0.08$.

In this structure (Fig. VI,35), N(1) functions as the nitro-nitrogen with N–O = 1.18 A. and O–N–O of the nitro group = 129°. The N(1)–N(2) separation is 1.40 A. The molecules are flat and lie in sheets separated by more than 3 A. from one another both in and between the sheets.

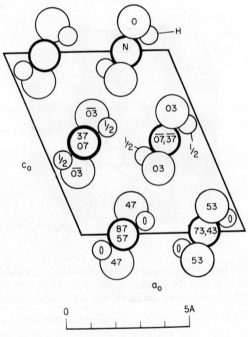

Fig. VI,35. The monoclinic structure of NH_2NO_2 viewed along its b_0 axis. Origin in lower left.

VI,39.   Crystals of *sulfamide*, $SO_2(NH_2)_2$, are orthorhombic, with an eight molecule cell of the dimensions:

$$a_0 = 9.14 \text{ A.}; \quad b_0 = 16.85 \text{ A.}; \quad c_0 = 4.580 \text{ A.}$$

The space group is C_{2v}^{19} (*Fdd2*), with sulfur atoms in

S: (8a) $00u; \; {}^1/_4, {}^1/_4, u + {}^1/_4$; F.C.   with $u = 0$ (arbitrary)

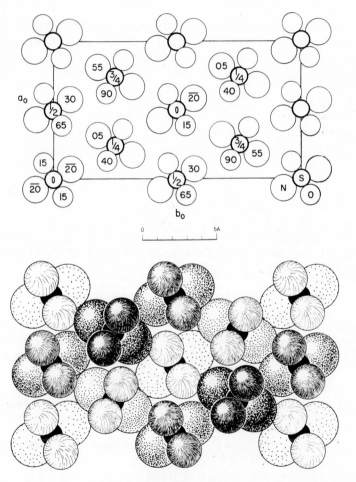

Fig. VI,36a (top). The orthorhombic structure of $SO_2(NH_2)_2$ projected along its c_0 axis. Origin in lower left.

Fig. VI,36b (bottom). A packing drawing of the $SO_2(NH_2)_2$ arrangement seen along its c_0 axis. The sulfur atoms are black, the NH_2 radicals are dotted.

All other atoms are in the general positions:

(8b) $xyz;\ \bar{x}\bar{y}z;\ {}^1/_4-x,y+{}^1/_4,z+{}^1/_4;\ x+{}^1/_4,{}^1/_4-y,z+{}^1/_4;$ F.C.

with the following parameters:

Atom	x	y	z
O	0.8844	0.0339	0.1533
N	0.0697	0.0691	0.8050

The x-ray data gave no evidence for the positions of the hydrogen atoms.

The way the tetrahedral $SO_2(NH_2)_2$ molecules pack in the crystal is shown in Figure VI,36. In a single molecule, S–O = 1.391 A. and S–N = 1.600 A. The angles are O–S–O = 119°24′, O–S–N = 106°12′ and 106°36′, and N–S–N = 112°6′. The shortest intermolecular distance is an O–N = 3.02 A.; others range upwards from 3.48 A.

VI,40. The phase ZrTaNO appears as a compound of definite composition. It is hexagonal, with a unimolecular cell of the dimensions:

$$a_0 = 3.645 \text{ A.}, \qquad c_0 = 3.881 \text{ A.}$$

The Zr atom is in (1a) 000 of D_{3h}^1 (P6m2) and the Ta in (1d) ${}^1/_3\ {}^2/_3\ {}^1/_2$. It is suggested that O is in (1c) ${}^1/_3\ {}^2/_3\ 0$ and N in (1f) ${}^2/_3\ {}^1/_3\ {}^1/_2$.

In the arrangement that results, the separation Zr–Ta = 2.84 A. Other atomic distances are N–O = 2.863 A., Zr–O = 2.11 A., Ta–O = 1.94 A., and Ta–N = 2.11 A.

Sulfides, etc.

VI,41. At temperatures below ca. 120°C., *silver bismuth diselenide*, $AgBiSe_2$, is no longer rhombohedral (see **VI,12**) but truly hexagonal, though its unit is little changed. The trimolecular hexagonal cell has the edges:

$$a_0 = 4.18 \text{ A.}, \qquad c_0 = 19.67 \text{ A.}$$

Atoms are considered to be in the following positions of D_{3d}^3 ($P\bar{3}m1$)

Ag(1):	(1a)	000	
Ag(2):	(2d)	$\pm({}^1/_3\ {}^2/_3\ u)$	with $u = -0.328$
Bi(1):	(1b)	$0\ 0\ {}^1/_2$	
Bi(2):	(2d)	with $u = 0.163$	
Se(1):	(2c)	$\pm(00u)$	with $u = 0.253$
Se(2):	(2d)	$u = -0.074$	
Se(3):	(2d)	with $u = 0.406$	

As can be seen by comparing Figure VI,37 with Figure VI,12 for Na-FeO$_2$, these represent only minor displacements of the atoms from their positions in the rhombohedral structure.

The sulfide AgBiS$_2$ appears also to have this hexagonal structure at room temperature. For it,

$$a_0 = 4.07 \text{ A.,} \qquad c_0 = 19.06 \text{ A.}$$

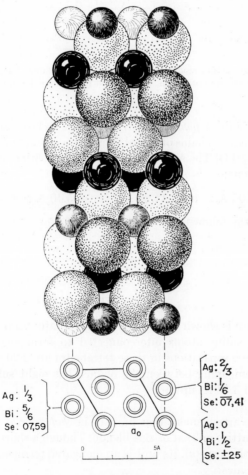

Fig. VI,37. Two projections of the hexagonal structure of low-temperature AgBiSe$_2$. The smallest circles are silver, the bismuth atoms are slightly larger and black.

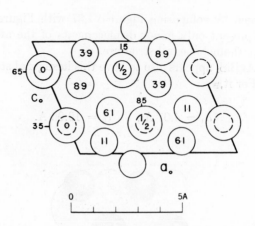

Fig. VI,38a. A projection along the b_0 axis of the monoclinic structure of KFeS₂. The largest circles are potassium, the smallest iron atoms. Origin in lower left.

VI,42.   Crystals of *potassium ferric disulfide*, KFeS₂, are built up of the same kind of endless chains of linked tetrahedra of sulfur atoms that characterize SiS₂ **(IV,e14)**. The symmetry is monoclinic with a tetramolecular cell of the dimensions:

$$a_0 = 7.05 \text{ A.}; \quad b_0 = 11.28 \text{ A.}; \quad c_0 = 5.40 \text{ A.}; \quad \beta = 112°30'$$

Atoms are in the following positions of C_{2h}^6 $(C2/c)$:

Fe: $(4e)$ $\pm(0 \ u \ ^1/_4; \ ^1/_2, u+^1/_2, ^1/_4)$ with $u = -0.008$

K: $(4e)$ with $u = 0.355$

S: $(8f)$ $\pm(xyz; \ x,\bar{y},z+^1/_2; \ x+^1/_2,y+^1/_2,z; \ x+^1/_2,^1/_2-y,z+^1/_2)$

with $x = 0.195, \ y = 0.111, \ z = 0.10$

The structure is shown in Figure VI,38. Iron atoms are at the centers of tetrahedra of sulfur atoms interconnected to form chains as in Figure VI,39. The Fe–S separations in these tetrahedra are 2.20 and 2.28 A. Each potassium atom, lying between the chains, has eight sulfur neighbors at distances from 3.33 A. upward.

VI,43.   The arrangement that prevails in *chalcopyrite*, CuFeS₂, is essentially a superlattice on that of zinc blende. Though a shorter cell was earlier described for this crystal, its unit is an elongated tetragonal prism containing four molecules:

$$a_0 = 5.24 \text{ A.}, \quad c_0 = 10.30 \text{ A.}$$

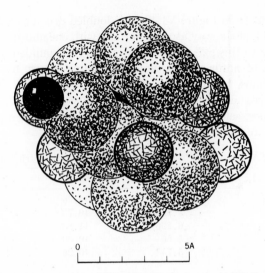

Fig. VI,38b. A packing drawing of the monoclinic structure of KFeS₂ viewed along its b_0 axis. The sulfur atoms are dotted, the potassium atoms are smaller and line-shaded.

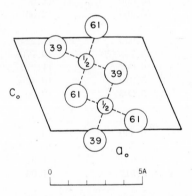

Fig. VI,39. A projection along b_0 of the cell of KFeS₂ showing a chain of linked FeS₂ tetrahedra.

Atoms are in special positions of V_d^{12} $(I\overline{4}2d)$:

Cu: (4a) $000; 0\,^1/_2\,^1/_4$; B.C.

Fe: (4b) $00^1/_2; 0\,^1/_2\,^3/_4$; B.C.

S: (8d) $u\,^1/_4\,^1/_8; \bar{u}\,^3/_4\,^1/_8; ^3/_4\,u\,^7/_8; ^1/_4\,\bar{u}\,^7/_8$; B.C.   with $u =$ ca. $^1/_4$

As is evident from Figure VI,40, the doubled c_0 compared to the a_0 axis is an expression of the way atoms alternate with one another in the metallic planes normal to this axis. Metallic atoms are surrounded here, as in ZnS, by tetrahedra of sulfur atoms, while each sulfur in turn has a tetrahedron of metallic atoms, two of which are copper and two iron. The significant interatomic distances are Fe–S = 2.20 A., Cu–S = 2.32 A., S–S = 3.56 A.

Other sulfides, selenides, tellurides, phosphides, and arsenides with this arrangement are listed in Table VI,6.

TABLE VI,6. Crystals with the Chalcopyrite Structure

Compound	a_0, A.	c_0, A.	u
AgAlS$_2$	5.695	10.26	0.30
AgAlSe$_2$	5.956	10.75	0.27
AgAlTe$_2$	6.296	11.83	0.26
AgFeS$_2$	5.66	10.30	0.25
AgGaS$_2$	5.743	10.26	0.28
AgGaSe$_2$	5.973	10.88	0.27
AgGaTe$_2$	6.288	11.94	0.26
AgInS$_2$[a]	5.816	11.17	0.25
AgInSe$_2$	6.090	11.67	0.25
AgInTe$_2$	6.406	12.56	0.25
CdGeAs$_2$	5.942	11.224	0.285
CdGeP$_2$	5.738	10.776	—
CdSnAs$_2$	6.092	11.922	—
CuAlS$_2$	5.312	10.42	0.27
CuAlSe$_2$	5.606	10.90	0.26
CuAlTe$_2$	5.964	11.78	0.25
CuGaS$_2$	5.349	10.47	0.25
CuGaSe$_2$	5.607	10.99	0.25
CuGaTe$_2$	5.994	11.91	0.25
CuInS$_2$	5.517	11.06	0.20
CuInSe$_2$	5.773	11.55	0.22
CuInTe$_2$	6.167	12.34	0.225
CuTlS$_2$	5.580	11.17	0.19
CuTlSe$_2$	5.832	11.63	0.23
ZnGeAs$_2$	5.670	11.153	—
ZnGeP$_2$	5.46	10.76	—
ZnSiAs$_2$	5.608	10.89	—
ZnSiP$_2$	5.398	10.44	—

[a] AgInS$_2$, tempered above 700°C. and quenched, gives a ZnO-like structure with a_0 = 4.121 A., c_0 = 6.674 A.

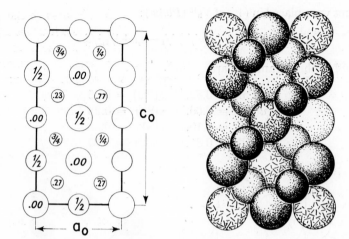

Fig. VI,40a (left). A projection of the tetragonal structure of chalcopyrite, $CuFeS_2$, upon an a face. Copper, iron, and sulfur atoms are represented by circles of decreasing size.
Fig. VI,40b (right). A packing drawing of the $CuFeS_2$ structure in which the atoms have been given their neutral radii. The line-shaded spheres are the copper atoms.

Zinc tin diarsenide, $ZnSnAs_2$, resembles chalcopyrite but lacks the segregation of its metal atoms. As a result, the structure is cubic, with $a_0 = 5.851$ A. The metallic atoms are to be considered as haphazardly distributed among the zinc positions in ZnS (**III,c1**). The same thing appears to be true of $MgGeP_2$, the cubic cell of which has the edge length:

$$a_0 = 5.652 \text{ A.}$$

This is also said to be the case for the following two semiconductors of a different composition:

$$Ge_2CuP_3: \quad a_0 = 5.375 \text{ A.}$$
$$Si_2CuP_3: \quad a_0 = 5.25 \text{ A.}$$

VI,44. The two copper minerals *wolfsbergite*, $CuSbS_2$, and *emplectite*, $CuBiS_2$, have been given a structure which bears some resemblance to that of the hexagonal ZnS, wurtzite (**III,c2**). They are orthorhombic, with the tetramolecular units:

$$CuSbS_2: \quad a_0 = 14.456 \text{ A.}; \quad b_0 = 6.008 \text{ A.}; \quad c_0 = 3.784 \text{ A.}$$
$$CuBiS_2: \quad a_0 = 14.512 \text{ A.}; \quad b_0 = 6.125 \text{ A.}; \quad c_0 = 3.890 \text{ A.}$$

All atoms were placed in $(4c)$ of V_h^{16} $(Pbnm)$:

$$(4c) \quad \pm (u\,v\,{}^1/_4;\; {}^1/_2-u,v+{}^1/_2,{}^1/_4)$$

For $CuSbS_2$ the parameters are

$$
\begin{aligned}
u(Cu) &= 0.82, & v(Cu) &= 0.25 \\
u(Sb) &= 0.06, & v(Sb) &= 0.23 \\
u(S1) &= 0.10, & v(S1) &= 0.62 \\
u(S2) &= 0.82, & v(S2) &= 0.87
\end{aligned}
$$

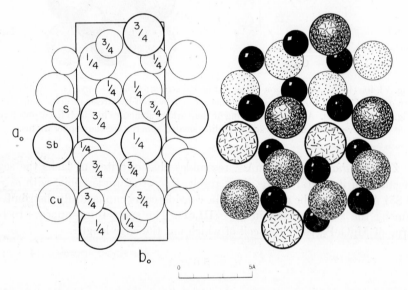

Fig. VI,41a (left). A projection along the c_0 axis of the orthorhombic structure of Cu-SbS$_2$. Origin in lower left.

Fig. VI,41b (right). A packing drawing of the orthorhombic CuSbS$_2$ structure viewed along its c_0 axis. The atoms have been given their neutral radii with the copper atoms dotted and the sulfur atoms black.

In this grouping (Fig. VI,41), copper has about it a slightly distorted tetrahedron of sulfur atoms at distances varying between 2.25 and 2.33 A., while antimony has three sulfur neighbors at distances between 2.44 and 2.57 A.

Some recent work has assigned to emplectite and wolfsbergite compositions approaching Bi_2CuS_4 and Sb_2CuS_4, respectively. If this should be correct the structure given above cannot, of course, be valid.

TABLE VI,7. Parameters of the Atoms in TlAsS₂

Atom	x	y	z
Tl(1)	0.051	0.313	0.160
Tl(2)	0.101	0.056	0.732
As(1)	0.190	0.820	0.237
As(2)	0.151	0.585	0.554
S(1)	0.125	0.320	0.750
S(2)	0.150	0.580	0.200
S(3)	0.125	0.790	0.510
S(4)	0.200	0.030	0.200

VI,45. The mineral *lorandite*, TlAsS₂, is monoclinic, with a unit containing eight molecules and having the dimensions:

$$a_0 = 12.27 \text{ A.}; \quad b_0 = 11.33 \text{ A.}; \quad c_0 = 6.11 \text{ A.}; \quad \beta = 104°12'.$$

All atoms have been placed in general positions of C_{2h}^5 ($P2_1/a$):

$$(4e) \quad \pm(xyz; \; x+{}^1/_2,{}^1/_2-y,z)$$

with the parameters of Table VI,7.

The structure is shown in Figure VI,42. The crystal contains AsS₂ chains running in the b_0 direction. Each chain consists of flat AsS₃ trigonal pyramids linked together by sharing sulfur corners. The two crystallographically different pyramids are essentially alike, with As–S = 2.05–2.39 A. in one case and 2.15–2.35 A. in the other. Thallium atoms lie between the chains and presumably bind them together in the crystal, with each thallium having three closest sulfur neighbors at distances between 2.84 and 3.25 A.

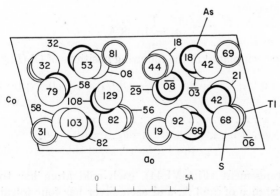

Fig. VI,42. The monoclinic structure of TlAsS₂ projected along its b_0 axis. Origin in lower left.

342 CRYSTAL STRUCTURES

VI,46. An electron diffraction study of *thallous antimony diselenide*, TlSbSe$_2$, has given it an orthorhombic structure with the bimolecular unit:

$$a_0 = 4.18 \text{ A.;} \quad b_0 = 4.50 \text{ A.;} \quad c_0 = 12.00 \text{ A.}$$

An arrangement has been described based on V_h^{19} (*Cmmm*), with atoms in the following positions:

Tl: (2b) $\quad {}^1/_2\,{}^1/_2\,0;\, 0\,{}^1/_2\,{}^1/_2$
Sb: (2a) $\quad 000;\, {}^1/_2\,0\,{}^1/_2$
Se: (4h) $\quad \pm(0uv;\, {}^1/_2,u,{}^1/_2-v)$ $\qquad$ with $u = 0.5$, $v = 0.272$

This is a structure (Fig. VI,43) in which there are zigzag chains of selenium atoms with Se–Se = 2.15 A. and Se–Se–Se = 152°. The shortest Tl–Se = 2.74 A. and Tl–Sb = 3.07 A.

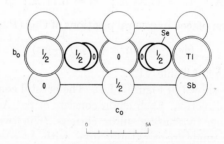

Fig. VI,43. The orthorhombic structure proposed for TlSbSe$_2$ projected along its a_0 axis. Origin in lower left.

VI,47. *Petzite*, Ag$_3$AuTe$_2$, is cubic with a cell containing eight molecules and having the edge:

$$a_0 = 10.38 \text{ A.}$$

Atoms have been placed in the following positions of O^8 ($I4_132$):

Ag: (24f) $\quad u\,0\,{}^1/_4;\, {}^1/_2-u,0,{}^3/_4;\, {}^1/_4-u,0,{}^1/_4;\, u+{}^1/_4,0,{}^3/_4;\, \text{tr; B.C.}$
$\qquad\qquad\qquad\qquad\qquad\qquad\qquad\qquad\qquad\qquad$ with $u = 0.365$
Au: (8a) $\quad {}^1/_8\,{}^1/_8\,{}^1/_8;\, {}^5/_8\,{}^3/_8\,{}^7/_8;\, {}^7/_8\,{}^5/_8\,{}^3/_8;\, {}^3/_8\,{}^7/_8\,{}^5/_8;\, \text{B.C.}$
Te: (16e) $\quad uuu;\, u+{}^1/_2,{}^1/_2-u,\bar{u};\, {}^1/_4-u,{}^1/_4-u,{}^1/_4-u;$
$\qquad\qquad\quad {}^3/_4-u,u+{}^1/_4,u+{}^3/_4;\, \text{tr; B.C.}$ $\qquad$ with $u = 0.266$

In this arrangement (Fig. VI,44), each gold atom has two tellurium atoms at a distance of 2.53 A. and each silver has four tetrahedrally distributed tellurium atoms 2.90 and 2.95 A. away. The close Te–Te is 3.91 A.

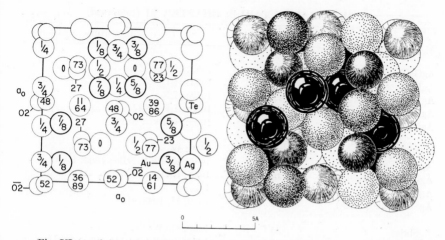

Fig. VI,44a (left). The cubic structure of Ag_3AuTe_2 viewed along a cube axis.
Fig. VI,44b (right). A packing drawing of the cubic Ag_3AuTe_2 structure seen along a cube axis. The gold atoms are black, the tellurium atoms fine-line shaded.

BIBLIOGRAPHY TABLE, CHAPTER VI

Compound	Paragraph	Literature
AgAlS$_2$	43	1953: H,F,K,M&S
AgAlSe$_2$	43	1953: H,F,K,M&S
AgAlTe$_2$	43	1953: H,F,K,M&S
Ag$_3$AuTe$_2$ (petzite)	48	1959: F
AgBiS$_2$	23,41	1936: R; 1951: G; 1959: G&W; S&M
AgBiSe$_2$	12,23,41	1959: G&W; S&M
AgBiTe$_2$	23	1959: S&M
AgClO$_2$	20	1931: L&S; 1957: C,R&L; 1961: C&M
AgCrO$_2$	12	1957: H&dL
AgCrS$_2$	12	1957: H&dL
AgCrSe$_2$	12	1957: H&dL
AgFeO$_2$	12	1936: K,E,G&K; 1960: F&M
AgFeS$_2$	43	1944: B
AgGaS$_2$	43	1953: H,F,K,M&S
AgGaSe$_2$	43	1953: H,F,K,M&S

(continued)

BIBLIOGRAPHY TABLE, CHAPTER VI (*continued*)

Compound	Paragraph	Literature
$AgGaTe_2$	43	1953: H,F,K,M&S
$AgInS_2$	43	1953: H,F,K,M&S
$AgInSe_2$	43	1953: H,F,K,M&S
$AgInTe_2$	43	1953: H,F,K,M&S
AgN_3	3	1935: B; 1962: M&S
$AgNCO$	5	1959: W
$AgNO_2$	17	1936: K; 1954: NBS; 1962: L&M
Ag_2PbO_2	28	1950: B&E
$AgSCN$	7	1957: L
$AgSbS_2$ (miargyrite)	23	1932: H; 1938: H
$AgSbSe_2$	23	1959: G&W; S&M
$AgSbTe_2$	23	1959: G&W; S&M; 1961: P&H; 1962: NBS
$BaCdO_2$	27	1962: S
$BaCoO_2$	25	1960: S
$BaNiO_2$	26	1951: L
$Ba(UO_2)O_2$	30	1947: S&S
$BaZnO_2$	25	1960: S; S,H&Z
$Ca(BO_2)_2$	37	1931: Z; 1932: Z&Z
$CaCN_2$	12	1927: D; 1942: B; 1958: Y,K,T&Y
$Ca(H_2PO_2)_2$	22	1958: L
$Ca(UO_2)O_2$	31	1948: Z; 1961: R
$CdGeAs_2$	43	1958: P
$CdGeP_2$	43	1961: F&P
$CdSnAs_2$	43	1958: P
$Cd(UO_2)O_2$	32	1961: K,P,I,S&S; K,P,S&I
$CsCNO$	1	1959: W
$CsHF_2$	1,23	1956: K,F&ME
CsI_3	14	1922: C&D; 1923: C&D; 1925: B&P
$CsIBr_2$	14	1922: C&D; 1923: C&D; 1925: B&P
$CsICl_2$	12	1920: W; 1952: NBS
$CuAlO_2$	12	1955: H&dL; 1956: D
$CsNO_2$	23	1957: F,C&T
$CuAlS_2$	43	1953: H,F,K,M&S
$CuAlSe_2$	43	1953: H,F,K,M&S
$CuAlTe_2$	43	1953: H,F,K,M&S
$CuBiS_2$ (emplectite)	44	1932: H; 1933: H; 1952: N
$CuBiSe_2$	23	1959: S&M

(*continued*)

BIBLIOGRAPHY TABLE, CHAPTER VI (*continued*)

Compound	Paragraph	Literature
CuCoO$_2$	12	1956: D; 1958: D
CuCrO$_2$	12	1949: S; 1955: D&V; 1956: D; 1961: B&D
CuCrS$_2$	12	1957: H&dL
CuCrSe$_2$	12	1957: H&dL
CuFeO$_2$	12	1934: W&T; 1935: S&T; 1938: P; 1946: P; 1956: D; 1961: B&D
CuFeS$_2$ (chalcopyrite)	43	1917: B&E; 1923: G&G; 1932: P&B; 1933: OD; 1934: K&T
CuGaO$_2$	12	1955: H&dL
CuGaS$_2$	43	1953: H,F,K,M&S; 1958: S,G&S
CuGaSe$_2$	43	1953: H,F,K,M&S
CuGaTe$_2$	43	1953: H,F,K,M&S
CuInS$_2$	43	1953: H,F,K,M&S
CuInSe$_2$	43	1953: H,F,K,M&S
CuInTe$_2$	43	1953: H,F,K,M&S
CuMnO$_2$	16	1958: K
CuRhO$_2$	12	1961: B&D
CuSbS$_2$ (wolfsbergite)	44	1932: H; 1933: H
CuTlS$_2$	43	1953: H,F,K,M&S
CuTlSe$_2$	43	1953: H,F,K,M&S
DCrO$_2$	12	1962: H&I
HCrO$_2$	12	1957: D; 1962: H&I
KAg(CN)$_2$	8	1933: H
KAlO$_2$	34	1935: B
KAu(CN)$_2$	10	1959: R&C
K(AmO$_2$)F$_2$	31	1954: A,E&Z
KBO$_2$	35	1935: C,S&A; 1937: Z
KBiS$_2$	23	1944: B; 1955: G&F; 1959: S&M
KBiSe$_2$	23	1955: G&Z
KCNO	1	1925: H&P; 1959: W
KCNS	5	1933: K; 1934: G,B&T; 1958: NBS
KCrS$_2$	12	1943: R&S
KCu(CN)$_2$	9	1957: C
KFeO$_2$	34	1933: H&L; 1935: B; 1950: B&B
KFeS$_2$	42	1933: OD; 1942: B&MG
KHF$_2$	1,23	1923: B; 1939: H&R; 1952: P&L; 1956: K,F&ME

(*continued*)

BIBLIOGRAPHY TABLE, CHAPTER VI (*continued*)

Compound	Paragraph	Literature
KN_3	1	1925: H&P; 1936: F
KNO_2	18	1936: Z
$KTlO_2$	12	1961: H&W
$LiAlO_2$	12,15	1961: T,L,B&C; 1962: L&C
$LiBiS_2$	23	1955: G&F; G&Z
$LiCoO_2$	12	1958: J,H&S; 1961: B&D
$LiCrO_2$	12	1954: R&B; 1961: B&D
$LiErO_2$	25	1962: B&G
$LiFeO_2$	23,24	1931: P&B; 1938: H; 1944: B,B&N; 1955: C; 1958: B&C; K; 1960: B
$LiHF_2$	12	1962: F&R
$LiInO_2$	24	1958: H&S; 1961: H&R
$LiLuO_2$	24	1962: B&G
$LiNO_2$	12	1954: D,B&S
$LiNiO_2$	12	1961: B&D
$LiPN_2$	24	1960: E,L,M&R
$LiRhO_2$	12	1961: B&D
$LiScO_2$	25	1961: R
$LiTiO_2$	23	1962: L
$LiTlO_2$	23,24	1961: H&W
$LiTmO_2$	25	1962: B&G
$Li_2(UO_2)O_2$	33	1961: K,I,S&S; S,I,S&K
$LiVO_2$	12	1954: R&B
$LiYO_2$	25	1962: B&G
$LiYbO_2$	25	1962: B&G
$Li(Co_{0.5}Mn_{0.5})O_2$	23	1960: B
$Li(Co_{0.5}Ti_{0.5})O_2$	23	1960: B
$Li(Fe_{0.5}Mn_{0.5})O_2$	23	1960: B
$Li(Fe_{0.5}Ti_{0.5})O_2$	23	1960: B
$Li(Mn_{0.5}Ti_{0.5})O_2$	23	1960: B
$Li(Ni_{0.5}Mn_{0.5})O_2$	23	1960: B
$Li(Ni_{0.5}Ti_{0.5})O_2$	23	1960: B
$MgGeP_2$	43	1961: F&P
$Mg(UO_2)O_2$	29	1954: R&P; Z
$N(CH_3)_4Cl_2I$	13	1939: M
$NH_4Ag(SCN)_2$	11	1957: L&S
NH_4BrClI	14	1935: M; 1937: M

(*continued*)

BIBLIOGRAPHY TABLE, CHAPTER VI *(continued)*

Compound	Paragraph	Literature
NH_4ClO_2	18	1931: L&S; 1959: G,S&T
NH_4HF_2	2	1932: H&L; 1933: P; 1940: R&H; 1960: MD
$NH_4H_2PO_2$	21	1934: Z&M
NH_4I_3	14	1935: M
NH_2NO_2	38	1937: B&TD
NH_4N_3	2	1936: F
NH_4SCN	6	1949: Z&Z
$(NH_2)_2SO_2$	39	1956: T&M
$NaAlO_2$	15	1959: B&S; 1961: T,B&C; T,L,B&C; 1962: L&C
$NaAu(CN)_2$	8	1948: S&Z
$NaBO_2$	35	1935: C,S&A; 1937: F; C,S&A; 1938: F
$NaBiS_2$	23	1944: B; 1955: G&F; 1959: S&M
$NaBiSe_2$	23	1955: G&Z
$NaCNO$	12	1961: NBS
$NaCrO_2$	12	1954: R&B
$NaCrS_2$	12	1942: B&MG; 1943: R&S
$NaCrSe_2$	12	1961: S
$NaFeO_2$	12,15	1933: G; 1935: G; 1939: T&S; 1948: R,R&S; 1954: B&B; 1961: T,B&C; 1962: L&C
$NaHF_2$	12	1923: R,H&L; 1926: A&H; 1954: NBS
$NaInO_2$	12	1958: H&S
$NaInS_2$	12	1961: H,L&F
$NaInSe_2$	12	1961: H,L&F
NaN_3	12	1925: H&P; 1934: W; 1936: F; 1939: B
$NaNO_2$	17	1931: Z; 1943: S&MG; 1946: S&MG; 1951: T; 1952: C; 1953: NBS; 1954: T; 1955: C; 1961: H&S; K&F; 1962: K,F&U
$NaNiO_2$	12,16	1954: D,B&S
$NaOCN$ (isocyanate)	12	1934: W; 1938: B; 1943: B; 1959: W
$NaTiO_2$	12	1962: H,L&O
$NaTlO_2$	12	1961: H&W
$Na_2(UO_2)O_2$	33	1958: K,I,S&S; 1961: K,I,S&S; K,P,I,-S&S; S,I,S&K
$NaVO_2$	12	1954: R&B
$(PCl_4)ICl_2$	13	1952: Z&B

(continued)

BIBLIOGRAPHY TABLE, CHAPTER VI (*continued*)

Compound	Paragraph	Literature
$Pb(UO_2)O_2$	30	1958: F&B; 1961: K,P,S&I
RbCNO	1	1959: W
$RbCrS_2$	12	1948: R,R&S
$RbCrSe_2$	12	1948: R,R&S; 1961: S
$RbHF_2$	1,23	1956: K,F&ME
RbN_3	1	1930: G,P&R; P; 1931: B,R&G
$RbTlO_2$	12	1961: H&W
$Sr(N_3)_2$	4	1947: L&W
$Sr(UO_2)O_2$	31	1948: Z
$SrZnO_2$	27	1960: vS&H; 1961: S&H
$TlAsS_2$ (lorandite)	45	1932: H; 1958: Z&Z; 1959: Z&Z
$TlBiS_2$	23	1959: S&M
$TlBiTe_2$	12	1961: H&W
TlCNO	1	1959: W
TlCNS	5	1934: B,G&T; S; 1958: NBS
$TlNO_2$	23	1955: C,N&B
$TlSbS_2$	23	1959: S&M
$TlSbSe_2$	46	1956: P,S&B
$TlSbTe_2$	12	1961: H&W
$ZnGeAs_2$	43	1957: G; 1958: P
$ZnGeP_2$	43	1958: P
$ZnSnAs_2$	43	1960: F&P
$Zn_4O(BO_2)_6$	37	1961: S,G,B&R
ZrTaNO	40	1954: S

BIBLIOGRAPHY, CHAPTER VI

1917

Burdick, C. L., and Ellis, J. H., "X-Ray Examination of the Crystal Structure of Chalcopyrite," *Proc. Natl. Acad. Sci.*, **3**, 644; *J. Am. Chem. Soc.*, **39**, 2518.

1920

Wyckoff, R. W. G., "The Crystal Structure of Caesium Dichloriodide," *J. Am. Chem. Soc.*, **42**, 1100.

1922

Clark, G. L., and Duane, W., "A Study of Secondary Valence by Means of X-Rays," *Phys. Rev.*, **20**, 85.

1923

Bozorth, R. M., "The Crystal Structure of Potassium Hydrogen Fluoride," *J. Am. Chem. Soc.*, **45**, 2128.

Clark, G. L., and Duane, W., "A New Method of Crystal Analysis and the Reflection of Characteristic X-Rays," *J. Opt. Soc. Am.*, **7**, 455.

Clark, G. L., and Duane, W., "The Reflection by a Crystal of X-Rays Characteristic of Chemical Elements in It," *Proc. Natl. Acad. Sci.*, **9**, 126.

Gross, R., and Gross, N., "The Atomic Arrangement of Chalcopyrite and the Structure of the Contact Faces of Regularly Overgrown Crystals," *Neues Jahrb. Mineral.*, **48**, 113.

Rinne, F., Hentschel, H., and Leonhardt, J., "Investigations of the Construction of Sodium Hydrogen Fluoride Employing the Idea of Atomic Domains and on the X-Ray Study of this Compound," *Z. Krist.*, **58**, 629.

1925

Bozorth, R. M., and Pauling, L., "The Crystal Structures of Cesium Triiodide and Cesium Dibromoiodide," *J. Am. Chem. Soc.*, **47**, 1561.

Hendricks, S. B., and Pauling, L., "The Crystal Structures of Sodium and Potassium Trinitrides and Potassium Cyanate and the Nature of the Trinitride Group," *J. Am. Chem. Soc.*, **47**, 2904.

1926

Andersen, C. C., and Hassel, O., "The Structure of Crystalline Sodium Hydrofluoride and the Form of the Ion HF_2," *Z. Physik. Chem.*, **123**, 151.

Yardley, K., "X-Ray Examination of Aramayoite," *Mineral. Mag.*, **21**, 163.

1927

Dehlinger, U., "The Space Group of $(CN_2H_2)_2$ and the Crystal Structure of $CaCN_2$," *Z. Krist.*, **65**, 286.

1930

Günther, P., Porger, J., and Rosbaud, P., "The Crystal Structure and Percussion Sensitivity of Rubidium Trinitride and Barium Trinitride," *Z. Physik. Chem.*, **6B**, 459.

Pauling, L., "The Crystal Structure of Rubidium Trinitride," *Z. Physik. Chem.*, **8B**, 326; cf. Günther, P., and Rosbaud, P., *ibid.*, **8B**, 329.

Stackelberg, M. v., "Investigations on Carbides I. The Crystal Structure of Carbides MC_2," *Z. Physik. Chem.*, **9B**, 437; *Naturwiss.*, **18**, 305.

1931

Büssem, W., Rosbaud, P., and Günther, P., "Crystal Structure of Rubidium Azide," *Z. Physik. Chem.*, **15B**, 58.

Levi, G. R., and Scherillo, A., "Crystallographic Investigation of the Salts of Chlorous Acid," *Z. Krist.*, **76**, 431.

Miles, F. D., "The Formation and Characteristics of Crystals of Lead Azide and of some other Initiating Explosives," *J. Chem. Soc.*, **1931**, 2532.

Posnjak, E., and Barth, T. F. W., "A New Type of Crystal Fine-Structure: Lithium Ferrite $(Li_2O \cdot Fe_2O_3)$," *Phys. Rev.*, **38**, 2234.

Zachariasen, W. H., "The Crystal Lattice of Calcium Metaborate," *Proc. Natl. Acad. Sci.*, **17**, 617.
Ziegler, G. E., "Crystal Structure of Sodium Nitrite," *Phys. Rev.*, **38**, 1040.

1932

Hassel, O., and Luzanski, N., "X-Ray Investigation of Ammonium Bifluoride," *Z. Krist.*, **83**, 448.
Hofmann, W., "Structural and Morphological Relations Between Ores of the Formula Type ABC_2," *Fortschr. Mineral.*, **17**, 422.
Pauling, L., and Brockway, L. O., "The Crystal Structure of Chalcopyrite," *Z. Krist.*, **82**, 188.
Zachariasen, W. H., and Ziegler, G. E., "The Crystal Structure of Calcium Metaborate," *Z. Krist.*, **83**, 354.

1933

Goldsztaub, S., "Crystal Structure of Sodium Ferrite," *Compt. Rend.*, **196**, 280.
Hilpert, S., and Lindner, A., "Ferrites II. The Alkaline, Alkaline Earth and Lead Ferrites," *Z. Physik. Chem.*, **22B**, 395.
Hoard, J. L., "The Crystal Structure of Potassium Silver Cyanide," *Z. Krist.*, **84**, 231.
Hofmann, W., "Structural and Morphological Relationships of Ores of the Type ABC_2. I. The Structures of Wolfsbergite and Emplectite and their Relation to Stibnite," *Z. Krist.*, **84**, 177.
Klug, H. P., "The Crystal Structure of Potassium Thiocyanate," *Z. Krist.*, **85**, 214.
O'Daniel, H., "$KFeS_2$ and $CuFeS_2$," *Z. Krist.*, **86**, 192.
Pauling, L., "The Crystal Structure of Ammonium Hydrogen Fluoride," *Z. Krist.*, **85**, 380.

1934

Büssem, W., Günther, P., and Tubin, R., "The Structure of Thallium and Potassium Thiocyanates," *Z. Physik. Chem.*, **24B**, 1.
Gossner, B., and Kraus, O., "The Crystal Form and Chemical Composition of Polybasite," *Centr. Mineral. Geol.*, **1934A**, 1.
Kozu, S., and Takané, K., "The Crystal Structure of Chalcopyrite," *Proc. Imp. Acad. Tokyo*, **10**, 498.
Strada, M., "Investigations on the Structure of Pseudo-halogens and their Compounds I. Thallium Thiocyanate," *Gazz. Chim. Ital.*, **64**, 400.
Wartmann, F. S., and Thompson, A. J., "Progress Reports—Metallurgical Division 3. Studies in the Metallurgy of Copper. Preparation and Properties of Copper Ferrite," *Bur. Mines, Rept. Invest.*, **No. 3228**, 15.
West, C. D., "The Crystal Structures of some Alkali Hydrosulfides and Monosulfides," *Z. Krist.*, **88**, 97.
Zachariasen, W. H., and Mooney, R. C. L., "The Structure of the Hypophosphite Group as Determined from the Crystal Lattice of Ammonium Hypophosphite," *J. Chem. Phys.*, **2**, 34.

1935

Barth, T. F. W., "Non-Silicates with Cristobalite-like Structure," *J. Chem. Phys.*, **3**, 323.

Bassière, M., "The Crystal Structure of Silver Trinitride, AgN₃," *Compt. Rend.*, **201**, 735.

Cole, S. S., Scholes, S. R., and Amberg, C. R., "The System $R_2O-B_2O_3$. II. Properties of Anhydrous and Hydrated Metaborates of Sodium and Potassium," *J. Am. Ceram. Soc.*, **18**, 58.

Goldsztaub, S., "The Study of Several Derivatives of Ferric Oxide; Determinations of their Structures," *Bull. Soc. Franc. Mineral. Crist.*, **58**, 6.

Mooney, R. C. L., "The Structure of Ammonium Tri-Iodide, NH_4I_3, " *Z. Krist.*, **90A**, 143.

Mooney, R. C. L., "The Crystal Structure of Ammonium Chlorobrom-Iodide, NH_4-ClBrI," *Phys. Rev.*, **47**, 807.

Soller, W., and Thompson, A. J., "The Crystal Structure of $CuFeO_2$," *Bull. Am. Phys. Soc.*, **10**, 17.

1936

Frevel, L. K., "The Configuration of the Trinitride, N_3, Ion," *J. Am. Chem. Soc.*, **58**, 779.

Frevel, L. K., "The Crystal Structure of Ammonium Trinitride, NH_4N_3," *Z. Krist.*, **94A**, 197.

Ketelaar, J. A. A., "The Crystal Structure of Silver Nitrite, $AgNO_2$," *Z. Krist.*, **95A**, 383.

Krause, A., Ernst, Z., Gawrych, S., and Kocay, W., "X-Ray Structure and Catalytic Properties of Silver Ferrites. Amorphous and Crystalline Oxyhydrates and Oxides and Hydrous Oxides, XXVIII," *Z. Anorg. Allgem. Chem.*, **228**, 352.

Ramdohr, P., "Galena, Schapbachite and Matildite," *Fortschr. Mineral.*, **22**, 56.

Ziegler, B. E., "The Crystal Structure of Potassium Nitrite, KNO_2," *Z. Krist.*, **94A**, 491.

1937

Bassière, M., "On the Structure of $Cd(N_3)_2$," *Compt. Rend.*, **204**, 1573.

Cole, S. S., Scholes, S. R., and Amberg, C. R., "Correction in Specific Gravity and Unit Cell Size of $Na_2O·B_2O_3$," *J. Am. Ceram. Soc.*, **20**, 215.

Fang, S. M., "On the Crystal Structure of Sodium Metaborate, $NaBO_2$," *J. Am. Ceram. Soc.*, **20**, 214.

Mooney, R. C. L., "The Crystal Structure of NH_4ClBrI," *Z. Krist.*, **98A**, 324.

Zachariasen, W. H., "The Crystal Structure of KBO_2," *J. Chem. Phys.*, **5**, 919.

1938

Bassière, M., "On the Structure of the Isocyanate of Sodium," *Compt. Rend.*, **206**, 1309.

Fang, S. M., "The Crystal Structure of $NaBO_2$," *Z. Krist.*, **99A**, 1.

Hoffman, A., "The Crystal Chemistry of Lithium Ferrites," *Naturwiss.*, **26**, 431.

Hofmann, W., "The Structure of Miargyrite, $AgSbS_2$," *Ber. Preuss. Akad. Wiss.*, **1938**, 111.

Pabst, A., "The Crystal Structure of Delafossite," *Am. Mineralogist*, **23**, 175.

1939

Bassière, M., "The Crystal Structure of the Azides: The Constitution of the Trinitride Ion," *Compt. Rend.*, **208**, 659; *J. Chim. Phys.*, **36**, 71.

Berman, H., and Wolfe, C. W., "The Structure of Aramayoite, $Ag(Sb,Bi)S_2$," *Mineral. Mag.*, **25**, 466.

Helmholz, L., and Rogers, M. T., "A Redetermination of the F—F Distance in KHF_2," *J. Am. Chem. Soc.*, **61**, 2590.

Mooney, R. C. L., "An X-Ray Determination of the Structure of Tetramethyl Ammonium Dichloroiodide Crystals, $N(CH_3)_4ICl_2$," *Z. Krist.*, **A100**, 519.

Toropow, N. A., and Shishacow, N. A., "The Binary System Sodium Ferrite–Sodium Aluminate," *Acta Physicochim., URSS*, **11**, 277.

1940

Rogers, M. T., and Helmholz, L., "A Redetermination of the Parameters of NH_4HF_2," *J. Am. Chem. Soc.*, **62**, 1533.

Tazaki, H., "The Structure of Rhombic Metaboric Acid," *J. Sci. Hiroshima Univ., Ser. A*, **10**, 55.

1942

Boon, J. W., and MacGillavry, C. H., "The Crystal Structure of $KFeS_2$ and $NaCrS_2$," *Rec. Trav. Chim.*, **61**, 910.

Bredig, M. A., "The Crystal Structure of $CaCN_2$," *J. Am. Chem. Soc.*, **64**, 1730.

Tanaka, Y., "Reactions between Solid Oxides at Higher Temperatures. IV. The Reaction between Calcium Oxide and Stannic Oxide," *Bull. Chem. Soc. Japan*, **17**, 70.

1943

Bassière, M., "The Crystal Structure of Sodium Isocyanate," *Mém. Serv. Chim. État (Paris)*, **30**, 30.

Rüdorff, W., and Stegemann, K., "The Crystal Structure and Magnetic Behavior of the Alkali Thiochromites," *Z. Anorg. Allgem. Chem.*, **251**, 376.

Straumanis, M., and Circulis, A., "The Preparation and Properties of $Cu(N_3)_2$," *Z. Anorg. Allgem. Chem.*, **251**, 315.

Strijk, B., and MacGillavry, C. H., "A High-Temperature Modification of $NaNO_2$," *Rec. Trav. Chim.*, **62**, 705.

1944

Barblan, F., Brandenberger, E., and Niggli, P., "Ordered and Disordered Structures of Titanates and Ferrites and Ordered Transformations of the TiO_2 Modification," *Helv. Chim. Acta*, **27**, 88.

Boon, J. W., "Crystal Structure of $NaBiS_2$ and $KBiS_2$," *Rec. Trav. Chim.*, **63**, 32.

Boon, J. W., "The Crystal Structure of Chalcopyrite ($CuFeS_2$) and $AgFeS_2$. The Permutoidic Reactions $KFeS_2 \rightarrow CuFeS_2$ and $KFeS_2 \rightarrow AgFeS_2$," *Rec. Trav. Chim.*, **63**, 69.

1946

Pabst, A., "Notes on the Structure of Delafossite," *Am. Mineralogist*, **31**, 539.

Peacock, M. A., "Crystallography of Artificial and Natural Smithite," *Univ. Toronto Studies, Geol. Ser.*, **50**, 81.

Strijk, B., and MacGillavry, C. H., "Rectification of 'A High-Temperature Modification of Sodium Nitrite'," *Rec. Trav. Chim.*, **65**, 127.

1947

Llewellyn, F. J., and Whitmore, F. E., "The Crystal Structure of Strontium Azide," *J. Chem. Soc.*, 1947, 881.

Samson, S., and Sillen, L. G., "The Crystal Structure of Barium Uranate; the Nonexistence of the UO_4 Group," *Arkiv Kemi*, 25A, 16 pp.

1948

Pfefferkorn, G., "The Structure of Lead Azide," *Z. Naturforsch.*, 3A, 364.

Rudorff, W., Ruston, W. R., and Scherhaufer, A., "The Crystal Structure of Sodium Selenochromite, $NaCrSe_2$, and Preliminary Investigations on Related Compounds," *Acta Cryst.*, 1, 196.

Shugam, E. A., and Zhdanov, G. S., "X-Ray Investigation of Crystal Structure of Na-$Au(CN)_2$," *Tr. Inst. Kristallogr. Akad. Nauk, SSSR*, 4, 179.

Zachariasen, W. H., "Crystal Chemical Studies of the 5*f*-Series of Elements. IV. The Crystal Structure of $Ca(UO_2)O_2$ and $Sr(UO_2)O_2$," *Acta Cryst.*, 1, 281.

1949

Stroupe, J. D., "An X-Ray Diffraction Study of the Copper Chromites and of the 'Copper–Chromium Oxide' Catalyst," *J. Am. Chem. Soc.*, 71, 569.

Zvonkova, Z. V., and Zhdanov, G. S., "X-Ray Study of Ammonium Thiocyanate," *Zh. Fiz. Khim.*, 23, 1495.

1950

Burdese, A., and Brisi, C., "Solid Solutions between Potassium Ferrite and Aluminate. Magnetic and X-Ray Researches," *Atti Accad. Sci. Torino, Classe Sci. Fis. Mat. Nat.*, 85, 231.

Byström, A., and Evers, L., "The Crystal Structures of Ag_2PbO_2 and $Ag_5Pb_2O_6$," *Acta Chem. Scand.*, 4, 613.

1951

Graham, A. R., "Matildite, Aramayoite, Miargyrite," *Am. Mineralogist*, 36, 436.

Lander, J. J., "The Crystal Structures of $NiO \cdot BaO$, etc.," *Acta Cryst.*, 4, 148.

Truter, M. R., "Dimensions of the Nitrite Ion," *Nature*, 168, 344.

1952

Carpenter, G. B., "The Crystal Structure of Sodium Nitrite," *Acta Cryst.*, 5, 132.

Juza, R., and Schulz, W., "Preparation and Properties of Li_3AlP_2 and Li_3AlAs_2," *Z. Anorg. Allgem. Chem.*, 269, 1.

Nuffield, E. W., "Mineral Thiosalts. XVI. Cuprobismuthite," *Am. Mineralogist*, 37, 447.

Peterson, S. W., and Levy, H. A., "A Single Crystal Neutron Diffraction Determination of the Hydrogen Position in KHF_2," *J. Chem. Phys.*, 20, 704.

Zelezny, W. F., and Baenziger, N. C., "The Crystal Structure of Tetrachlorophosphonium Dichloroiodide," *J. Am. Chem. Soc.*, 74, 6151.

1953

Cavalca, L., Nardelli, M., and Braibanti, A., "Sodium Argentonitrite," *Gazz. Chim. Ital.*, **83**, 476.

Hahn, H., Frank, G., Klingler, W., Meyer, A. D., and Storger, G., "Ternary Chalcogenides with Chalcopyrite Structures," *Z. Anorg. Allgem. Chem.*, **271**, 153.

1954

Asprey, L. B., Ellinger, F. H., and Zachariasen, W. H., "Preparation, Identification and Crystal Structure of a Quinquevalent Americium Compound, K(AmO₂)F₂," *J. Am. Chem. Soc.*, **76**, 5235.

Bertaut, F., and Blum, P., "Structure of a New Variety of Sodium Ferrite (NaFeO₂)," *Compt. Rend.*, **239**, 429.

Dyer, L. D., Borie, B. S., Jr., and Smith, G. P., "Alkali Metal Nickel Oxides of the Type MNiO₂," *J. Am. Chem. Soc.*, **76**, 1499.

Juza, R., and Schulz, W., "Ternary Phosphides and Arsenides of Lithium with Elements of Groups III and IV," *Z. Anorg. Allgem. Chem.*, **275**, 65.

Rüdorff, W., and Becker, H., "The Structures of LiVO₂, NaVO₂, LiCrO₂, and NaCrO₂," *Z. Naturforsch.*, **9B**, 614.

Rüdorff, W., and Pfister, F., "Alkaline Earth Uranates(VI) and their Reduction Products," *Z. Naturforsch.*, **9B**, 568.

Schönberg, N., "Structure of the Metallic Quaternary Phase ZrTaNO," *Acta Chem. Scand.*, **8**, 627.

Truter, M. R., "Refinement of a Non-Centrosymmetrical Structure: Sodium Nitrite," *Acta Cryst.*, **7**, 73.

Zachariasen, W. H., "The Crystal Structure of Magnesium Orthouranate," *Acta Cryst.*, **7**, 788.

1955

Carpenter, G. B., "Further Least-Squares Refinement of the Crystal Structure of Sodium Nitrite," *Acta Cryst.*, **8**, 852.

Cavalca, L., Nardelli, M., and Bassi, I. W., "The Structure of Thallous Nitrite," *Gazz. Chim. Ital.*, **85**, 153.

Collongues, R., "The Properties of Lithium Ferrite, FeLiO₂," *Compt. Rend.*, **241**, 1577.

Dannhauser, W., and Vaughan, P. A., "The Crystal Structure of Cuprous Chromite," *J. Am. Chem. Soc.*, **77**, 896.

Gattow, G., and Zemann, J., "The Crystal Chemistry of Alkali Bismuth Sulfides and Alkali Bismuth Selenides," *Z. Anorg. Allgem. Chem.*, **279**, 324.

Glemser, O., and Filcek, M., "Alkali Thiobismuthates(III)," *Z. Anorg. Allgem. Chem.*, **279**, 321.

Hahn, H., and Lorent, C. de, "Ternary Chalcogenides. VII. Synthesis of Ternary Oxides of Aluminum, Gallium and Indium with Univalent Copper and Silver," *Z. Anorg. Allgem. Chem.*, **279**, 281.

1956

Delorme, C., "Some Compounds of the Type M₂O₃–Cu₂O," *Acta Cryst.*, **9**, 200.

Kruh, R., Fuwa, K., and McEver, T. E., "The Crystal Structures of Alkali Metal Bifluorides," *J. Am. Chem. Soc.*, **78**, 4256.

Pinsker, Z. G., Semiletov, S. A., and Belova, E. N., "Electron Diffraction Determination of the Structure of Tl₂Sb₂Se₄," *Dokl. Akad. Nauk SSSR*, **106**, 1003.

Trueblood, K. M., and Mayer, S. W., "The Crystal Structure of Sulfamide," *Acta Cryst.*, **9**, 628.

1957

Beevers, C. A., and Trotman-Dickenson, A. F., "The Crystal Structure of Nitramide, NH₂NO₂," *Acta Cryst.*, **10**, 34.

Cromer, D. T., "The Crystal Structure of KCu(CN)₂," *J. Phys. Chem.*, **61**, 1388.

Curti, R., Riganti, V., and Locchi, S., "The Crystal Structure of AgClO₂," *Acta Cryst.*, **10**, 687.

Douglass, R. M., "The Crystal Structure of HCrO₂," *Acta Cryst.*, **10**, 423.

Ferrari, A., Cavalca, L., and Tani, M. E., "The Structure of Cesium Nitrite," *Gazz. Chim. Ital.*, **87**, 310.

Goodman, C. H. L., "A New Group of Compounds with Diamond-type (Chalcopyrite) Structure," *Nature*, **179**, 828.

Hahn, H., and Lorent, C. de, "Ternary Chalcogenides. XII. Ternary Chalcogenides of Chromium with Univalent Copper and Silver," *Z. Anorg. Allgem. Chem.*, **290**, 68.

Lindqvist, I., "On the Crystal Structure of Silver Thiocyanate," *Acta Cryst.*, **10**, 29.

Lindqvist, I., and Strandberg, B., "The Crystal Structure of Ammonium Silver Dithiocyanate," *Acta Cryst.*, **10**, 173.

1958

Behar, I., and Collongues, R., "Application of the Micrographic Method to the Study of Transformations of Ferrites," *Congr. Intern. Chim. Pure Appl. 16ᵉ, Paris, 1957, Mém. Sect. Chim. Minérale*, p. 109.

Delorme, C., "The Asymmetry of the Bivalent Copper Ion in Combinations of the NaCl Type and of the Spinel Type. II. Study of the M₂CuO₄ Spinels and their Solid Solutions," *Bull. Soc. Franc. Mineral. Crist.*, **81**, 79.

Frondel, C., and Barnes, I., "Structural Relations of UO₂, Isometric PbUO₄, and Orthorhombic PbUO₄," *Acta Cryst.*, **11**, 562.

Hoppe, R., and Schepers, B., "Alkali Indates: LiInO₂ and NaInO₂," *Z. Anorg. Allgem. Chem.*, **295**, 233.

Johnston, W. D., Heikes, R. R., and Sestrich, D., "The Preparation, Crystallography and Magnetic Properties of the LiₓCo₍₁₋ₓ₎O System," *Phys. Chem. Solids*, **7**, 1.

Kato, E., "Phase Transitions of the Li₂O–Fe₂O₃ System. I. Thermal and Electric Properties of Lithium Ferrite, LiFeO₂," *Bull. Chem. Soc. Japan*, **31**, 108.

Kondrashev, Y. D., "The Crystal Structure and Composition of Crednerite, CuMnO₂," *Kristallografiya*, **3**, 696.

Kovba, L. M., Ippolitova, E. A., Simanov, Y. P., and Spitsyn, V. I., "X-Ray Diffraction of Alkali Metal Uranates," *Dokl. Akad. Nauk SSSR*, **120**, 1042.

Loopstra, B. O., "X-Ray and Neutron Diffraction Study of Calcium Hypophosphite and Phosphorous Acid," *JENER (Joint Estab. Nucl. Energy Res.)*, Publ. 15, 64 pp.

Pfister, H., "Crystal Structure of Ternary Alloys of the Type AᴵᴵBᴵⱽC₂ⱽ," *Acta Cryst.*, **11**, 221.

Strunz, H., Geier, B. H., and Seeliger, E., "Gallite, CuGaS₂, the First Independent Gallium Mineral and its Distribution in the Ores of the Tsumeb and Kipushi Mines," *Neues Jahrb. Mineral., Monatsh.*, **1958**, 241.

Yamamoto, Y., Kinoshita, K., Tamaru, K., and Yamanaka, T., "Redetermination of the Crystal Structure of Calcium Cyanamide," *Bull. Chem. Soc. Japan*, **31**, 501.

Zemann, A., and Zemann, J., "The Structural Type of Lorandite, TlAsS₂," *Naturwiss.*, **45**, 488.

1959

Beletskii, M. S., and Saksonov, Y. G., "X-Ray Diffraction Studies of Polymorphic Transition Formations in Sodium Aluminate," *Zh. Neorgan. Khim.*, **4**, 972.

Frueh, A. J., Jr., "The Crystallography of Petzite, Ag_3AuTe_2," *Am. Mineralogist*, **44**, 693.

Geller, S., and Wernick, J. H., "Ternary Semiconducting Compounds with Sodium Chloride-like Structures: $AgSbSe_2$, $AgSbTe_2$, $AgBiS_2$, $AgBiSe_2$," *Acta Cryst.*, **12**, 46.

Gillespie, R. B., Sparks, R. A., and Trueblood, K. N., "The Crystal Structure of Ammonium Chlorite at $-35°C.$," *Acta Cryst.*, **12**, 867.

Rosenzweig, A., and Cromer, D. T., "The Crystal Structure of $KAu(CN)_2$," *Acta Cryst.*, **12**, 709.

Semiletov, S. A., and Man, L. I., "Electron Diffraction Determination of the Structure of Thin Films of $TlBiSe_2$ and $TlSbS_2$," *Kristallografiya*, **4**, 414.

Waddington, T. C., "Lattice Parameters and Infrared Spectra of Some Inorganic Cyanates," *J. Chem. Soc.*, **1959**, 2499.

Zemann, A., and Zemann, J., "Crystal Structure of Lorandite, $TlAsS_2$," *Acta Cryst.*, **12**, 1002.

1960

Brixner, L. H., "Preparation, Structure and Electrical Properties of some Substituted Lithium-Oxo-Metallates," *J. Inorg. Nucl. Chem.*, **16**, 162.

Eckerlin, P., Langereis, C., Maak, I., and Rabenau, A., "$LiPN_2$" *Angew. Chem.*, **72**, 268.

Feitknecht, W., and Moser, K., "Formation of Silver Iron Oxide from Iron Hydroxide and Silver Oxide in Alkaline Solution," *Z. Anorg. Allgem. Chem.*, **304**, 181.

Folberth, O. G., and Pfister, H., "The Crystal Structure of $ZnSnAs_2$," *Acta Cryst.*, **13**, 199.

Hoppe, R., and Werding, G., "Alkali Metal Oxothallates(III)," *Naturwiss.*, **47**, 203.

McDonald, T. R. R., "The Electron Density Distribution in Ammonium Bifluoride," *Acta Cryst.*, **13**, 113.

Schnering, H. G. v., and Hoppe, R., "The Crystal Structure of $SrZnO_2$," *Naturwiss.*, **47**, 467.

Schnering, H. G. v., Hoppe, R., and Zemann, J., "The Crystal Structure of $BaZnO_2$," *Z. Anorg. Allgem. Chem.*, **305**, 241.

Spitsbergen, U., "The Crystal Structures of $BaZnO_2$, $BaCoO_2$, and $BaMnO_2$," *Acta Cryst.*, **13**, 197.

1961

Bertaut, E. F., and Dulac, J., "On the Isomorphism of the Ternary Oxides of Trivalent Chromium and Rhodium," *Phys. Chem. Solids*, **21**, 118.

Cooper, J., and Marsh, R. E., "On the Structure of $AgClO_2$," *Acta Cryst.*, **14**, 202.

Folberth, O. G., and Pfister, H., "New Ternary Semiconducting Phosphides, $MgGeP_2$, $CuSi_2P_3$, and $CuGe_2P_3$," *Acta Cryst.*, **14**, 325.

Hockings, E. F., and White, J. G., "The Crystal Structures of $TlSbTe_2$ and $TlBiTe_2$," *Acta Cryst.*, **14**, 328.

Hoppe, R., and Röhrborn, H.-J., "Crystal Structure of $LiInO_2$," *Naturwiss.*, **48**, 452.

Hoppe, R., and Werding, G., "Oxythallates of the Alkali Metals," *Z. Anorg. Allgem. Chem.*, **307**, 174.

Hoppe, R., Lidecke, W., and Frorath, F. C., "Sodium Thioindate and Sodium Selenoindate," *Z. Anorg. Allgem. Chem.*, **309**, 49.

Hoshino, S., and Shibuya, I., "Anomalous Temperature Dependence of Lattice Constants of Ferroelectric Sodium Nitrite," *J. Phys. Soc. Japan*, **16**, 1254.

Kay, M. I., and Frazer, B. C., "A Neutron Diffraction Refinement of the Low Temperature Phase of $NaNO_2$," *Acta Cryst.*, **14**, 56.

Kovba, L. M., Ippolitova, E. A., Simanov, Y. P., and Spitsyn, V. I., "Crystalline Structure of Uranates. I. Uranates with Tetragonal Layers $(UO_2)O_2$," *Zh. Fiz. Khim.*, **35**, 563.

Kovba, L. M., Polunina, G. P., Simanov, Y. P., and Ippolitova, E. A., "X-Ray Diffraction Studies of Some Uranates," *Issled. v Obl., Khim. Urana, Sb. Statei*, **1961**, 15.

Kovba, L. M., Polunina, G. P., Ippolitova, E. A., Simanov, Y. P., and Spitsyn, V. I., "Crystal Structure of Uranates. II. Uranates Containing Uranyl–Oxygen Chains," *Zh. Fiz. Khim.*, **35**, 719.

McGeachin, H. M., and Tromans, F. R., "Phosphonitrilic Derivatives. VII. Crystal Structures of Tetrameric Phosphonitrilic Fluoride," *J. Chem. Soc.*, **1961**, 4777.

Plust, H. G., and Hugi, W., "Electron Microscope Investigation of Some New Polymorphic Thermoelectric Compounds," *Schweiz. Arch. Angew. Wiss. Tech.*, **27**, 458.

Recker, K., "The Entry of Uranium into Calcium Fluoride," *Fortschr. Mineral.*, **39**, 69.

Rooymans, C. J. M., "The Crystal Structure of $LiScO_2$," *Z. Anorg. Allgem. Chem.*, **313**, 234.

Scherhaufer, A. M., "Relation between Lattice Constants and Conductivity of Alkali Selenochromites," *Oesterr. Chem. Ztg.*, **62**, 18.

Schnering, H. G., and Hoppe, R., "The Crystal Structure of $SrZnO_2$," *Z. Anorg. Allgem. Chem.*, **312**, 87.

Smith, P., Garcia-Blanco, S., and Rivoir, L., "Structure of Zinc Metaborate, Zn_4O-$(BO_2)_6$," *Anal. Real Soc. Espan. Fis. Quim. (Madrid)*, Ser. A, **57**, 263.

Smith, P., Garcia-Blanco, S., and Rivoir, L., "A New Structural Type of Metaborate Anion," *Z. Krist.*, **115**, 460.

Spitsyn, V. I., Ippolitova, E. A., Simanov, Y. P., and Kovba, L. M., "Normal Uranates of Alkali Metals," *Issled. Obl. v Khim. Urana, Sb. Statei*, **1961**, 5.

Théry, J., Briançon, D., and Collongues, R., "The Structure and Properties of Sodium Aluminate $NaAlO_2$ and its Solid Solutions with the Ferrite $NaFeO_2$," *Compt. Rend.*, **252**, 1475.

Théry, J., Lejus, A.-M., Briançon, D., and Collongues, R., "Structure and Properties of Alkaline Aluminates," *Bull. Soc. Chim. France*, **1961**, 973.

1962

Bertaut, F., and Gondrand, M., "A Study of Combinations of Rare Earth Oxides and Lithium of the Type $TLiO_2$," *Compt. Rend.*, **255**, 1135.

Frevel, L. K., and Rinn, H. W., "The Crystal Structure of $LiHF_2$," *Acta Cryst.*, **15**, 286.

Hagenmuller, P., Lecerf, A., and Onillon, M., "A New Oxide of Trivalent Titanium, $NaTiO_2$," *Compt. Rend.*, **255**, 928.

Hamilton, W. C., and Ibers, J. A., "A Neutron Diffraction Study of Polycrystalline $HCrO_2$ and $DCrO_2$," *J. Phys. Soc. Japan, Suppl. B-II*, **17**, 383.

Hazekamp, R., Migchelsen, T., and Vos, A., "Refinement of the Structure of Metastable Phosphonitrilic Chloride $(PNCl_2)_4$," *Acta Cryst.*, **15**, 539.

Kay, M. I., Frazer, B. C., and Ueda, R., "The Disordered Structure of $NaNO_2$ at 185°C.," *Acta Cryst.*, **15**, 506; *J. Phys. Soc. Japan, Suppl. B-II*, **17**, 389.

Lecerf, A., "A Double Oxide of Lithium and Trivalent Titanium, $LiTiO_2$," *Compt. Rend.*, **254**, 2003.

Lejus, A.-M., and Collongues, R., "The Structure and Properties of Lithium Aluminates," *Compt. Rend.*, **254**, 2005.

Long, R. E., and Marsh, R. E., "A Reinvestigation of the Crystal Structure of Silver Nitrite," *Acta Cryst.*, **15**, 448.

Marr, H. E., III, and Stanford, R. H., Jr., "The Unit-Cell Dimensions of Silver Azide," *Acta Cryst.*, **15**, 1313.

Schnering, H. G., "The Crystal Structure of $BaCdO_2$," *Z. Anorg. Allgem. Chem.*, **314**, 144.

Chapter VII

COMPOUNDS OF THE TYPE $R_n(MX_3)_p$

The same difficulties are encountered in trying to find a rational arrangement of the compounds $R_n(MX_3)_p$ as were presented by the chemically simpler crystals of the preceding chapters. Many of the crystals to be treated here are predominantly ionic, with MX_3 groups functioning as anions, and a number can be considered as distortions of common structures already described. Many others, however, have a complexity that puts them outside a scheme of classification based on these analogies, and consequently it has seemed better to follow a looser classification based on chemical composition.

The chapter is divided into two sections. In the first are crystals with the composition RMX_3; in the second are all compounds possessed of less simple formulas. Each section begins with compounds like the carbonates, nitrates, and chlorates which are pronouncedly ionic, and these are followed by other oxygen-containing compounds where the close-packing of individual atoms or ions appears important in determining the structure that occurs. They are followed by halides and compounds in which atomic chains are present; and the section is concluded with any, miscellaneous, substances that do not naturally belong in one or another of the foregoing groups.

A. COMPOUNDS RMX_3

Nitrates, Carbonates, etc.

VII,a1. *Sodium nitrate*, $NaNO_3$, and the crystals isomorphous with it are NaCl-like arrangements of R^+ and MX_3 ions distorted by the spatial requirements of its complex anions. The symmetry is rhombohedral, with a bimolecular unit which for $NaNO_3$ has the dimensions:

$$a_0 = 6.3247 \text{ A.}, \qquad \alpha = 47°16'$$

Atoms were early found to be in the following positions of D_{3d}^6 ($R\bar{3}c$):

Na: (2b) $\pm(\frac{1}{4} \frac{1}{4} \frac{1}{4})$
N: (2a)   000; $\frac{1}{2} \frac{1}{2} \frac{1}{2}$
O: (6e) $u\bar{u}0$; $\bar{u}0u$; $0u\bar{u}$;
             $\frac{1}{2}-u, u+\frac{1}{2}, \frac{1}{2}$; $u+\frac{1}{2}, \frac{1}{2}, \frac{1}{2}-u$; $\frac{1}{2}, \frac{1}{2}-u, u+\frac{1}{2}$

with a parameter that has recently been redetermined as 0.2401. The corresponding hexagonal unit containing six molecules has the edge lengths:

$$a_0' = 5.0708 \text{ A.,} \qquad c_0' = 16.818 \text{ A. } (23°\text{C.})$$

In this cell the atomic coordinates are

Na: (6b) $\pm(0\ 0\ \frac{1}{4})$; rh
N: (6a)   000; $0\ 0\ \frac{1}{2}$; rh
O: (18e) $u00$; $0u0$; $\bar{u}\bar{u}0$;
             $\bar{u}\ 0\ \frac{1}{2}$; $0\ \bar{u}\ \frac{1}{2}$; $u\ u\ \frac{1}{2}$; rh   with $u = 0.2401$

The origin used in this description lies not in a center of symmetry but in one of the intersections of twofold axes midway between centers along the principal axis. If one chooses to employ coordinates referred to an origin in a center, as is done in the *International Tables*, the z coordinates of the foregoing hexagonal description will be displaced by $\frac{1}{4}$.

The arrangement of the atoms in terms of the elongated rhombohedral unit and viewed down its threefold axis is shown in Figure VIIA,1. Referred to the hexagonal axes the structure is shown in Figure VIIA,2. The nitrate ion is a planar equilateral triangle with nitrogen at the center and N–O = 1.217 A.

This substance and all others with the same structure show a very pronounced and perfect rhombohedral cleavage. The axes of the cleavage rhombohedron thus formed are those of a tetramolecular pseudo-unit which has its ions distributed as are the atoms in NaCl; its relation to the true unit is indicated in Figure VIIA,3.

Lithium nitrate, $LiNO_3$, has this structure at room temperature, and other nitrates have it as a high-temperature modification. It is also possessed by the *calcite* form of $CaCO_3$, by other divalent carbonates, and by certain borates. The cell data and values of the oxygen parameters as established for these various crystals are listed in Table VIIA,1.

At elevated temperatures the x-ray reflections from $NaNO_3$ due only to the oxygen atoms become weaker until above 275°C. they have disappeared. This undoubtedly is due to disorder in the orientations of the nitrate ions, and detailed studies have been made in the region between ca. 150° and 275°C. where this disorder seems to be developing. These observa-

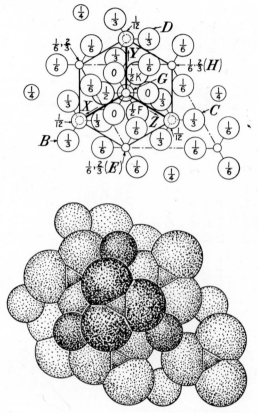

Fig. VIIA,1a (top) A projection of a portion of the NaNO₃ structure on a plane normal to the threefold axes and passing through the apex of the cleavage rhombohedron. Distances in this figure are given below the plane. The smallest circles are the nitrogen, the largest are the oxygen atoms. The letters refer to correspondingly marked atoms in Figure VIIA,3.

Fig. VIIA,1b (bottom) A packing drawing of the atoms of NaNO₃ shown in Figure VIIA, 1a.

tions are thought to indicate that above 275 °C. the nitrate ions are either in a free rotation about their threefold axes or are randomly distributed between their low temperature position and one that is turned 30° (or 180°) to this; in the transition zone below this temperature the disorder would be partial.

The structure described in **VII,b7** for dolomite, $CaMg(CO_3)_2$, differs from this arrangement only in the lower symmetry and minor distortions involved in accommodating its two cations of different sizes.

TABLE VIIA,1. Crystals with the Rhombohedral $NaNO_3$ Arrangement

Crystal	Unit cell		u	Hexagonal pseudo-cell		Cleavage rhombohedron	
	a_0, A.	α		a_0'', A.	c_0'', A.	a_0', A.	α'
AgNO₃	6.342	47°49'	—	5.168	16.903 (150°C.)	—	—
	6.411	47°36'	—	5.174	17.018 (210°C.)	—	—
	6.427	47°45'	—	5.203	17.045 (164°C.)	—	—
AgNO₃(II)	6.361	46°6'	0.2593	4.99008	17.05951 (18°C.)	6.412	101°55'
CaCO₃ (calcite)	—	—	—	5.00011	17.09498	—	—
CdCO₃	6.1306	47°19'	ca. 0.25	4.9204	16.298	6.28	102°48'
CoCO₃	5.66505	48°33'	—	4.6581	14.958 (26°C.)	5.91	103°22'
FeCO₃	5.795	47°45'	0.27	4.626	15.288	6.02	103°5'
InBO₃	5.841	48°10'	—	4.77	15.45	—	—
	5.856	48°38'	—	4.823	15.456	—	—
KNO₃	7.148	44°35'	—	5.423	19.277 (128°C.)	—	—
	7.542	42°4'	—	5.414	20.590 (335°C.)	—	—
LiNO₃	5.747	48°11'	0.264	4.692	15.206 (20°C.)	5.90	103°14'
	5.866	47°33'	—	4.729	15.577 (251°C.)	—	—
LuBO₃	6.104	47°28'	—	4.913	16.214	—	—
MgCO₃	5.6752	48°12'	0.27	4.6330	15.013 (26°C.)	6.064	102°58'
MnCO₃	5.9049	47°43'	—	4.7768	15.664	6.01	102°50'
NaNO₃	6.3247	47°16'	0.2402	5.0708	16.818 (23°C.)	6.48	102°40'
	6.535	45°42'	—	5.075	17.523 (275°C.)	—	—
	6.460	45°31'	—	5.075	17.607 (310°C.)	—	—
NiCO₃	5.5795	48°40'	—	4.5975	14.723	—	—
RbNO₃	7.807	41°7'	—	5.483	21.410 (250°C.)	—	—
ScBO₃	5.781	48°28'	—	4.75	15.27	—	—
YBO₃	6.44	46°17'	—	5.06	17.21	—	—
ZnCO₃	5.6833	48°20'	—	4.6528	15.025	5.928	103°27'

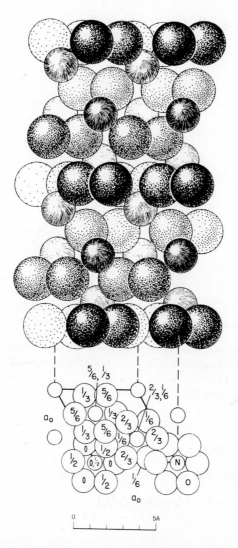

Fig. VIIA,2. Two projections in terms of the hexagonal cell of the structure of NaNO₃. In the packing drawing of the structure viewed normal to a prism face the oxygen atoms are the large circles, the nitrogen atoms do not show.

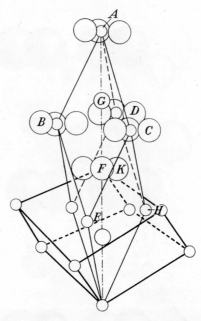

Fig. VIIA,3. A drawing to show the relation between the correct unit rhombohedron of the sodium nitrate arrangement and its cleavage rhombohedron. The bimolecular unit is the elongated cell, the cleavage pseudocell is that outlined by the thicker lines. The atoms marked with capitals refer to those similarly designated in Figure VIIA,1a.

VII,a2. *Aragonite*, the modification of $CaCO_3$ that is thermodynamically stable at room temperature, is typical of a number of crystals having orthorhombic symmetry. Its tetramolecular unit has the dimensions:

$$a_0 = 7.968 \text{ A.}; \quad b_0 = 5.741 \text{ A.}; \quad c_0 = 4.959 \text{ A. } (26°C.)$$

The space group is V_h^{16} (*Pbnm*), with atoms in the positions:

(4c) $\pm (u\, v\, ^1/_4; \; ^1/_2-u, v+^1/_2, ^1/_4)$

(8d) $\pm (xyz; \; x,y,^1/_2-z; \; ^1/_2-x,y+^1/_2,z; \; ^1/_2-x,y+^1/_2,^1/_2-z)$

Approximate parameters established many years ago are

Atom	Position	x	y	z
Ca	(4c)	0.417	0.750	$^1/_4$
C	(4c)	0.75	−0.083	$^1/_4$
O(1)	(4c)	0.917	−0.083	$^1/_4$
O(2)	(8d)	0.67	−0.083	0.48

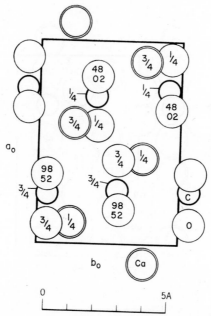

Fig. VIIA,4. A projection of the orthorhombic structure of the aragonite modification of $CaCO_3$ viewed along its c_0 axis. Origin in the lower left.

In this structure (Fig. VIIA,4) as in other carbonates and nitrates, the anions are planar equilateral triangles with C–O = ca. N–O = ca. 1.25 A.

This structure is pseudohexagonal (Fig. VIIA,5), and the distribution of its ions is much like that of the nickel and arsenic atoms in NiAs (III,d1). The extent of this similarity is made clear by the fact that in terms of an orthorhombic pseudocell NiAs would have the axial ratio 1:0.943:0.577 (compared with 1:0.721:0.619 for aragonite).

Determinations including approximate parameters have been made for two other compounds with this structure. For *potassium nitrate*, KNO_3, the cell dimensions are

$$a_0 = 0.1709 \text{ A.}; \quad b_0 = 6.4255 \text{ A.}; \quad c_0 = 5.4175 \text{ A.}$$

Its approximate parameters are

Atom	Position	x	y	z
K	(4c)	0.416	0.750	$1/4$
N	(4c)	0.75	−0.083	$1/4$
O(1)	(4c)	0.883	−0.083	$1/4$
O(2)	(8d)	0.686	−0.083	0.444

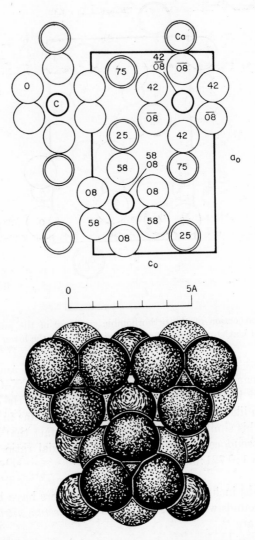

Fig. VIIA,5a (top). A projection of the orthorhombic aragonite structure along its b_0 axis. Viewed in this direction the pseudohexagonal character of the arrangement is evident. Origin in lower right.

Fig. VIIA,5b (bottom). A packing drawing of the aragonite structure as shown in Figure VIIA,5a. The line shadowed spheres are the calcium atoms. The carbon atoms appear as little triangles.

For *lead carbonate*, $PbCO_3$, the cell dimensions are

$$a_0 = 8.468 \text{ A.}; \; b_0 = 6.146 \text{ A.}; \; c_0 = 5.166 \text{ A.}$$

Its approximate parameters are similar to those of the preceding two crystals:

Atom	Position	x	y	z
Pb	(4c)	0.417	0.750	$1/4$
C	(4c)	0.764	−0.097	$1/4$
O(1)	(4c)	0.909	−0.097	$1/4$
O(2)	(8d)	0.691	−0.097	0.455

The following carbonates have been shown to have this structure, but atomic positions have not been established. Their cell dimensions are

$BaCO_3$: $a_0 = 8.8345$ A.; $b_0 = 6.5490$ A.; $c_0 = 5.2556$ A.
$SrCO_3$: $a_0 = 8.414$ A.; $\;\; b_0 = 6.029$ A.; $\;\; c_0 = 5.107$ A. (26 °C.)
$(Ca,Pb)CO_3$,
tarnowitzite: $a_0 = 8.015$ A.; $\;\; b_0 = 5.79$ A.; $\;\;\; c_0 = 4.97$ A.
$CaBa(CO_3)_2$,
alstonite: $a_0 = 8.77$ A.; $\;\;\; b_0 = 6.11$ A.; $\;\;\; c_0 = 4.99$ A.

If alstonite is not a solid solution with calcium and barium together filling the special positions (4c), its correct symmetry presumably is less than holohedral and its atoms are somewhat displaced from the positions prevailing in aragonite. Its relation to aragonite would then be like that of dolomite to calcite. Another mineral with the composition of alstonite, barytocalcite, has the different structure described in **VII,b8**.

Two rare-earth borates are known to have this structure, and further work may well reveal others. For these two, the cells are

$LaBO_3$: $a_0 = 8.252$ A.; $b_0 = 5.872$ A.; $c_0 = 5.104$ A.
$NdBO_3$: $a_0 = 7.968$ A.; $b_0 = 5.741$ A.; $c_0 = 4.959$ A.

VII,a3.   A structure has been suggested, though not yet adequately demonstrated, for the unstable μ-form of calcium carbonate $CaCO_3$, *vaterite*. It is, like aragonite, orthorhombic and pseudohexagonal but it has a different tetramolecular cell with the dimensions:

$$a_0 = 4.13 \text{ A.}; \;\;\; b_0 = 7.15 \text{ A.} \; (= \sqrt{3} \, a_0); \;\;\; c_0 = 8.48 \text{ A.}$$

The space group is stated as V_h^{16} ($Pbnm$), with atoms in the following positions:

Ca: (4a) $000; 0\,0\,^1/_2; ^1/_2\,^1/_2\,0; ^1/_2\,^1/_2\,^1/_2$

C: (4c) $\pm(u\,v\,^1/_4; ^1/_2-u,v+^1/_2,^1/_4)$ with $u = 0.157$, $v = 0.67$

O(1): (4c) with $u = 0.471$, $v = 0.67$

O(2): (8d) $\pm(xyz; x+^1/_2,^1/_2-y,z+^1/_2; x,y,^1/_2-z; ^1/_2-x,y+^1/_2,z)$

with $x = 0.00$, $y = 0.67$, $z = 0.118$

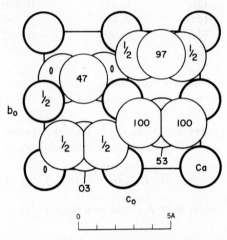

Fig. VIIA,6. A projection along its a_0 axis of the orthorhombic structure proposed for the μ-form of CaCO₃. The carbon atoms do not show within their triangular CO₃ ions.

This is an arrangement (Fig. VIIA,6) that can be considered as an end member of the series of which bästnasite is the simplest example (VII,b11).

Several rare-earth borates (Table VIIA,2) seem to be isomorphous with vaterite, though the data from them are compatible with an exact hexagonal symmetry ($b_0 = \sqrt{3}\,a_0$). It is important to determine if vaterite and these other crystals actually are orthorhombic and if the atomic arrangement as described above is correct.

VII,a4. *Ammonium nitrate*, NH_4NO_3, is unusually pleomorphic, with not less than five modifications.

The form occurring just below the melting point in the temperature range between 169° and 125°C. is cubic and gives the very simple x-ray pattern of the CsCl arrangement (III,b1). For this I modification,

$$a_0 = 4.40 \text{ A}.$$

TABLE VIIA,2
Crystals with the Vaterite Structure

Atom	a_0, A.	b_0, A.	c_0, A.
DyBO$_3$	3.791	6.566	8.84
ErBO$_3$	3.761	6.514	8.79
EuBO$_3$	3.845	6.660	8.94
GdBO$_3$	3.829	6.632	8.89
HoBO$_3$	3.776	6.540	8.80
LuBO$_3$	3.725	6.452	8.71
SmBO$_3$	3.858	6.682	8.96
TmBO$_3$	3.748	6.491	8.76
YBO$_3$	3.777	6.542	8.81
YbBO$_3$	3.732	6.464	8.74

It would appear that the centers of the NH$_4$ and NO$_3$ groups are at 000 and $1/2$ $1/2$ $1/2$ but that there is no fixed orientation for their hydrogen and oxygen atoms, i.e., the ions are freely "rotating."

The II form of NH$_4$NO$_3$ stable between ca. 125° and 84°C. is tetragonal, with an arrangement that differs from the foregoing mainly in the fact that entirely free rotation no longer can occur. Its unit is bimolecular, with the dimensions:

$$a_0 = 5.75 \text{ A.}, \quad c_0 = 5.00 \text{ A.}$$

The structure that has been described involves an unusual kind of disorder. Making use of the space group C_{4v}^2 ($P4bm$), it places the centers of the ammonium groups in ($2a$) 000; $1/2$ $1/2$ 0. There are two kinds of nitrate ions. For one, N(1) is in either $1/2$ 0 u or $1/2$ 0 $\bar{u}$; O(1) is in $1/2$ 0 v or $1/2$ 0 $\bar{v}$ and O(1') in $t+1/2,t,w$; $1/2-t,\bar{t},w$ or $t+1/2,t,\bar{w}$; $1/2-t,\bar{t},\bar{w}$, with $u =$ 0.53, $v = 0.75$, $t = 0.12$, $w = 0.42$. For the other, N(2) is in either 0 $1/2$ u or 0 $1/2$ $\bar{u}$; O(2) in 0 $1/2$ v or 0 $1/2$ $\bar{v}$ and O(2') in $\bar{t},t+1/2,w$; $t,1/2-t,w$ or $\bar{t},t+1/2,\bar{w}$; $t,1/2-t,\bar{w}$, with the same parameters.

VII,a5. The III modification of *ammonium nitrate*, NH$_4$NO$_3$, stable between 84° and 32°C. is built on a different principle. Its structure, like that of aragonite (**VII,a2**), resembles NiAs rather than CsCl. The orthorhombic cell containing four molecules has the edges:

$$a_0 = 7.18 \text{ A.}; \quad b_0 = 7.71 \text{ A.}; \quad c_0 = 5.83 \text{ A.}$$

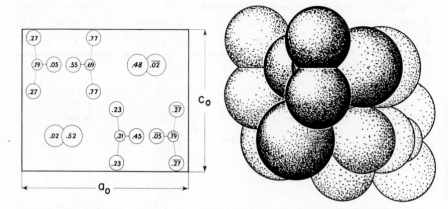

Fig. VIIA,7a (left). Atoms in the orthorhombic unit of modification III of NH_4NO_3 projected along its b_0 axis. Atoms of the NO_3 group are connected by light lines.
Fig. VIIA,7b (right). A packing drawing of the NH_4NO_3 atoms of Figure VIIA,7a. The largest spheres are NH_4 ions, the smallest nitrogen atoms.

Atoms are in the following positions of V_h^{16} (*Pbnm*):

(4c) $\pm(u\ v\ ^1/_4;\ ^1/_2-u,v+^1/_2,^1/_4)$

(8d) $\pm(xyz;\ x,y,^1/_2-z;\ ^1/_2-x,y+^1/_2,z;\ ^1/_2-x,y+^1/_2,^1/_2-z)$

with the approximate parameters:

Atom	Position	x	y	z
NH_4	(4c)	0.30	0.52	$^1/_4$
N	(4c)	−0.09	−0.19	$^1/_4$
O(1)	(4c)	−0.19	−0.05	$^1/_4$
O(2)	(8d)	−0.07	−0.27	0.06

The planar nitrate ions that result have N–O = 1.28 A. and O–O = 2.91 A.

In this crystal (Fig. VIIA,7), a pseudohexagonal axis is along b_0; and the extent of the distortion from the NiAs structure may be gauged from the fact that the NH_4 and NO_3 ions would be in Ni and As positions if the axial ratio were 0.943:1:0.577 instead of 0.922:1:0.778 and if $u = \ ^1/_4$ and $v = \ ^1/_2$ for NH_4 and $u' = -^1/_6$ and $v' = -^1/_4$ for the nitrogen atom of NO_3.

VII,a6. The two lowest temperature forms of *ammonium nitrate*, NH_4NO_3, revert to the CsCl-pattern of arrangement that prevails in the high-temperature forms. Of these, the *room-temperature*, IV, modification, stable between 32° and $-18°C$. is orthorhombic, with a bimolecular cell of the dimensions:

$$a_0 = 5.757 \text{ A.}; \quad b_0 = 5.451 \text{ A.}; \quad c_0 = 4.935 \text{ A.}$$

Atoms have been assigned the following positions of $V_h{}^{13}$ (*Pmmn*):

NH₄: (2b) $0\,{}^1/_2\,u;\ {}^1/_2\,0\,\bar{u}$ with $u = 0.57$
  N: (2a) $00u;\ {}^1/_2\,{}^1/_2\,u$ with $u = 0.03$
O(1):   (2a)   with $u = 0.28$
O(2):   (4f) $u0v;\ \bar{u}0v;\ u+{}^1/_2,{}^1/_2,\bar{v};\ {}^1/_2-u,{}^1/_2,\bar{v}$
                                with $u = 0.19, v = -0.095$

This arrangement (Fig. VIIA,8) results in planar NO_3 groups which are nearly regular and have N–O = 1.24 and 1.26 A. The nearest approach between NH_4 groups and oxygen atoms and between the oxygen atoms of adjacent NO_3 ions is the same, 2.95 A.

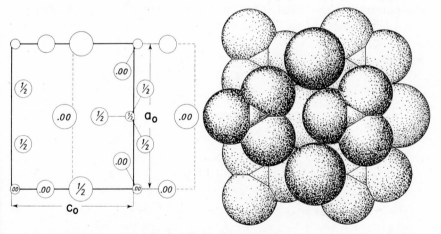

Fig. VIIA,8a (left). A projection of the orthorhombic structure of the room temperature, IV, form of NH_4NO_3 on its b_0 axis. Axes of the unit are marked by full lines. The smallest circles are nitrogen, the largest are NH_4 radicals.
Fig. VIIA,8b (right). A drawing to show the way the $NO_3{}^-$ and $NH_4{}^+$ ions pack in form IV of NH_4NO_3. The largest circles are the NH_4 ions.

The resemblance to CsCl is brought out by choosing a unimolecular pseudocell which has the same c_0 axis, and a_0' and b_0' axes that are half the face diagonals of the true unit. Alternate NO_3 ions at the corners of this pseudocell point in opposite directions; its elongation along c_0 results from the coincidence of this axis with one of the N–O bonds.

VII,a7. The *lowest-temperature*, V, form of *ammonium nitrate*, NH_4NO_3, has recently been described as tetragonal rather than hexagonal as previously supposed. Its unit containing eight molecules has the following dimensions:

$$a_0 = 7.98 \text{ A.}, \qquad c_0 = 9.78 \text{ A. (at } -150°\text{C.)}$$

The space group is C_4^3 ($P4_2$), with atoms in the positions:

(2a) $00u;\ 0,0,u+{}^1/_2$
(2b) ${}^1/_2\,{}^1/_2\,u;\ {}^1/_2,{}^1/_2,u+{}^1/_2$
(2c) $0\ {}^1/_2\,u;\ {}^1/_2,0,u+{}^1/_2$
(4d) $xyz;\ \bar{x}\bar{y}z;\ \bar{y},x,z+{}^1/_2;\ y,\bar{x},z+{}^1/_2$

The parameters are listed in Table VIIA,3. The hydrogen positions as stated here are based on limited experimental evidence supplemented by the assumption that N–H = 1.04 A.

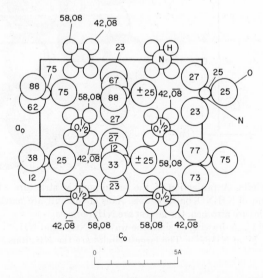

Fig. VIIA,9a. A projection of the tetragonal low temperature, V, modification of NH_4NO_3 viewed along an a_0 axis. Origin in lower left.

TABLE VIIA,3
Positions and Parameters in the Lowest Temperature (V) Form of NH_4NO_3

Atom	Position	x	y	z
N(1)	(4d)	0.250	0.250	0.518
O(1)	(4d)	0.120	0.270	0.456
O(2)	(4d)	0.330	0.230	0.456
O(3)	(4d)	0.250	0.250	0.642
N(2)	(4d)	0.250	0.250	0.018
O(4)	(4d)	0.120	0.230	0.956
O(5)	(4d)	0.380	0.270	0.956
O(6)	(4d)	0.250	0.250	0.142
N(3)	(2a)	0	0	0.250
H(1)	(4d)	0.076	0.076	0.190
H(2)	(4d)	0.076	−0.076	0.310
N(4)	(2c)	0	$1/2$	−0.250
H(3)	(4d)	0.576	0.076	0.190
H(4)	(4d)	0.424	0.076	0.310
N(5)	(2b)	$1/2$	$1/2$	0.250
H(5)	(4d)	0.424	0.424	0.310
H(6)	(4d)	0.424	0.576	0.190
N(6)	(2c)	0	$1/2$	0.250
H(7)	(4d)	−0.076	0.424	0.310
H(8)	(4d)	−0.076	0.576	0.190

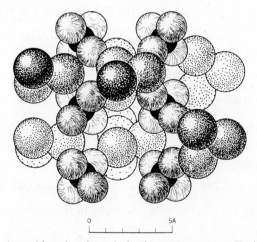

0   5A

Fig. VIIA,9b. A packing drawing of the low temperature, V, form of NH_4NO_3. The hydrogen atoms of the NH_4 ions are fine-line shaded, the central nitrogen atoms are black. Nitrate nitrogens do not show.

The structure, as shown in Figure VIIA,9, is a clearly CsCl-like grouping of NO_3 and NH_4 ions, each ion having eight of the opposite sort around it. Between ions, N–O = 2.88–3.17 A. There are two short H–O = 1.91 and 2.01 A.; the other two hydrogen–oxygen separations are H(4)–O(6) = 2.42 A. and H(3)–O(6) = 2.82 A.

VII,a8. High-temperature forms of two alkali nitrates which originally were thought to have small unimolecular cubic units like that of NH_4NO_3(I) (**VII,a4**) have more recently been described in terms of larger cells that contain eight molecules. Their dimensions are

High-RbNO$_3$: a_0 = 8.74 A. (at ca. 190 °C.)
High-CsNO$_3$: a_0 = 8.980 A. (above 161 °C.)

Structures described for them, based on $T_h{}^6$ (*Pa*3), place the atoms in the following positions:

Rb [or Cs] (1): (4a) 000; F.C.
Rb [or Cs] (2): (4b) $^1/_2$ $^1/_2$ $^1/_2$; F.C.
       N: (8c) $\pm (uuu; u+^1/_2,^1/_2-u,\bar{u}; \bar{u},u+^1/_2,^1/_2-u;$
                 $^1/_2-u,\bar{u},u+^1/_2)$
                 with u(RbNO$_3$) = 0.285 and u(CsNO$_3$) = 0.275
       O: (24d) $\pm (xyz; x+^1/_2,^1/_2-y,\bar{z}; z+^1/_2,^1/_2-x,\bar{y};$
                 $y+^1/_2,^1/_2-z,\bar{x})$; tr
                 with, for RbNO$_3$, $x = y$ = 0.278, z = 0.399
                 and, for CsNO$_3$, $x = y$ = 0.219, z = 0.386

Below 161 °C., *cesium nitrate*, CsNO$_3$, is reportedly rhombohedral, with a hexagonal cell of the dimensions:

$$a_0 = 10.74 \text{ A.}, \qquad c_0 = 7.68 \text{ A.}$$

The atomic arrangement has not been established.

VII,a9. In the structure found for the room-temperature form of *thallous nitrate*, TlNO$_3$, atoms are in an orthorhombic cell containing eight molecules and having the dimensions:

$$a_0 = 6.287 \text{ A.}; \quad b_0 = 12.31 \text{ A.}; \quad c_0 = 8.001 \text{ A.}$$

The space group is $V_h{}^{16}$ (*Pbnm*) with atoms in the positions:

(4c) $\pm (u\, v\,^1/_4; ^1/_2-u,v+^1/_2,^1/_4)$
(8d) $\pm (xyz; ^1/_2-x,y+^1/_2,^1/_2-z; x,y,^1/_2-z; x+^1/_2,^1/_2-y.\bar{z})$

with the parameters of Table VIIA,4.

TABLE VIIA,4. Positions and Parameters of the Atoms in Low-TlNO₃

Atom	Position	x	y	z
Tl	(8d)	0.213	0.125	0.500
N(1)	(4c)	0.297	0.371	$1/4$
N(2)	(4c)	0.829	0.108	$1/4$
O(1)	(8d)	0.198	0.375	0.117
O(2)	(4c)	0.516	0.361	$1/4$
O(3)	(4c)	0.022	0.080	$1/4$
O(4)	(4c)	0.685	0.036	$1/4$
O(5)	(4c)	0.782	0.207	$1/4$

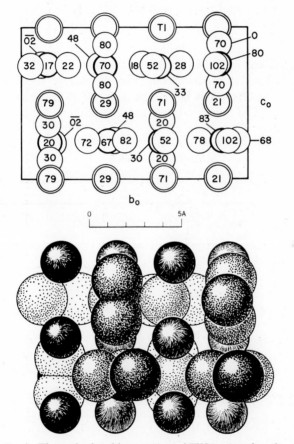

Fig. VIIA,10a (top). The orthorhombic structure of TlNO₃ seen along its a_0 axis. Origin in lower right.

Fig. VIIA,10b (bottom). A packing drawing of the TlNO₃ arrangement viewed along its a_0 axis. The thallium atoms are fine-line shaded; some of the nitrogen atoms show as small black segments.

In deducing this arrangement, the nitrate ions were assigned the size and shape found for them in $NaNO_3$. The structure that results is illustrated in Figure VIIA,10. In it eight oxygens are unsymmetrically distributed about the thallium atoms with the shortest Tl–O = 2.4 A.

VII,a10. *Anhydrous nitric acid*, HNO_3, which crystallizes at −41.6°C., has below this temperature a pseudoorthorhombic monoclinic cell that contains 16 molecules and has the edge lengths:

$$a_0 = 16.23 \text{ A.}; \quad b_0 = 8.57 \text{ A.}; \quad c_0 = 6.31 \text{ A.}; \quad \beta = 90°$$

Atoms are in the general positions of C_{2h}^5 $(P2_1/a)$:

$$(4e) \quad \pm(xyz; \; x+{}^1/_2, {}^1/_2-y, z)$$

with the parameters of Table VIIA,5.

In the structure thus described (Fig. VIIA,11), the intermolecular oxygen separations have not been such as to indicate the positions of the hydrogen atoms. The four crystallographically different nitrate groups are

TABLE VIIA,5
Parameters of the Atoms in HNO_3

Atom	x	y	z
N(1)	0.008	0.257	0.027
N(2)	0.131	0.077	0.479
N(3)	0.252	0.413	−0.032
N(4)	0.381	0.073	0.474
O(1)	−0.055	0 248	−0.100
O(2)	0.036	0.140	0.083
O(3)	0.043	0.372	0.095
O(4)	0.071	0.102	0.591
O(5)	0.164	−0.033	0.400
O(6)	0.168	0.205	0.420
O(7)	0.308	0.433	0.083
O(8)	0.212	0.302	−0.120
O(9)	0.208	0.531	−0.100
O(10)	0.320	0.090	0.591
O(11)	0.418	−0.042	0.395
O(12)	0.417	0.197	0.420

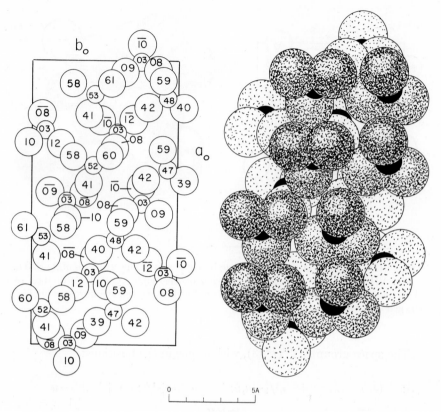

Fig. VIIA,11a (left). A projection along the c_0 axis of the monoclinic pseudoortho-rhombic structure of HNO_3. Only nitrate groups are shown, the nitrogens being the smaller circles. Origin in upper right.

Fig. VIIA,11b (right). A packing drawing of the nitrate groups in monoclinic HNO_3 viewed along its c_0 axis. The small black circles are nitrogen.

of similar but not identical dimensions, N–O varying between 1.19 and 1.33 A. and the O–N–O angle between 101° and 140°.

The diffraction data point to the existence of appreciable disorder in this structure.

VII,a11. The unusual compound Ag_3SNO_3 forms cubic crystals whose tetramolecular cells have the edge length:

$$a_0 = 7.929 \text{ A.}$$

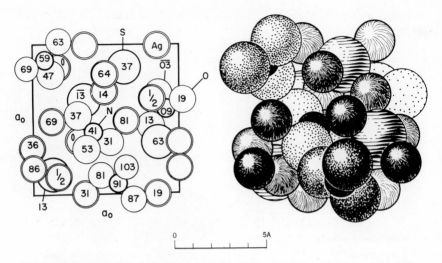

Fig. VIIA,12a (left). A projection along its cubic axis of the structure of Ag_3SNO_3.
Fig. VIIA,12b (right). A packing drawing of the cubic Ag_3SNO_3 structure seen along an axis of the unit. The silver atoms are fine-line shaded, the nitrogens are black. Oxygen atoms are dotted.

The space group is T^4 ($P2_13$), with atoms in the positions:

S: (4a) $uuu;\ u+\frac{1}{2},\frac{1}{2}-u,\bar{u};\ \frac{1}{2}-u,\bar{u},u+\frac{1}{2};\ \bar{u},u+\frac{1}{2},\frac{1}{2}-u$
with $u = 0.133$

N: (4a) with $u = 0.411$

Ag: (12b) $xyz;\ x+\frac{1}{2},\frac{1}{2}-y,\bar{z};\ y+\frac{1}{2},\frac{1}{2}-z,\bar{x};\ z+\frac{1}{2},\frac{1}{2}-x,\bar{y};$ tr
with $x = 0.310,\ y = 0.997,\ z = 0.364$

O: (12b) with $x = 0.310,\ y = 0.373,\ z = 0.532$

This structure (Fig. VIIA,12) consists of NO_3 groups of the usual dimensions embedded in endless Ag_3S chains. In these chains each sulfur atom is surrounded by six silver atoms at a distance of either 2.41 or 2.55 A. The chains are formed from these Ag_6S clusters tied together by sharing a silver atom.

VII,a12. *Uranyl carbonate,* UO_2CO_3, as the mineral rutherfordine is orthorhombic, with a bimolecular cell of the dimensions:

$$a_0 = 4.845 \text{ A.};\quad b_0 = 9.205 \text{ A.};\quad c_0 = 4.296 \text{ A.}$$

Atoms are in the following positions of the space group V_h^{13} ($Pmmn$) (origin shifted to $1/4$ $1/4$ 0 with respect to the usual description):

U: (2b) $\pm(1/4\ 3/4\ u)$ with $u = 0.750$

C: (2a) $\pm(1/4\ 1/4\ u)$ with $u = 0.601$

O(1): (2a) with $u = 0.303$

O(2): (4e) $\pm(1/4\ u\ v;\ 1/4, 1/2-u, v)$ with $u = 0.960,\ v = 0.750$

O(3): (4f) $\pm(u\ 1/4\ v;\ 1/2-u, 1/4, v)$ with $u = 0.021,\ v = 0.750$

This determination makes use of the assumption that the CO_3 and UO_2 ions have their usual sizes, with C–O = 1.28 A. and U–O = 1.93 A.

The structure itself (Fig. VIIA,13) is thought to be a more or less idealized one inasmuch as the x-ray patterns show a notable streaking. The disorder this expresses has been imagined as due to indefiniteness in the orientation of the carbonate ions.

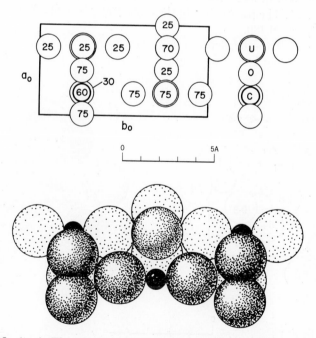

Fig. VIIA,13a (top). The orthorhombic structure of UO_2CO_3 viewed along its c_0 axis. Origin in lower left.

Fig. VIIA,13b (bottom). A packing drawing of the UO_2CO_3 arrangement seen along its c_0 axis. The uranium atoms are small and black, the carbons do not show.

VII,a13. *Sodium chlorate*, $NaClO_3$, has an NaCl arrangement of its ions distorted to accommodate the ClO_3 anions in such a way that the symmetry remains cubic. The four-molecule unit has the edge length:

$$a_0 = 6.5756 \text{ A. } (25°\text{C.})$$

Atoms are in the following positions of T^4 $(P2_13)$:

Na and Cl: $(4a)$ uuu; $u+\frac{1}{2},\frac{1}{2}-u,\bar{u}$; $\bar{u},u+\frac{1}{2},\frac{1}{2}-u$; $\frac{1}{2}-u,\bar{u},u+\frac{1}{2}$
with $u(\text{Na}) = 0.0659$ and $u(\text{Cl}) = 0.4168$

O: $(12b)$ xyz; $x+\frac{1}{2},\frac{1}{2}-y,\bar{z}$; $\bar{x},y+\frac{1}{2},\frac{1}{2}-z$; $\frac{1}{2}-x,\bar{y},z+\frac{1}{2}$;
zxy; $\bar{z},x+\frac{1}{2},\frac{1}{2}-y$; $\frac{1}{2}-z,\bar{x},y+\frac{1}{2}$; $z+\frac{1}{2},\frac{1}{2}-x,\bar{y}$;
yzx; $\frac{1}{2}-y,\bar{z},x+\frac{1}{2}$; $y+\frac{1}{2},\frac{1}{2}-z,\bar{x}$; $\bar{y},z+\frac{1}{2},\frac{1}{2}-x$
with $x = 0.3052$, $y = 0.5921$, $z = 0.4936$

Sodium bromate, $NaBrO_3$, also has this structure, with

$$a_0 = 6.705 \text{ A. } (25°\text{C.})$$

For it, the sodium and bromine atoms in $(4a)$ have $u(\text{Na}) = 0.075$ and $u(\text{Br}) = 0.405$. The parameters of the oxygen atoms in $(12b)$ are $x = 0.258$, $y = 0.614$, $z = 0.480$.

The extent of the departure of this structure (Fig. VIIA,14) from the NaCl arrangement is indicated by the differences of $u(\text{Na})$ and $u(\text{Cl})$ from

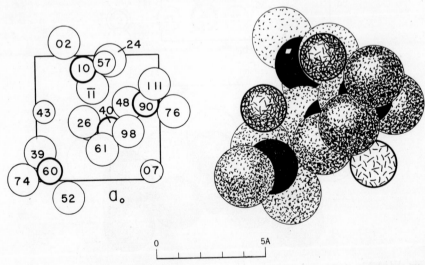

Fig. VIIA,14a (left). A projection along a cubic axis of the structure of $NaBrO_3$. The small heavy circles are bromine, the small light circles sodium atoms. Origin in lower right.

Fig. VIIA,14b (right). A packing drawing of the $NaBrO_3$ structure viewed along a cubic axis. The sodium atoms are line-shaded, the bromine atoms large and black.

zero and one half. The ClO_3 ion is a flat trigonal pyramid having the three oxygen atoms of its base 2.38 A. apart. The chlorine atom is 0.48 A. above this base and distant 1.455 A. from the atoms in it. In the BrO_3 ion, Br–O = 1.78 A. and O–O = 2.95 A. The Na–O separation found in $NaClO_3$ is 2.48 A.; in $NaBrO_3$ it is 2.38 A. These metal-to-oxygen distances are substantially the sum of ionic radii and the same as the corresponding distance in $NaNO_3$.

According to a recent determination of the absolute configuration of this optically active substance, a dextrorotatory crystal will have the foregoing arrangement referred to right-handed axes.

VII,a14. The structure of *potassium bromate*, $KBrO_3$, is analogous to that of $NaHF_2$ (**VI,12**) in being a rhombohedral distortion of either the NaCl or the CsCl arrangement, depending on the angle of its rhombohedral cell. In this it is also like the $NaNO_3$ arrangement (**VII,a1**), the difference being that in $KBrO_3$ all the MX_3 ions have the same (rather than alternate) orientations viewed down the principal axis. The unit is unimolecular, with

$$a_0 = 4.413 \text{ A.}, \qquad \alpha = 85°48'$$

TABLE VIIA,6
Additional Crystals with the $KBrO_3$ Structure

Compound	a_0, A.	α	$u(M)$	$u(O)$	$v(O)$
	Rhombohedral Unit				
$RbClO_3$	4.440	86°38′	0.513	0.583	0.197
$TlBrO_3$	4.45	87°24′	0.492	0.568	0.126
$TlClO_3$	4.43	86°20′	0.49	0.55	0.17
$TlIO_3$	4.44	89°6′	—	—	—
KNO_3 (high pressure)	4.365	76°56′	0.44	0.55	0.22
NH_4ClO_3	4.444	86°24′	0.46	0.566	0.136

Compound	a_0', A.	c_0', A.	$u'(M)$	$u'(O)$	$v'(O)$
	Hexagonal Unit				
$RbClO_3$	6.092	8.129	0.513	0.129	0.454
$TlBrO_3$	6.149	8.049	0.492	0.147	0.421
$TlClO_3$	6.061	8.149	0.49	0.127	0.423
$TlIO_3$	6.230	7.810	—	—	—
KNO_3 (high pressure)	5.430	9.110	0.44	0.11	0.44
NH_4ClO_3	6.084	8.166	0.46	0.143	0.423

In this rhombohedron, atoms are in the following special positions of C_{3v}^5 ($R3m$):

    K: ($1a$) uuu with $u = 0$
    Br: ($1a$)   with $u = 0.50$
    O: ($3b$) $uuv; uvu; vuu$ with $u = 0.58, v = 0.12$

In the corresponding trimolecular hexagonal cell with

$$a_0' = 6.015 \text{ A.}, \qquad c_0' = 8.142 \text{ A.}$$

K and Br are in ($3a$) $00u$; rh, with $u(\text{K}) = 0$ and $u(\text{Br}) = 0.50$, and the oxygens are in

    ($9b$) $u\bar{u}v; u\,2u\,v; 2\bar{u}\,\bar{u}\,v$; rh   with $u' = 0.153, v' = 0.427$

This arrangement (Fig. VIIA,15) is closer to the CsCl than to the NaCl structure, the $\alpha = 85°48'$ of its unimolecular cell being nearer to 90° than is the 107°30' that characterizes its larger NaCl-like rhombohedron. The flat BrO_3 pyramids have the same shape as in other bromates and chlorates, with Br–O = 1.68 A. and O–O = 2.73 A. In the isomorphous NH_4ClO_3, Cl–O = 1.45 A. and O–Cl–O = 106°.

Several other chlorates, bromates, and iodates of large cations have this structure. Their cell dimensions and the atomic parameters found for them are listed in Table VIIA,6. It was early supposed that at temperatures above ca. 230°C. the nitrates of the heavier alkalis and of thallium have this arrangement, but as **VII,a8** indicates, recent work points to a more complicated structure.

The formal analogy between this arrangement and that of the cubic perewskite, $CaTiO_3$, (**VII,a21**) is so close that some iodates which very possibly have the $KBrO_3$ arrangement have been classified as perewskites. An example is *potassium iodate*, KIO_3, with a unit rhombohedron generally given as

$$a_0 = 4.410 \text{ A.}, \qquad \alpha = 89°25'$$

According to a recent x-ray and neutron study not yet reported in detail (1962: O,R,P,R&Z), however, the true unit is larger, contains eight molecules, and has $a_0 = 8.938$ A. In this case its structure is perhaps more closely related to that of the high-temperature $RbNO_3$ discussed in **VII,a8**.

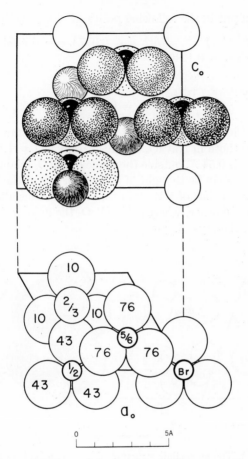

Fig. VIIA,15. Two projections of the hexagonal cell corresponding to the rhombohedral unit of KBrO$_3$. In the shaded projection the bromine atoms are small and black,
the potassium atoms are fine-line shaded.

VII,a15. *Potassium chlorate*, KClO$_3$, has a structure which in spite of its
monoclinic symmetry is closely related to that of the bromate, **VII,a14.**
Its unit contains two molecules and has the dimensions:

$a_0 = 4.6569$ A.; $b_0 = 5.59089$ A.; $c_0 = 7.0991$ A.; $\beta = 109°38.9'$ (25°C.)

Atoms have been put in the following positions of $C_{2h}{}^2$ ($P2_1m$):

K: (2e) $\pm(u\,{}^1/_4\,v)$ with $u = 0.3573$, $v = 0.7104$
Cl: (2e)   with $u = 0.1240$, $v = 0.1792$
O(1): (2e)   with $u = 0.3897$, $v = 0.1178$
O(2): (4f) $\pm(xyz;\ \bar{x},y+{}^1/_2,\bar{z})$

with $x = 0.1406$, $y = 0.459$, and $z = 0.3104$

The resulting ClO_3 ions are trigonal pyramids of substantially the same shape and size as those in $NaClO_3$. In this case, $Cl\text{–}O = 1.46\text{–}1.49$ A. and the chlorine atom is 0.54 A. outside the plane of the O_3 triangle. This structure (Fig. VIIA,16) can be thought of as a distortion of the NaCl arrangement. It differs from that of $KBrO_3$ in that its ClO_3 ions alternate in their orientation to one another, whereas in $KBrO_3$ the anion orientation is the same throughout the crystal.

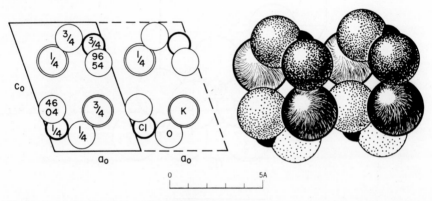

Fig. VIIA,16a (left). The monoclinic structure of room temperature $KClO_3$ seen along its b_0 axis. Adjacent cells in the direction of a_0 are shown. Origin in the lower left.

Fig. VIIA,16b (right). A packing drawing of room temperature $KClO_3$ viewed along its principal, b_0, axis. As in Figure VIIA,16a the contents of two cells are shown. The potassium atoms are large and fine-line shaded, the chlorine atoms are black.

VII,a16. At 250°C. *potassium chlorate*, $KClO_3$, becomes orthorhombic but continues to give diffraction data closely resembling those from the low-temperature, monoclinic, modification. At 280°C. its tetramolecular cell has the dimensions:

$$a_0 = 4.74\ \text{A.};\quad b_0 = 5.64\ \text{A.};\quad c_0 = 13.80\ \text{A.}$$

The relation between these and the monoclinic axes of the room temperature form is shown in Figure VIIA,17.

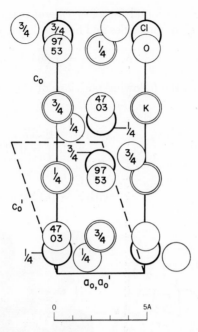

Fig. VIIA,17. A projection of the orthorhombic, high temperature form of KClO₃ viewed along its b_0 axis. The relation between its axes (full lines) and those of the lower temperature monoclinic form (dashed lines) is shown.

The space group is V_h^{16} (*Pcmn*), with atoms in the positions:

(4c) $\pm(u\ ^1/_4\ v;\ ^1/_2-u,^1/_4,v+^1/_2)$

(8d) $\pm(xyz;\ ^1/_2-x,^1/_2-y,z+^1/_2;\ \bar{x},y+^1/_2,\bar{z};\ x+^1/_2,\bar{y},^1/_2-z)$

The chosen atomic positions and parameters, not accurately established by x-ray data, are taken to be the following:

Atom	Position	x	y	z
K	(4c)	0.50	$^1/_4$	−0.135
Cl	(4c)	0.00	$^1/_4$	0.085
O(1)	(4c)	0.345	$^1/_4$	0.055
O(2)	(8d)	0.00	0.47	0.14

As can be seen by comparing Figure VIIA,17 with Figure VIIA,16 for the low-KClO₃, the two arrangements differ chiefly in the repetition of atomic layers in successive planes parallel to a_0b_0.

VII,a17. The structure found for *silver chlorate*, $AgClO_3$, is tetragonal, with a cell containing eight molecules and having the edges:

$$a_0 = 8.486 \text{ A.}, \qquad c_0 = 7.894 \text{ A.}$$

Atoms have been given the following positions of C_{4h}^5 $(I4/m)$:

Ag(1):   (4d) $\pm (0 \; ^1/_2 \; ^1/_4)$; B.C.
Ag(2):   (4e) $\pm (00u)$; B.C.   with $u = 0.277$
  Cl: (8h) $\pm (uv0; \bar{v}u0)$; B.C.   with $u = 0.215, v = 0.235$
O(1):   (8h)   with $u = 0.340, v = 0.350$
O(2):   (16i) $\pm (xyz; \bar{y}xz; \bar{x}\bar{y}z; y\bar{x}z)$; B.C.
                       with $x = 0.120, y = 0.250, z = 0.152$

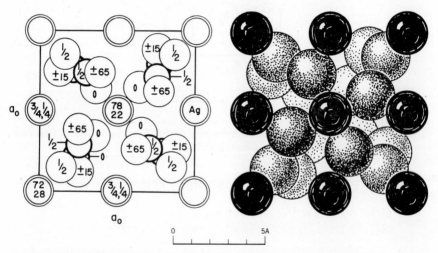

Fig. VIIA,18a (left). The tetragonal structure of $AgClO_3$ projected along its c_0 axis. Doubly-ringed atoms are silver; heavily ringed atoms are chlorine.
Fig. VIIA,18b (right). A packing drawing of the $AgClO_3$ arrangement as shown in Figure VIIA,18a. The silver atoms are black, the chlorine atoms do not show.

In developing this structure (Fig. VIIA,18), the assumption was made that the ClO_3 ion would have the same size and shape as in $KClO_3$ and other chlorates. It yields Ag–O distances of 2.47–2.55 A. and O–O separations between adjacent ions from 2.70 A. upwards.

Silver bromate, $AgBrO_3$, has the same arrangement, with

$$a_0 = 8.609 \text{ A.}, \qquad c_0 = 8.092 \text{ A. } (25°C.)$$

VII,a18. The structure deduced many years ago for *lithium iodate*, $LiIO_3$, differs fundamentally from that of other alkali iodates (**VII,a20**) in not showing separate IO_3 ions; instead, it surrounds each iodine atom by six oxygen atoms, each of which is equidistant from two iodines. The symmetry is hexagonal and the bimolecular cell has the dimensions:

$$a_0 = 5.469 \text{ A.}, \qquad c_0 = 5.155 \text{ A.}$$

Iodine atoms are in $(2c)$ $\pm(\frac{1}{3}\,\frac{2}{3}\,\frac{1}{4})$, and oxygen atoms have been placed in the following positions of D_6^6 ($C6_32$):

$$(6g) \quad uu0;\ \bar{u}\ \bar{u}\ \frac{1}{2};\ 0\bar{u}0;\ 0\ u\ \frac{1}{2};\ \bar{u}00;\ u\ 0\ \frac{1}{2}$$

with $u = \frac{1}{3}$. The lithium could not be located from the x-ray data, but it was pointed out that a reasonable Li–O = 2.23 A. results if they are in $(2b)$ $0\ 0\ \frac{1}{4};\ 0\ 0\ \frac{3}{4}$. Evidently, a quantitative reexamination should be made of this structure (Fig. VIIA,19) to be sure of the non-existence in it of iodate ions.

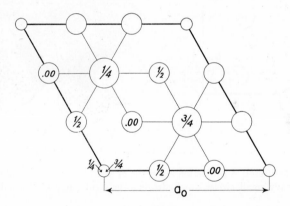

Fig. VIIA,19. A basal projection of the hexagonal structure described for $LiIO_3$. The largest circles are iodine, the smallest lithium atoms.

VII,a19. Crystals of *iodic acid*, HIO_3, are orthorhombic, with a tetra-molecular cell of the dimensions:

$$a_0 = 5.5379 \text{ A.}; \quad b_0 = 5.8878 \text{ A.}; \quad c_0 = 7.7333 \text{ A. (26°C.)}$$

All atoms are in general positions of V^4 ($P2_12_12_1$):

$$(4a) \quad xyz;\ \frac{1}{2}-x,\bar{y},z+\frac{1}{2};\ x+\frac{1}{2},\frac{1}{2}-y,\bar{z};\ \bar{x},y+\frac{1}{2},\frac{1}{2}-z$$

388 CRYSTAL STRUCTURES

A study using neutrons has confirmed an earlier x-ray investigation and has led to more accurate parameters and a definition of the hydrogen positions. These parameters are as follows:

Atom	x	y	z
I	0.2036	0.0861	0.1580
O(1)	0.0712	0.5225	0.2429
O(2)	0.3340	0.1978	0.0864
O(3)	0.4043	0.1566	0.4481
H	0.2340	0.3220	0.1350

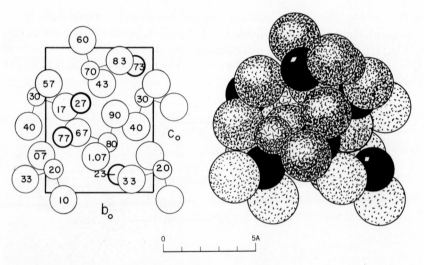

Fig. VIIA,20a (left). A projection along the a_0 axis of the orthorhombic structure of HIO$_3$. The heavy small circles are hydrogen, the light small circles iodine. Origin in lower right.
Fig. VIIA,20b (right). The orthorhombic structure of HIO$_3$ viewed along the a_0 axis. The iodine atoms are black, the hydrogen atoms line shaded.

In this structure (Fig. VIIA,20), the IO$_3$ group is a trigonal pyramid with iodine at the apex and I–O separations lying between 1.78 and 1.90 A. The hydrogen is attached to O(2) at a distance of 0.99 A.; it comes within 1.70 A. of the O(1) atom of an adjacent molecule. Also, between O(1) and the iodine of a neighboring molecule there is the short separation of 2.50 A. This has the effect of tying the molecules into chains running in the b_0 direction.

VII,a20. The two iodates, $NaIO_3$ and NH_4IO_3, yield x-ray data so much like those from $CaTiO_3$ that they were originally thought to be only slightly distorted perewskite structures (**VII,a21**). The atomic arrangement developed by careful analysis is, however, a different one.

The true unit for *sodium iodate*, $NaIO_3$, is tetramolecular, with the dimensions:

$$a_0 = 5.74 \text{ A.}; \quad b_0 = 6.37 \text{ A.}; \quad c_0 = 8.11 \text{ A.}$$

Its atoms are in the following positions of V_h^{16} (*Pbnm*):

Na: (4*b*) $^1/_2\,0\,0;\ ^1/_2\,0\,^1/_2;\ 0\,^1/_2\,0;\ 0\,^1/_2\,^1/_2$

I: (4*c*) $\pm(u\,v\,^1/_4;\ ^1/_2-u,v+^1/_2,^1/_4)$ with $u = 0.008,\ v = 0.013$

O(1): (4*c*)   with $u = 0.715,\ v = 0.110$

O(2): (8*d*) $\pm(xyz;\ ^1/_2-x,y+^1/_2,^1/_2-z;\ x,y,^1/_2-z;\ ^1/_2-x,y+^1/_2,z)$

with $x = 0.140,\ y = 0.160,\ z = 0.08$. As can be seen from Figure VIIA,21, the iodate ions that result have the same general shape as the ClO_3 and BrO_3 ions already described. In IO_3, $I-O = 1.80$ and 1.83 A. and $O-O = 2.65$ and 2.81 A. The CsCl-like nature of this arrangement is apparent when it is viewed along the c_0 axis (rather than along the a_0 of the figure).

The x-ray data from *ammonium iodate*, NH_4IO_3, are very nearly those of a cubic crystal. Its tetramolecular unit has, however, the dimensions:

$$a_0 = 6.41 \text{ A.}; \quad b_0 = 6.38 \text{ A.}; \quad c_0 = 9.25 \text{ A.}$$

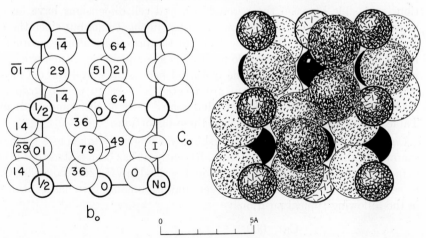

Fig. VIIA,21a (left). A projection along the a_0 axis of the orthorhombic structure of $NaIO_3$. Origin in lower right.

Fig. VIIA,21b (right). A packing drawing of the orthorhombic structure of $NaIO_3$ viewed along its a_0 axis. The sodium atoms are line shaded, the oxygen atoms dotted.

This crystal is undoubtedly like $NaIO_3$, and the following approximate parameters have been suggested for it:

For I: $u = v = 0$

For O: $u = 0.72$, $v = 0.13$, $x = y = 0.14$, $z = 0.13$

Lithium bromate, $LiBrO_3$, probably has this structure, with

$$a_0 = 5.06 \text{ A.}; \quad b_0 = 5.99 \text{ A.}; \quad c_0 = 7.86 \text{ A.}$$

Atomic parameters were not established.

Perewskite-Like Compounds

VII,a21. The largest group of RMX_3 compounds is that typified by *perewskite*, $CaTiO_3$. Many of the crystals with this arrangement (Table VIIA,7) give exceedingly simple diffraction patterns that can be accounted for in terms of a cubic cell containing one molecule. Others listed in the Table are doubly refracting under the polarizing microscope; and it has not always been clear whether this was due to strain or to small departures from a perfect cubic symmetry. Still others are certainly of less symmetry and probably have appreciably distorted atomic arrangements, often with larger unit cells. As can be seen from Table VIIA,7, the data from many of these substances are conflicting and it frequently is impossible to decide from the existing information which unit is correct. Widespread solid solution is possible between the perewskites, and cell dimensions have been measured for many of them; a few are listed in the Table, but since these results contribute little to a knowledge of the prevailing atomic arrangements no attempt has been made to provide a complete list. Much work is now being done on several of the distorted perewskites, and the detailed structures worked out for them are treated in subsequent paragraphs. Probably, future studies will show that many more of the compounds in Table VIIA,7 have one or another of these distortions.

Calcium titanate, $CaTiO_3$, as perewskite, was originally taken as a typical example of the simple cubic structure. Recently, it has been stated that it is in fact orthorhombic, with the distortion described in **VII,a26.** Above 900°C. it seems to become truly cubic, however, and presumably then has its atoms in the following typical positions:

Ca: 000

Ti: $^1/_2\,^1/_2\,^1/_2$

O: $^1/_2\,^1/_2\,0$; $^1/_2\,0\,^1/_2$; $0\,^1/_2\,^1/_2$

TABLE VIIA,7
Crystals with the CaTiO₃ Structure

Crystal	Cubic or Pseudocubic a_0, A.	Remarks
$AgZnF_3$	3.98	—
$BaCeO_3$	4.397	1 molecule
	8.754	8 mol. monoclinic
$BaFeO_3$	4.012	—
$BaMoO_3$	4.0404	—
$BaPbO_3$	4.273	—
$BaPrO_3$	4.354	1 mol.
	8.708	8 mol. monoclinic
$BaPuO_3$	4.39	—
$BaSnO_3$	4.1168	—
$BaThO_3$	4.480	1 mol.
	8.985	8 mol. monoclinic
$BaTiO_3$	4.0118 (201°C.) ⎱	Remains cubic through
	4.0783 (1372°C.) ⎰	this interval
$BaUO_3$	4.40	—
$BaZrO_3$	4.1929 (26°C.)	—
$CaCeO_3$	7.70	8 mol.
$CaMnO_3$	7.465	90% Mn^{4+}, see monoclinic
$CaSnO_3$	3.92	—
$CaThO_3$	8.74	8 mol. monoclinic pseudocubic
$CaTiO_3$ (perewskite)	3.84	—
$CaVO_3$	3.76	—
$CaZrO_3$	4.020	Pseudocubic
$CdCeO_3$	7.65	8 mol.
$CdSnO_3$	7.80	8 mol. monoclinic
$CdThO_3$	8.74	8 mol. monoclinic pseudocubic
$CdTiO_3$	7.50	8 mol. monoclinic pseudocubic
$CeAlO_3$	3.772	—
$CeCrO_3$	3.866	—
$CeFeO_3$	3.900	Pseudocubic
$CeGaO_3$	3.879	—
$CeVO_3$	3.90	—
$CsCaF_3$	4.522	—
$CsCdBr_3$	5.33	1 mol.
	10.70	8 mol. monoclinic
$CsCdCl_3$	5.20	1 mol.
	10.40	8 mol. monoclinic

(*continued*)

TABLE VIIA,7 (*continued*)

Crystal	Cubic or Pseudocubic a_0, A.	Remarks
CsHgBr₃	5.77	1 mol.
	11.54	8 mol. monoclinic
CsHgCl₃	5.44	1 mol.
	10.88	8 mol. monoclinic
CsIO₃	4.66	1 mol.
	9.324	8 mol. monoclinic
CsPbBr₃	5.874 (>13°C.)	—
CsPbCl₃	5.605 (>47°C.)	—
CsPbF₃ (high)	4.81	—
DyMnO₃	3.70	—
EuAlO₃	3.725	—
EuCrO₃	3.803	—
EuFeO₃	3.836	—
EuTiO₃	3.897	—
FeBiO₃	3.965	$\alpha = 89°28'$, Bi at 0
	7.64	8 mol.
GdAlO₃	3.71	—
GdCrO₃	3.795	—
GdFeO₃	3.820	—
GdMnO₃	3.82	—
HgNiF₃	8.15	8 mol. pseudocubic
KCaF₃	8.742	8 mol. monoclinic, $\beta = 92°36'$
KCdF₃	4.293	—
KCoF₃	4.069 (298°K.)	—
KFeF₃	4.122 (298°K.)	Cubic
	4.108 (78°K.)	$\alpha = 89°51'$
KIO₃	4.410	$\alpha = 89°24'$
	8.92	8 mol. monoclinic
KMgF₃	3.973	1 mol.
	8.02	8 mol. monoclinic, $\beta = 91°18'$
KMnF₃	4.190 (298°K.)	—
KNbO₃	4.007 (ca. 500°C.)	—
	4.016 (−140°C.)	$\alpha = 89°50'$
KNiF₃	4.012	1 mol.
	4.002 (78°K.)	—
	8.02	8 mol. monoclinic
KTaO₃	3.9885	—

(*continued*

TABLE VIIA,7 (*continued*)

Crystal	Cubic or Pseudocubic a_0, A.	Remarks
$KZnF_3$	4.055	1 mol.
	8.112 (25°C.)	8 mol. monoclinic
$LaAlO_3$	3.778	1 mol.
	7.58	8 mol. monoclinic
$LaCoO_3$	3.824	Rhombohedral, $\alpha = 90°42'$
	7.651	8 mol.
$LaCrO_3$	3.874	—
$LaFeO_3$	3.920	1 mol.
	7.852	8 mol.
$LaGaO_3$	3.875	—
$LaNiO_3$	7.676	8 mol., $\alpha = 90°41'$
$LaRhO_3$	3.94	—
$LaTiO_3$	3.92	—
$LaVO_3$	3.99	1 mol.
	7.842	8 mol.
$LiBaF_3$	3.996	—
Li_xWO_3	3.72	—
$MgCeO_3$	8.54	8 mol. monoclinic
NH_4CoF_3	4.129	—
NH_4MnF_3	4.241	—
NH_4NiF_3	8.15	8 mol. pseudocubic
$NaAlO_3$	3.73	Pseudocubic
$NaMgF_3$	3.955 (>900°C.)	—
$NaNbO_3$	3.915	Monoclinic, $\beta = 90°40'$
	7.78	8 mol.
$NaTaO_3$	3.881	1 mol.
	7.76	8 mol. monoclinic
$NaWO_3$	3.8622	—
$NdAlO_3$	3.752	—
$NdCoO_3$	3.777	—
$NdCrO_3$	3.835	—
$NdFeO_3$	3.870	—
$NdGaO_3$	3.851	Pseudocubic
$NdMnO_3$	3.80	—
$NdVO_3$	3.89	—
$PbCeO_3$	7.62	8 mol.
$PbThO_3$	8.95	8 mol. monoclinic

(*continued*)

TABLE VIIA,7 (*continued*)

Crystal	Cubic or Pseudocubic a_0, A.	Remarks
α-PbTiO$_3$	3.960 (535°C.)	—
PbZrO$_3$	9.28	8 mol.
PrAlO$_3$	3.757	1 mol.
	5.31	Rhombohedral, $\alpha = 60°20'$
PrCoO$_3$	3.787	Rhombohedral, $\alpha = 90°13'$
PrCrO$_3$	3.852	—
PrFeO$_3$	3.887	—
PrGaO$_3$	3.863	—
PrMnO$_3$	3.82	—
PrVO$_3$	3.89	—
PuAlO$_3$	5.33	Rhombohedral, $\alpha = 56°4'$
PuMnO$_3$	3.86	Pseudocubic
RbCaF$_3$	4.452	—
RbCoF$_3$	4.062	—
RbIO$_3$	4.52	1 mol.
	9.04	8 mol. monoclinic
RbMgF$_3$	8.19	8 mol. monoclinic, $\beta = 98°30'$
RbMnF$_3$	4.250	—
SmAlO$_3$	3.734	—
SmCoO$_3$	3.75	—
SmCrO$_3$	3.812	—
SmFeO$_3$	3.845	—
SmVO$_3$	3.89	—
SrCeO$_3$	4.27	1 mol.
	8.54	8 mol. monoclinic
SrCoO$_3$	7.725	8 mol.
SrFeO$_3$	3.869	—
SrHfO$_3$	4.069	1 mol.
	8.138	8 mol. monoclinic
SrMoO$_3$	3.9751	—
SrSnO$_3$	4.0334	1 mol.
	8.070	8 mol.
SrThO$_3$	8.84	8 mol. monoclinic
SrTiO$_3$	3.9051 (25°C.)	—

(*continued*)

TABLE VIIA,7 (*continued*)

Crystal	Cubic or Pseudocubic a_0, A.	Remarks
$SrZrO_3$	4.101	1 mol.
	8.218	8 mol.
$TaSnO_3$	3.880	—
$TlCoF_3$	4.138	—
$TlIO_3$	4.510	Rhombohedral, $\alpha = 89°22'$
$YAlO_3$	3.68	1 mol.
	7.34	8 mol. monoclinic
$YCrO_3$	3.768	—
$YFeO_3$	3.785	—
$BaNi_{0.33}Nb_{0.67}O_3$	4.065	—
$BaCa_{0.5}W_{0.5}O_3$	8.390	—
$BaCe_{0.5}Nb_{0.5}O_3$	4.293	—
$BaCo_{0.33}Ta_{0.67}O_3$	4.086	Pseudocubic
$BaCo_{0.5}W_{0.5}O_3$	4.050	1 mol.
	8.098	8 mol.
$BaDy_{0.5}Nb_{0.5}O_3$	4.224	—
$BaEr_{0.5}Nb_{0.5}O_3$	4.208	—
$BaEu_{0.5}Nb_{0.5}O_3$	4.243	—
$BaFe_{0.5}Nb_{0.5}O_3$	4.057	—
$BaFe_{0.5}Ta_{0.5}O_3$	4.08	—
$BaFe_{0.5}W_{0.5}O_3$	8.133	8 mol.
$BaGd_{0.5}Nb_{0.5}O_3$	4.242	—
$BaHo_{0.5}Nb_{0.5}O_3$	4.216	—
$BaIn_{0.5}Nb_{0.5}O_3$	4.1454	—
$BaLi_{0.5}Re_{0.5}O_3$	8.1183	8 mol.
$BaLu_{0.5}Nb_{0.5}O_3$	4.187	—
$BaMg_{0.5}W_{0.5}O_3$	8.099	8 mol.
$BaMo_{0.5}Co_{0.5}O_3$	4.0429	—
$BaMo_{0.5}Ni_{0.5}O_3$	4.0225	—
$BaNa_{0.5}Re_{0.5}O_3$	8.2963	8 mol.
$BaNd_{0.5}Nb_{0.5}O_3$	4.277	—
$BaNi_{0.33}Ta_{0.67}O_3$	4.075	Pseudocubic
$BaNi_{0.5}W_{0.5}O_3$	4.0326	1 mol.
	8.066	8 mol.
$BaPr_{0.5}Nb_{0.5}O_3$	4.285	—

(*continued*)

TABLE VIIA,7 (*continued*)

Crystal	Cubic or Pseudocubic a_0, A.	Remarks
$BaSc_{0.5}Nb_{0.5}O_3$	4.129	—
$BaSc_{0.5}Ta_{0.5}O_3$	4.12	—
$BaSm_{0.5}Nb_{0.5}O_3$	4.248	—
$BaTb_{0.5}Nb_{0.5}O_3$	4.229	—
$BaTm_{0.5}Nb_{0.5}O_3$	4.201	—
$BaY_{0.5}Nb_{0.5}O_3$	4.180	—
$BaYb_{0.5}Nb_{0.5}O_3$	4.192	—
$BaYb_{0.5}Ta_{0.5}O_3$	4.17	—
$BaZn_{0.33}Nb_{0.67}O_3$	4.07	—
$BaZn_{0.5}W_{0.5}O_3$	8.116	8 mol.
$BaCaZrGeO_6$	4.172	Tetragonal
$CaNi_{0.33}Nb_{0.67}O_3$	3.89	—
$CaNi_{0.33}Ta_{0.67}O_3$	3.93	Orthorhombic
$(Ca,Na)(Ti,Nb)O_3$ (dysanalyte)	3.826	—
$CsCd(NO_2)_3$	5.390	—
$CsHg(NO_2)_3$	5.475	—
$K_{0.5}Bi_{0.5}TiO_3$	3.95 ($>410°C.$)	—
$KCd(NO_2)_3$	5.325	—
$K_{0.5}Ce_{0.5}TiO_3$	3.90	—
$K_{0.5}La_{0.5}TiO_3$	3.914	Pseudocubic
$K_{0.5}Nd_{0.5}TiO_3$	3.874	Pseudocubic
$KBaTiNbO_6$	4.016	—
$KBaCaTiZrNbO_9$	4.07	Pseudocubic
$K_2CeLaTi_4O_{12}$	3.90	—
$La_{0.6}Ba_{0.4}MnO_3$	3.91	Pseudocubic
$La_{0.6}Ca_{0.4}MnO_3$	3.84	Pseudocubic
$La_{0.6}Sr_{0.4}MnO_3$	3.88	Pseudocubic
$LaMg_{0.5}Ge_{0.5}O_3$	3.90	—
$LaMg_{0.5}Ti_{0.5}O_3$	3.932	—
$LaNi_{0.5}Ti_{0.5}O_3$	3.93	—
$LaZr_{0.5}Ca_{0.5}O_3$	4.174	—
$LaZr_{0.5}Mg_{0.5}O_3$	4.06	—
$LaMnO_{3.07-3.15}$	3.904–3.889	—
$LaMnO_{3.10-3.23}$	3.880–3.878	Rhombohedral, $\alpha = 90°32'$
$NH_4Cd(NO_2)_3$	5.355	—

(*continued*)

TABLE VIIA,7 (*continued*)

Crystal	Cubic or Pseudocubic a_0, A.	Remarks
$Na_{0.5}Bi_{0.5}TiO_3$	3.891	Rhombohedral, $\alpha = 89°36'$
$Na_{0.5}La_{0.5}TiO_3$	3.86	—
$NdMg_{0.5}Ti_{0.5}O_3$	3.90	Orthorhombic
$(Na,Ca,Ce)(Ti,Nb)O_3$ (loparite)	3.854	—
$NaBaTiNbO_6$	3.97	—
$PbMg_{0.33}Nb_{0.67}O_3$	4.041(22°C.)	—
$PbNi_{0.33}Nb_{0.67}O_3$	4.025	—
$PbFe_{0.67}W_{0.33}O_3$	3.98	—
$PbFe_{0.5}Nb_{0.5}O_3$	4.01	—
$PbFe_{0.5}Ta_{0.5}O_3$	4.01	—
$PbSc_{0.5}Nb_{0.5}O_3$	4.08	—
$PbSc_{0.5}Ta_{0.5}O_3$	4.08	—
$PbYb_{0.5}Nb_{0.5}O_3$	4.16	—
$PbYb_{0.5}Ta_{0.5}O_3$	4.14	—
$(Pr,Nd)MnO_3$	3.86	Pseudocubic
$(Pr,Nd)_{0.6}Ba_{0.4}MnO_3$	3.89	Pseudocubic
$(Pr,Nd)_{0.6}Sr_{0.4}MnO_3$	3.86	Pseudocubic
$RbCd(NO_2)_3$	5.375	—
$RbHg(NO_2)_3$	5.450	—
$SrMg_{0.33}Ta_{0.67}O_3$	4.000	Pseudocubic
$SrNi_{0.33}Nb_{0.67}O_3$	3.99	—
$SrCr_{0.5}Nb_{0.5}O_3$	3.9421	—
$SrCr_{0.5}Ta_{0.5}O_3$	3.94	—
$SrGa_{0.5}Nb_{0.5}O_3$	3.9477	—
$SrIn_{0.5}Nb_{0.5}O_3$	4.0569	—
$SrMo_{0.5}Co_{0.5}O_3$	7.918(>320°C.)	—
$SrMo_{0.5}Ni_{0.5}O_3$	7.878 (>230°C.)	—
$SrMo_{0.5}Zn_{0.5}O_3$	7.954 (>420°C.)	—
$SrW_{0.5}Co_{0.5}O_3$	7.904 (>400°C.)	—
$SrW_{0.5}Ni_{0.5}O_3$	7.908 (>300°C.)	—
$SrW_{0.5}Zn_{0.5}O_3$	7.956 (430°C.)	—
$SrLaNdTiCrAlO_9$	3.90	Orthorhombic
$TlCd(NO_2)_3$	5.340	—
$TlHg(NO_2)_3$	5.385	—

(*continued*)

TABLE VIIA,7 (*continued*)

Crystal	a_0, A.	Tetragonal c_0, A.	Remarks
AlBiO$_3$	7.61	7.94	8 mol.
BaFeO$_3$	3.98	4.01	—
BaTiO$_3$	3.9947	4.0336 (20°C.)	—
BaZrS$_3$	4.990	5.088	—
CeAlO$_3$	3.767	3.794	—
CrBiO$_3$	7.77	8.08	8 mol.
CsMgF$_3$	9.39	8.72	8 mol.
CsPbCl$_3$	5.590	5.630	—
CsZnF$_3$	9.90	9.05	8 mol.
KCoF$_3$	4.057	4.049 (78°K.)	—
KCrF$_3$	4.274	4.019	—
KCuF$_3$	4.140	3.926 (298°K.)	—
	4.121	3.913 (78°K.)	—
KNbO$_3$	4.00	4.07 (260°C.)	—
KZnF$_3$	8.51	8.11	8 mol.
LaVO$_3$	5.546	7.827	4 mol.
LiUO$_3$	5.406	7.506	—
NaMgF$_3$	3.942	3.933 (760–900°C.)	1 mol.
	7.66	7.31	8 mol.
NaNbO$_3$	7.856	15.712 (425°C.)	—
NaZnF$_3$	7.76	8.15	8 mol.
PbSnO$_3$	7.86	8.13	8 mol.
β-PbTiO$_3$	3.99330	4.03030	—
PbZrO$_3$	4.150	4.099 (23°C.)	—
RbZnF$_3$	8.71	8.03	8 mol.
BaDy$_{0.5}$Nb$_{0.5}$O$_3$	8.607	8.437	8 mol.
BaEr$_{0.5}$Nb$_{0.5}$O$_3$	8.607	8.427	8 mol.
BaEu$_{0.5}$Nb$_{0.5}$O$_3$	8.607	8.507	8 mol.
BaGd$_{0.5}$Nb$_{0.5}$O$_3$	8.607	8.496	8 mol.
BaHo$_{0.5}$Nb$_{0.5}$O$_3$	8.607	8.434	8 mol.

(*continued*)

TABLE VIIA,7 (*continued*)

Crystal	Tetragonal a_0, A.	c_0, A.	Remarks
$BaIn_{0.5}Nb_{0.5}O_3$	8.607	8.279	8 mol.
$BaLa_{0.5}Nb_{0.5}O_3$	8.607	8.690	8 mol.
$BaLu_{0.5}Nb_{0.5}O_3$	8.607	8.364	8 mol.
$BaNd_{0.5}Nb_{0.5}O_3$	8.607	8.540	8 mol.
$BaSm_{0.5}Nb_{0.5}O_3$	8.607	8.518	8 mol.
$BaTm_{0.5}Nb_{0.5}O_3$	8 607	8.408	8 mol.
$BaYb_{0.5}Nb_{0.5}O_3$	8.607	8.374	8 mol.
$K_{0.5}Bi_{0.5}TiO_3$	3.913	3.993	Cubic above 410°C., with $a_0 = 3.95$ A.
$SrMo_{0.5}Co_{0.5}O_3$	3.9367	3.9764	1 mol.
	5.581	7.940	4 mol.
$SrMo_{0.5}Ni_{0.5}O_3$	3.9237	3.9474	1 mol.
	5.560	7.886	4 mol.
$SrMo_{0.5}Zn_{0.5}O_3$	5.561	7.966	4 mol.
$SrW_{0.5}Co_{0.5}O_3$	3.9502	3.9746	1 mol.
	5.596	7.978	4 mol.
$SrW_{0.5}Ni_{0.5}O_3$	3.9310	3.9592	1 mol.
	5.575	7.918	4 mol.
$SrW_{0.5}Zn_{0.5}O_3$	5.579	7.976	4 mol.

Crystal	Monoclinic a_0, A.	b_0, A.	c_0, A.	β	Remarks
$CaMnO_3$	7.481	7.449	7.481	91°7′	—
	10.683	7.449	10.476	—	Orthorhombic
$CaMoO_3$	7.80	7.77	7.80	91°23′	—
$KMnF_3$	4.168	4.171	4.185	89°51′	78°K.
$LaCrO_3$	7.777	7.750	7.777	90°15′	—
$LaMnO_3$	7.960	7.698	7.960	91°52′	—
	11.439	7.698	11.072	—	Orthorhombic
$NaMnF_3$	3.997	3.992	3.997	91°57′	—
$YCrO_3$	7.61	7.54	7.61	92°56′	—

(*continued*)

TABLE VIIA,7 (*continued*)

Crystal	Orthorhombic a_0, A.	b_0, A.	c_0, A.	Remarks
CaSnO$_3$	5.519	7.884	5.668	—
	11.506	7.882	11.325	—
Monoclinic pseudocell	3.952	3.941	3.952	$\beta = 91°30'$
CaTiO$_3$	10.761	7.645	10.886	—
Monoclinic pseudocell	3.827	3.823	3.827	$\beta = 90°40'$
CaZrO$_3$	11.174	8.010	11.515	—
Monoclinic pseudocell	4.011	4.005	4.011	$\beta = 91°43'$
CdSnO$_3$	5.547	5.577	7.867	—
CdTiO$_3$	10.615	7.615	10.834	—
Monoclinic pseudocell	3.791	3.808	3.791	$\beta = 91°10'$
DyAlO$_3$	5.21	5.31	7.40	—
DyFeO$_3$	5.30	5.60	7.62	—
GdCoO$_3$	3.732	3.807	3.676	—
LaCrO$_3$	5.487	7.750	5.511	—
NaMgF$_3$	5.363	7.676	5.503	—
NaNbO$_3$	5.5052	5.5682	3.8795	—
(see **VII,a24**)	15.5180	5.5682	5.5052	—
	5.512	5.577	4×3.885	—
SmCoO$_3$	3.747	3.803	3.728	—
SrCeO$_3$	6.011	8.588	6.156	1250–1650°C.
SrPbO$_3$	5.864	5.949	8.336	—

Here (Fig. VIIA,22), the calcium and titanium atoms are distributed exactly as are the cesium and chlorine atoms in CsCl. There is no reason to believe that definite TiO$_3$ groups exist in the crystal, however, and therefore it is better to look upon it as determined by the packing requirements of the large oxygen and calcium ions. These four atoms taken together are in a cubic close-packing with the small titanium ions in holes. As this packing picture would predict, oxides with the perewskite structure all have unit cells of about the same size. The same thing is true of the fluorides and chlorides, each as a group. In CaTiO$_3$ the significant atomic separations are: O–O = 2.70 A., Ca–O and Ti–O = 1.90 A. In fluorides, F–F = ca. 2.85 A. and in chlorides, Cl–Cl = ca. 3.75 A.

Among the non-cubic perewskite-like compounds, some do not require larger units. Thus for the tetragonal *lead titanate*, PbTiO$_3$, with its unimolecular cell having

$$a_0 = 3.904 \text{ A.}, \qquad c_0 = 4.150 \text{ A.}$$

x-ray and neutron diffraction work indicates that the atoms are in the following positions:

$$Pb: \quad 000$$
$$Ti: \quad 1/2 \; 1/2 \; 0.541$$
$$O(1): \quad 1/2 \; 1/2 \; 0.112$$
$$O(2): \quad 1/2 \, 0 \, 0.612 ; \, 0 \, 1/2 \, 0.612$$

The room-temperature form of *barium titanate*, $BaTiO_3$, has a tetragonal cell, with

$$a_0 = 3.9947 \text{ A.}, \qquad c_0 = 4.0336 \text{ A.}$$

It is thought to have its atoms in the positions (similar to those in $PbTiO_3$):

$$Ba: \quad 000$$
$$Ti: \quad 1/2 \; 1/2 \; 0.512$$
$$O(1): \quad 1/2 \; 1/2 \; 0.023$$
$$O(2): \quad 1/2 \, 0 \, 0.486 ; \, 0 \, 1/2 \, 0.486$$

Recently, a detailed attempt to define the atomic parameters by a variety of methods failed to yield appreciably different results.

In common with other compounds of this sort, $BaTiO_3$ forms many solid solutions, some of which shift the cell dimensions until they become equal and the crystal appears cubic. This is the case, for instance, with a preparation containing ca. 16 mole-% $CaSnO_3$.

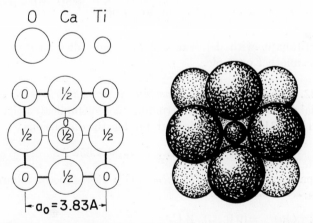

Fig. VIIA,22a (left). The unit of the simple cubic perewskite structure projected along an axis. The largest circles are the oxygen, the smallest are the titanium atoms.

Fig. VIIA,22b (right). A packing drawing of the atoms of the perewskite arrangement. oxygens are the largest circles, the titanium of $CaTiO_3$ the smallest.

Cesium lead trichloride, $CsPbCl_3$, is another tetragonal perewskite which seems to show the same kind of atomic displacements as $PbTiO_3$. For it,

$$a_0 = 5.590 \text{ A.}, \qquad c_0 = 5.630 \text{ A.}$$

and the displacement of the chlorine atoms is ca. 0.06. Above 47°C. this compound is truly cubic, with $a_0 = 5.599$ A.

In view of its approximate x-ray data, $NaIO_3$, described in the preceding paragraph (**VII,a20**), was originally thought to have the perewskite arrangement. It is very possible that with better data other crystals listed in Table VIIA,7 would be found to be like this iodate.

Many crystals that are perewskite-like at sufficiently high temperatures lose symmetry at lower temperatures. Often, several modifications are then found which do not, however, represent profound alterations in atomic arrangement. An example of this type of pleomorphism involving forms that remain pseudocubic is *sodium niobate,* $NaNbO_3$, for which the following modifications exist:

Cubic: $a_0 = 3.942$ A. at 640°C.
Tetragonal: $a_0 = 7.866$ A. (2 × 3.933 A.)
   $c_0 = 15.760$ A. (4 × 3.940 A.) at 560°C.
Pseudotetragonal: $a_0 = 7.848$ A. (2 × 3.924 A.)
   $c_0 = 15.696$ A. (4 × 3.924 A.) at 420°C.
Monoclinic: $a_0 = c_0 = 7.828$ A. (2 × 3.914 A.)
   $b_0 = 15.524$ A. (4 × 3.881 A.)
   $\beta = 90°39'$ at 20°C.

Barium titanate, with detailed structures described in succeeding paragraphs, is another example.

VII,a22. The compound *potassium cupric fluoride,* $KCuF_3$, has a tetragonal distorted perewskite structure more complicated than those discussed in **VII,a21**. Its tetramolecular unit has the dimensions:

$$a_0 = 5.855 \text{ A.}, \qquad c_0 = 7.852 \text{ A.}$$

The space group is D_{4h}^{18} ($I4/mcm$), with atoms in the following special positions:

K: (4a) $\pm(0\ 0\ 1/4)$; B.C.
Cu: (4d) $0\ 1/2\ 0$; $1/2\ 0\ 0$; B.C.
F(1):   (4b) $0\ 1/2\ 1/4$; $1/2\ 0\ 1/4$; B.C.
F(2):   (8h) $\pm(u,u+1/2,0;\ 1/2-u,u,0)$; B.C.   with $u = 0.228$

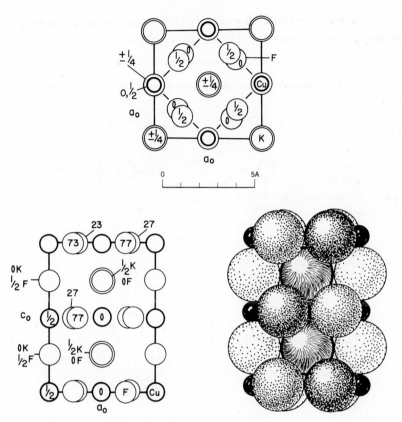

Fig. VIIA,23a (top). A projection of the tetragonal $KCuF_3$ distortion of the perewskite arrangement seen along its c_0 axis.

Fig. VIIA,23b (bottom left). The tetragonal $KCuF_3$ arrangement projected along an a_0 axis.

Fig. VIIA,23c (bottom right). A packing drawing of the $KCuF_3$ structure seen along an a_0 axis. The potassium atoms are large and fine-line shaded, the copper are small and black.

In this structure (Fig. VIIA,23), each copper atom has four fluorine neighbors with Cu–F = 1.89 or 1.96 A. and two more at a distance of 2.25 A.

The relation to the simple perewskite arrangement and the type of distortion that prevails here are evident from the basal projection of Figure VIIA,23a. Two of the axes of the undistorted unimolecular cell would, as the dashed lines indicate, follow the diagonals of the base of the true unit, the other would halve the c_0 axis of this unit.

Below the Neél temperature (243 °K.), the structure of this compound is the same as above. At 78 °K. the unit cell has

$$a_0 = 5.828 \text{ A.}, \qquad c_0 = 7.822 \text{ A.}$$

Probably, *potassium chromium trifluoride*, $KCrF_3$, also has this structure, with

$$a_0 = 6.03 \text{ A.}, \qquad c_0 = 8.02 \text{ A.}$$

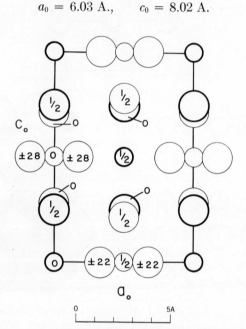

Fig. VIIA,24. A projection along an a_0 axis of the tetragonal structure of $Cs_2AgAuCl_6$. The large heavy circles are cesium, the small heavy circles gold atoms. Of the light circles the smaller are silver.

VII,a23. Structures found for preparations of $Cs_2AgAuCl_6$ and $Cs_2Au_2Cl_6$ that have been subjected to prolonged heating are "superlattices" on the perewskite arrangement. As ordinarily made, these salts give the cubic diffraction patterns and undoubtedly have the simple grouping ascribed to perewskite (**VII,a21**). After heating to 350°C. for several days their symmetry becomes tetragonal and they assume an atomic arrangement which has the bimolecular cells:

$$Cs_2AgAuCl_6: \quad a_0 = 7.38 \text{ A.}, \, c_0 = 11.01 \text{ A.}$$
$$Cs_2Au_2Cl_6: \quad a_0 = 7.49 \text{ A.}, \, c_0 = 10.87 \text{ A.}$$

Atoms in $Cs_2AgAuCl_6$ have been put in the following special positions of D_{4h}^{17} $(I4/mmm)$:

Cs: $(4d)$ $0\ ^1/_2\ ^1/_4;\ ^1/_2\ 0\ ^1/_4;$ B.C.
Ag: $(2b)$ $0\ 0\ ^1/_2;\ ^1/_2\ ^1/_2\ 0$
Au: $(2a)$ $000;\ ^1/_2\ ^1/_2\ ^1/_2$
Cl(1): $(4e)$ $\pm(00u);$ B.C.   with $u = 0.282$
Cl(2): $(8h)$ $\pm(vv0;\ v\bar{v}0);$ B.C.   with $v = 0.220$

For $Cs_2Au_2Cl_6$, with silver replaced by gold, nearly the same parameters apply: $u = 0.288$ and $v = 0.228$.

In these arrangements (Fig. VIIA,24), the atomic positions have been displaced from those in perewskite until in $Cs_2AgAuCl_6$ each gold atom is surrounded by a square of chlorine atoms at a distance of 2.30 A., while each silver atom is most closely associated with a dumbbell of two chlorine atoms at a distance of 2.36 A. The corresponding separations in the gold compound are 2.41 and 2.30 A.

VII,a24.   On cooling through 5°C., the symmetry of $BaTiO_3$ changes from tetragonal to orthorhombic. This orthorhombic form has the bimolecular unit:

$$a_0 = 3.990 \text{ A.;}\quad b_0 = 5.669 \text{ A.;}\quad c_0 = 5.682 \text{ A.}$$

Its atoms have been placed in the following positions of C_{2v}^{14} $(Amm2)$:

Ba: $(2a)$ $00u;\ 0,^1/_2,u+^1/_2$ with $u = 0$
Ti: $(2b)$ $^1/_2\ 0\ u;\ ^1/_2,^1/_2,u+^1/_2$ with $u = 0.510$
O(1): $(2a)$ with $u = 0.490$
O(2): $(4e)$ $^1/_2\ u\ v;\ ^1/_2\ \bar{u}\ v;\ ^1/_2,u+^1/_2,v+^1/_2;\ ^1/_2,^1/_2-u,v+^1/_2$
                                with $u = 0.253,\ v = 0.237$

As Figure VIIA,25 shows, this represents only a small shift from the tetragonal structure as given in **VII,a21**. These orthorhombic b_0 and c_0 axes are diagonal to and $\sqrt{2}$ greater than those of the unimolecular tetragonal cell described there.

Measured cell dimensions make it probable that the following compounds also have this arrangement:

NaTaO$_3$: $a_0 = 3.8831$ A.; $b_0 = 5.4778$ A.; $c_0 = 5.5239$ A.
KNbO$_3$: $a_0 = 3.9714$ A.; $b_0 = 5.6946$ A.; $c_0 = 5.7203$ A.
KTaO$_3$: $a_0 = 3.9714$ A.; $b_0 = 5.6946$ A.; $c_0 = 5.7203$ A.

Other data have given $KTaO_3$ as cubic (Table VIIA,7), but it is not clear if two forms actually exist.

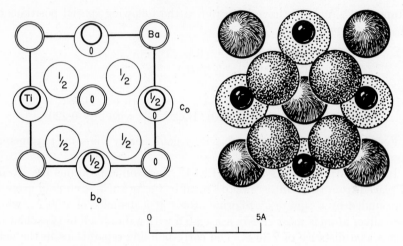

Fig. VIIA,25a (left). A projection along its a_0 axis of the orthorhombic low temperature modification of BaTiO$_3$. Origin in lower right.

Fig. VIIA,25b (right). A packing drawing of the projection along a_0 of the low temperature tetragonal modification of BaTiO$_3$. The barium atoms are fine-line shaded, the titanium atoms are small and black.

VII,a25. The distortion of the perewskite structure found by neutron and x-ray studies for *lead zirconate*, PbZrO$_3$, has orthorhombic symmetry, with the eight-molecule cell:

$$a_0 = 5.884 \text{ A.}; \quad b_0 = 11.768 \text{ A.}; \quad c_0 = 8.220 \text{ A.}$$

Its space group has been chosen as $C_{2v}{}^8$ (*Pba2*), with atoms in the positions:

(2a) $00u;\ {}^1/_2\,{}^1/_2\,u$
(2b) $0\,{}^1/_2\,u;\ {}^1/_2\,0\,u$
(4c) $xyz;\ \bar{x}\bar{y}z;\ {}^1/_2-x,y+{}^1/_2,z;\ x+{}^1/_2,{}^1/_2-y,z$

Parameters are stated in Table VIIIA,8 and the resulting arrangement is shown in Figure VIIA,26.

Originally, the symmetry of this substance was described as tetragonal, with the cell dimensions:

$$a_0' = 4.159 \text{ A.}, \qquad c_0' = 4.109 \text{ A.}$$

The orthorhombic cell is connected to this by the relations:

$$a_0 = a_0'\sqrt{2};\ \ b_0 = 2a_0'\sqrt{2};\ \ c_0 = 2c_0'$$

TABLE VIIA,8
Parameters of the Atoms in PbZrO$_3$

Atom	Position	x	y	z
Pb(1)	(4c)	0.706	0.127	0.000
Pb(2)	(4c)	0.706	0.127	0.500
Zr(1)	(4c)	0.243	0.124	0.250
Zr(2)	(4c)	0.243	0.124	0.750
O(1)	(4c)	0.270	0.150	0.980
O(2)	(4c)	0.270	0.150	0.480
O(3)	(4c)	0.040	0.270	0.300
O(4)	(4c)	0.040	0.270	0.750
O(5)	(2b)	0.000	0.500	0.250
O(6)	(2b)	0.000	0.500	0.800
O(7)	(2a)	0.000	0.000	0.250
O(8)	(2a)	0.000	0.000	0.800

Perhaps with further work other substances listed in Table VIIA,7 will prove to have this structure.

VII,a26. *Gadolinium ferrite*, GdFeO$_3$, is typical of a large group of orthorhombic distorted perewskite-like compounds. Four molecules are contained in a unit with the edges:

$$a_0 = 5.346 \text{ A.}; \quad b_0 = 5.616 \text{ A.}; \quad c_0 = 7.668 \text{ A.}$$

The space group is V_h^{16} (*Pbnm*), and atoms are in the following of its equivalent positions:

Gd: (4c) $\pm(u\,v\,{}^1/_4;\,{}^1/_2-u,v+{}^1/_2,{}^1/_4)$
with $u = -0.018$, $v = 0.060$
Fe: (4b) ${}^1/_2\,0\,0;\,{}^1/_2\,0\,{}^1/_2;\,0\,{}^1/_2\,0;\,0\,{}^1/_2\,{}^1/_2$
O(1): (4c) with $u = 0.05$, $v = 0.47$
O(2): (8d) $\pm(xyz;\,{}^1/_2-x,y+{}^1/_2,{}^1/_2-z;\,\bar{x},\bar{y},z+{}^1/_2;\,x+{}^1/_2,{}^1/_2-y,\bar{z})$
with $x = -0.29$, $y = 0.275$, $z = 0.05$

The structure as a whole is shown in Figure VIIA,27. As can be seen from Figure VIIA,28, each cell is built up of four distorted perewskite units oriented diagonally to the a_0 and b_0 axes. For GdFeO$_3$ they have the dimensions:

$$a_0' = c_0' = 3.877 \text{ A.}; \quad b_0' = 3.834 \text{ A.}; \quad \beta' = 92°48'$$

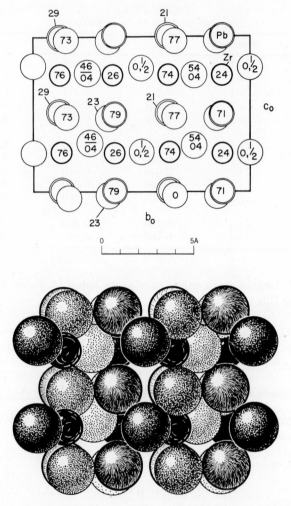

Fig. VIIA,26a (top). The orthorhombic PbZrO₃ distortion of the perewskite structure viewed along its a_0 axis. Origin in lower right.

Fig. VIIA,26b (bottom). A packing drawing of the orthorhombic PbZrO₃ arrangement seen along its a_0 axis. The zirconium atoms are black, the lead atoms fine-line shaded.

The compounds of Table VIIA,9, some of which were originally described in terms of other units, are isomorphous with GdFeO₃. It is likely that certain of the monoclinic perewskites listed in Table VIIA,7 will also be found to have this arrangement.

TABLE VIIA,9
Additional Compounds with the $GdFeO_3$ Structure

Compound	a_0, A.	b_0, A.	c_0, A.
$CaRuO_3$	5.36	5.53	7.67
$EuAlO_3$	5.271	5.292	7.458
$EuFeO_3$	5.371	5.611	7.686
$GdAlO_3$	5.247	5.304	7.447
$GdCrO_3$	5.312	5.514	7.611
$GdScO_3$	5.487	5.756	7.925
$GdVO_3$	5.345	5.623	7.638
$LaCrO_3$	5.477	5.514	7.755
$LaFeO_3$	5.556	5.565	7.862
$LaGaO_3$	5.496	5.524	7.787
$LaScO_3$	5.678	5.787	8.098
$NaCoF_3$	5.420	5.603	7.793
$NaMgF_3$	5.350	5.474	7.652
$NaNiF_3$	5.360	5.525	7.705
$NaZnF_3$	5.400	5.569	7.756
$NdCrO_3$	5.412	5.494	7.695
$NdFeO_3$	5.441	5.573	7.753
$NdGaO_3$	5.426	5.502	7.706
$NdScO_3$	5.574	5.771	7.998
$NdVO_3$	5.440	5.589	7.733
$PrCrO_3$	5.444	5.484	7.710
	5.44	5.48	7.71
$PrFeO_3$	5.495	5.578	7.810
$PrGaO_3$	5.465	5.495	7.729
$PrScO_3$	5.615	5.776	8.027
$PrVO_3$	5.477	5.545	7.759
	5.48	5.59	7.76
$PuCrO_3$	5.46	5.51	7.76
$PuVO_3$	5.48	5.61	7.78
$SmAlO_3$	5.285	5.290	7.473
$SmCrO_3$	5.372	5.502	7.650
$SmFeO_3$	5.394	5.592	7.711
$SrRuO_3$	5.53	5.57	7.85
$YAlO_3$	5.179	5.329	7.370
$YCrO_3$	5.247	5.518	7.540
$YFeO_3$	5.302	5.589	7.622
$YScO_3$	5.431	5.712	7.894

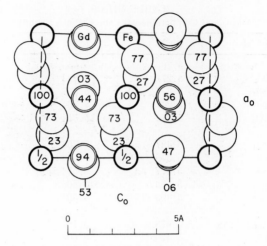

Fig. VIIA,27a. The orthorhombic structure of GdFeO$_3$ projected along its b_0 axis. Origin in lower right.

Systems of solid solutions of the type La(M$_x$Mn$_{1-x}$)O$_3$ (with M = Cr, $0.33 < x \leqq 1$ and M = Fe, $0.5 < x \leqq 1$) have this structure. Another example, more thoroughly studied, is the substance La(Co$_{0.2}$Mn$_{0.8}$)O$_3$. For it,

$$a_0 = 5.525 \text{ A.}; \quad b_0 = 5.530 \text{ A.}; \quad c_0 = 7.819 \text{ A.}$$

Lanthanum atoms are in (4c) with $u = -0.01$, $v = 0.023$; (Co,Mn) atoms are in (4b); and the oxygen atoms are in

O(1):   (4c)   with $u = 0.01$, $v = 0.50$
O(2):   (8d)   with $x = -0.29$, $y = 0.27$, $z = 0.06$

Perewskite itself, CaTiO$_3$, has recently been assigned this distortion of the simple arrangement described in **VII,a21**. Its orthorhombic unit in the *Pbnm* orientation has the dimensions:

$$a_0 = 5.37 \text{ A.}; \quad b_0 = 5.44 \text{ A.}; \quad c_0 = 7.64 \text{ A.}$$

The titanium atoms, like the iron atoms of GdFeO$_3$, are in special positions (4a). The calcium atoms in (4c) have $u = 0.00$, $v = 0.030$. For O(1) in (4c), $u = 0.037$, $v = 0.482$; for O(2) in (8d), $x = -0.268$, $y = 0.268$, $z = 0.026$.

Though approaching cubic symmetry as the temperature is raised, perewskite remains orthorhombic up to 900°C.

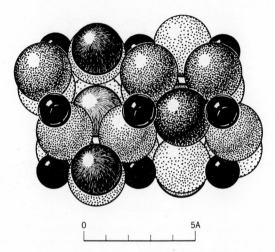

Fig. VIIA,27b. A packing drawing of the orthorhombic GdFeO₃ structure seen along its b_0 axis. The iron atoms are black, the gadolinium atoms are fine-line shaded.

The compound KMnF₃ which at room temperature is cubic with the perewskite structure (**VII,a21**) and $a_0 = 4.186$ A., has this GdFeO₃ arrangement below 184°K. Its orthorhombic cell then has the dimensions:

$$a_0 = 5.885 \text{ A.}; \quad b_0 = 5.885 \text{ A.}; \quad c_0 = 8.376 \text{ A. (95°K.)}$$
$$a_0 = 5.900 \text{ A.}; \quad b_0 = 5.900 \text{ A.}; \quad c_0 = 8.330 \text{ A. (65°K.)}$$

The manganese atoms are in ($4b$) of $V_h{}^{16}$ (*Pbnm*), the potassium and one set of fluorine atoms in ($4c$) and the other fluorine atoms in ($4d$). Parameters have been described for these atoms at three different temperatures. They are

$$
\begin{aligned}
\text{K:} \quad & u = -0.005, \ v = 0.02 \ (65°\text{K.}) \\
& u = -0.01, \quad v = 0.02 \ (84°\text{K.}) \\
& u = -0.01, \quad v = 0.02 \ (95°\text{K.}) \\
\text{F(1):} \quad & u = 0.035, \quad v = 0.50 \ (65°\text{K.}) \\
& u = 0.038, \quad v = 0.49 \ (84°\text{K.}) \\
& u = 0.023, \quad v = 0.49 \ (95°\text{K.}) \\
\text{F(2):} \quad & x = -0.280; \ y = 0.28; \ z = 0.030 \ (65°\text{K.}) \\
& x = -0.268; \ y = 0.268; \ z = 0.018 \ (84°\text{K.}) \\
& x = -0.25; \quad y = 0.25; \quad z = 0.034 \ (95°\text{K.})
\end{aligned}
$$

At elevated temperatures, *lanthanum gallium trioxide*, LaGaO₃, and *samarium aluminum trioxide*, SmAlO₃, invert from this GdFeO₃ grouping

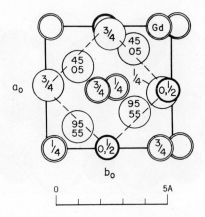

Fig. VIIA,28. The orthorhombic $GdFeO_3$ arrangement projected along its c_0 axis to show the relation of its unit to that of the simple cubic perewskite structure. The iron atoms are heavily ringed. Origin in lower left.

to a rhombohedral form which, though its structure has not yet been established, probably is not very different. Their bimolecular rhombohedra have the dimensions:

$$LaGaO_3: \quad a_0 = 5.544 \text{ A., } \alpha = 60°25' \text{ (900°C.)}$$
$$SmAlO_3: \quad a_0 = 5.316 \text{ A., } \alpha = 60°19' \text{ (850°C.)}$$

Three other aluminates are isomorphous. For them,

$$LaAlO_3: \quad a_0 = 5.357 \text{ A., } \alpha = 60°6'$$
$$NdAlO_3: \quad a_0 = 5.286 \text{ A., } \alpha = 60°25'$$
$$PrAlO_3: \quad a_0 = 5.307 \text{ A., } \alpha = 60°20'$$

The hexagonal cells of these compounds containing six molecules have the edge lengths:

$$LaAlO_3: \quad a_0 = 5.365 \text{ A., } c_0 = 13.11 \text{ A.}$$
$$LaGaO_3: \quad a_0 = 5.579 \text{ A., } c_0 = 13.54 \text{ A.}$$
$$NdAlO_3: \quad a_0 = 5.319 \text{ A., } c_0 = 12.91 \text{ A.}$$
$$PrAlO_3: \quad a_0 = 5.334 \text{ A., } c_0 = 12.97 \text{ A.}$$
$$SmAlO_3: \quad a_0 = 5.341 \text{ A., } c_0 = 12.99 \text{ A.}$$

VII,a27. A structure somewhat different from that of the preceding $GdFeO_3$ (**VII,a26**) has been found for the similarly distorted *cadmium titanate*, $CdTiO_3$. Its tetramolecular orthorhombic unit has the dimensions:

$$a_0 = 5.348 \text{ A.; } \quad b_0 = 7.615 \text{ A.; } \quad c_0 = 5.417 \text{ A.}$$

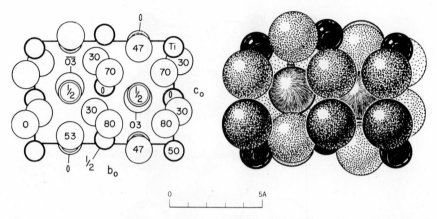

Fig. VIIA,29a (left). A projection of the orthorhombic CdTiO₃ arrangement viewed along its a_0 axis. Origin in lower right.

Fig. VIIA,29b (right). A packing drawing of the orthorhombic CdTiO₃ structure seen along its a_0 axis. The titanium atoms are black, the cadmium atoms fine-line shaded.

These axes follow a different sequence, and the chosen space group is the lower symmetry C_{2v}^9 $(Pc2_1n)$. All atoms have been placed in its general positions:

$$(4a) \quad xyz;\ \bar{x},y+{}^1\!/_2,\bar{z};\ x+{}^1\!/_2,y+{}^1\!/_2,{}^1\!/_2-z;\ {}^1\!/_2-x,y,z+{}^1\!/_2$$

The determined parameters and those for *sodium tantalate*, NaTaO₃, which has been given the same structure, are as follows:

Atom	x	y	z
Cd(Na)	0.006 (−0.01)	0.75 (0.78)	0.016 (0.02)
Ti(Ta)	0.505 (0.50)	0.00 (0.00)	−0.065 (0.00)
O(1)	−0.03 (−0.02)	0.75 (0.76)	0.55 (0.52)
O(2)	0.30 (0.29)	−0.03 (−0.03)	0.31 (0.29)
O(3)	0.30 (0.29)	0.57 (0.56)	0.31 (0.29)

The arrangement is illustrated in Figure VIIA,29. A comparison with Figure VIIA,28 makes clear the shifts in oxygen positions that differentiate this grouping from that of GdFeO₃.

The unit cell for NaTaO₃ has the dimensions:

$$a_0 = 5.4941 \text{ A.};\quad b_0 = 3.8754 \text{ A.};\quad c_0 = 5.5130 \text{ A.}$$

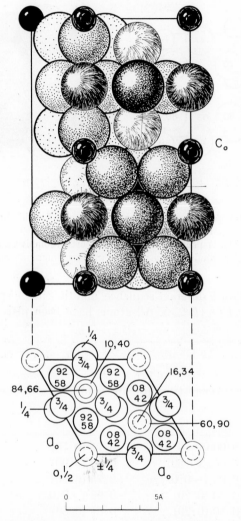

Fig. VIIA,30. Two projections of the structure of the hexagonal modification of BaTiO₃. In the upper shaded projection the titanium atoms are black, the bariums fine-line shaded.

VII,a28. In addition to its cubic and tetragonal ferroelectric forms, *barium titanate*, BaTiO₃, has a stable hexagonal modification. Its unit contains six molecules and has the edge lengths:

$$a_c = 5.735 \text{ A.}, \qquad c_0 = 14.05 \text{ A.}$$

Atoms are in the following special positions of D_{6h}^4 $(C6/mmc)$:

Ba(1): $(2b)$ $0\;0\;^1/_4;0\;0\;^3/_4$
Ba(2): $(4f)$ $\pm(^1/_3\;^2/_3\;u;\;^2/_3,^1/_3,u+^1/_2)$ with $u = 0.097$
Ti(1): $(2a)$ $000;0\;0\;^1/_2$
Ti(2): $(4f)$ with $u = 0.845$
O(1): $(6h)$ $\pm(u\;2u\;^1/_4;\;2\bar{u}\;\bar{u}\;^1/_4;\;u\;\bar{u}\;^1/_4)$ with $u = 0.522$
O(2): $(12k)$ $\pm(u\;2u\;v;\qquad 2\bar{u}\;\bar{u}\;v;\qquad u\bar{u}v;$
                    $u,2u,^1/_2-v;\;2\bar{u},\bar{u},^1/_2-v;\;u,\bar{u},^1/_2-v)$
                          with $u = 0.836,\;v = 0.076$

This structure, like the cubic form, is a stacking of close-packed layers of barium plus oxygen atoms (Fig. VIIA,30). In cubic $BaTiO_3$ there is repetition every three layers; in this hexagonal form the repetition occurs along the c_0 axis at every sixth layer. Of the several ways such a repetition could be achieved, the one actually found is that to be designated as $ABCACB$ in Figure VIIA,31. Titanium atoms lie in octahedral holes of the Ba+O packing, but two-thirds of the TiO_6 octahedra share faces to form Ti_2O_9 groups while the other third do not. Within these Ti_2O_9 coordination groups the Ti–Ti separation is 2.67 A. and the Ti–O distances are 1.96 and 2.02 A. In the TiO_6 octahedra the Ti–O distance is 1.96 A. Each barium atom has twelve oxygen neighbors. For Ba(1) there are six oxygen atoms at a distance of 2.89 A. and six more at 2.94 A.; for Ba(2) there are three oxygen atoms at 2.78 A., six at 2.88 A., and three more at 2.96 A.

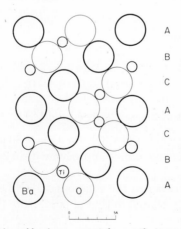

Fig. VIIA,31. The repetition of barium–oxygen layers that prevails in hexagonal $BaTiO_3$

This hexagonal form of $BaTiO_3$, which is not ferroelectric, is probably identical with the modification that has previously been described as rhombohedral.

Cesium manganous fluoride, $CsMnF_3$, has this arrangement, with

$$a_0 = 6.213 \text{ A.}, \qquad c_0 = 15.074 \text{ A.}$$

The cesium atoms are in $(2b)$ and $(4f)$ with $u = 0.0986$ and the manganese in $(2a)$ and $(4f)$ with $u = 0.8498$. For the fluorine atoms in $(6h)$, $u = 0.522$ and for those in $(12k)$ $u = 0.835$, $v = 0.078$. The parameters are all very similar to the values found for the corresponding atoms in $BaTiO_3$. Each cesium has 12 fluorine neighbors at distances between 3.06 and 3.22 A. Manganese atoms on the other hand have but six close fluorines each, with $Mn-F = 2.12-2.16$ A. The closest approach of manganese atoms to one another is 3.004 A. and of manganese to cesium 3.67 A.

The mixed compound $BaIr_{0.33}Ti_{0.67}O_3$ has this structure, with

$$a_0 = 5.74 \text{ A.}, \qquad c_0 = 14.2 \text{ A.}$$

It has also been found for an oxygen-deficient preparation having the composition $BaTi_{0.48}^{III}Ti_{0.52}^{IV}O_{2.76}$.

VII,a29. At elevated temperatures, *barium manganite*, $BaMnO_3$, is also hexagonal but with the shorter tetramolecular cell:

$$a_0 = 5.669 \text{ A.}, \qquad c_0 = 9.375 \text{ A.}$$

The space group is D_{6h}^4 $(P6_3/mmc)$, and atoms are reported to be in the positions:

Ba(1): $(2a)$ $000; 0\ 0\ ^1/_2$
Ba(2): $(2c)$ $\pm(^1/_3\ ^2/_3\ ^1/_4)$
Mn: $(4f)$ $\pm(^1/_3\ ^2/_3\ u; \ ^2/_3, ^1/_3, u+^1/_2)$ with $u = 0.61$
O(1): $(6g)$ $^1/_2 0\ 0;\ 0\ ^1/_2 0; ^1/_2\ ^1/_2 0;$
                  $^1/_2 0\ ^1/_2; 0\ ^1/_2\ ^1/_2; ^1/_2\ ^1/_2\ ^1/_2$
O(2): $(6h)$ $\pm(u\ 2u\ ^1/_4; 2\bar{u}\ \bar{u}\ ^1/_4; u\ \bar{u}\ ^1/_4)$ with $u = \ ^5/_6$

In this arrangement (Fig. VIIA,32), Ba–O = 2.83 and 2.86 A., Mn–O = 1.94 and 2.10 A. Manganese atoms approach within 2.62 A. of one another.

VII,a30. Several years ago the compound $BaNiO_3$ was assigned a hexagonal structure closely related to that of perewskite (**VII,a21**). Its bimolecular cell has the edges:

$$a_0 = 5.580 \text{ A.}, \qquad c_0 = 4.832 \text{ A.}$$

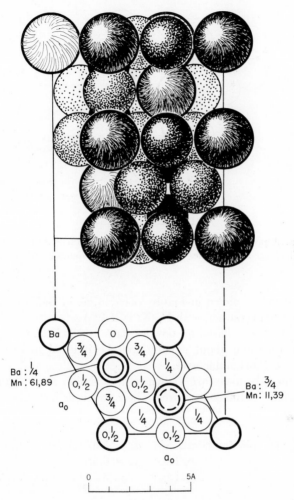

Fig. VIIA,32. Two projections of the hexagonal BaMnO₃ structure. In the upper shaded projection the manganese atoms are black, the bariums are large and fine-line shaded.

Atoms were placed in the following positions of C_{6v}^4 ($C6mc$):

Ba: (2*b*) $^2/_3\,^1/_3\,^3/_4$; $^1/_3\,^2/_3\,^1/_4$

Ni: (2*a*) 000; 0 0 $^1/_2$

O: (6*c*) $u\bar{u}v$; $u\,2u\,v$; $2\bar{u}\,\bar{u}\,v$;

  $\bar{u},u,v+^1/_2$; $\bar{u},2\bar{u},v+^1/_2$; $2u,u,v+^1/_2$ with $u = \,^1/_6$, $v = \,^1/_4$

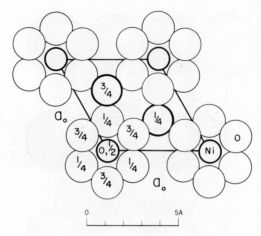

Fig. VIIA,33a. A projection along the c_0 axis of the hexagonal structure of BaNiO$_3$. The large heavy circles are barium atoms. Origin in lower left.

In the arrangement shown in Figure VIIA,33, as in BaNiO$_2$ (**VI,a26**), the nickel atoms are in contact, with Ni–Ni = 2.42 A.

This structure was based on powder data and it has since been shown that the pattern can about equally well be indexed in terms of either tetragonal or orthorhombic axes.

Three other substances are known to give similar x-ray data and hence probably to have similar structures. Their unit cells in terms of hexagonal axes are

$$\text{BaMnO}_3 \text{ (low-temperature)}: \quad a_0 = 5.672 \text{ A.}, \; c_0 = 4.71 \text{ A.}$$
$$\text{SrTiS}_3: \quad a_0 = 6.590 \text{ A.}, \; c_0 = 5.707 \text{ A.}$$
$$\text{BaTiS}_3: \quad a_0 = 6.730 \text{ A.}, \; c_0 = 5.829 \text{ A.}$$

The intensity data from these sulfides are reported to give no more than indifferent agreement with the hexagonal atomic arrangement chosen for BaNiO$_3$, and it has therefore been concluded that the correct structure may rather be orthorhombic. Evidently, more work should be carried out on all these crystals.

VII,a31. The oxygen atoms in *lithium antimony trioxide*, LiSbO$_3$, are close-packed like the large ions in perewskite (**VII,a21**), but in this case the

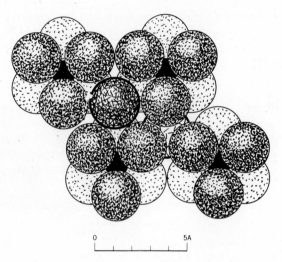

0   5A

Fig. VIIA,33b. A packing drawing of the hexagonal $BaNiO_3$ structure viewed along its c_0 axis. The barium atoms are line shaded, the nickel atoms are small and black.

packing is hexagonal rather than cubic. The symmetry is orthorhombic, with a tetramolecular cell of the dimensions:

$$a_0 = 4.893 \text{ A.}; \quad b_0 = 8.491 \text{ A.}; \quad c_0 = 5.183 \text{ A.}$$

Its atoms have been placed in the following positions of V_h^6 ($Pncn$):

(4c) $\pm (0 \; u \; ^1/_4; \; ^1/_2, ^1/_2 - u, ^1/_4)$
(4d) $\pm (^1/_4 \; ^1/_4 \; u; \; ^1/_4, ^3/_4, u + ^1/_2)$
(8e) $\pm (xyz; \; x, \bar{y}, z + ^1/_2; \; x + ^1/_2, y + ^1/_2, \bar{z}; \; ^1/_2 - x, y + ^1/_2, z + ^1/_2)$

with the parameters:

Atom	Position	x	y	z
Li	(4c)	0	0.730	$^1/_4$
Sb	(4c)	0	0.097	$^1/_4$
O(1)	(4d)	$^1/_4$	$^1/_4$	0.444
O(2)	(8e)	0.209	0.081	0.907

The resulting structure is illustrated in Figure VIIA,34. Around each antimony atom is an octahedron of oxygen atoms with Sb–O = 2.00–2.05 A. Along the c_0 axis these octahedra are joined into strings by sharing edges, the lengths of these O–O edges being from 2.61 A. upwards. The lithium atoms lie in holes within the approximately hexagonal close-packing of the oxygens; they are thus also octahedrally surrounded by six oxygens, with Li–O = 2.01–2.07 A.

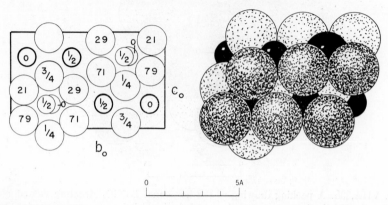

Fig. VIIA,34a (left). A projection along the a_0 axis of the orthorhombic structure of LiSbO₃. The antimony are the heavy, the lithium the light small circles. Origin in lower right.

Fig. VIIA,34b (right). The orthorhombic structure of LiSbO₃ viewed along its a_0 axis. The oxygen atoms are dotted; the antimony are the smaller of the black circles.

VII,a32. The mineral *ilmenite*, $FeTiO_3$, and several other compounds isomorphous with it have an arrangement almost identical with the Cr_2O_3 grouping described in **V,a3**. It differs only in having half the metallic atoms unlike the other half and in the lower symmetry this entails. There are two molecules in the unit rhombohedron having the dimensions:

$$a_0 = 5.538 \text{ A.}, \qquad \alpha = 54°41$$

The space group is $C_{3i}{}^2$ ($R\overline{3}$), with atoms in the positions:

Fe: (2c) $\pm(uuu)$ with $u = 0.358$
Ti: (2c)   with $u = 0.142$
O: (6f) $\pm(xyz; zxy; yzx)$ with $x = 0.555, y = -0.040, z = 0.235$

The hexamolecular cell referred to hexagonal axes has the edges:

$$a_0' = 5.082 \text{ A.}, \qquad c_0' = 14.026 \text{ A.}$$

The iron and titanium atoms in (6c) $\pm(00u)$; rh have $u(\text{Fe}) = 0.358$, $u(\text{Ti}) = 0.142$, as before. The oxygen atoms with the coordinates:

$$(18f) \qquad \pm(xyz; \bar{y}, x-y, z; y-x, \bar{x}, z); \text{rh}$$

have $x = 0.305, y = 0.015, z = 0.250$.

Like the simple oxide arrangement, this is a structure that is best looked on as a nearly perfect hexagonal close-packing of the oxygen atoms with the metallic atoms in interstices. The distribution of these atoms is shown in Figure VIIA,35. As would be expected from their similar structures, there is extensive solid solution possible between compounds with the oxide and ilmenite arrangements.

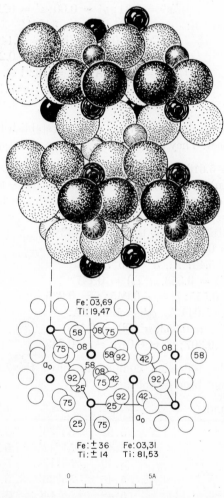

Fig. VIIA,35. Two projections of the rhombohedral structure of FeTiO₃ in terms of its hexagonal axes. In the packing drawing the iron atoms are black, the titanium atoms small and fine-line shaded.

Other compounds for which complete determinations have been made are

$$NiMnO_3: \quad a_0 = 5.343 \text{ A.}, \ \alpha = 54°39'$$
$$u(Ni) = 0.352, \ u(Mn) = 0.148$$
$$x = 0.56; \ y = -0.06; \ z = 0.25$$
$$a_0' = 4.905 \text{ A.}, \ c_0' = 13.59 \text{ A.}$$
$$u'(Ni) = 0.352, \ u'(Mn) = 0.148$$
$$x' = 0.31; \ y' = 0.00; \ z' = 0.25$$
$$CoMnO_3: \quad a_0 = 5.385 \text{ A.}, \ \alpha = 54°31'$$
$$u(Co) = 0.354, \ u(Mn) = 0.146$$
$$x = 0.57; \ y = -0.07; \ z = 0.25$$
$$a_0' = 4.933 \text{ A.}, \ c_0' = 13.91 \text{ A.}$$
$$u'(Co) = 0.354, \ u'(Mn) = 0.146$$
$$x' = 0.32; \ y' = 0.00; \ z' = 0.25$$

Neutron diffraction data have been used to establish the parameters of two more substances. They are

$$MnTiO_3: \quad a_0 = 5.610 \text{ A.}, \ \alpha = 54°30'$$
$$u(Mn) = 0.357; \ u(Ti) \text{ presumably is } 1/2 - u(Mn) = 0.143$$
For oxygen, $x = 0.560, \ y = -0.050, \ z = 0.220.$
$$a_0' = 5.137 \text{ A.}, \ c_0' = 14.283 \text{ A.}$$
$$u'(Mn) = 0.357, \ u'(Ti) = 0.143$$
$$x' = 0.317; \ y' = 0.023; \ z' = 0.243$$
$$NiTiO_3: \quad a_0 = 5.437 \text{ A.}, \ \alpha = 55°7'$$
$$u(Ni) = 0.353, \ u(Ti) = 0.147$$
$$x = 0.555; \ y = -0.045; \ z = 0.235$$
$$a_0' = 5.044 \text{ A.}, \ c_0' = 13.819 \text{ A.}$$
$$u'(Ni) = 0.353, \ u'(Ti) = 0.147$$
$$x' = 0.307; \ y' = 0.013; \ z' = 0.248$$

In addition, the following compounds have this structure, though parameters have not been established:

Compound	a_0, A.	α	a_0', A.	c_0', A.
CdTiO₃ (low)	5.82	53°36'	5.248	14.906
CoTiO₃	5.49	54°42'	5.044	13.961
CrRhO₃	5.42	55°12'	5.022	13.737
FeRhO₃	5.46	55°26'	5.079	13.817
FeVO₃	5.42	55°14'	5.026	13.735
LiNbO₃	5.47	55°43'	5.112	13.816
MgGeO₃ (high-pressure)	5.40	54°24'	4.936	13.76
MgTiO₃	5.54	54°39'	5.054	13.898

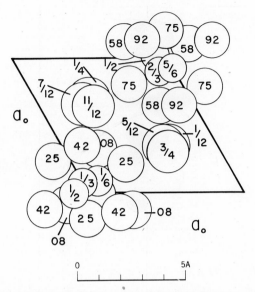

Fig. VIIA,36. A projection along the c_0 axis of the hexagonal structure of CsCuCl₃. The largest circles are cesium, the smallest circles copper atoms.

Halides, etc.

VII,a33.   Crystals of *cesium cupric chloride*, CsCuCl₃, are hexagonal, with a unit containing six molecules and having the edges:

$$a_0 = 7.20 \text{ A.,} \qquad c_0 = 18.00 \text{ A.}$$

The atoms have been found to be in the following positions of $D_6{}^2$ ($C6_12$):

Cu: (6a) $u\,0\,0;\ 0\,u\,{}^1/_3;\ \bar{u}\,\bar{u}\,{}^2/_3;$
$\bar{u}\,0\,{}^1/_2; 0\,\bar{u}\,{}^5/_6; u\,u\,{}^1/_6$ with $u = 0.07$

Cs: (6b) $u\,2u\,{}^1/_4; 2\bar{u}\,\bar{u}\,{}^7/_{12}; u\,\bar{u}\,{}^{11}/_{12};$
$\bar{u}\,2\bar{u}\,{}^3/_4; 2u\,u\,{}^1/_{12}; \bar{u}\,u\,{}^5/_{12}$ with $u = 0.345$

Cl(1):   (6b)   with $u = 0.90$
Cl(2):   (12c) $xyz;\qquad \bar{y},x-y,z+{}^1/_3; y-x,\bar{x},z+{}^2/_3;$
$\bar{x},\bar{y},z+{}^1/_2; y,y-x,z+{}^5/_6; x-y,x,z+{}^1/_6;$
$y,x,{}^1/_3-z; \bar{x},y-x,{}^2/_3-z; x-y,\bar{y},\bar{z};$
$\bar{y},\bar{x},{}^5/_6-z; x,x-y,{}^1/_6-z; y-x,x,{}^1/_2-z$
with $x = 0.35,\ y = 0.22,\ z = 0.25$

In this structure (Fig. VIIA,36) the cesium and chlorine atoms together are in approximately close-packed layers, with the copper atoms lying between successive layers. The Cs–Cl separations range from 3.40 to 3.82 A.

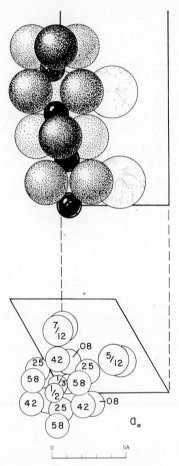

Fig. VIIA,37. Two projections of a chain of copper and chlorine atoms and its nearby cesium atoms in the CsCuCl₃ structure. In the upper shaded portion the cesium atoms are large and fine-line shaded, the copper are black.

Each copper atom is surrounded by two chlorine atoms at a distance of 2.27 A., by two others nearly as close (2.30 A.), and by two more distant (2.65 A.). This association of copper and chlorine atoms is such as to suggest that they form a chain ion of the sort shown in Figure VIIA,37.

Cesium nickel trichloride, $CsNiCl_3$, probably has this structure, with

$$a_0 = 7.18 \text{ A.}, \quad c_0 = 17.79 \text{ A.}$$

On the basis of electron diffraction measurements, however, both it and the cupric salt have been ascribed a smaller unit one-third as high. For the bimolecular cell that results:

$$a_0 = 7.18 \text{ A.}, \quad c_0 = 5.93 \text{ A.}$$

In the simplified structure described for it and based on D_{6h}^4 ($C6/mmc$), atoms are in the positions:

Ni: (2a) $000; 0\,0\,^1/_2$
Cs: (2d) $\pm(^2/_3\,^1/_3\,^1/_4)$
Cl: (6h) $\pm(u\,2u\,^1/_4; 2\bar{u}\,\bar{u}\,^1/_4; u\,\bar{u}\,^1/_4)$ with $u = 0.156$

In this arrangement (Fig. VIIA,38), the closest Ni–Cl = 2.43 A., and Cs–Cl = 3.59 A.

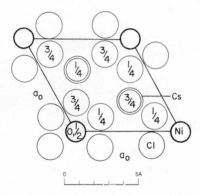

Fig. VIIA,38. A basal projection of the simpler structure ascribed to $CsNiCl_3$.

This simpler structure has also been ascribed to the following:

$CsVCl_3$: $a_0 = 7.23$ A., $c_0 = 6.03$ A.
$KVCl_3$: $a_0 = 6.90$ A., $c_0 = 5.98$ A.

It would be important through further work to determine if both these closely related arrangements actually exist as structures for these compounds.

VII,a34.   The tetragonal *ammonium mercuric chloride*, NH_4HgCl_3, has the unimolecular cell:

$$a_0 = 4.19 \text{ A.}, \qquad c_0 = 7.94 \text{ A.}$$

Atoms have been assigned the following special positions of D_{4h}^1 ($P4/mmm$):

Hg:	($1a$)	000
NH_4:	($1d$)	$\frac{1}{2}\,\frac{1}{2}\,\frac{1}{2}$
Cl(1):	($1c$)	$\frac{1}{2}\,\frac{1}{2}\,0$
Cl(2):	($2g$)	$\pm(00u)$ with $u = 0.294$

In such a grouping (Fig. VIIA,39), each mercury atom will have about it a distorted octahedron of chlorine atoms, two of which are shared with no other mercury atom [Hg–Cl(2) = 2.34 A.], the other four being shared by four different mercury atoms [Hg–Cl(1) = 2.96 A.]. Each ammonium ion has eight chlorine neighbors with NH_4–Cl(2) = 3.38 A. The Cl(2) atoms are distant from one another by the unusually short 3.26 A.

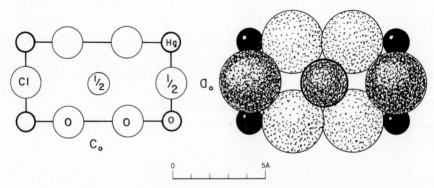

Fig. VIIA,39a (left). A projection along an a_0 axis of the tetragonal structure of NH_4-$HgCl_3$. Origin in lower right.

Fig. VIIA,39b (right). The tetragonal structure of NH_4HgCl_3 viewed along an a_0 axis. The ammonium group is line shaded, the chlorine atoms are dotted.

VII,a35.   Crystals of *potassium mercury isothiocyanate*, $KHg(SCN)_3$, and of the corresponding *ammonium* salt, $NH_4Hg(SCN)_3$, are monoclinic with the bimolecular cells:

$$KHg(SCN)_3: \quad a_0 = 11.081 \text{ A.}; \, b_0 = 4.07 \text{ A.}; \, c_0 = 10.915 \text{ A.}$$
$$\beta = 114°45'$$
$$NH_4Hg(SCN)_3: \quad a_0 = 11.185 \text{ A.}; \, b_0 = 4.08 \text{ A.}; \, c_0 = 10.935 \text{ A.}$$
$$\beta = 114°55'$$

The space group is C_{2h}^2 ($P2_1/m$) with all atoms in the special positions (2e): $\pm(u\,^1/_4\,v)$. Parameters assigned the atoms in the two compounds are listed in Table VIIA,10.

The resulting structure is shown in Figure VIIA,40. In it the mercury atoms make contact with the sulfur atoms of the isocyanate radicals with

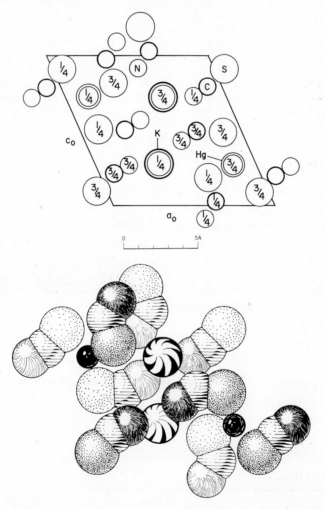

Fig. VIIA,40a (top). A projection along its b_0 axis of the monoclinic structure of KHg-(SCN)₃. Origin in lower left.

Fig. VIIA,40b (bottom). A shaded drawing of the projection of KHg(SCN)₃ shown in Figure VIIA,40a. The mercury atoms are small and black. In the SCN radicals the sulfur atoms are dotted and the nitrogen atoms fine-line shaded.

TABLE VIIA,10
Parameters of the Atoms in KHg(SCN)$_3$ and NH$_4$Hg(SCN)$_3$

Atom	KHg(SCN)$_3$		NH$_4$Hg(SCN)$_3$	
	u	v	u	v
Hg	0.146	0.727	0.147	0.723
K or NH$_4$	0.400	0.278	0.415	0.273
S(1)	0.120	0.512	0.122	0.509
S(2)	0.667	0.187	0.669	0.193
S(3)	0.063	0.901	0.071	0.904
N(1)	0.398	0.567	0.393	0.566
N(2)	0.610	0.914	0.611	0.924
N(3)	0.795	0.731	0.805	0.729
C(1)	0.282	0.549	0.280	0.552
C(2)	0.646	0.028	0.643	0.033
C(3)	0.911	0.791	0.917	0.812

Hg–S(1,3) = 2.51 and 2.45 A. and Hg–S(2) = 2.78 A. There is no Hg-(SCN)$_3$ ion, but instead each Hg has two S(2) atoms at the 2.78 A. distance as well as the two nearer S(1) and S(3) atoms. The distance between the alkali ions and the nitrogen atoms of the SCN radicals lies between 2.90 and 3.18 A.

VII,a36. *Ammonium cadmium chloride,* NH$_4$CdCl$_3$, and its rubidium analogue, RbCdCl$_3$, are orthorhombic. Their tetramolecular cells have the dimensions:

NH$_4$CdCl$_3$:　a_0 = 14.90 A.; b_0 = 9.00 A.; c_0 = 3.96 A.
RbCdCl$_3$:　a_0 = 14.93 A.; b_0 = 9.01 A.; c_0 = 4.01 A.

The space group is V$_h^{16}$ (*Pbnm*) and it has been concluded that all atoms are in special positions:

$$(4c) \quad \pm (u\,v\,^1/_4;\ ^1/_2-u,v+^1/_2,^1/_4)$$

with the following parameters for NH$_4$CdCl$_3$:

Atom	u	v
NH$_4$	0.828	0.425
Cd	0.057	0.164
Cl(1)	0.208	0.286
Cl(2)	0.494	0.172
Cl(3)	0.926	0.021

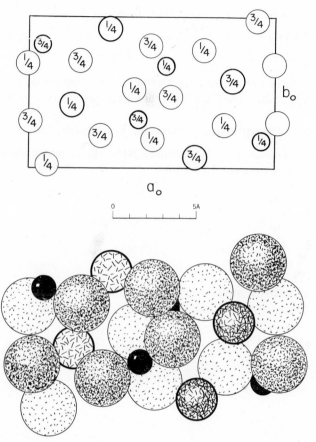

Fig. VIIA,41a (top). A projection along the c_0 axis of the orthorhombic structure of NH$_4$CdCl$_3$. The larger heavy circles are ammonium ions, the small heavy circles cadmium atoms. Origin in lower right.

Fig. VIIA,41b (bottom). The orthorhombic structure of NH$_4$CdCl$_3$ viewed along its c_0 axis. The ammonium groups are line shaded, the chlorine atoms dotted.

In this arrangement (Fig. VIIA,41), each cadmium atom is surrounded by a somewhat distorted octahedron of chlorine atoms at distances between 2.60 and 2.72 A. The structure as a whole can be represented as strings of pairs of these octahedra extending along the c_0 axis, the individual octahedra in a string sharing edges with one another. On this basis the strings themselves are held together to form the crystal by the large alkali ions that lie between them. The coordination of the ammonium ions is not very clearcut, but each has about it nine chlorine atoms at distances between 3.27 and 3.82 A.

Two other salts have this structure. They are

$$KCdCl_3: \quad a_0 = 14.56 \text{ A.}; \, b_0 = 8.78 \text{ A.}; \, c_0 = 3.99 \text{ A.}$$
$$NaHgCl_3: \quad a_0 = 18.3 \text{ A.}; \quad b_0 = 9.48 \text{ A.}; \, c_0 = 4.2 \text{ A.}$$

Parameters have been stated for $NaHgCl_3$, but they are so different from the confirmed values for NH_4CdCl_3 that a further study seems necessary.

Phosphates, Arsenates, Vanadates, etc.

VII,a37. *Ammonium metavanadate,* NH_4VO_3, is orthorhombic with a tetramolecular unit of the dimensions:

$$a_0 = 4.902 \text{ A.}; \, b_0 = 11.79 \text{ A.}; \, c_0 = 5.827 \text{ A.}$$

The space group is V_h^{11} (*Pbcm*), with atoms in the positions:

$$(4c) \quad \pm (u \, ^1/_4 \, 0; \, u \, ^1/_4 \, ^1/_2)$$
$$(4d) \quad \pm (u \, v \, ^1/_4; \, u, ^1/_2 - v, ^3/_4)$$

For this repeatedly studied crystal, the most recent parameters are as follows:

Atom	Position	x	y	z
V	(4d)	0.4636	0.1736	$^1/_4$
N	(4d)	−0.063	0.412	$^1/_4$
O(1)	(4c)	0.578	$^1/_4$	0
O(2)	(4d)	0.123	0.167	$^1/_4$
O(3)	(4d)	0.572	0.044	$^1/_4$

The resulting structure, as shown in Figure VIIA,42, is one that surrounds each vanadium atom by a tetrahedron of oxygen atoms with V–O = 1.64–1.80 A. Around each NH_4 ion are six closest oxygens at distances between 2.83 and 3.42 A. This is an arrangement of tetrahedral atomic strings which in some way resembles that of the pyroxenes (**XII,b1** loose-leaf). Among the silicates it is possessed by the high-temperature form of *barium metasilicate*, $BaSiO_3$, for which

$$a_0 = 4.54 \text{ A.}; \quad b_0 = 12.27 \text{ A.}; \quad c_0 = 5.56 \text{ A.}$$

The high-temperature modification of *barium metagermanate*, $BaGeO_3$, is isomorphous, with

$$a_0 = 4.58 \text{ A.}; \quad b_0 = 12.76 \text{ A.}; \quad c_0 = 5.68 \text{ A.}$$

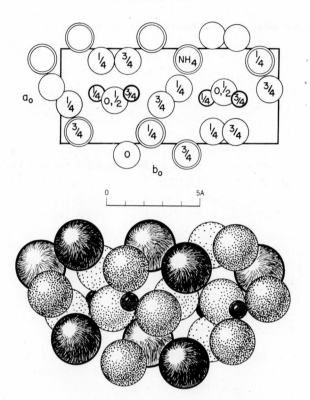

Fig. VIIA,42a (top). A projection along c_0 of the orthorhombic NH_4VO_3 arrangement. Origin in lower left.
Fig. VIIA,42b (bottom). A packing drawing of NH_4VO_3 seen along its c_0 axis. The vanadium atoms are small and black, the NH_4 ions are fine-line shaded.

Potassium metavanadate, KVO_3, with this structure has, like the ammonium salt, been thoroughly studied. For it, the unit has the edge lengths:

$$a_0 = 5.22 \text{ A.}; \quad b_0 = 10.82 \text{ A.}; \quad c_0 = 5.70 \text{ A.}$$

and the parameters:

Atom	Position	x	y	z
K	(4d)	0.4768	0.1624	$1/4$
V	(4d)	-0.0629	0.3948	$1/4$
O(1)	(4c)	0.587	$1/4$	0
O(2)	(4d)	0.160	0.148	$1/4$
O(3)	(4d)	0.606	0.023	$1/4$

Two other alkali metavanadates with this structure are

$$RbVO_3: \quad a_0 = 4.99 \text{ A.}; \quad b_0 = 11.98 \text{ A.}; c_0 = 5.654 \text{ A.}$$
$$CsVO_3: \quad a_0 = 5.385 \text{ A.}; \quad b_0 = 12.20 \text{ A.}; c_0 = 5.788 \text{ A.}$$

The corresponding four *alkali beryllium trifluorides* are crystallographically isomorphous with these metavanadates, their cells having the dimensions:

$$NH_4BeF_3: \quad a_0 = 4.610 \text{ A.}; b_0 = 12.85 \text{ A.}; c_0 = 5.777 \text{ A.}$$
$$KBeF_3: \quad a_0 = 4.53 \text{ A.}; \quad b_0 = 12.22 \text{ A.}; c_0 = 5.53 \text{ A.}$$
$$RbBeF_3: \quad a_0 = 4.53 \text{ A.}; \quad b_0 = 12.55 \text{ A.}; c_0 = 5.81 \text{ A.}$$
$$CsBeF_3: \quad a_0 = 4.84 \text{ A.}; \quad b_0 = 12.82 \text{ A.}; c_0 = 6.06 \text{ A.}$$

It would be natural to conclude that these four substances have the NH_4VO_3 structure. Nevertheless, it has recently been stated that NH_4BeF_3 actually is monoclinic and that each of the others, though orthorhombic, has a structure based on a different space group. For *rubidium beryllium trifluoride*, $RbBeF_3$, this space group is V^4 ($P2_12_12_1$), with all atoms in the general positions:

$$(4a) \quad xyz; \; {}^1/_2-x,\bar{y},z+{}^1/_2; \; x+{}^1/_2,{}^1/_2-y,\bar{z}; \; \bar{x},y+{}^1/_2,{}^1/_2-z$$

The parameters assigned these atoms are

Atom	x	y	z
Rb	0.25	0.143	0.047
F(1)	0.25	0.457	0.024
F(2)	0.16	0.149	0.340
F(3)	−0.20	0.399	0.263

Positions were not given the beryllium atoms, but it was considered probable that they, too, are near the centers of tetrahedra formed by the fluorines.

It seems evident that additional work is needed to determine if these substances actually do not have the same structure as the vanadates.

VII,a38. *Ammonium tetrametaphosphate*, $(NH_4)_4P_4O_{12}$, is orthorhombic, with a unit containing four of these molecules and having the edge lengths:

$$a_0 = 10.42 \text{ A.}; \quad b_0 = 10.82 \text{ A.}; \quad c_0 = 12.78 \text{ A.}$$

TABLE VIIA,11

Positions and Parameters of the Atoms in $(NH_4)_4P_4O_{12}$

Atom	Position	x	y	z
P(1)	(8f)	0	0.036	0.150
P(2)	(8d)	0.211	0	0
O(1)	(8f)	0	−0.100	0.154
O(2)	(8f)	0	0.121	0.236
O(3)	(16g)	0.283	−0.095	0.064
O(4)	(16g)	0.120	0.078	0.075
NH_4(1)	(8e)	$\frac{1}{4}$	0.272	$\frac{1}{4}$
NH_4(2)	(8f)	0	0.319	−0.044

Atoms are in the following special and general positions of V_h^{18} $(Cmca)$:

(8d) $\pm (u00; u \, ^1/_2 \, ^1/_2; u+^1/_2,^1/_2,0; u+^1/_2,0,^1/_2)$

(8f) $\pm (0uv; \, ^1/_2,u,^1/_2-v; \, ^1/_2,u+^1/_2,v; \, 0,u+^1/_2,^1/_2-v)$

(8e) $\pm (^1/_4 \, u \, ^1/_4; \, ^3/_4 \, u \, ^1/_4; \, ^3/_4,u+^1/_2,^1/_4; \, ^1/_4,u+^1/_2,^1/_4)$

(16g) $\pm (xyz; \, x,^1/_2-y,z+^1/_2; \, x+^1/_2,y+^1/_2,z; \, x+^1/_2,\bar{y},z+^1/_2;$
    $\bar{x}yz; \, \bar{x},^1/_2-y,z+^1/_2; \, ^1/_2-x,y+^1/_2,z; \, ^1/_2-x,\bar{y},z+^1/_2)$

The atomic positions and parameters as determined are listed in Table VIIA,11.

They lead to a structure (Fig. VIIA,43) made up of P_4O_{12} ions in a face-centered array and interleaved with layers of NH_4 cations. The P_4O_{12} ions are groups of four PO_4 tetrahedra sharing an oxygen with one another (Fig. VIIA,44). The oxygen tetrahedra are regular; the phosphorus atoms, however, are not central but farther from the shared oxygen [O(4)]. In the two crystallographically different tetrahedra that exist, P(1)–O(4) = 1.63 A., P(1)–O(1) = 1.48 A., P(1)–O(2) = 1.44 A., P(2)–O(4) = 1.59 A., and P(2)–O(3) = 1.51 A. Each NH_4 ion has four especially near oxygen neighbors [NH_4(1)–O = 2.79 A., 3.03 A. and NH_4(2)–O = 2.80 A., 2.82 A., 2.85 A.] and these suggest that there may be some hydrogen bonding. Other oxygens are not much more distant; thus NH_4(1) has four more neighbors at 3.14 and 3.38 A. and NH_4(2) two more at 3.25 A.

VII,a39. Crystals of *lithium arsenate*, $(LiAsO_3)_n$, are monoclinic, with a unit containing eight molecules and having the dimensions:

$$a_0 = 10.18 \text{ A.}; \quad b_0 = 9.43 \text{ A.}; \quad c_0 = 5.25 \text{ A.}; \quad \beta = 110°32'$$

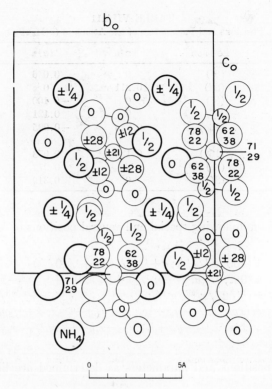

Fig. VIIA,43a. A projection along the a_0 axis of the orthorhombic structure of $(NH_4)_4P_4O_{12}$. The smallest circles are phosphorus. Origin in lower right.

Atoms are in the following positions of C_{2h}^6 $(C2/c)$:

$(4e)$ $\pm(0\ u\ ^1/_4;\ ^1/_2,u+^1/_2,^1/_4)$

$(8f)$ $\pm(xyz;\ \bar{x},y,^1/_2-z;\ x+^1/_2,y+^1/_2,z;\ ^1/_2-x,y+^1/_2,^1/_2-z)$

TABLE VIIA,12

Positions and Parameters of the Atoms in $(LiAsO_3)_n$

Atom	Position	x	y	z
Li(1)	$(4e)$	0	−0.111	$^1/_4$
Li(2)	$(4e)$	0	0.220	$^1/_4$
As	$(8f)$	0.206	0.404	0.732
O(1)	$(8f)$	0.372	0.409	0.881
O(2)	$(8f)$	0.125	0.255	0.650
O(3)	$(8f)$	0.131	0.475	0.952

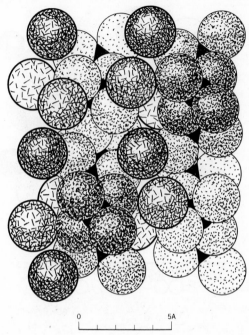

0   5A

Fig. VIIA,43b. The orthorhombic structure of $(NH_4)_4P_4O_{12}$ viewed along its a_0 axis. The ammonium groups are line shaded, the oxygen atoms dotted.

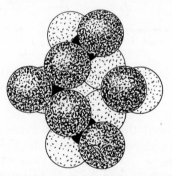

Fig. VIIA,44. A P_4O_{12} group (viewed along the b_0 axis) showing its general shape.

The chosen parameters are those of Table VIIA,12. This is a structure (Fig. VIIA,45) closely resembling that of diopside, $CaMg(SiO_3)_2$. The AsO_3 chains extending indefinitely along the c_0 direction are made up of AsO_4 tetrahedra that share corners. The As–O separations in the tetrahedra lie between 1.60 and 1.81 A. The close Li–O distances are 1.93, 2.04, and 2.06 A.

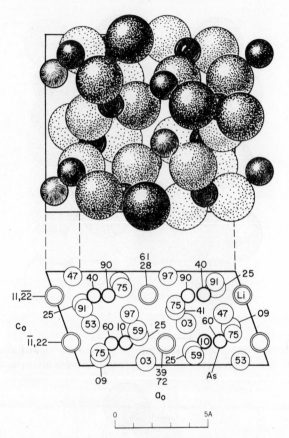

Fig. VIIA,45. Two projections of the monoclinic $(LiAsO_3)_n$ structure viewed along its b_0 axis. In the shaded upper projection the arsenic atoms are black, the lithium small and fine-line shaded.

It is very possible that this is the same type of structure as that of klino-enstatite, $MgSiO_3$ and of the isomorphous $MgGeO_3$. For the latter,

$$a_0 = 9.60 \text{ A.}; \quad b_0 = 8.92 \text{ A.}; c_0 = 5.16 \text{ A.}; \quad \beta = 100°49'$$

The other form of $MgGeO_3$ is enstatite-like.

VII,a40. *Sodium metaphosphate*, $(NaPO_3)_n$, in the modification that is one of the two forms of Kurrol's salt is monoclinic, with an eight-molecule cell of the dimensions:

$$a_0 = 12.12 \text{ A.}; \quad b_0 = 6.20 \text{ A.}; \quad c_0 = 6.99 \text{ A.}; \quad \beta = 92°$$

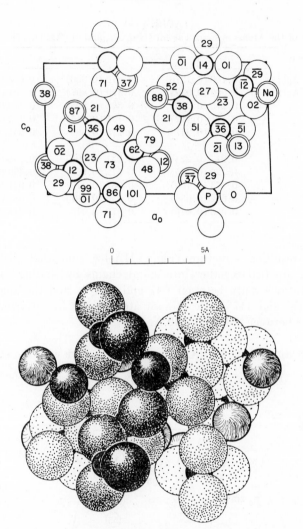

Fig. VIIA,46a (top). A projection along b_0 of the monoclinic structure of $(NaPO_3)_n$.
Origin in lower left.
Fig. VIIA,46b (bottom). A shaded drawing of the basal projection of $(NaPO_3)_n$ shown
in Figure VIIA,46a. The phosphorus atoms are black, the sodium are fine-line shaded.

The space group is C_{2h}^5 ($P2_1/n$) and all atoms are in its general positions:

$$(4e) \quad \pm(xyz; \; x+\tfrac{1}{2}, \tfrac{1}{2}-y, z+\tfrac{1}{2})$$

with the parameters of Table VIIA,13.

TABLE VIIA,13
Parameters of the Atoms in Kurrol's Salt $(NaPO_3)_n$ and of $(AgPO_3)_n$ (in parentheses)

Atom	x	y	z
Na or Ag(1)	0.130 (0.1276)	0.870 (0.336)	0.629 (0.613)
Na or Ag(2)	0.006 (0.0305)	0.379 (0.897)	0.764 (0.770)
P(1)	0.215 (0.226)	0.362 (0.825)	0.489 (0.486)
P(2)	0.111 (0.112)	0.125 (0.615)	0.182 (0.180)
O(1)	0.228 (0.247)	0.215 (0.669)	0.654 (0.634)
O(2)	0.217 (0.224)	0.231 (0.695)	0.286 (0.298)
O(3)	0.120 (0.125)	0.512 (0.963)	0.499 (0.489)
O(4)	0.173 (0.160)	0.987 (0.470)	0.021 (0.027)
O(5)	0.042 (0.053)	0.290 (0.797)	0.087 (0.088)
O(6)	0.056 (0.049)	0.978 (0.479)	0.314 (0.301)

In the resulting structure (Fig. VIIA,46) somewhat distorted PO_4 tetrahedra are tied together into spiral chains by sharing corners. The P–O separations range between 1.47 and 1.64 A. and the angles O–P–O between 99° and 117°. The sodium atoms have five oxygen neighbors with Na–O between 2.35 and 2.49 A.

Silver metaphosphate, $(AgPO_3)_n$, has this structure with the cell:

$$a_0 = 11.86 \text{ A.}; \quad b_0 = 6.06 \text{ A.}; \quad c_0 = 7.31 \text{ A.}; \quad \beta = 93°30'$$

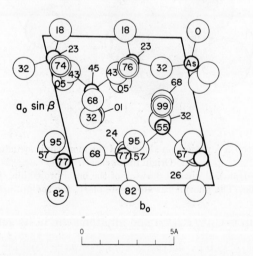

Fig. VIIA,47a. A projection along its c_0 axis of the triclinic structure of $(NaAsO_3)_n$. Origin in lower left.

The parameters are given in parentheses in Table VIIA,13. They are similar to those for the sodium salt except for a displacement of $1/2$ along the b_0 axis.

VII,a41. *Sodium arsenate*, $(NaAsO_3)_n$, is triclinic, with a unit containing six molecules and having the dimensions:

$$a_0 = 8.07 \text{ A.}; \; b_0 = 7.44 \text{ A.}; \; c_0 = 7.32 \text{ A.}$$
$$\alpha = 90°; \quad \beta = 91°30'; \; \gamma = 104°$$

The atoms, in general positions $(2i)$ $\pm(xyz)$ of C_i^1 $(P\bar{1})$, have the parameters of Table VIIA,14.

As Figure VIIA,47 indicates, there are endless chains of AsO_4 tetrahedra linked together by sharing corners running parallel to the b_0 axis. The orientation of the tetrahedra in a chain repeats itself every fourth AsO_4—in contrast to the $(MX_3)_n$ chains in the diopside structure, where the repetition is every third tetrahedron.

In the AsO_4 tetrahedra, As–O lies between 1.67 and 1.79 A. The sodium atoms are octahedrally surrounded by oxygens at distances that range from 2.28 to 2.61 A.

This structure differs from the one determined some time ago for β-wollastonite, $CaSiO_3$, though the two substances resemble one another closely. Undoubtedly, a restudy should be made of the silicate.

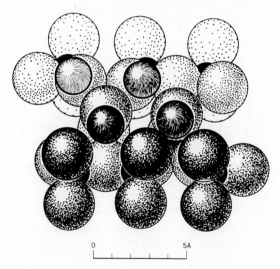

Fig. VIIA,47b. A packing drawing of the $(NaAsO_3)_n$ arrangement viewed along its c_0 axis. The arsenic atoms are black, the sodiums fine-line shaded.

TABLE VIIA,14
Parameters of the Atoms in $(NaAsO_3)_n$

Atom	x	y	z
As(1)	0.1710	0.387	0.7708
As(2)	0.1710	0.956	0.7708
As(3)	0.3904	0.723	0.5480
O(1)	0.291	0.495	0.946
O(2)	0.291	0.911	0.946
O(3)	−0.033	0.370	0.819
O(4)	−0.033	0.865	0.819
O(5)	0.214	0.182	0.684
O(6)	0.242	0.507	0.568
O(7)	0.242	0.865	0.568
O(8)	0.424	0.732	0.318
O(9)	0.568	0.770	0.685
Na(1)	0.221	0.426	0.243
Na(2)	0.204	0.932	0.255
Na(3)	0.528	0.759	0.992

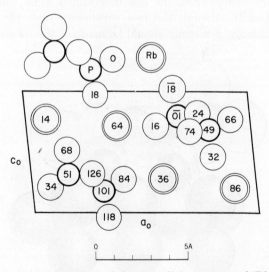

Fig. VIIA,48a. A projection along b_0 of the monoclinic structure of $(RbPO_3)_n$. Origin in lower left.

VII,a42. Crystals of *rubidium metaphosphate*, $(RbPO_3)_n$, are monoclinic, with a tetramolecular unit having the dimensions:

$$a_0 = 12.123 \text{ A.}; \quad b_0 = 4.228 \text{ A.}; \quad c_0 = 6.479 \text{ A.}; \quad \beta = 96°19'$$

Atoms are in the general positions of C_{2h}^5 $(P2_1/n)$:

$$(4e) \quad \pm(xyz; \ x+{}^1/_2,{}^1/_2-y,z+{}^1/_2)$$

with the parameters:

Atom	x	y	z
Rb	0.093	0.137	0.777
P	0.183	0.510	0.314
O(1)	0.223	0.76	0.166
O(2)	0.081	0.34	0.215
O(3)	0.172	0.68	0.520

The structure is shown in Figure VIIA,48. In it, the $(PO_3)_n$ tetrahedra form endless chains parallel to the b_0 axis with the distances and angles of

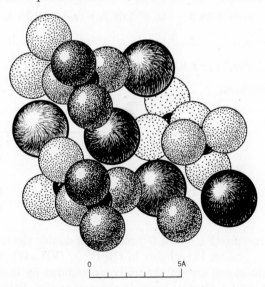

0   5A

Fig. VIIA,48b. A packing drawing of the projection along b_0 of the $(RbPO_3)_n$ arrangement. The phosphorus atoms are black, the rubidium atoms large and fine-line shaded.

Figure VIIA,49. The rubidium atoms which hold these chains together in the crystal have six oxygen neighbors with Rb–O $= 2.90$–3.02 A. and a seventh with Rb–O $= 3.20$ A. The $(PO_3)_n$ chains in this crystal are very similar to the $(SO_3)_n$ chains in one form of SO_3 and somewhat like the $(SiO_3)_n$ chains in metasilicates (Chapter XII).

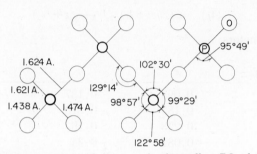

Fig. VIIA,49. The bond angles and distances in the endless PO_3 chains in $(RbPO_3)_n$.

VII,a43. The high-temperature form of *barium germanate*, $BaGeO_3$, is orthorhombic, with the tetramolecular unit:

$$a_0 = 4.58 \text{ A.}; \quad b_0 = 5.68 \text{ A.}; \quad c_0 = 12.76 \text{ A.}$$

The space group is V^4 ($P2_12_12_1$), with all atoms in the positions:

$$(4a) \quad xyz; \; {}^1\!/_2-x,\bar{y},z+{}^1\!/_2; \; x+{}^1\!/_2,{}^1\!/_2-y,\bar{z}; \; \bar{x},y+{}^1\!/_2,{}^1\!/_2-z$$

with the parameters:

Atom	x	y	z
Ba	0.224	0.187	0.359
Ge	0.675	0.690	0.425
O(1)	0.267	0.730	0.440
O(2)	0.738	0.885	0.326
O(3)	0.730	0.394	0.410

This is a structure (Fig. VIIA,50) containing double GeO_3 chains that are like the corresponding PO_3 chains in $(RbPO_3)_n$ **(VII,a42)**. Within a chain the germanium atoms are tetrahedrally surrounded by four oxygens with Ge–O $= 1.70$–1.89 A. As the figure demonstrates, the chains are bound together in the crystal with each barium atom surrounded by seven oxygens at distances between 2.62 and 2.99 A.

It is probable that the high-temperature modification of $BaSiO_3$ also has this arrangement.

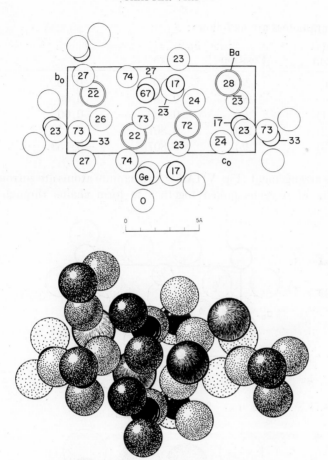

Fig. VIIA,50a (top). The orthorhombic structure of BaGeO₃ projected along its a_0 axis. Origin in lower left.

Fig. VIIA,50b (bottom). A packing drawing of the BaGeO₃ arrangement seen along its a_0 axis. The germanium atoms are black, the barium atoms fine-line shaded.

VII,a44. Crystals of *copper germanium trioxide*, $CuGeO_3$, are orthorhombic, with a bimolecular unit of the approximate dimensions:

$$a_0 = 4.8 \text{ A.}; \quad b_0 = 8.5 \text{ A.}; \quad c_0 = 2.94 \text{ A.}$$

The space group has been chosen as C_{2v}^2 ($Pb2_1m$), with atoms in the positions:

$$(2a) \quad uv0; \ \bar{u},v+{}^1/_2,0$$
$$(2b) \quad u\,v\,{}^1/_2; \ \bar{u},v+{}^1/_2,{}^1/_2$$

The parameters are as follows:

Atom	Position	x	y	z
Cu	$(2a)$	0.50	0.25	0
Ge	$(2b)$	0.08	0.00	$^1/_2$
O(1)	$(2a)$	0.302	0.00	0
O(2)	$(2b)$	0.807	0.153	$^1/_2$
O(3)	$(2b)$	0.807	−0.153	$^1/_2$

In this arrangement (Fig. VIIA,51), germanium atoms are surrounded by tetrahedra of oxygens linked together to form chains through sharing

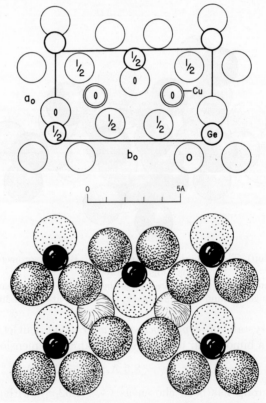

Fig. VIIA,51a (top). The orthorhombic structure of $CuGeO_3$ projected along its c_0 axis. Origin in lower left.

Fig. VIIA,51b (bottom). A packing drawing of the $CuGeO_3$ arrangement seen along its c_0 axis. The germanium atoms are black, the copper atoms fine-line shaded.

corners. The separation Ge–O = 1.86–1.87 A. Atoms of copper are octahedrally surrounded by six oxygens, with Cu–O = 2.24 A.

VII,a45. *Silver antimony trioxide*, $AgSbO_3$, is cubic with 16 molecules in a unit of the edge length:

$$a_0 = 10.32 \text{ A.}$$

The powder diagram of this substance is very close to that given by pyrochlor, $CaNaSb_2O_6(OH)$ [**XI,d1** loose-leaf] and it is considered that the structures are essentially the same. On this basis the space group is O_h^7 (*Fd3m*) and atoms are in the positions:

Ag: (16c) $\quad 1/8\ 1/8\ 1/8;\ 1/8\ 3/8\ 3/8;\ 3/8\ 1/8\ 3/8;\ 3/8\ 3/8\ 1/8;$ F.C.

Sb: (16d) $\quad 5/8\ 5/8\ 5/8;\ 5/8\ 7/8\ 7/8;\ 7/8\ 5/8\ 7/8;\ 7/8\ 7/8\ 5/8;$ F.C.

O: (48f) $\quad \pm(u00;\quad 0u0;\quad 00u;$
$\qquad u+1/4,1/4,1/4;\ 1/4,u+1/4,1/4;\ 1/4,1/4,u+1/4);$ F.C.

with $u = 0.29$

Sodium antimony trioxide, $NaSbO_3$, has been reported to occur occasionally with this structure, though in a poorly crystallized condition. For it,

$$a_0 = 10.20 \text{ A.}$$

The more common form of this substance has the $FeTiO_3$ arrangement (**VII,a32**), with $a_0 = 6.14$ A., $\alpha = 51°18$.

VII,a46. According to a preliminary note, *sodium zinc hydroxide*, $Na[Zn(OH)_3]$, forms excellent tetragonal crystals, with

$$a_0 = 10.86 \text{ A.,} \qquad c_0 = 5.35 \text{ A.}$$

The eight molecules in this cell have been placed in special positions:

(8h) $\quad \pm(uv0;\ u+1/2,1/2-v,0;\ \bar{v}\,u\,1/2;\ v+1/2,u+1/2,1/2)$

of D_{4h}^{13} (*P4/mbc*), with the parameters:

Atom	u	v
Zn	0.059	0.307
Na	0.377	0.404
O(1)	0.173	0.449
O(2)	0.096	0.128
O(3)	0.379	0.183

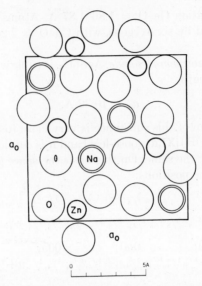

Fig. VIIA,52. The tetragonal structure of NaZn(OH)₃ projected along its c_0 axis. All the atoms shown here are in the plane with $z = 0$; the next atomic plane at $z = \frac{1}{2}$ is just like this but turned 90° about an a_0 axis.

As Figure VIIA,52 shows, each zinc atom has around it three oxygen atoms, with Zn–O = 1.96 and 2.00 A. The next nearest oxygen neighbor is 2.65 A. away. The shortest Na–O = 2.27 A.

BIBLIOGRAPHY TABLE, CHAPTER VIIA

Compound	Paragraph	Literature
AgBrO₃	a17	1928: H; 1954: NBS
AgClO₃	a17	1927: F&F; 1928: H; Z; 1940: P; 1942: NS&P
AgNO₃	a1	1928: Z; 1955: M; 1956: F; 1959: N; 1961: P
AgPO₃	a40	1958: J; 1961: J
Ag₃SNO₃	a11	1959: B
AgSbO₃	a45	1938: S
AgZnF₃	a21	1953: DV&R
AlBiO₃	a21	1947: NS
BaCO₃	a2	1928: W; Z; 1931: C&G; 1935: C&L; 1953: NBS; 1959: NBS
BaCa₀.₅W₀.₅O₃	a21	1959: F,K&W

(continued)

BIBLIOGRAPHY TABLE, CHAPTER VIIA *(continued)*

Compound	Paragraph	Literature
$BaCaZrGeO_6$	a21	1954: R
$BaCeO_3$	a21	1934: H; 1935: H; 1947: NS; 1960: S&W
$BaCe_{0.5}Nb_{0.5}O_3$	a21	1960: B
$BaCo_{0.33}Ta_{0.67}O_3$	a21	1954: R
$BaCo_{0.5}W_{0.5}O_3$	a21	1959: F,K&W; 1960: B
$BaDy_{0.5}Nb_{0.5}O_3$	a21	1960: B
$BaEr_{0.5}Nb_{0.5}O_3$	a21	1960: B
$BaEu_{0.5}Nb_{0.5}O_3$	a21	1960: B
$BaFeO_3$	a21	1954: M&K; 1961: D,F&S
$BaFe_{0.5}Nb_{0.5}O_3$	a21	1960: A
$BaFe_{0.5}Ta_{0.5}O_3$	a21	1960: A
$BaFe_{0.5}W_{0.5}O_3$	a21	1959: F,K&W
$BaGd_{0.5}Nb_{0.5}O_3$	a21	1960: B
$BaGeO_3$	a37, a43	1960: L; 1962: H
$BaHo_{0.5}Nb_{0.5}O_3$	a21	1960: B
$BaIn_{0.5}Nb_{0.5}O_3$	a21	1960: B
$BaIr_{0.33}Ti_{0.67}O_3$	a28	1961: D,K&W
$BaLa_{0.5}Nb_{0.5}O_3$	a21	1960: B
$(Ba,La)TiO_3$	a21	1961: J&S
$BaLi_{0.5}Re_{0.5}O_3$	a21	1961: S&W
$BaLu_{0.5}Nb_{0.5}O_3$	a21	1960: B
$BaMg_{0.5}W_{0.5}O_3$	a21	1959: F,K&W
$BaMnO_3$	a29, a30	1962: H
$BaMoO_3$	a21	1960: B
$BaMo_{0.5}Co_{0.5}O_3$	a21	1960: B
$BaMo_{0.5}Ni_{0.5}O_3$	a21	1960: B
$BaNa_{0.5}Re_{0.5}O_3$	a21	1961: S&W
$BaNd_{0.5}Nb_{0.5}O_3$	a21	1960: B
$BaNiO_3$	a30	1951: L
$BaNi_{0.33}Nb_{0.67}O_3$	a21	1960: A
$BaNi_{0.33}Ta_{0.67}O_3$	a21	1954: R
$BaNi_{0.5}W_{0.5}O_3$	a21	1959: F,K&W; 1960: A; B
$BaPbO_3$	a21	1958: W; 1959: W&B
$BaPrO_3$	a21	1935: H; 1947: NS
$BaPr_{0.5}Nb_{0.5}O_3$	a21	1960: B
$BaPuO_3$	a21	1960: R,H&B

(continued)

BIBLIOGRAPHY TABLE, CHAPTER VIIA (*continued*)

Compound	Paragraph	Literature
$BaSc_{0.5}Nb_{0.5}O_3$	a21	1960: A; B
$BaSc_{0.5}Ta_{0.5}O_3$	a21	1960: A
$BaSiO_3$	a37, a43	1960: L
$BaSm_{0.5}Nb_{0.5}O_3$	a21	1960: B
$BaSnO_3$	a21	1943: NS; 1946: M; 1947: NS; 1959: W&B; 1960: S&W
$BaTb_{0.5}Nb_{0.5}O_3$	a21	1960: B
$BaThO_3$	a21	1934: H; 1935: H; 1947: NS; 1960: S&W
$BaTiO_3$	a21, a24, a28	1926: G; 1943: NS; 1945: M; 1946: M; 1947: K&R; NS; 1948: B&E; E&B; V&G; 1951: E; E,S&J; R; V; W; 1952: D,K&B; NBS; 1953: E; F; 1955: F,D&P; 1957: S,D&P; 1958: D,M&C; V,Z,V&S; 1961: D,K&W; E; J&S
$BaTiS_3$	a30	1956: H&M
$BaTm_{0.5}Nb_{0.5}O_3$	a21	1960: B
$BaUO_3$	a21	1954: R&P; 1960: R,H&B
$BaY_{0.5}Nb_{0.5}O_3$	a21	1960: A; B
$BaYb_{0.5}Nb_{0.5}O_3$	a21	1960: B
$BaYb_{0.5}Ta_{0.5}O_3$	a21	1960: A
$BaZn_{0.33}Nb_{0.67}O_3$	a21	1960: A
$BaZn_{0.5}W_{0.5}O_3$	a21	1959: F,K&W
$BaZrO_3$	a21	1934: H; 1935: H; 1943: NS; 1946: M; 1947: NS; 1954: NBS; 1958: V,Z,V&S
$BaZrS_3$	a21	1956: H&M
$CaCO_3$	a1, a2, a3	1914: B; 1915: B; 1919: S; 1920: W; 1923: M; 1924: B; H; R; 1925: B; C,B&DF; G,W&M; T; W; 1927: T; 1929: M; 1930: K; OD; 1931: B; F; 1934: W&S; 1935: S; 1936: F&C; 1938: S; 1940: I&S; T; 1952: NBS; 1953: B; 1954: NBS; 1957: S,V&D; 1959: M; 1961: G; L,R&M
$CaCeO_3$	a21	1943: NS; 1947: NS
$CaMnO_3$	a21	1950: J&vS; 1955: Y
$CaMoO_3$	a21	1956: MC,K&W

(*continued*)

BIBLIOGRAPHY TABLE, CHAPTER VIIA (*continued*)

Compound	Paragraph	Literature
$(Ca,Na)(Ti,Nb)O_3$, dysanalyte	a21	1925: B; 1936: Z; 1939: Z
$CaNi_{0.33}Nb_{0.67}O_3$	a21	1960: A
$CaNi_{0.33}Ta_{0.67}O_3$	a21	1954: R
$(Ca,Pb)CO_3$, tarnowitzite	a2	1930: S
$CaRuO_3$	a26	1959: R&W
$CaSnO_3$	a21	1926: G; 1945: R; 1946: M; 1947: NS
$CaThO_3$	a21	1947: NS
$CaTiO_3$	a21, a26	1925: B; L&N; 1926: G; 1936: Z; 1939: Z; 1943: NS; Z; 1946: M; 1947: NS; 1957: K&B
$CaVO_3$	a21	1954: R&B
$CaZrO_3$	a21	1926: G; 1929: R,E&S; 1946: M; 1947: NS; 1956: R
$CdCO_3$	a1	1928: Z; 1936: F&C; 1961: G
$CdCeO_3$	a21	1943: NS; 1947: NS
$CdSnO_3$	a21	1943: NS; 1947: NS; 1960: S
$CdThO_3$	a21	1947: NS
$CdTiO_3$	a21, a27, a32	1928: Z; 1934: P&B; 1946: M; 1947: NS; 1957: K&M
$CeAlO_3$	a21	1949: Z; 1955: R&F
$CeCrO_3$	a21	1952: W,W,G&R; 1954: K&R; W&W; 1955: R&F
$CeFeO_3$	a21	1954: K&R; 1955: R&F
$CeGaO_3$	a21	1954: K&R; R&F
$CeVO_3$	a21	1952: W,W,G&R; 1954: R&B; W&W
$CoCO_3$	a1	1929: F&C; 1932: B; 1936: F&C; 1959: NBS; 1961: G
$CoMnO_3$	a32	1958: C; S,T&V
$CoTiO_3$	a32	1934: B&P; P&B
$CrBiO_3$	a21	1947: NS
$CrRhO_3$	a32	1961: K
$Cs(Ag,Au)Cl_3$	a21	1937: F
$Cs_2AgAuCl_6$	a23	1934: E; 1938: E&P
$CsAuCl_3$	a21, a23	1934: E; 1937: F; 1938: E&P
$CsBeF_3$	a37	1960: L; 1962: M&PK

(*continued*)

BIBLIOGRAPHY TABLE, CHAPTER VIIA (*continued*)

Compound	Paragraph	Literature
CsCaF₃	a21	1952: L&W
CsCdBr₃	a21	1928: N&P; 1947: NS
CsCdCl₃	a21	1927: F&B; 1928: N&P; 1947: NS
CsCd(NO₂)₃	a21	1935: F&C
CsCuCl₃	a33	1946: K&S; 1947: W
CsHgBr₃	a21	1928: N&P; 1947: NS
CsHgCl₃	a21	1927: N; 1928: N&P; 1947: NS
CsHg(NO₂)₃	a21	1935: F&C
CsIO₃	a21	1928: Z; 1947: NS
CsMgF₃	a21	1952: L&W
CsMnF₃	a28	1962: Z,L&T
CsNO₃	a8	1934: W&MC; 1937: F&H; 1953: K; 1957: F,S&O
CsNiCl₃	a33	1955: T
CsPbBr₃	a21	1959: M
CsPbCl₃	a21	1957: M; 1959: M
CsPbF₃	a21	1956: SD,B&H
CsVCl₃	a33	1959: S&E
CsVO₃	a37	1960: E
CsZnF₃	a21	1952: L&W
CuGeO₃	a44	1954: G
DyAlO₃	a21	1960: D&W
DyBO₃	a3	1961: L,R&M
DyFeO₃	a21	1960: D&W
DyMnO₃	a21	1957: V&K
ErBO₃	a3	1961: L,R&M
EuAlO₃	a21, a26	1955: R&F; 1956: G&B
EuBO₃	a3	1961: L,R&M
EuCrO₃	a21	1955: R&F
EuFeO₃	a21, a26	1955: R&F; 1956: G&W
EuTiO₃	a21	1953: B,F&B
FeBiO₃	a21	1960: F,S,F&B; Z&T; 1961: F,V,Z&S
FeCO₃, siderite	a1	1914: B; 1920: W; 1932: G&H; 1935: S; 1960: S; 1961: G
FeLaO₃	a21	1951: GG
FePrO₃	a21	1951: GG

(*continued*)

BIBLIOGRAPHY TABLE, CHAPTER VIIA (*continued*)

Compound	Paragraph	Literature
$FeRhO_3$	a32	1961: K
$FeTiO_3$	a32	1934: B&P; P&B; 1959: S,P&I
$FeVO_3$	a32	1960: B&C
$GdAlO_3$	a21, a26	1955: R&F; 1956: G&B; 1960: D&W
$GdBO_3$	a3	1961: L,R&M
$GdCoO_3$	a21	1954: R&F
$GdCrO_3$	a21, a26	1955: R&F; 1957: G
$GdFeO_3$	a21, a26	1955: R&F; 1956: G; G&W; 1960: D&W
$GdMnO_3$	a21	1950: J&vS
$GdScO_3$	a26	1957: G
$GdVO_3$	a26	1957: G
HIO_3	a19	1928: Z; 1941: R&H; 1954: G; NBS
HNO_3	a10	1949: L; 1950: L; 1951: L
$HgNiF_3$	a21	1959: R,K,L&B
$HoBO_3$	a3	1961: L,R&M
$InBO_3$	a1	1932: G&H; 1961: L,R&M
$KBaTiNbO_6$	a21	1954: R
$KBaCaTiZrNbO_9$	a21	1954: R
$KBeF_3$	a37	1960: L
$K_{0.5}Bi_{0.5}TiO_3$	a21	1960: A; 1962: I,K,V&Z
$KBrO_3$	a14	1925: S; 1928: Z; 1955: M
$KCaF_3$	a21	1952: B; L&W
$KCdCl_3$	a36	1947: B
$KCdF_3$	a21	1952: B
$KCd(NO_2)_3$	a21	1935: F&C
$K_{0.5}Ce_{0.5}TiO_3$	a21	1954: R
$K_2CeLaTi_4O_{12}$	a21	1954: R
$KClO_3$	a15, a16	1925: S; 1928: Z; 1929: Z; 1930: S; 1953: I&O; 1957: R&L; 1958: A; 1959: B,S&T
$KCoF_3$	a21	1959: O,S&F; R,K,L&B; 1961: K; O&S; 1962: O&S
$KCrF_3$	a21, a22	1959: E&P; 1961: K
$KCuF_3$	a21, a22	1959: E&P; O,S&F; 1961: K; O&S
$KFeF_3$	a21	1959: O,S&F; 1961: K; O&S; 1962: O&S
$KHg(SCN)_3$	a35	1952: Z&S

(*continued*)

BIBLIOGRAPHY TABLE, CHAPTER VIIA (*continued*)

Compound	Paragraph	Literature
KIO_3	a14, a21	1925: S; 1926: G; 1947: NS; 1960: S&W; 1961: NS&K; 1962: O,R,P, R&Z
$K_{0.5}La_{0.5}TiO_3$	a21	1954: R
$KMgF_3$	a21	1926: G; 1947: NS; 1952: L&W; 1953: DV&R; 1956: R&H
$KMnF_3$	a21, a26	1957: S,B&K; 1959: O,S&F; 1961: B&K; H,L&D; K; O&S; 1962: O&S
KNO_3	a1, a2, a14	1928: Z; 1931: E; 1939: B; 1947: T; 1952: NBS; 1956: F
$KNbO_3$	a21, a24	1926: G; 1932: Q; 1951: V; W; 1952: L; 1954: S,D,P&P
$K_{0.5}Nd_{0.5}TiO_3$	a21	1954: R
$KNiF_3$	a21	1926: G; 1947: NS; 1959: O,S&F; R,K,L&B; 1961: K; O&S; 1962: O&S
$KTaO_3$	a21, a24	1932: Q; 1947: NS; 1951: V
$KVCl_3$	a33	1959: S&E
KVO_3	a37	1954: E&B; 1958: P,M&H; 1960: E
$KZnF_3$	a21	1926: G; 1947: NS; 1952: L&W; 1954: NBS; 1961: K
$LaAlO_3$	a21, a26	1926: G; 1947: NS; 1955: R&F; 1956: G&B
$LaBO_3$	a2	1932: G&H; 1960: NBS; 1961: L,R&M
$(La,Ba)MnO_3$	a21	1950: J&vS
$(La,Ca)MnO_3$	a21	1950: J&vS; 1955: Y
$LaCoO_3$	a21	1950: A,F&W; 1952: W,W,G&R; 1954: R&F; W&W; 1955: Y; 1957: K&W; W,P&B; 1962: vS
$LaCrO_3$	a21, a26	1943: NS; 1952: W,W,G&R; 1954: W&W; 1955: R&F; 1957: G; K&W
$LaFeO_3$	a21, a26	1943: NS; 1947: NS; 1955: R&F; Y; 1956: G&W; 1957: K&W
$LaGaO_3$	a21, a26	1926: G; 1954: K&R; R&F; 1957: G
$LaMg_{0.5}Ge_{0.5}O_3$	a21	1954: R
$LaMg_{0.5}Ti_{0.5}O_3$	a21	1954: R; 1960: A
$La(Mn,Co)O_3$	a26	1957: G

(*continued*)

BIBLIOGRAPHY TABLE, CHAPTER VIIA (*continued*)

Compound	Paragraph	Literature
$LaMnO_{3+x}$	a21	1943: NS; 1947: NS; 1955: H; Y; 1957: K&W
$LaNiO_3$	a21	1957: K&W; 1960: F,M&L
$LaNi_{0.5}Ti_{0.5}O_3$	a21	1954: R
$LaRhO_3$	a21	1957: W,P&B; 1961: K
$LaScO_3$	a26	1957: G
$(La,Sr)MnO_3$	a21	1950: J&vS; 1955: Y
$LaTiO_3$	a21	1954: K&W; 1961: J&S
$LaVO_3$	a21	1952: W,W,G&R; 1954: R&B; W&W; 1957: K,D&W
$LaZr_{0.5}Ca_{0.5}O_3$	a21	1956: R; 1960: B
$LaZr_{0.5}Mg_{0.5}O_3$	a21	1956: R; 1960: B
$(LiAsO_3)_n$	a39	1956: H&DS; 1957: H
$LiBaF_3$	a21	1952: L&W
$LiBrO_3$	a20	1962: B
$LiIO_3$	a18	1931: Z&B
$LiNO_3$	a1	1928: Z; 1956: F
$LiNbO_3$	a32	1928: Z
$LiSbO_3$	a31	1954: E&I
$LiUO_3$	a21	1960: K&G
Li_xWO_3	a21	1950: M&N
$LuBO_3$	a1, a3	1961: L,R&M
$MgCO_3$, magnesite	a1	1924: L&F; 1927: B&D; 1935: S; 1958: G&G; 1961: G
$MgCeO_3$	a21	1947: NS
$MgGeO_3$	a32, a39	1957: R; 1959: R&L; 1962: R&S
$MgSiO_3$	a39	1943: B; 1950: I
$MgTiO_3$, geikielite	a32	1928: Z; 1934: P&B; 1954: NBS; 1959: W
$MnCO_3$	a1	1914: B; 1920: W; 1929: B&A; 1930: K; 1936: F&C; 1957: G&G; 1961: G
$MnTiO_3$	a32	1934: P&B; 1959: S,P&I
NH_4BeF_3	a37	1960: L; 1962: M&PK
NH_4CdCl_3	a36	1938: B&P; 1939: MG,N,D&K
$NH_4Cd(NO_2)_3$	a21	1935: F&C
NH_4ClO_3	a14	1962: G,G&T
NH_4CoF_3	a21	1959: R,K,L&B
NH_4HgCl_3	a34	1938: H

(*continued*)

BIBLIOGRAPHY TABLE, CHAPTER VIIA (*continued*)

Compound	Paragraph	Literature
$NH_4Hg(SCN)_3$	a35	1952: Z&S
NH_4IO_3	a20	1932: G; 1943: MG&E; 1947: NS
NH_4MnF_3	a21	1957: S,B&K; 1961: H,L&D
NH_4NO_3	a4, a5, a6, a7	1931: K,H&P; 1932: H,P&K; W; 1937: F&H; 1947: G&W; 1955: M; 1958: AP,A&C; 1959: S; 1960: AP; AP&B; 1962: AP,A&C
NH_4NiF_3	a21	1959: R,K,L&B
$(NH_4PO_3)_4$	a38	1949: R,K&MG; 1950: A&F; 1951: R,K&MG; 1952: P,MC,Z&D
NH_4VO_3	a37	1950: L; 1954: S&H; 1959: NBS; 1960: E
$NaAlO_3$	a21	1954: K&R
$(NaAsO_3)_n$	a41	1954: DS,L&T; 1956: L
$NaBaTiNbO_6$	a21	1954: R
$NaBeF_3$	[XII]	1943: OD&T; 1945: OD&T
$Na_{0.5}Bi_{0.5}TiO_3$	a21	1960: A; 1962: I,K,V&Z
$NaBrO_3$	a13	1921: D&G; K,B&K; 1922: V; 1923: K; K,B&K; 1938: H; 1940: H; 1954: NBS
$(Na,Ca,Ce)(Nb,Ti)O_3$, loparite	a21	1930: vG
$NaClO_3$	a13	1921: D&G; K,B&K; 1922: V; W; 1923: K; K,B&K; 1929: Z; 1940: H; 1952: NBS; 1957: R&C; 1959: A; B,S&T
$NaCoF_3$	a26	1959: R,K,L&B
$NaHgCl_3$	a36	1954: W&D
$NaIO_3$	a20, a21	1928: Z; 1943: MG&vE; 1947: NS&N
$Na_{0.5}La_{0.5}TiO_3$	a21	1960: A
$NaMgF_3$	a21, a26	1952: L&W; 1959: R,K,L&B; 1961: C,E,S&M
$NaMnF_3$	a21	1957: S,B&K
$NaNO_3$	a1	1914: B; 1920: W; 1931: K,P&H; 1932: B&K; 1933: W; 1934: S&M; 1945: K&S; 1947: T; 1955: M; 1957: S,V&D; 1960: I; K&V; 1962: KS

(*continued*)

BIBLIOGRAPHY TABLE, CHAPTER VIIA (*continued*)

Compound	Paragraph	Literature
$NaNbO_3$	a21	1925: B; 1926: G; 1932: Q; 1947: NS; 1951: V; W; 1956: F; 1961: W&M
$NaNiF_3$	a26	1959: R,K,L&B
$Na(PO_3)_n$	a40	1951: D&K; P&W; 1954: DS,L&T; 1959: J; 1961: J
$NaSbO_3$	a45	1938: S
$NaTaO_3$	a21, a24, a27	1932: Q; 1947: NS; 1951: V; 1957: K&M
$NaVO_3$	[XII]	1943: S
$NaWO_3$	a21	1932: dJ; 1935: H; 1942: D; 1943: NS; 1949: M; 1951: B,B,L&J; N&K; 1954: B&B; 1960: A&R
$NaZnF_3$	a21, a26	1952: L&W; 1959: R,K,L&B
$NaZn(OH)_3$	a46	1961: vS
$NdAlO_3$	a21, a26	1955: R&F; 1956: G&B; 1960: D&W
$NdBO_3$	a2	1960: NBS; 1961: L,R&M
$NdCoO_3$	a21	1952: W,W,G&R; 1954: R&F; W&W
$NdCrO_3$	a21, a26	1952: W,W,G&R; 1954: K&R; W&W; 1955: R&F; 1957: G
$NdFeO_3$	a21, a26	1955: R&F; 1956: G&W; 1960: D&W
$NdGaO_3$	a21, a26	1954: K&R; 1957: G
$NdMg_{0.5}Ti_{0.5}O_3$	a21	1954: R
$NdMnO_3$	a21	1957: V&K
$NdScO_3$	a26	1957: G
$NdVO_3$	a21, a26	1952: W,W,G&R; 1954: W&W; 1957: G
$NiCO_3$	a1	1952: L; 1959: P; 1960: NBS; 1961: G
$NiMnO_3$	a32	1958: C; S,T&V
$NiTiO_3$	a32	1930: T; 1934: P&B; 1959: S,P&I
$PbCO_3$, cerussite	a2	1928: Z; 1933: C&LC; 1938: L&H; S
$PbCeO_3$	a21	1943: NS; 1947: NS
$PbFe_{0.5}Nb_{0.5}O_3$	a21	1960: A
$PbFe_{0.5}Ta_{0.5}O_3$	a21	1960: A
$Pb(Fe,W)O_3$	a21	1960: A; I

(*continued*)

BIBLIOGRAPHY TABLE, CHAPTER VIIA (*continued*)

Compound	Paragraph	Literature
$Pb(Mg,Nb)O_3$	a21	1960: A; I
$Pb(Ni,Nb)O_3$	a21	1960: A; I
$PbSc_{0.5}Nb_{0.5}O_3$	a21	1960: A
$PbSc_{0.5}Ta_{0.5}O_3$	a21	1960: A
$PbSnO_3$	a21	1943: NS; 1947: NS
$PbThO_3$	a21	1947: NS
$PbTiO_3$	a21, a24	1937: C&E; 1943: NS; 1946: M; 1947: NS; 1950: S,H&S; 1951: S&H; 1955: I; S,P&F; 1956: S,P&F
$PbYb_{0.5}Nb_{0.5}O_3$	a21	1960: A
$PbYb_{0.5}Ta_{0.5}O_3$	a21	1960: A
$PbZrO_3$	a21, a25	1935: H; 1943: NS; 1946: M; 1947: NS; 1951: S,M&H; U&S; 1952: S; 1957: J,S,M&P
$PrAlO_3$	a21, a26	1955: R&F; 1956: G&B; 1960: R,H&B
$PrCoO_3$	a21	1952: W,W,G&R; 1954: R&F; W&W
$PrCrO_3$	a21, a26	1952: W,W,G&R; 1954: W&W; 1955: R&F; 1957: G; 1960: R,H&B
$PrFeO_3$	a21, a26	1955: R&F; 1956: G&W; 1960: R,H&B
$PrGaO_3$	a21, a26	1954: R&F; 1957: G
$PrMnO_3$	a21	1957: V&K
$(Pr,Nd)MnO_3$	a21	1950: J&vS
$(Pr,Nd,Ba)MnO_3$	a21	1950: J&vS
$(Pr,Nd,Sr)MnO_3$	a21	1950: J&vS
$PrScO_3$	a26	1957: G
$PrVO_3$	a21, a26	1952: W,W,G&R; 1954: W&W; 1957: G; 1960: R,H&B
$PuAlO_3$	a21	1960: R,H&B
$PuCrO_3$	a26	1960: R,H&B
$PuFeO_3$	a21	1960: R,H&B
$PuMnO_3$	a21	1960: R,H&B
$PuVO_3$	a26	1960: R,H&B
$RbBeF_3$	a37	1960: L; 1962: M&PK
$RbCaF_3$	a21	1952: L&W

(*continued*)

BIBLIOGRAPHY TABLE, CHAPTER VIIA *(continued)*

Compound	Paragraph	Literature
$RbCdCl_3$	a36	1939: MG,N,D&K
$RbCd(NO_2)_3$	a21	1935: F&C
$RbClO_3$	a14	1951: G&GB
$RbCoF_3$	a21	1959: R,K,L&B
$RbHg(NO_2)_3$	a21	1935: F&C
$RbIO_3$	a21	1926: G; 1947: NS
$RbMgF_3$	a21	1952: L&W
$RbMnF_3$	a21	1957: S,B&K; 1961: H,L&D
$RbNO_3$	a1, a8, a14	1928: Z; 1933: P&S; 1937: F&H; 1951: K
$(RbPO_3)_n$	a42	1956: C
$RbVO_3$	a37	1960: E
$RbZnF_3$	a21	1952: L&W
$ScBO_3$	a1	1932: G&H
$SmAlO_3$	a21, a26	1954: K&R; 1955: R&F; 1956: G&B; 1957: G
$SmBO_3$	a3	1961: L,R&M
$SmCoO_3$	a21	1952: W,W,G&R; 1954: R&F; W&W
$SmCrO_3$	a21, a26	1952: W,W,G&R; 1954: W&W; 1955: R&F; 1957: G
$SmFeO_3$	a21, a26	1955: R&F; 1956: G&W
$SmVO_3$	a21	1952: W,W,G&R; 1954: W&W
$SrCO_3$	a2	1928: W; Z; 1952: NBS
$SrCeO_3$	a21	1935: H; 1947: NS
$SrCoO_3$	a21	1955: Y
$SrCr_{0.5}Nb_{0.5}O_3$	a21	1960: B
$SrCr_{0.5}Ta_{0.5}O_3$	a21	1954: R
$SrFeO_3$	a21	1955: Y
$SrGa_{0.5}Nb_{0.5}O_3$	a21	1960: B
$SrHfO_3$	a21	1933: H; 1935: H; 1947: NS
$SrIn_{0.5}Nb_{0.5}O_3$	a21	1960: B
$SrLaNdTiCrAlO_9$	a21	1954: R
$SrMg_{0.33}Ta_{0.67}O_3$	a21	1954: R
$SrMoO_3$	a21	1960: B
$SrMo_{0.5}Co_{0.5}O_3$	a21	1960: B; 1962: K&F
$SrMo_{0.5}Ni_{0.5}O_3$	a21	1960: B; 1962: K&F
$SrMo_{0.5}Zn_{0.5}O_3$	a21	1962: K&F

(continued)

BIBLIOGRAPHY TABLE, CHAPTER VIIA (*continued*)

Compound	Paragraph	Literature
$SrNi_{0.33}Nb_{0.67}O_3$	a21	1960: A
$SrPbO_3$	a21	1958: W
$SrRuO_3$	a26	1959: R&W
$SrSnO_3$	a21	1935: H; 1946: M; 1947: NS; 1960: S&W
$SrThO_3$	a21	1947: NS
$SrTiO_3$	a21	1926: G; 1935: H; 1945: R; 1946: M; 1947: NS; 1952: NBS; 1953: B,F&B
$SrTiS_3$	a30	1956: H&M
$SrW_{0.5}Co_{0.5}O_3$	a21	1959: F,K&W; 1960: B; 1962: K&F
$SrW_{0.5}Ni_{0.5}O_3$	a21	1959: F,K&W; 1960: B; 1962: K&F
$SrW_{0.5}Zn_{0.5}O_3$	a21	1959: F,K&W; 1960: B; 1962: K&F
$SrZrO_3$	a21	1928: Z; 1933: H; 1935: H; 1946: M; 1947: NS; 1960: S&W
$TaSnO_3$	a21	1955: G
$TlBrO_3$	a14	1948: R&A
$TlCd(NO_2)_3$	a21	1935: F&C
$TlClO_3$	a14	1951: S&MC
$TlCoF_3$	a21	1959: R,K,L&B
$TlHg(NO_2)_3$	a21	1935: F&C
$TlIO_3$	a14, a21	1947: R&A; 1948: S; 1960: S&W
$TlNO_3$	a9	1937: F&H; 1943: R&A; 1950: F,C&T; 1957: H&K; 1961: K&P
$TmBO_3$	a3	1961: L,R&M
UO_2CO_3	a12	1955: C,C&E
$YAlO_3$	a21, a26	1926: G; 1947: NS; 1955: R&F; 1956: G&B; G&W
YBO_3	a1, a3	1932: G&H; 1961: L,R&M
$YCrO_3$	a21, a26	1954: L&K; 1955: R&F; 1956: G&W; 1957: G
$YFeO_3$	a21, a26	1955: R&F; 1956: G&W
$YScO_3$	a26	1957: G
$YbBO_3$	a3	1961: L,R&M
$ZnCO_3$	a1	1924: L&F; 1929: B&A; 1932: G&H; 1936: F&C; 1961: G

B. COMPOUNDS $R_n(MX_3)_m$

Carbonates and Nitrates

VII,b1. A redetermination has recently been made of the structure of *sodium bicarbonate*, $NaHCO_3$, which changes somewhat the parameters established many years ago. The cell initially employed for this monoclinic crystal was tetramolecular, with the dimensions:

$$a_0 = 7.51 \text{ A.}; \quad b_0 = 9.70 \text{ A.}; \quad c_0 = 3.53 \text{ A.}; \quad \beta = 93°19'$$

For the new determination, a different but equivalent tetramolecular cell was chosen. For it (Fig. VIIB,1), the a_0 and c_0 axes were interchanged and the non-common axis chosen so that the orientation of the space group C_{2h}^5 would be $P2_1/c$ instead of the $P2_1/n$ used before. For this new cell,

$$a_0' = c_0 = 3.51 \text{ A.}; b_0' = b_0 = 9.71 \text{ A.}; c_0' = 8.05 \text{ A.}; \beta' = 111°51'$$

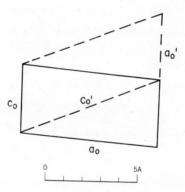

Fig. VIIB,1. The relation between the axes of the two units used to describe the structure of $NaHCO_3$.

Continuing to use the old axes, however, and placing the atoms in the general positions according to $P2_1/n$:

$$(4e) \quad \pm (xyz; \, x+^1/_2,\,^1/_2-y,z+^1/_2)$$

the revised and original (in parentheses) parameters are as follows:

Atom	x	y	z
Na	0.2855 (0.278)	0.0047 (0.000)	0.7127 (0.708)
C	0.0768 (0.069)	0.2370 (0.236)	0.2866 (0.314)
O(1)	0.0709 (0.069)	0.3668 (0.367)	0.2605 (0.314)
O(2)	0.2054 (0.200)	0.1629 (0.169)	0.1937 (0.183)
O(3)	0.9400 (0.939)	0.1707 (0.169)	0.4358 (0.444)

The parameters $(x'y'z')$ according to the new more oblique axes $(a_0'b_0'c_0')$ convert to the parameters used above (xyz), applying to the original axes $(a_0b_0c_0)$ through the transformation.

$$x = z'; \; y = y'; \; z = x'-z'$$

This arrangement in terms of the original axes is shown in Figure VIIB,2. It consists of layers of carbonate ions interspersed with layers of sodium atoms. The hydrogen position has not been found experimentally, but it is considered to be closely bonded to $O(3)$, thus creating a hydrogen bond between this oxygen and an $O(2)$ atom of an adjacent carbonate ion with $O(3)$–H. . .$O(2) = 2.595$ A. This CO_3 ion is not symmetrical, C–$O(3) = 1.346$ A. being considerably longer than the 1.263 and 1.264 A. that apply to the two other oxygens. The angle $O(1)$–C–$O(2) = 124°58'$ makes the other two angles of the planar CO_3 ion smaller than 120°. The sodium atoms are octahedrally surrounded by six oxygens at distances between 2.389 and 2.471 A.

VII,b2. Crystals of *potassium acid carbonate*, $KHCO_3$, are monoclinic with a tetramolecular cell for which there are two recent, rather poorly agreeing sets of dimensions:

$$a_0 = 15.176 \; (15.04) \text{ A.}; \; b_0 = 5.630 \; (5.51) \text{ A.}; \; c_0 = 3.708 \; (3.68) \text{ A.};$$

$$\beta = 104°31' \; (104°30')$$

Atoms have been placed in general positions of C_{2h}^5 $(P2_1/a)$:

$$(4e) \quad \pm(xyz; \; x+{}^1/_2, {}^1/_2-y, z)$$

with the following parameters:

Atom*	x	y	z
K	0.1633 (0.166)	0.021 (0.025)	0.301 (0.300)
C	0.1177 (0.122)	0.529 (0.525)	−0.150 (−0.160)
O(1)	0.1960 (0.195)	0.541 (0.525)	0.078 (0.090)
O(2)	0.0813 (0.082)	0.725 (0.725)	−0.317 (−0.285)
O(3)	0.0789 (0.082)	0.320 (0.325)	−0.233 (−0.285)

* Parameters according to 1952: H are in parentheses, with z values shifted by $^1/_2$.

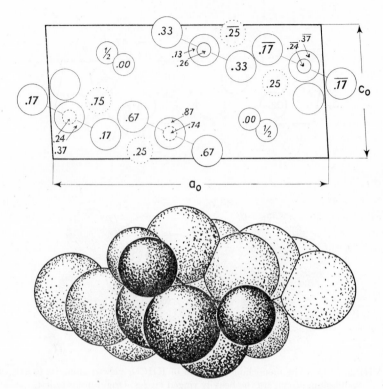

Fig. VIIB,2a (top). Atoms in the monoclinic unit of NaHCO₃ projected on its b_0 face. The large circles are the oxygen, the smallest circles are carbon atoms; they are joined by light lines to form the CO₃ groups. Positions once proposed for the hydrogen atoms are given by the dotted circles.

Fig. VIIB,2b (bottom). A packing drawing of the NaHCO₃ arrangement viewed along its b_0 axis. The oxygen atoms are the large, the sodium atoms the small circles. Carbon atoms do not show.

The structure is illustrated in Figure VIIB,3. Though the two determinations lead to the same general arrangement, differences in the parameters are sufficient to render uncertain such details of the structure as the exact dimensions of the carbonate ion [C–O = 1.24–1.28 A. according to one structure and 1.28–1.33 A. according to the other]. Hydrogen positions were not determined, but it is considered that there are hydrogen bonds which in one case would have the length O–H–O = 2.57 A., and in the other 2.61 A.

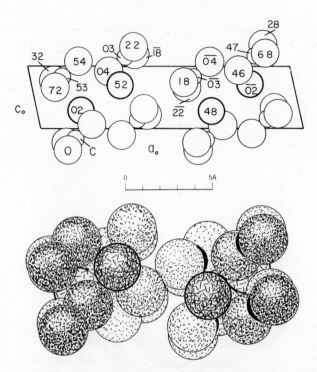

Fig. VIIB,3a (top). The monoclinic structure of KHCO₃ viewed along its b_0 axis. The potassium atoms are the heavily ringed circles. Origin in lower left.

Fig. VIIB,3b (bottom). A packing drawing to show the monoclinic KHCO₃ structure seen along its b_0 axis. The potassium atoms are line-shaded; oxygens are dotted.

VII,b3. According to a preliminary announcement published many years ago, crystals of *ammonium bicarbonate*, NH_4HCO_3, are orthorhombic, with a unit containing eight molecules and having the dimensions:

$$a_0 = 7.30 \text{ A.}; \quad b_0 = 10.81 \text{ A.}; \quad c_0 = 8.78 \text{ A.}$$

The space group is V_h^{10} (*Pccn*), with all atoms in the general positions:

$$(8e) \quad \pm(xyz; \ x+1/2,y+1/2,\bar{z}; \ 1/2-x,y,z+1/2; \ x,1/2-y,z+1/2)$$

The following parameters were considered to be approximately correct:

Atom	x	y	z
N	0.25	0.01	0.41
C	0.00	0.24	0.16
O(1)	−0.03	0.19	0.03
O(2)	−0.03	0.19	0.28
O(3)	0.06	0.35	0.17

They provide an arrangement of NH_4 and HCO_3 ions which is a distortion of the CsCl structure (Fig. VIIB,4). Viewed from another standpoint, the hydrogen atoms can be thought of as providing bonds that tie the carbonate groups into strings running parallel to the c_0 axis.

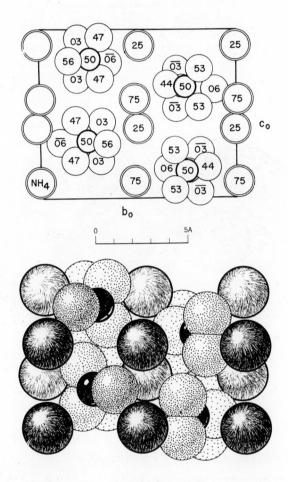

Fig. VIIB,4a (top). The orthorhombic NH_4HCO_3 structure projected along its a_0 axis. Carbon atoms are heavily ringed. Origin in lower right.
Fig. VIIB,4b (bottom). A packing drawing of the NH_4HCO_3 arrangement viewed along its a_0 axis. The carbon atoms are black, the oxygens are dotted.

VII,b4. Anhydrous *lithium carbonate*, Li_2CO_3, is monoclinic, with the tetramolecular cell:

$$a_0 = 8.39 \text{ A.}; \quad b_0 = 5.00 \text{ A.}; \quad c_0 = 6.21 \text{ A.}; \quad \beta = 114°30'$$

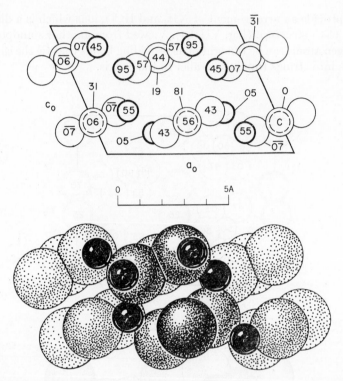

Fig. VIIB,5a (top). The monoclinic structure of Li_2CO_3 seen along its b_0 axis. The lithium atoms are the small heavily ringed circles; oxygens are the large circles. Origin in lower left.

Fig. VIIB,5b (bottom). A packing drawing of the monoclinic Li_2CO_3 structure viewed along its b_0 axis. The lithium atoms are black; the carbon atoms do not show.

Atoms have been placed in the following special and general positions of C_{2h}^6 $(C2/c)$:

C: (4e) $\pm(0\ u\ ^1/_4;\ ^1/_2,u+^1/_2,^1/_4)$ with $u = 0.057$

Li: (8f) $\pm(xyz;\ x,\bar{y},z+^1/_2;\ x+^1/_2,y+^1/_2,z;\ x+^1/_2,^1/_2-y,z+^1/_2)$

with $x = 0.203$, $y = 0.450$, $z = 0.840$

O(1): (4e) with $u = 0.313$
O(2): (8f) with $x = 0.145$, $y = -0.067$, $z = 0.320$

In this structure (Fig. VIIB,5) the carbonate ions have their usual planar shape, with C–O = 1.27 and 1.28 A. Each lithium atom has four oxygens tetrahedrally arranged about it at distances between 1.96 and 2.00 A.

VII,b5. The orthorhombic structure given *potassium silver carbonate*, $KAgCO_3$, has a unit cell with the edge lengths:

$$a_0 = 20.23 \text{ A.}; \quad b_0 = 5.75 \text{ A.}; \quad c_0 = 5.95 \text{ A.}$$

The eight molecules per cell are in the following positions of V_h^{27} (*Ibca*):

Ag:	(8e)	$\pm(0 \ ^1/_4 \ u; \ ^1/_2 \ ^3/_4 \ u)$; B.C.	with $u = 0.128$
K:	(8c)	$\pm(u \ 0 \ ^1/_4; \ u \ ^1/_2 \ ^3/_4)$; B.C.	with $u = -0.182$
C:	(8c)	with $u = 0.122$	
O(1):	(8c)	with $u = 0.187$	
O(2):	(16f)	$\pm(xyz; \ x,\bar{y},^1/_2-z; \ ^1/_2-x,y,\bar{z}; \ \bar{x},^1/_2-y,z)$; B.C.	

with $x = 0.089$, $y = 0.146$, $z = 0.378$

In the structure that results (Fig. VIIB,6), silver, potassium, and carbonate ions are distributed in a way that bears little relation to other structures. The carbonate ions are of the usual size and shape. Each silver ion has four oxygen atoms at a distance of 2.42 A.; each potassium atom has nine nearest oxygen neighbors, one at 2.65 A., four at 2.88 A., and four at 3.00 A.

VII,b6. *Potassium plutonyl carbonate*, $KPuO_2CO_3$, and analogous ammonium and rubidium compounds are hexagonal, with bimolecular cells of the dimensions:

$$KPuO_2CO_3: \quad a_0 = 5.09 \text{ A.}, \ c_0 = 9.83 \text{ A.}$$
$$NH_4PuO_2CO_3: \quad a_0 = 5.09 \text{ A.}, \ c_0 = 10.39 \text{ A.}$$
$$RbAmO_2CO_3: \quad a_0 = 5.12 \text{ A.}, \ c_0 = 10.46 \text{ A.}$$

Atoms have been found to be in the following special positions of D_{6h}^4 (*C6/mmc*):

K, NH$_4$, or Rb:	(2a)	$000; \ 0 \ 0 \ ^1/_2$
Pu or Am:	(2c)	$\pm(^1/_3 \ ^2/_3 \ ^1/_4)$
C:	(2d)	$\pm(^2/_3 \ ^1/_3 \ ^1/_4)$
O(1):	(6h)	$\pm(u \ 2u \ ^1/_4; \ 2\bar{u} \ \bar{u} \ ^1/_4; \ u \ \bar{u} \ ^1/_4)$
O(2):	(4f)	$\pm(^1/_3 \ ^2/_3 \ v; \ ^2/_3,^1/_3,v+^1/_2)$

For the potassium, ammonium, and rubidium compounds, $u = 0.812$, 0.812, and 0.811; for the same compounds, $v = 0.447$, 0.437, and 0.435.

The resulting arrangement is shown in Figure VIIB,7. In deducing it, the C–O distance was assumed to be 1.28 A. and the Pu–O(2) separation 1.94 A. The alkali atoms have 12 oxygen neighbors with, for $KPuO_2CO_3$,

K–O = 2.96 or 2.98 A. The plutonium atoms are distant from six carbonate oxygens with Pu–O(1) = 2.55 A.

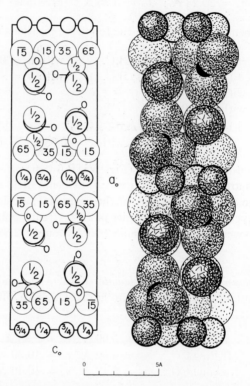

Fig. VIIB,6a (left). The orthorhombic structure of $KAgCO_3$ projected along the b_0 axis. The potassium are the large, the silver the small heavily ringed circles. The smaller light circles are carbon. Origin in lower right.

Fig. VIIB,6b (right). A packing drawing of the orthorhombic structure of $KAgCO_3$ viewed along its b_0 axis. The potassium atoms are line-shaded, the silver atoms are heavily ringed and dotted. The carbon atoms are black.

VII,b7. The mineral *dolomite*, $CaMg(CO_3)_2$, is a definite compound and not a solid solution of its two component carbonates, though its unit rhombohedral cell is very like theirs. This unimolecular rhombohedron has the dimensions:

$$a_0 = 6.0154 \text{ A.}, \qquad \alpha = 47°7'$$

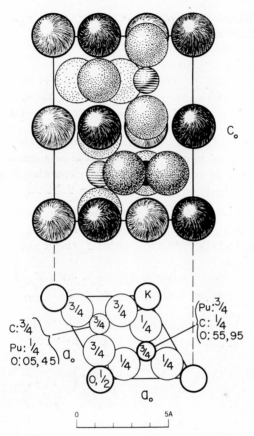

Fig. VIIB,7. Two projections of the hexagonal structure of $KPuO_2CO_3$. In the upper projection the carbon atom is black and the oxygen atoms dotted. The potassium atoms are large and fine-line shaded.

The symmetry is less than holohedral, with atoms in the following positions of C_{3i}^2 $(R\bar{3})$:

$$
\begin{array}{lll}
\text{Ca:} & (1a) & 000 \\
\text{Mg:} & (1b) & {}^1/_2\,{}^1/_2\,{}^1/_2 \\
\text{C:} & (2c) & \pm(uuu) \\
\text{O:} & (6f) & \pm(xyz;\ zxy;\ yzx)
\end{array}
$$

with $x = 0.4814$, $y = -0.0281$, $z = 0.2787$

Expressed in terms of the trimolecular hexagonal cell, with

$$a_0' = 4.8079 \text{ A.}, \qquad c_0' = 16.010 \text{ A.}$$

atoms are distributed as follows:

$$\text{Ca:} \quad (3a) \quad 000; \text{rh}$$
$$\text{Mg:} \quad (3b) \quad 0\,0\,{}^1/_2; \text{rh}$$
$$\text{C:} \quad (6c) \quad \pm(00u); \text{rh} \qquad \text{with } u = \text{ca. } {}^1/_4$$
$$\text{O:} \quad (18f) \quad \pm(xyz; \bar{y},x-y,z; y-x,\bar{x},z)$$
$$\text{with } x = 0.2374, \, y = -0.0347, \, z = 0.2440$$

Except for the small displacement of the oxygen atoms required by the presence of cations of two different sizes, the arrangement is practically identical with that of calcite (**VII,a1**). Its close atomic separations are C–O = 1.283 A., Mg–O = 2.095 A., Ca–O = 2.390 A.

In some dolomites, significant amounts of the magnesium atoms are replaced by ferrous iron with little change in cell dimensions. Thus in the mineral *ankerite*, $Ca(Mg,Fe)(CO_3)_2$, with 12% FeO:

$$a_0 = 6.062 \text{ A.}, \qquad \alpha = 46° \, 58'$$
$$a_0' = 4.832 \text{ A.}, \qquad c_0' = 16.14 \text{ A.}$$

The following three synthetic compounds with the dolomite arrangement are perhaps typical of others that could be synthesized:

$$CdMg(CO_3)_2: \quad a_0 = 5.898 \text{ A.}, \qquad \alpha = 47°47'$$
$$a_0' = 4.7770 \text{ A.}, \qquad c_0' = 15.641 \text{ A.}$$
$$CaMn(CO_3)_2: \quad a_0 = 6.140 \text{ A.}, \qquad \alpha = 46°49'$$
$$a_0' = 4.8797 \text{ A.}, \qquad c_0' = 16.367 \text{ A.}$$

It is considered that in this substance there is disorder in the cation distribution.

$$CaSn(BO_3)_2: \quad a_0 = 6.001 \text{ A.}, \qquad \alpha = 47°42'$$
$$a_0' = 4.853 \text{ A.}, \qquad c_0' = 15.920 \text{ A.}$$

VII,b8. The double carbonate of barium and calcium *barytocalcite*, $BaCa(CO_3)_2$, is monoclinic with the bimolecular cell:

$$a_0 = 8.15 \text{ A.}; \quad b_0 = 5.22 \text{ A.}; \quad c_0 = 6.58 \text{ A.}; \quad \beta = 106°8'$$

Atoms have been placed in the general positions of C_2^2 ($P2_1$):

$$(2a) \quad xyz; \bar{x}.y+{}^1/_2,\bar{z}$$

with the parameters of Table VIIB,1.

TABLE VIIB,1
Parameters of the Atoms in Barytocalcite

Atom	x	y	z
Ba	0.142	0.250	0.283
Ca	0.383	0.734	0.800
C(1)	0 083	0.205	0.770
C(2)	0.358	0.740	0.299
O(1)	0.012	0.150	0.893
O(2)	0.105	0.444	0.697
O(3)	0.133	0.020	0.639
O(4)	0.287	0.680	0.422
O(5)	0.379	0.974	0.226
O(6)	0.407	0.550	0.168

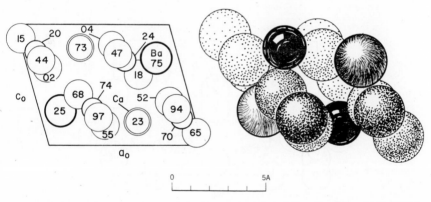

Fig. VIIB,8a (left). The simple monoclinic structure of barytocalcite, $BaCa(CO_3)_2$, projected along its b_0 axis. Origin in lower left.

Fig. VIIB,8b (right). A packing drawing of the barytocalcite arrangement seen along its b_0 axis. The calcium atoms are black, the barium atoms are fine-line shaded. Carbon atoms do not show.

The resulting structure (Fig. VIIB,8) is a composite of CO_3 groups of the usual size interspersed with calcium and barium atoms. The atoms of calcium have six oxygen neighbors at Ca–O $= 2.35$–2.62 A.; for barium the coordination is less definite, but there are six oxygens for which Ba–O lies within the range 2.56–2.64 A., with three more having Ba–O < 3.00 A.

The other mineral with this composition, alstonite, has the aragonite structure (VII,a2).

VII,b9. A structure has been proposed for the double carbonate *shortite*, $Na_2Ca_2(CO_3)_3$. Its symmetry is orthorhombic and it has a bimolecular unit of the dimensions:

$$a_0 = 4.99 \text{ A.}; \quad b_0 = 10.99 \text{ A.}; \quad c_0 = 7.11 \text{ A.}$$

Atoms have been placed in the following positions of C_{2v}^{14} (*Amm2*):

(2a) $00u; 0,^1/_2,u+^1/_2$

(2b) $^1/_2\,0\,u; \,^1/_2,^1/_2,u+^1/_2$

(4d) $0uv; 0\bar{u}v; 0,u+^1/_2,v+^1/_2; 0,^1/_2-u,v+^1/_2$

(4e) $^1/_2\,u\,v; \,^1/_2\,\bar{u}\,v; \,^1/_2.u+^1/_2,v+^1/_2; \,^1/_2,^1/_2-u,v+^1/_2$

(8f) $xyz; \bar{x}\bar{y}z; x,y+^1/_2,z+^1/_2; \bar{x},^1/_2-y,z+^1/_2;$
        $\bar{x}yz; x\bar{y}z; \bar{x},y+^1/_2,z+^1/_2; x.^1/_2-y,z+^1/_2$

with the positions and parameters of Table VIIB,2.

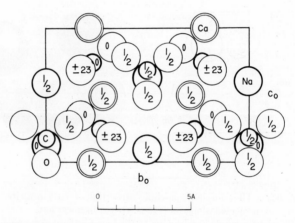

Fig. VIIB,9. The orthorhombic structure of shortite, $Na_2Ca_2(CO_3)_3$, projected along its a_0 axis. Origin in lower right.

The atomic arrangement to which this gives rise is shown in Figure VIIB,9. In it there are the usual carbonate ions, with C–O = 1.31 A. The coordination of oxygens about the metallic atoms is somewhat indefinite. Calcium has seven oxygen neighbors at distances between 2.32 and 2.52 A. There are four oxygens around the Na(1) atoms at distances of 2.48 and 2.66 A. and four more 2.92 A. away. Around the Na(2) ions are six oxygens at distances of 2.34 or 2.44 A. and one more at 2.56 A.

Undoubtedly, additional work should be done on this substance.

TABLE VIIB,2
Positions and Parameters of the Atoms in Shortite

Atom	Position	x	y	z
Ca	(4e)	$1/2$	0.213	−0.014
Na(1)	(2a)	0	0	0.130
Na(2)	(2b)	$1/2$	0	0.600
C(1)	(4d)	0	0.226	0 743
C(2)	(2b)	$1/2$	0	0.185
O(1)	(4d)	0	0.325	0.850
O(2)	(8f)	0.230	0.177	0.690
O(3)	(2b)	$1/2$	0	0.000
O(4)	(4e)	$1/2$	0 104	0 276

VII,b10. The double carbonate *huntite*, $Mg_3Ca(CO_3)_4$, is rhombohedral like the simpler carbonates, with a unimolecular cell of the dimensions:

$$a_0' = 6.075 \text{ A.}, \qquad \alpha = 102°56'$$

The trimolecular hexagonal cell has the edges:

$$a_0 = 9.505 \text{ A.}, \qquad c_0 = 7.821 \text{ A.}$$

What is considered to be a reasonably accurate structure based on D_3^7 (*R*32) places atoms in the following positions referred to the hexagonal axes:

Ca: (3a)   000; rh

Mg: (9d) u00; 0u0; $\bar{u}\bar{u}$0; rh   with u = 0.541

C(1):   (3b)   0 0 $1/2$; rh

C(2):   (9e) u 0 $1/2$; 0 u $1/2$; $\bar{u}$ $\bar{u}$ $1/2$; rh   with u = 0.461

O(1):   (9e)   with u = −0.135

O(2):   (9e)   with u = −0.404

O(3):   (18f) xyz; $\bar{y},x-y,z$; $y-x,\bar{x},z$;
                $yx\bar{z}$; $\bar{x},y-x,\bar{z}$; $x-y,\bar{y},\bar{z}$
                        with x = 0.461, y = 0.135, z = 0.506

In this structure (Fig. VIIB,10) one carbonate ion lies in the basal plane and the other three are slightly tilted with respect to it. The unique CO_3 has six magnesium neighbors with Mg–O = 2.09 or 2.10 A.; the others have two calciums and four magnesiums at Mg–O = 2.10 A. and Ca–O = 2.35 A. The C–O separation is 1.28 A.

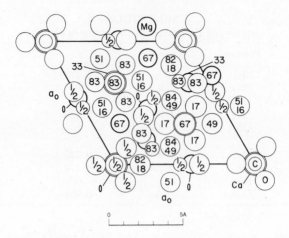

Fig. VIIB,10. A projection of the hexagonal (rhombohedral) structure of huntite, $Mg_3Ca(CO_3)_4$, along its principal axis.

VII,b11. The mineral *bastnäsite*, $(Ce,La)FCO_3$, is hexagonal, with a six molecule unit having the dimensions:

$$a_0 = 7.094 \text{ A.}, \qquad c_0 = 9.718 \text{ A.}$$

The following structure based on D_{3h}^4 $(C\bar{6}2c)$ has been proposed but not firmly established:

(Ce,La):   (6g) $\bar{u}\bar{u}0;\quad 0u0;\quad u00;$
                   $\bar{u}\;\bar{u}\;^1/_2;\; 0\;u\;^1/_2;\; u\;0\;^1/_2$ with $u = {}^1/_3$

F(1):   (2a) $000;\; 0\;0\;^1/_2$

F(2):   (4f) $\pm({}^1/_3\;{}^2/_3\;u;\; {}^2/_3,{}^1/_3,u+^1/_2$ with $u = 0$

C: (6h) $u\;v\;^1/_4;\; \bar{v},u-v,^1/_4;\; v-u,\bar{u},^1/_4;$
            $v\;u\;^3/_4;\; \bar{u},v-u,^3/_4;\; u-v,\bar{v},^3/_4$

                                        with $u = {}^1/_3,\; v = 0.245$

O(1):   (6h)   with $u = {}^1/_3,\; v = 0.07$

O(2):   (12i) $xyz;\qquad \bar{y},x-y,z;\qquad y-x,\bar{x},z;$
                $x,y,^1/_2-z;\; \bar{y},x-y,^1/_2-z;\; y-x,\bar{x},^1/_2-z;$
                $y,x,z+^1/_2;\; \bar{x},y-x,z+^1/_2;\; x-y,\bar{y},z+^1/_2;$
                $yx\bar{z};\qquad \bar{x},y-x,\bar{z};\qquad x-y,\bar{y},\bar{z}$

                                  with $x = {}^1/_3,\; y = {}^1/_3,\; z = 0.14$

The resulting arrangement is indicated in Figure VIIB,11.

Bastnäsite can be considered as the simplest member of a series of double carbonates which are either hexagonal or pseudohexagonal and have units of related sizes. Atomic arrangements have been suggested but not determined, and this may be difficult because some of these minerals yield x-ray patterns indicative of a considerable degree of disorder.

These minerals are:

Synchisite: $CeFCO_3 \cdot CaCO_3$. This mineral is orthorhombic or monoclinic pseudohexagonal, with a hexagonal pseudocell of the dimensions:

$$a_0' = 7.107 \text{ A.}, \qquad c_0' = 18.24 \text{ A.}$$

There is some evidence that in the fully ordered condition the true c_0 may be three times greater.

Parisite: $2CeFCO_3 \cdot CaCO_3$. This is rhombohedral with a cell which, in terms of hexagonal axes, has the dimensions:

$$a_0 = 7.176 \text{ A.}, \qquad c_0 = 28.04 \text{ A.}$$

or perhaps the still longer $c_0' = 84.11$ A.

Roentgenite: $3CeFCO_3 \cdot 2CaCO_3$. This also is rhombohedral, with the hexagonal cell:

$$a_0 = 7.131 \text{ A.}, \qquad c_0 = 23.14 \text{ A.}$$

or perhaps the longer 69.42 A.

A related barium compound *cordylite*, $2CeFCO_3 \cdot BaCO_3$, has been given the hexagonal cell:

$$a_0' = 7.53 \text{ A.}, \qquad c_0' = 22.8 \text{ A.}$$

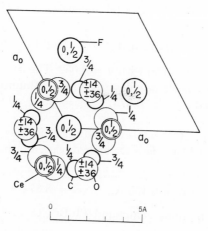

Fig. VIIB,11. A projection along its c_0 axis of some of the atoms in the unit cell of hexagonal (Ce,La)FCO_3.

VII,b12.   The structure assigned to *northupite*, $Na_2Mg(CO_3)_2 \cdot NaCl$, and the corresponding brominated compound, is based on the cubic space group T_h^4 (*Fd3*). Their unit cubes with

$$a_0 = 14.01 \text{ A. for } Na_2Mg(CO_3)_2 \cdot NaCl$$

and

$$a_0 = 14.20 \text{ A. for } Na_2Mg(CO_3)_2 \cdot NaBr$$

contain 16 molecules, and atoms have been placed in the following positions (for northupite):

Mg: (16*d*) $^5/_8\,^5/_8\,^5/_8$; $^5/_8\,^7/_8\,^7/_8$; $^7/_8\,^5/_8\,^7/_8$; $^7/_8\,^7/_8\,^5/_8$; F.C.

Cl: (16*c*) $^1/_8\,^1/_8\,^1/_8$; $^1/_8\,^3/_8\,^3/_8$; $^3/_8\,^1/_8\,^3/_8$; $^3/_8\,^3/_8\,^1/_8$; F.C.

C: (32*e*) uuu; $u\bar{u}\bar{u}$; $^1/_4-u,^1/_4-u,^1/_4-u$; $^1/_4-u,u+^1/_4,u+^1/_4$;

$\bar{u}u\bar{u}$; $\bar{u}\bar{u}u$; $u+^1/_4,^1/_4-u,u+^1/_4$; $u+^1/_4,u+^1/_4,^1/_4-u$; F.C.

with $u = 0.405$

Na: (48*f*) $u00$; $\bar{u}00$; $u+^1/_4,^1/_4,^1/_4$; $^1/_4-u,^1/_4,^1/_4$; tr; F.C.

with $u = 0.225$

O: (96*g*) xyz; $x\bar{y}\bar{z}$; $^1/_4-x,^1/_4-y,^1/_4-z$; $^1/_4-x,y+^1/_4,z+^1/_4$;

$\bar{x}y\bar{z}$; $\bar{x}\bar{y}z$; $x+^1/_4,^1/_4-y,z+^1/_4$; $x+^1/_4,y+^1/_4,^1/_4-z$; tr; F.C.

with $x = 0.392$, $y = 0.348$, $z = 0.475$

Essentially the same arrangement has been given *tychite*, $Na_4Mg_2(CO_3)_4 \cdot Na_2SO_4$. The eight molecules in its unit cube, with

$$a_0 = 13.93 \text{ A.}$$

have analogous positions except that in place of the 16 chlorine atoms there are eight sulfurs in

(8*a*) 000; $^1/_4\,^1/_4\,^1/_4$; F.C.

and 32 more oxygens in (32*e*) with $u = 0.062$. The parameters for the other atoms, in the positions already described for northupite, are $u(C) = 0.400$, $u(Na) = 0.225$, and $x(O) = 0.375$, $y(O) = 0.352$, $z(O) = 0.473$.

VII,b13.   There have been several studies of *phosgenite*, the chloro-carbonate of lead, $Pb_2Cl_2CO_3$. It is tetragonal, with a tetramolecular cell of the edge lengths:

$$a_0 = 8.155 \text{ A.,} \qquad c_0 = 8.874 \text{ A.}$$

The analogous bromine compound $Pb_2Br_2CO_3$ has the same structure, with

$$a_0 = 8.354 \text{ A.,} \qquad c_0 = 9.074 \text{ A.}$$

The space group is D_{4h}^5 $(P4/mbm)$, with atoms in the following positions:

(4e) $\pm (00u; \, {}^1/_2 \, {}^1/_2 \, u)$

(4g) $\pm (u,u+{}^1/_2,0; \, u+{}^1/_2,\bar{u},0)$

(4h) $\pm (u,u+{}^1/_2,{}^1/_2; \, u+{}^1/_2,\bar{u},{}^1/_2)$

(8k) $\pm (u,u+{}^1/_2,v; \, \bar{u},{}^1/_2-u,v; \, u+{}^1/_2,\bar{u},v; \, {}^1/_2-u,u,v)$

It would appear that the correct assignment of positions and parameters is that of Table VIIB,3.

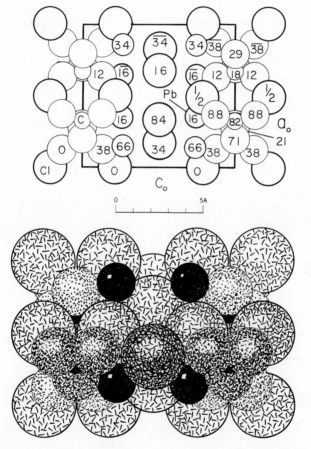

Fig. VIIB,12a (top). The tetragonal structure of phosgenite, $Pb_2Cl_2CO_3$, seen along an a_0 axis. Origin in lower right.

Fig. VIIB,12b (bottom). A packing drawing of the tetragonal structure of $Pb_2Cl_2CO_2$ seen along an a_0 axis. The larger black circles are lead, the smaller black carbon atoms scarcely show. The chlorine atoms are line-shaded.

TABLE VIIB,3
Positions and Parameters of the Atoms in $Pb_2Cl_2CO_3$ and $Pb_2Br_2CO_3$[a]

Atom	Position	x	y	z
Pb	$(8k)$	0.165 (0.160)	0.665 (0.660)	0.260 (0.254)
Cl(1)	$(4e)$	0	0	0 204 (0.215)
Cl(2)	$(4h)$	0.342 (0.335)	0.842 (0.835)	$1/2$
O(1)	$(4g)$	0 215 (0.222)	0.715 (0.722)	0
O(2)	$(8k)$	0.378 (0.381)	0.878 (0.881)	0.123 (0.120)
C	$(4g)$	0.324 (0.328)	0.824 (0.828)	0

[a] Parameters for the bromide, where they differ from those in phosgenite, are given in parentheses.

In this structure (Fig. VIIB,12) the lead and halogen positions could be established from the x-ray data; the carbonate positions have been assigned by spatial and symmetry considerations. In the chloride, the shortest Pb–Cl = 2.95 A., Pb–O = 2.37 A., Cl–O = 3.31 A., and Cl–Cl = 3.62 A. In the bromide the analogous distances are Pb–Br = 3.03 A., Pb–O = 2.41 A., Br–O = 3.44 A., and Br–Br = 3.90 A.

VII,b14. A recent reconsideration of earlier work on the structure of *malachite*, $Cu_2(OH)_2CO_3$, together with some new data have led to a more satisfactory atomic arrangement. There are four molecules in a monoclinic unit of the dimensions:

$$a_0 = 9.48 \text{ A.}; \quad b_0 = 12.03 \text{ A.}; \quad c_0 = 3.21 \text{ A.}; \quad \beta = 98°$$

The space group is C_{2h}^5 ($P2_1/a$), with all atoms in the general positions:

$$(4e) \quad \pm(xyz; \ x+1/2, 1/2-y, z)$$

The chosen parameters are those of Table VIIB,4.

TABLE VIIB,4. Parameters of the Atoms in Malachite

Atom	x	y	z
Cu(1)	0.00	0.21	0.89
Cu(2)	0.235	0.39	0.38
O(1)	0.14	0 13	0.28
O(2)	0.34	0.24	0.50
O(3)	0.33	0.05	0.63
OH(1)	0.09	0.36	0.92
OH(2)	0.39	0.43	0.86
C	0.27	0.13	0.47

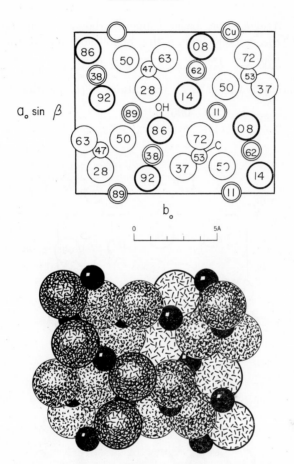

a₀ sin β

b₀

0   5A

Fig. VIIB,13a (top). The monoclinic structure of malachite, $Cu_2(OH)_2CO_3$, seen along its c_0 axis. Origin in lower left.

Fig. VIIB,13b (bottom). A packing drawing of the monoclinic malachite structure seen along its c_0 axis. Both the copper and the carbon atoms are black, the carbon atoms only partly showing at the center of their surrounding dotted oxygen atoms. The hydroxyl groups are line-shaded.

As can be seen from Figure VIIB,13, the structure is one in which each copper atom has two hydroxyl and two carbonate oxygen atoms as neighbors, approximately at the corners of a square; for Cu(1) these distances lie between 1.95 and 2.00 A., for Cu(2) they are 1.91–2.07 A. In addition, each Cu(1) has two more oxygens about 2.71 A. away and Cu(2) has two additional hydroxyls at a distance of ca. 2.41 A.

VII,b15. *Azurite*, the basic copper carbonate $2CuCO_3 \cdot Cu(OH)_2$ or $Cu_3(OH)_2(CO_3)_2$, is monoclinic, with a bimolecular unit of the dimensions:

$$a_0 = 5.00 \text{ A.}; \quad b_0 = 5.85 \text{ A.}; \quad c_0 = 10.35 \text{ A.}; \quad \beta = 92°20'$$

The space group is C_{2h}^5 $(P2_1/c)$, with atoms in the positions:

$$(2a) \quad 000; 0\, ^1/_2\, ^1/_2$$
$$(4e) \quad \pm(xyz; x, ^1/_2 - y, z + ^1/_2)$$

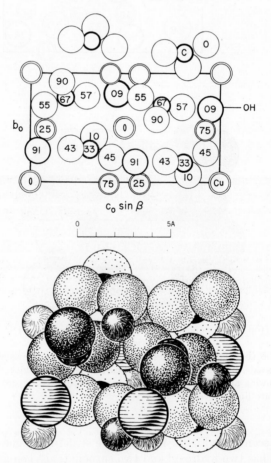

Fig. VIIB,14a (top). The monoclinic structure of azurite, $Cu_3(OH)_2(CO_3)_2$, projected along its a_0 axis. Origin in lower left.

Fig. VIIB,14b (bottom). A packing drawing of the azurite structure viewed along its a_0 axis. The carbon atoms are black and the small copper atoms are fine-line shaded. Oxygen atoms are dotted and the hydroxyls are heavily ringed.

TABLE VIIB,5. Positions and Parameters of the Atoms in Azurite

Atom	Position	x	y	z
Cu(1)	(2a)	0	0	0
Cu(2)	(4e)	0.252	0.495	0.085
OH	(4e)	0.092	0.812	0.444
C	(4e)	0.329	0.298	0.319
O(1)	(4e)	0.098	0.390	0.338
O(2)	(4e)	0.447	0.224	0.421
O(3)	(4e)	0.431	0.303	0.212

A recent redetermination has given results that conflict with those of an earlier study in all but the copper positions. According to this later and presumably correct structure, atoms have the positions and parameters listed in Table VIIB,5.

The atomic distribution in this arrangement is shown in Figure VIIB,14. The two copper atoms are differently coordinated. Each Cu(1) atom has around it an approximate square of oxygen atoms and hydroxyl radicals with Cu–2O = 1.88 A. and Cu–2OH = 1.98 A. Around each Cu(2) is a similar approximate square of 2O and 2OH with Cu–O(or OH) = 1.92–2.04 A.; there is also a third oxygen somewhat more distant (Cu–O = 2.38 A.). The carbonate groups have their usual triangular shapes with C–O = 1.24–1.30 A.

VII,b16. An approximate structure based on electron diffraction data has been described for white lead, the *basic lead carbonate*, $2PbCO_3 \cdot Pb(OH)_2$ or $Pb_3(OH)_2(CO_3)_2$. The observed symmetry is hexagonal with a nine-molecule unit of the dimensions:

$$a_0 = 9.06 \text{ A.}, \qquad c_0 = 24.8 \text{ A.}$$

It is concluded that the arrangement as built up along the c_0 axis consists of repetitions of two $PbCO_3$ layers with an interleaved $Pb(OH)_2$. The data indicate that this subunit has the dimensions:

$$a_0' = 9.06 \text{ A.}, \qquad c_0' = 8.27 \text{ A.}$$

In it, the atoms would be in the following special positions of C_{3v}^2 (P31m):

(1a) 00u

(2b) $^1/_3\,^2/_3\,u;\ ^2/_3\,^1/_3\,u$

(3c) $u0v;\ 0uv;\ \bar{u}\bar{u}v$

(6d) $xyz;\ \bar{y},x-y,z;\ y-x,\bar{x},z;$
      $yxz;\ \bar{x},y-x,z;\ x-y,\bar{y},z$

with the parameters of Table VIIB,6.

TABLE VIIB,6
Positions and Parameters of Atoms in the Subcell of $2PbCO_3 \cdot Pb(OH)_2$

Atom	Position	x	y	z
Pb(1)	(1a)	0	0	0.77
Pb(2)	(2b)	$1/_3$	$2/_3$	0.77
Pb(3)	(3c)	0.255	0	0.50
Pb(4)	(3c)	0.660	0	0.23
C(1)	(1a)	0	0	0.14
C(2)	(2b)	$1/_\circ$	$2/_\circ$	0.14
C(3)	(3c)	0.66	0	0.86
O(1)	(3c)	0.16	0	0.14
O(2)	(6d)	0.53	0 16	0 14
O(3)	(3c)	0.51	0	0.86
O(4)	(6d)	0.34	0.18	0.86
OH(1)	(1a)	0	0	0.42
OH(2)	(2b)	$1/_3$	$2/_3$	0.42
OH(3)	(3c)	0.70	0	0.58

There is indefiniteness in the way these subcells are stacked one above another. The stacking is considered to be approximately but not exactly that of the rhombohedral repetition; and the existence of this irregularity is supported by the fact that the x-ray patterns from white lead are very poor in lines.

In a subunit as defined above (Fig. VIIB,15), two thirds of the lead atoms are surrounded by six carbonate oxygens on one side (Pb–O = 2.75 A.) and by four hydroxyls on the other (Pb–O = 2.75 and 3.1 A.). The nearest oxygens to the other lead atoms all belong to hydroxyl groups, with Pb–O = 2.4 and 2.6 A. Best agreement with the data was obtained by giving the C–O separations in the carbonate ions the unusually large value of 1.45 A.

VII,b17. *Calcium nitrate*, $Ca(NO_3)_2$, and three crystals isomorphous with it have a structure related to that of fluorite (**IV,a1**). They are cubic, with tetramolecular cells of the edge lengths:

$$Ca(NO_3)_2: \quad a_0 = 7.590 \text{ A.}$$
$$Sr(NO_3)_2: \quad a_0 = 7.7798 \text{ A. } (20°C.)$$
$$Ba(NO_3)_2: \quad a_0 = 8.11 \text{ A.}$$
$$Pb(NO_3)_2: \quad a_0 = 7.853 \text{ A.}$$

Atoms are in the following positions of T_h^6 ($Pa3$):

R: (4a) $000; \, ^1/_2\,^1/_2\,0; \, ^1/_2\,0\,^1/_2; \, 0\,^1/_2\,^1/_2$

N: (8c) $\pm\,(uuu; \, u+^1/_2,^1/_2-u,\bar{u}; \, \bar{u},u+^1/_2,^1/_2-u; \, ^1/_2-u,\bar{u},u+^1/_2)$

O: (24d) $\pm\,(xyz; \qquad\qquad zxy; \qquad\qquad yzx;$
$\qquad\qquad x+^1/_2,^1/_2-y,\bar{z}; \, z+^1/_2,^1/_2-x,\bar{y}; \, y+^1/_2,^1/_2-z,\bar{x};$
$\qquad\qquad \bar{x},y+^1/_2,^1/_2-z; \, \bar{z},x+^1/_2,^1/_2-y; \, \bar{y},z+^1/_2,^1/_2-x;$
$\qquad\qquad ^1/_2-x,\bar{y},z+^1/_2; \, ^1/_2-z,\bar{x},y+^1/_2; \, ^1/_2-y,\bar{z},x+^1/_2)$

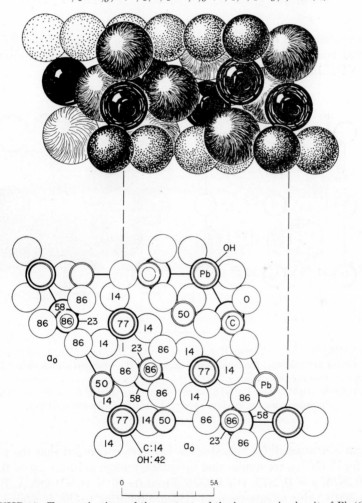

Fig. VIIB,15. Two projections of the contents of the hexagonal subunit of $Pb_3(OH)_2$-$(CO_3)_2$. In the upper projection oxygen atoms are dotted and OH groups are solid black. The lead atoms are large and fine-line shaded. Carbon atoms do not show.

Parameters assigned on the basis of x-ray data to three of these compounds are given below:

Crystal	u	x	y	z
$Ca(NO_3)_2$	0.339	0.253	0.293	0.467
$Sr(NO_3)_2$	0.341	0.264	0.291	0.468
$Ba(NO_3)_2$	0.350	0.280	0.296	0.474

These yield the expected planar nitrate ions displaced from the symmetrical positions of the CaF_2 structure by their packing requirements (Fig. VIIB,16).

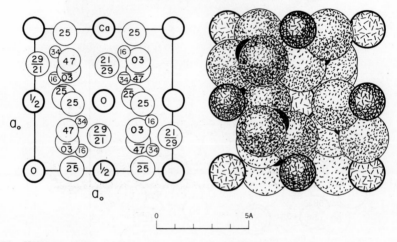

Fig. VIIB,16a (left).   A projection of the $Ca(NO_3)_2$ structure along a cubic axis. The heavy circles are calcium; the small light circles are the nitrogen atoms. Right-handed coordinates used for this drawing.

Fig. VIIB,16b (right). The $Ca(NO_3)_2$ structure viewed along a cubic axis. The line-shaded circles are calcium atoms, the black circles are nitrogen.

From a neutron diffraction study it has been concluded that the parameters for $Pb(NO_3)_2$ are similar and presumably more accurate: $u = 0.3535$, $x = 0.2875$, $y = 0.2875$, $z = 0.4856$. From them, N–O = 1.268 A. Heated up to 300°C. this nitrate gave no evidence pointing to the development of the type of hindered rotation of the nitrate ions that has been found in $NaNO_3$, for instance (**VII,a1**).

VII,b18.   Crystals of *ammonium trinitrate*, $NH_4NO_3 \cdot 2HNO_3$, are considered to be monoclinic, though, except for one weak reflection, their x-ray data are compatible with orthorhombic symmetry. The structure described for them has been based on this higher pseudosymmetry. Its unit contains two molecules and has the edge lengths:

$$a_0 = 6.57 \text{ A.}; \quad b_0 = 12.64 \text{ A.}; \quad c_0 = 4.56 \text{ A.}$$

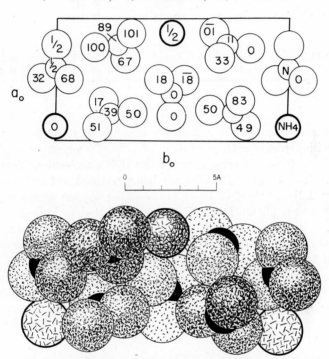

Fig. VIIB,17a (top). A projection of the orthorhombic structure of $NH_4NO_3 \cdot 2HNO_3$ along the c_0 axis. Origin in lower left.
    Fig. VIIB,17b (bottom). A packing drawing of the orthorhombic structure of NH_4-$NO_3 \cdot 2HNO_3$ viewed along the c_0 axis. The ammonium ions are line-shaded; the oxygens are dotted. Hydrogen atoms are not shown.

Choosing the space group as V^3 ($P22_12_1$), atoms have been placed in the following positions:

$$(2a) \quad u00; \; \bar{u} \, ^1/_2 \, ^1/_2$$
$$(2b) \quad v \, ^1/_2 \, 0; \; \bar{v} \, 0 \, ^1/_2$$
$$(4c) \quad xyz; \; x\bar{y}\bar{z}; \; \bar{x},^1/_2-y,z+^1/_2; \; \bar{x},y+^1/_2,^1/_2-z$$

The assigned positions and parameters are listed in Table VIIB,7.

TABLE VIIB,7

Positions and Parameters of the Atoms in $NH_4NO_3 \cdot 2HNO_3$

Atom	Position	x	y	z
NH_4	(2a)	0.120	0	0
N(1)	(2b)	0.389	$1/2$	0
O(1)	(2b)	0.208	$1/2$	0
O(2)	(4c)	0.489	0.447	0.185
N(2)	(4c)	0.223	0.234	0.386
O(3)	(4c)	0.324	0.213	0.170
O(4)	(4c)	0.107	0.177	0.515
O(5)	(4c)	0.237	0.331	0.496

The structure, as shown in Figure VIIB,17, is approximately one in which planes of ammonium and nitrate ions normal to the b_0 axis are interleaved with planes of nitric acid molecules. The NO_3 groups are planar but do not have threefold symmetry; in the HNO_3 molecules, N–O = 1.20, 1.22, and 1.33 A., the latter distance being ascribed to the N–OH separation. A relatively short distance between the oxygen atoms of a HNO_3 molecule and a nitrate ion (2.63 A.) is attributed to hydrogen bonding. The other close approaches of oxygen atoms range between 3.04 and 3.33 A. Each ammonium ion has 12 neighboring oxygen atoms at distances between 3.01 and 3.24 A.

VII,b19. *Cesium uranyl nitrate*, $Cs(UO_2)(NO_3)_3$, is rhombohedral, with a bimolecular cell of the dimensions:

$$a_0 = 8.56 \text{ A.,} \qquad \alpha = 68°30'$$

The corresponding hexagonal cell containing six molecules has

$$a_0' = 9.64 \text{ A.,} \qquad c_0' = 19.51 \text{ A.}$$

The assigned space group is D_{3d}^6 ($R\bar{3}c$), with atoms in the approximate positions:

Cs: (6a) $\pm(0\ 0\ 1/4)$; rh

U: (6b) $000; 0\ 0\ 1/2$; rh

N: (18e) $\pm(u\ 0\ 1/4; 0\ u\ 1/4; \bar{u}\ \bar{u}\ 1/4)$; rh   with $u = 0.30$–0.35

O(1): (18e) with $u = 0.43$–0.48

O(2): (36f) $\pm(xyz; \qquad \bar{y},x-y,z; \qquad y-x,\bar{x},z;$

               $\bar{y},\bar{x},z+1/2; x,x-y,z+1/2; y-x,y,z+1/2)$; rh

                     with $x = 0.30$–0.35, $y = 0.125$, $z = 0.25$

An equally approximate determination has been made of $Rb(UO_2)$-$(NO_3)_3$. For it,

$$a_0 = 8.30 \text{ A.,} \qquad \alpha = 68°42'$$
$$a_0' = 9.36 \text{ A.,} \qquad c_0' = 18.88 \text{ A.}$$

The selected atomic parameters are $u(N) = \frac{1}{3}$, $u(O1) = 0.46$, $x = \frac{1}{3}$, $y = 0.13$, $z = 0.25$.

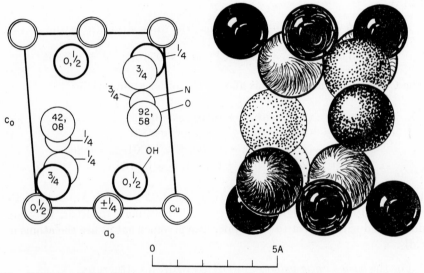

Fig. VIIB,18a (left). The monoclinic structure of $Cu_2(OH)_3NO_3$ projected along its b_0 axis. Origin in lower left.

Fig. VIIB,18b (right). A packing drawing of the $Cu_2(OH)_3NO_3$ arrangement viewed along its b_0 axis. The nitrogen atoms do not show. Copper atoms are black and nitrate oxygens are dotted. The hydroxyls are heavily outlined.

VII,b20. The *basic nitrate of copper*, $Cu_2(OH)_3NO_3$, is monoclinic, with a bimolecular cell of the dimensions:

$$a = 5.576 \text{ A.;} \quad b_0 = 6.050 \text{ A.;} \quad c_0 = 6.896 \text{ A.;} \quad \beta = 94°30'$$

The space group has been given as C_{2h}^2 ($P2_1/m$), with atoms in the positions:

Cu(1): (2a) 000; 0 $\frac{1}{2}$ 0
Cu(2): (2e) $\pm(u\,\frac{1}{4}\,v)$ with $u = 0.500$, $v = 0.000$
OH(1): (2e) with $u = 0.867$, $v = 0.833$
OH(2): (4f) $\pm(xyz; x,\frac{1}{2}-y,z)$
                with $x = 0.333$, $y = 0.00$, $z = 0.842$

$$N: \quad (2e) \quad \text{with } u = 0.200, v = 0.391$$
$$O(1): \quad (2e) \quad \text{with } u = 0.200, v = 0.217$$
$$O(2): \quad (4f) \quad \text{with } x = 0.200, y = 0.076, z = 0.479$$

In this arrangement, as shown in Figure VIIB,18, each copper atom has around it six oxygen and hydroxyl neighbors at distances between 2.00 and 2.35 A. The nitrate ions have their usual triangular shapes, with $N-O = 1.21$ A.

Basic cupric bromide, $Cu_2(OH)_3Br$, has the same structure, with

$$a_0 = 5.64 \text{ A.}; \quad b_0 = 6.139 \text{ A.}; \quad c_0 = 6.056 \text{ A.}; \quad \beta = 93°30'$$

The Cu(1) atoms are in $(2a)$ as above. The other atoms are in:

$$Cu(2): \quad (2e) \quad \text{with } u = 0.504, v = 0.003$$
$$OH(1): \quad (2e) \quad \text{with } u = 0.883, v = 0.837$$
$$OH(2): \quad (4f) \quad \text{with } x = 0.300, y = 0.004, z = 0.863$$
$$Br: \quad (2e) \quad \text{with } u = 0.2104, v = 0.3767$$

In this arrangement, Cu–OH lies between ca. 1.92 and 2.00 A., and Cu–Br = ca. 2.80 or ca. 2.99 A. Considering all the anions together, they are distributed with respect to the copper atoms somewhat as are the atoms in CdI_2.

For the *basic cupric iodide*, $Cu_2(OH)_3I$, with this structure:

$$a_0 = 5.653 \text{ A.}; \quad b_0 = 6.157 \text{ A.}; \quad c_0 = 6.560 \text{ A.}; \quad \beta = 95°10'$$

With atoms in the positions chosen for the analogous bromide, the parameters are:

$$Cu(2): \quad (2e) \quad \text{with } u = 0.509, v = 0.003$$
$$OH(1): \quad (2e) \quad \text{with } u = 0.881, v = 0.838$$
$$OH(2): \quad (4f) \quad \text{with } x = 0.311, y = -0.005, z = 0.861$$
$$I: \quad (2e) \quad \text{with } u = 0.2077, v = 0.3806$$

One of the several forms of the corresponding chloride, $CuCl_2 \cdot 3Cu(OH)_2$, has this structure. The dimensions of its unit are

$$a_0 = 5.65 \text{ A.}; \quad b_0 = 6.11 \text{ A.}; \quad c_0 = 5.73 \text{ A.}; \quad \beta = 93°45'$$

Other forms are treated in Chapter XI.

Sulfites, etc.

VII,b21. Crystals of *sodium sulfite*, Na_2SO_3, contain two molecules in the hexagonal unit:

$$a_0 = 5.441 \text{ A.}, \qquad c_0 = 6.133 \text{ A.}$$

Their atoms have been put in the following positions of C_{3i}^1 ($C\bar{3}$):

Na(1): (1a) 000
Na(2): (1b) $0\ 0\ {}^1/_2$
Na(3): (2d) $\quad {}^1/_3\,{}^2/_3\,u;\ {}^2/_3\,{}^1/_3\,\bar{u}$ with $u = 0.67$
   S: (2d)   with $u' = 0.17$
   O: (6g) $\pm(xyz;\ y-x,\bar{x},z;\ \bar{y},x-y,z)$
                           with $x = 0.14,\ y = 0.40,\ z = 0.25$

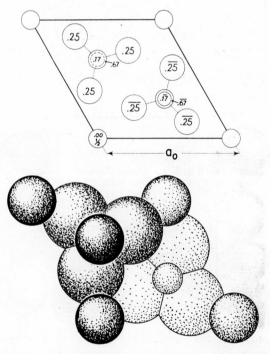

Fig. VIIB,19a (top). A basal projection of the hexagonal unit of the structure found for Na_2SO_3. The largest circles are the oxygen, the smallest the sulfur atoms. These atoms, forming SO_3^{2-} ions, are joined by light lines.

Fig. VIIB,19b (bottom). A packing drawing of the Na_2SO_3 arrangement seen along its c_0 axis. The sulfur atoms are the smallest, the oxygens the largest circles.

The SO_3 ions in this structure (Fig. VIIB,19) are like ClO_3 groups in being trigonal pyramids. The oxygen atoms that constitute the base of the SO_3 pyramids are 2.23 A. apart; sulfur, at its apex, is 1.39 A. distant from each oxygen atom. The closest approach of sodium and oxygen atoms is the expected 2.45 A.

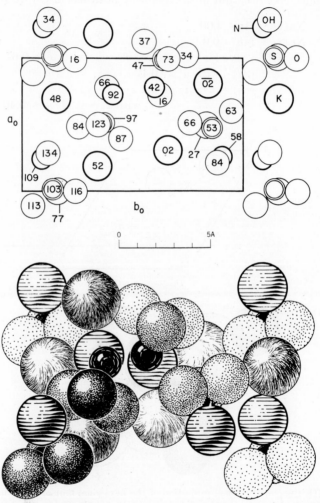

Fig. VIIB,20a (top). The orthorhombic structure of $KSO_3(NHOH)$ projected along its c_0 axis. Origin in lower left.

Fig. VIIB,20b (bottom). A packing drawing of the $KSO_3(NHOH)$ arrangement viewed along its c_0 axis. The nitrogen atoms are black, the potassium atoms large and fine-line shaded. Oxygen atoms are dotted. Sulfur atoms are black but scarcely show.

TABLE VIIB,8. Parameters of the Atoms in $KSO_3(NHOH)$

Atom	x	y	z
K	0.185	0.340	0.519
S	0.015	0.140	0.032
N	0.235	0.094	0.077
OH	0.284	0.115	0.336
O(2)	0.984	0.246	0.156
O(3)	0.013	0.153	0.772
O(4)	0.893	0.051	0.126

VII,b22. Crystals of *potassium hydroxyl amine-N-sulfonate*, KSO_3-(NHOH), have orthorhombic symmetry with a tetramolecular cell of the dimensions:

$$a_0 = 7.06 \text{ A.}; \quad b_0 = 12.02 \text{ A.}; \quad c_0 = 5.58 \text{ A.}$$

The space group is V^4 ($P2_12_12_1$), with all atoms in its general positions:

$$(4a) \quad xyz; \ {}^1/_2-x,\bar{y},z+{}^1/_2; \ x+{}^1/_2,{}^1/_2-y,\bar{z}; \ \bar{x},y+{}^1/_2,{}^1/_2-z$$

The determined atomic parameters are those of Table VIIB,8.

In the resulting structure (Fig. VIIB,20), the $SO_3(NHOH)$ anion has the dimensions of Figure VIIB,21, the sulfur atom being surrounded by a tetrahedron of three oxygen and one nitrogen atoms. Each potassium atom has eight oxygen neighbors at distances between 2.71 and 2.99 A.

It is considered that the anions in the structure are tied together by hydrogen bonds which are either of the type N—H· · ·O of length 2.86 A. or of type N· · ·H—O of length 2.90 A.

This structure is to be compared with that of potassium sulfamate, KSO_3NH_2 (Chapter VIII), and that of the amine disulfonate, $K_2NH(SO_3)_2$ (**VII,b23**).

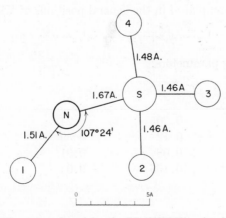

Fig. VIIB,21. Dimensions found for the $SO_3(NHOH)$ anion in $KSO_3(NHOH)$.

VII,b23. Crystals of *potassium amine disulfonate*, $K_2NH(SO_3)_2$, are monoclinic, with a tetramolecular cell of the dimensions:

$$a_0 = 12.430 \text{ A.}; \quad b_0 = 7.458 \text{ A.}; \quad c_0 = 7.175 \text{ A.}; \quad \beta = 91°11'$$

The space group is C_{2h}^6 $(C2/c)$ with the nitrogen atoms in

$$(4e) \quad \pm(0 \; u \; ^1/_4; \; ^1/_2, u+^1/_2, ^1/_4) \qquad \text{with } u = 0.4247$$

and all other atoms in general positions:

$$(8f) \quad \pm(xyz; \; x,\bar{y},z+^1/_2; \; x+^1/_2,y+^1/_2,z; \; x+^1/_2, ^1/_2-y, z+^1/_2)$$

with the following parameters:

Atom	x	y	z
K	0.3476	0.6414	0.6428
S	0.3985	0.1783	0.6397
O(1)	0.4422	0.2711	0.4810
O(2)	0.3276	0.0330	0.5879
O(3)	0.3522	0.3000	0.7721

The hydrogen positions were not determined.

This is a structure (Fig. VIIB,22) composed of K^+ and $NH(SO_3)_2^-$ ions. In the complex anions having the dimensions of Figure VIIB,23, sulfur atoms are tetrahedrally surrounded by three oxygen and the nitrogen atoms. K–O separations range upwards from 2.70 A.

VII,b24. *Selenious acid*, H_2SeO_3, forms orthorhombic crystals that have four molecules in a unit cell of the dimensions:

$$a_0 = 9.15 \text{ A.}; \quad b_0 = 6.00 \text{ A.}; \quad c_0 = 5.05 \text{ A.}$$

All atoms have been placed in the general positions of V^4 $(P2_12_12_1)$:

$$(4a) \quad xyz; \; ^1/_2-x,\bar{y},z+^1/_2; \; x+^1/_2, ^1/_2-y,\bar{z}; \; \bar{x},y+^1/_2, ^1/_2-z$$

with the following parameters:

Atom	x	y	z
Se	0.205	0.243	0.115
O(1)	0.12	0.39	0.365
O(2)	0.08	0.31	−0.135
O(3)	0.155	−0.04	ca. 0.133

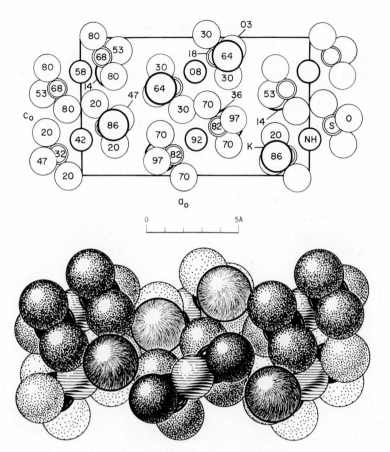

Fig. VIIB,22a (top). The monoclinic structure of $K_2NH(SO_3)_2$ projected along its b_0 axis. Origin in lower left.

Fig. VIIB,22b (bottom). A packing drawing of the $K_2NH(SO_3)_2$ arrangement viewed along its b_0 axis. The sulfur atoms are black and the oxygen atoms dotted. Potassium atoms are large and fine-line shaded.

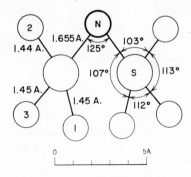

Fig. VIIB,23. Dimensions of the $NH(SO_3)_2$ anion in $K_2NH(SO_3)_2$.

This arrangement (Fig. VIIB,24) contains pyramidal SeO_3 groups which are linked together presumably through hydrogen bonds to form double layers parallel to the cleavage plane (100). Within an SeO_3 pyramid, Se–O = 1.72, 1.75, and 1.76 A. The shortest O–O separations between SeO_3 ions, through the presumed hydrogen bonds, are 2.56 and 2.60 A.

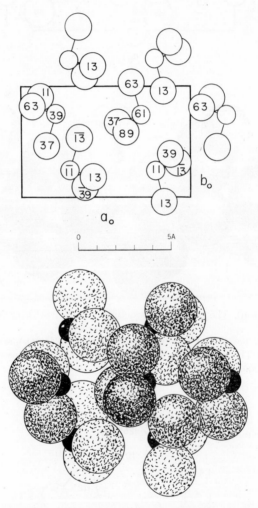

Fig. VIIB,24a (top). The orthorhombic structure of H_2SeO_3 viewed along its c_0 axis. The small circles are the selenium atoms. Origin in lower right.

Fig. VIIB,24b (bottom). The SeO_3 groups of the orthorhombic H_2SeO_3 structure viewed along its c_0 axis. The selenium atoms are black.

TABLE VIIB,9. Parameters of Atoms in $LiH_3(SeO_3)_2$

Atom	x	y	z
Se(1)	0.233	0.394	0.088
Se(2)	0.266	0.901	0.413
O(1)	0.210	0.229	0.863
O(2)	0.314	0.725	0.599
O(3)	0.305	0.268	0.353
O(4)	0.191	0.807	0.130
O(5)	0.485	0.466	0.061
O(6)	0.021	0.993	0.418

VII,b25. The ferroelectric substance *lithium acid selenite*, $LiH_3(SeO_3)_2$, is monoclinic, with a bimolecular cell of the dimensions:

$$a_0 = 6.258 \text{ A.}; \quad b_0 = 7.886 \text{ A.}; \quad c_0 = 5.433 \text{ A.}; \quad \beta = 105°12'$$

The space group is the low symmetry C_s^2 (Pn):

$$(2a) \quad xyz; \; x+\frac{1}{2},\bar{y},z+\frac{1}{2}$$

The cations could not be found from the x-ray measurements, but the other atoms have the parameters listed in Table VIIB,9.

The selenite ions, of substantially the same shape as those in H_2SeO_3 (**VII,b24**), have Se–O = 1.66–1.77 A. and O–Se–O = 96–110°. Their selenium atoms are ca. 0.80 A. outside the plane of the oxygens. It is considered that these anions are tied together in the crystal through O—H···O bonds having lengths of 2.52–2.57 A.

Iodates

VII,b26. *Basic copper iodate*, $Cu(OH)IO_3$, which occurs as the mineral *salesite*, is orthorhombic. Its tetramolecular unit has the dimensions:

$$a_0 = 10.80 \text{ A.}; \quad b_0 = 6.71 \text{ A.}; \quad c_0 = 4.79 \text{ A.}$$

Atoms are in the following positions of V_h^{16} $(Pnma)$:

Cu: (4a) $000; \; 0\,\frac{1}{2}\,0; \; \frac{1}{2}\,0\,\frac{1}{2}; \; \frac{1}{2}\,\frac{1}{2}\,\frac{1}{2}$

I: (4c) $\pm\,(u\,\frac{1}{4}\,v; \; u+\frac{1}{2},\frac{1}{4},\frac{1}{2}-v)$

                                with $u = 0.2443, v = -0.006$

O(1): (4c) with $u = 0.388, v = 0.174$

O(2): (8d) $\pm\,(xyz; \; \frac{1}{2}-x,y+\frac{1}{2},z+\frac{1}{2}; \; x,\frac{1}{2}-y,z; \; x+\frac{1}{2},y,\frac{1}{2}-z)$

                           with $x = 0.166, y = 0.049, z = 0.178$

OH: (4c) with $u = 0.029, v = -0.195$

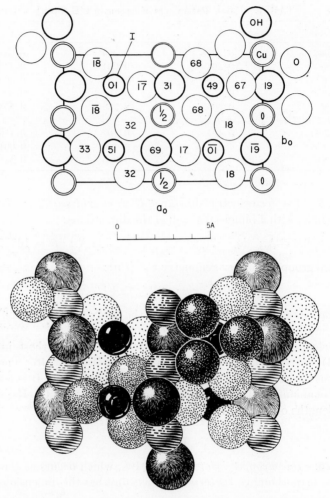

Fig. VIIB, 25a (top). A projection along its c_0 axis of the orthorhombic structure of Cu-(OH)IO₃. Origin in lower right.

Fig. VIIB,25b (bottom). A packing drawing of the structure of Cu(OH)IO₃ seen along its c_0 axis. The iodine atoms are black and the iodate oxygens dotted. Hydroxyls are fine-line shaded.

In this arrangement (Fig. VIIB,25), iodate ions are trigonal pyramids, with I–O = 1.78 or 1.82 A. and O–O = 2.70 or 2.75 A. Two oxygens and two hydroxyls form a square around each copper atom at distances of 1.95 and 2.01 A.; two additional oxygens at Cu–O(1) = 2.59 A. give it what can be considered a distorted octahedral environment.

VII,b27.   Crystals of *ceric iodate*, $Ce(IO_3)_4$, are tetragonal. There are two molecules in a cell of the edge lengths:

$$a_0 = 9.90 \text{ A.}, \qquad c_0 = 5.32 \text{ A.}$$

Atoms are in the following positions of C_{4h}^4 $(P4_2/n)$:

$$Ce: \quad (2a) \quad 000; \; {}^1/_2 \, {}^1/_2 \, {}^1/_2$$

and all others in the general positions:

$$(8g) \quad xyz; \; \bar{x}\bar{y}z; \; x+{}^1/_2, y+{}^1/_2, {}^1/_2-z; \; {}^1/_2-x, {}^1/_2-y, {}^1/_2-z;$$
$$\bar{y}x\bar{z}; \; y\bar{x}\bar{z}; \; {}^1/_2-y, x+{}^1/_2, z+{}^1/_2; \; y+{}^1/_2, {}^1/_2-x, z+{}^1/_2$$

with the parameters:

Atom	x	y	z
I	0.2769	0.0014	0.4633
O(1)	0.4140	0.1173	0.4816
O(2)	0.1900	0.0826	0.1984
O(3)	0.1504	0.0670	0.6827

In this arrangement (Fig. VIIB,26) the iodate ions are trigonal pyramids, with the iodine atom 0.88 A. outside the plane of the three oxygens. The I–O distances are 1.78, 1.83, and 1.84 A. Each cerium atom has eight oxygen neighbors, four at a distance of 2.31 A. and four more at 2.35 A.

VII,b28.   *Zirconium iodate*, $Zr(IO_3)_4$, has a structure different from that of the analogous cerium compound (**VII,b27**). It is, however, also tetragonal, with a bimolecular cell of the edge lengths:

$$a_0 = 8.38 \text{ A.}, \qquad c_0 = 7.49 \text{ A.}$$

The space group is C_{4h}^3 $(P4/n)$ and atoms are in the following positions:

$$(2c) \quad \pm ({}^1/_4 \, {}^1/_4 \, u)$$
$$(8g) \quad \pm (xyz; \; {}^1/_2-x, {}^1/_2-y, z; \; {}^1/_2-y, x, z; \; y, {}^1/_2-x, z)$$

The parameters have been established as:

Atom	Position	x	y	z
Zr	(2c)	${}^1/_4$	${}^1/_4$	0.0603
I	(8g)	-0.0061	-0.0302	0.2699
O(1)	(8g)	-0.1547	0.0022	-0.5578
O(2)	(8g)	0.0404	0.1787	0.2173
O(3)	(8g)	-0.1589	-0.0430	0.0929

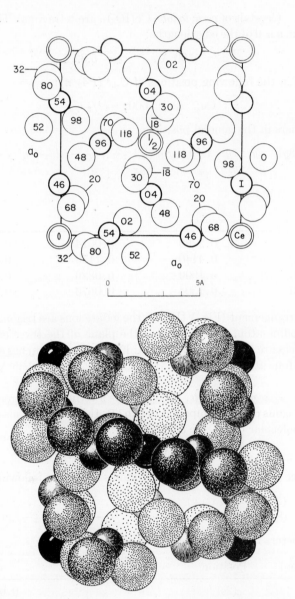

Fig. VIIB,26a (top). The tetragonal structure of Ce(IO₃)₄ projected along its c_0 axis.
Fig. VIIB,26b (bottom). A packing drawing of the Ce(IO₃)₄ arrangement seen along its c_0 axis. The cerium atoms are black, the iodine atoms small and fine-line shaded.

In the resulting arrangement (Fig. VIIB,27), the IO_3 anions are, as usual, low trigonal pyramids with iodine at the apices, but in this case the base is not quite regular. The separations I–O lie between 1.814 and 1.846 A., and the angles between the oxygens are 58°6', 58°53', and 63°1'. Each zirconium atom has eight oxygen neighbors at the corners of a nearly regular Archimedes antiprism, with Zr–O = 2.197 or 2.216 A. As has been found to be the case with other iodates, there are short separations between an iodine and an oxygen belonging to a neighboring IO_3; in this crystal one such I–O = 2.55 A.

Halides

VII,b29. *Disilver cesium iodide*, Ag_2CsI_3, is orthorhombic. Its tetramolecular unit has the edges:

$$a_0 = 11.08 \text{ A.}; \quad b_0 = 13.74 \text{ A.}; \quad c_0 = 6.23 \text{ A.}$$

The space group is V_h^{16} (*Pbnm*), with atoms in the positions:

(4c) $\pm (u\,v\,^1/_4;\ ^1/_2{-}u,v{+}^1/_2,^1/_4)$

(8d) $\pm (xyz;\ x{+}^1/_2,^1/_2{-}y,\bar{z};\ x,y,^1/_2{-}z;\ x{+}^1/_2,^1/_2{-}y,z{+}^1/_2)$

The chosen positions and parameters are as follows:

Atom	Position	x	y	z
Cs	(4c)	0.505	0.185	$^1/_4$
Ag	(8d)	0.342	0.499	0.000
I(1)	(4c)	0.214	0.366	$^1/_4$
I(2)	(4c)	0.495	0.622	$^1/_4$
I(3)	(4c)	0.802	0.372	$^1/_4$

The structure is represented in Figure VIIB,28. Both this arrangement and that of Cu_2CsCl_3 (**VII,b31**) can be thought of as built up of chains of tetrahedra and except that in these cases the chains are doubled, there are resemblances to the SiS_2 arrangement (**IV,e14**). In this crystal the silver atoms are at the centers of tetrahedra of iodine atoms, with Ag–I = 2.79–2.90 A. There are ten iodines around a cesium atom, with Cs–I = 3.93–4.18 A.

The unit cell of Ag_2CsI_3 is not very different in shape and size from that of Cs_2AgI_3 (**VII,b33**) if the a_0 and b_0 of one are interchanged. The space group, too, is the same but the halogen atoms are differently placed.

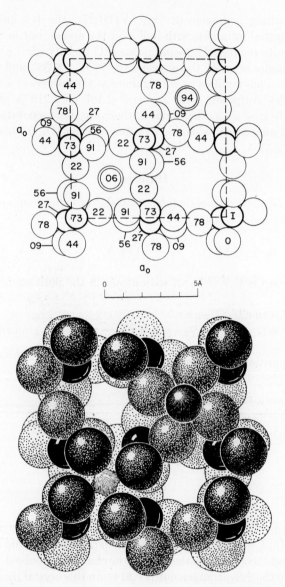

Fig. VIIB,27a (top). The tetragonal structure of $Zr(IO_3)_4$ projected along its c_0 axis. The zirconium atoms are doubly ringed, the iodine atoms are heavily ringed.

Fig. VIIB,27b (bottom). A packing drawing of the $Zr(IO_3)_4$ structure viewed along its c_0 axis. The iodine atoms are black, the zirconium atoms fine-line shaded.

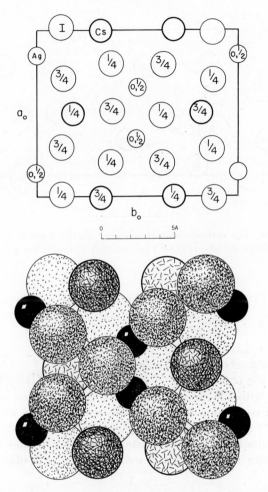

Fig. VIIB,28a (top). The orthorhombic structure of Ag_2CsI_3 viewed along its c_0 axis. Origin in lower left.

Fig. VIIB,28b (bottom). A packing drawing of the orthorhombic structure of Ag_2CsI_3 viewed along its c_0 axis. The iodine atoms are dotted, the cesium atoms line-shaded.

VII,b30. *Dipotassium silver iodide*, K_2AgI_3, is typical of a group of ortho-rhombic substances which, in contradiction to early crystallographic work, are not isomorphous with the analogous cesium compound Cs_2AgI_3 (**VII,b33**).

It has a tetramolecular cell of the dimensions:

$$a_0 = 19.52 \text{ A.}; \quad b_0 = 9.98 \text{ A.}; \quad c_0 = 4.74 \text{ A.}$$

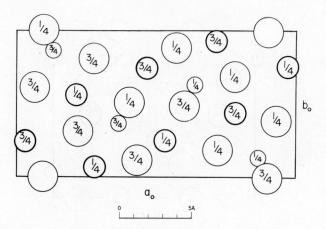

Fig. VIIB,29a. The orthorhombic structure of K₂AgI₃ projected along its c_0 axis. The smallest circles are silver; those that are heavily ringed are potassium. Origin in lower right.

All atoms have been found to be in the following special positions of V_h^{16} (*Pbnm*):

$$(4c) \quad \pm(u\,v\,{}^1\!/_4;\, {}^1\!/_2-u,v+{}^1\!/_2,{}^1\!/_4)$$

with the parameters listed in Table VIIB,10.

The resulting arrangement is shown in Figure VIIB,29. Though not isomorphous with the cesium compound, the structures are nevertheless similar, both containing tetrahedral strings parallel to their c_0 axes (cf. Figs. VIIB,32 and VIIB,29). In K₂AgI₃, silver atoms at the centers of these tetrahedra are almost equidistant from four iodines (2.82–2.85 A.). Each potassium atom has seven iodine neighbors at distances between 3.41 and 3.72 A.

TABLE VIIB,10
Parameters of the Atoms in K₂AgI₃

Atom	u	v
Ag	0.135	0.129
I(1)	0.278	0.187
I(2)	0.070	0.382
I(3)	−0.099	0.008
K(1)	0.714	0.070
K(2)	0.464	0.249

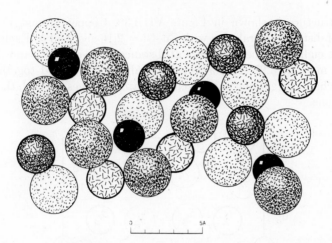

Fig. VIIB,29b. A view along its c_0 axis of the orthorhombic structure of K_2AgI_3. The silver atoms are black, the potassium atoms line-shaded.

Other crystals with this arrangement are

$$Rb_2AgI_3: \quad a_0 = 20.0 \text{ A.}; \; b_0 = 10.3 \text{ A.}; \; c_0 = 4.9 \text{ A.}$$
$$(NH_4)_2AgI_3: \quad a_0 = 21.3 \text{ A.}; \; b_0 = 10.9 \text{ A.}; \; c_0 = 4.6 \text{ A.}$$

VII,b31. *Dicuprous cesium chloride*, Cu_2CsCl_3, is orthorhombic, with a tetramolecular cell of the dimensions:

$$a_0 = 9.49 \text{ A.}; \quad b_0 = 11.88 \text{ A.}; \quad c_0 = 5.61 \text{ A.}$$

Atoms are in the following positions of V_h^{17} (*Cmcm*):

$(4c) \quad \pm(0\, u\, ^1/_4;\; ^1/_2, u+^1/_2, ^1/_4)$

$(8e) \quad \pm(u00;\; u\, 0\, ^1/_2;\; u+^1/_2, ^1/_2, 0;\; u+^1/_2, ^1/_2, ^1/_2)$

$(8g) \quad \pm(u\, v\, ^1/_4;\; \bar{u}\, v\, ^1/_4;\; u+^1/_2, v+^1/_2, ^1/_4;\; ^1/_2-u, v+^1/_2, ^1/_4)$

with the following positions and parameters:

Atom	Position	x	y	z
Cs	(4c)	0	-0.317	$^1/_4$
Cu	(8e)	-0.164	$^1/_2$	0
Cl(1)	(8g)	0.218	0.398	$^1/_4$
Cl(2)	(4c)	0	0.120	$^1/_4$

The structure is shown in Figure VIIB,30. Copper atoms are at the centers of distorted tetrahedra with Cu–Cl = 2.16 and 2.53 A. Each cesium atom has eight chlorine neighbors at distances between 3.62 and 3.70 A., and two more at 3.97 A. The CuCl$_4$ tetrahedra are linked together into doubled chains by sharing pairs of chlorine atoms (edges) in both the a_0 and c_0 directions.

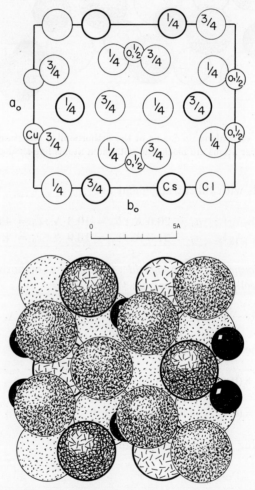

Fig. VIIB,30a (top). The orthorhombic structure of Cu$_2$CsCl$_3$ viewed along its c_0 axis. Origin in lower left.

Fig. VIIB,30b (bottom). A packing drawing of the orthorhombic Cu$_2$CsCl$_3$ structure viewed along its c_0 axis. The copper atoms are small and black; the cesium atoms are line-shaded.

TABLE VIIB,11
Positions and Parameters of the Atoms in $(NH_4)_2CuCl_3$

Atom	Position	x	y	z
Cu(1)	(4h)	0.278	0	$1/_2$
Cu(2)	(4j)	0	0.183	$1/_2$
Cl(1)	(4g)	0.187	0	0
Cl(2)	(4i)	0	0.119	0
Cl(3)	(8q)	0.373	0.085	$1/_2$
Cl(4)	(8q)	0.128	0.247	$1/_2$
NH$_4$(1)	(2b)	$1/_2$	0	0
NH$_4$(2)	(2d)	0	0	$1/_2$
NH$_4$(3)	(4i)	0	0.330	0
NH$_4$(4)	(8p)	0.227	0.152	0

VII,b32. *Diammonium cuprous chloride*, $(NH_4)_2CuCl_3$, has a structure that is orthorhombic in symmetry, like the other complex trihalides being considered here. Its large unit containing eight molecules has the edges:

$$a_0 = 14.71 \text{ A.}; \quad b_0 = 22.07 \text{ A.}; \quad c_0 = 4.08 \text{ A.}$$

Atoms are in the following special positions of V_h^{19} (*Cmmm*):

(2b) $1/_2\,0\,0;\,0\,1/_2\,0$ (2d) $0\,0\,1/_2;\,1/_2\,1/_2\,1/_2$
(4g) $\pm(u00;\,u+1/_2,1/_2,0)$ (4h) $\pm(u\,0\,1/_2;\,u+1/_2,1/_2,1/_2)$
(4i) $\pm(0u0;\,1/_2,u+1/_2,0)$ (4j) $\pm(0\,u\,1/_2;\,1/_2,u+1/_2,1/_2)$
(8p) $\pm(uv0;\,\bar{u}v0;\,u+1/_2,v+1/_2,0;\,1/_2-u,v+1/_2,0)$
(8q) $\pm(u\,v\,1/_2;\,\bar{u}\,v\,1/_2;\,u+1/_2,v+1/_2,1/_2;\,1/_2-u,v+1/_2,1/_2)$

The chosen positions and parameters are those of Table VIIB,11.

The structure thus obtained is illustrated in Figure VIIB,31. Copper atoms are at the centers of $CuCl_4$ tetrahedra which share corners to produce strings running along the c_0 axis. The distance Cu–Cl = 2.34–2.48 A. Each ammonium ion, lying between the chains, has eight chlorine neighbors, with NH$_4$–Cl = 3.27–3.68 A.

VII,b33. *Potassium cuprous chloride*, K_2CuCl_3, is orthorhombic with a tetramolecular cell of the dimensions:

$$a_0 = 12.00 \text{ A.}; \quad b_0 = 12.55 \text{ A.}; \quad c_0 = 4.20 \text{ A.}$$

The space group is V_h^{16} (*Pnam*), with all atoms in the special positions:

(4c) $\pm(u\,v\,1/_4;\,u+1/_2,1/_2-v,1/_4)$

The established parameters are listed in Table VIIB,12.

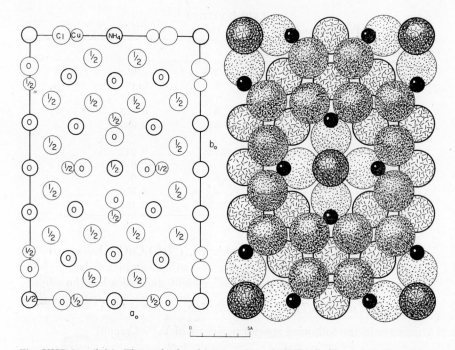

Fig. VIIB,31a (left). The orthorhombic structure of $(NH_4)_2CuCl_3$ viewed along its c_0 axis. Origin in lower right.

Fig. VIIB,31b (right). The orthorhombic structure of $(NH_4)_2CuCl_3$ viewed along its c_0 axis. The ammonium atoms are line-shaded, the copper atoms are small and black.

TABLE VIIB,12

Parameters of the Atoms in K_2CuCl_3

Atom	u	v
Cu	0.252	0.197
Cl(1)	0.133	0.052
Cl(2)	0.435	0.138
Cl(3)	0.277	0.791
K(1)	0.172	0.480
K(2)	−0.488	−0.327

These lead to a structure (Fig. VIIB,32) composed of potassium ions and $CuCl_4$ tetrahedra which, as with $(NH_4)_2CuCl_3$ (**VII,b32**), are linked into chains along the c_0 axis by sharing corners. Within one of these tetrahedra the Cu–Cl separations lie between 2.31 and 2.43 A. Each potassium atom, situated between the chains, has seven chlorine neighbors at distances that range between 3.12 and 3.27 A.

Several other complex halides have this structure. Their cell dimensions are:

$$(NH_4)_2CuBr_3: \quad a_0 = 13.1 \text{ A.}; \quad b_0 = 14.0 \text{ A.}; \quad c_0 = 4.4 \text{ A.}$$
$$Cs_2AgCl_3: \quad a_0 = 13.19 \text{ A.}; b_0 = 13.74 \text{ A.}; c_0 = 4.57 \text{ A.}$$
$$Cs_2AgI_3: \quad a_0 = 14.39 \text{ A.}; b_0 = 15.16 \text{ A.}; c_0 = 5.02 \text{ A.}$$

Atomic positions have not been determined for these compounds.

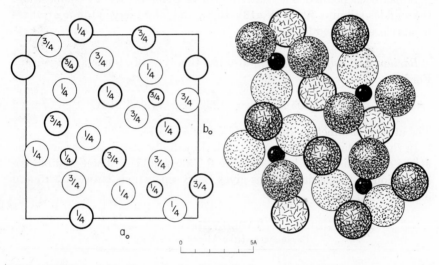

Fig. VIIB,32a (left). The orthorhombic K_2CuCl_3 structure viewed along its c_0 axis. The copper atoms are the small, the potassium atoms the large heavily ringed circles. Origin in lower right.

Fig. VIIB,32b (right). The orthorhombic K_2CuCl_3 structure viewed along its c_0 axis. The copper atoms are small and black, the chlorine atoms are dotted.

The mineral *aikinite*, $CuPbBiS_3$, has this structure. Rearranging the axes of the original description to conform with the sequence used above, it has the cell edges:

$$a_0 = 11.65 \text{ A.}; \quad b_0 = 11.30 \text{ A.}; \quad c_0 = 4.00 \text{ A.}$$

The atomic parameters, after making an appropriate shift of origin, are those listed in Table VIIB,13.

TABLE VIIB,13
Parameters of the Atoms in Aikinite

Atom	u	v
Cu	0.260	0.210
Bi(or Pb)	0.168	0.487
Pb(or Bi)	−0.517	−0.315
S(1)	0.130	0.045
S(2)	0.440	0.135
S(3)	0.270	0.810

It has been pointed out that, though of different overall composition, the sulfides stibnite, Sb_2S_3 (**V,a18**), Bi_2S_3, and Th_2S_3 have atomic arrangements closely related to this.

Barium zinc sulfide, Ba_2ZnS_3, also has this type of atomic arrangement. For it,

$$a_0 = 12.05 \text{ A.}; \quad b_0 = 12.65 \text{ A.}; \quad c_0 = 4.21 \text{ A.}$$

The atomic parameters are those of Table VIIB,14.

In this substance the sulfur atoms tetrahedrally distributed around zinc atoms are at distances lying between 2.33 and 2.45 A. The sevenfold coordination of the barium atoms gives Ba–S separations between 3.15 and 3.26 A.

TABLE VIIB,14
Parameters of the Atoms in Ba_2ZnS_3

Atom	u	v
Zn	0.253	0.197
Ba(1)	0.1735	0.4815
Ba(2)	−0.4885	−0.3255
S(1)	0.134	0.052
S(2)	0.441	0.140
S(3)	0.277	0.792

Sulfides

VII,b34. The structure that has been found for *cubanite*, Fe_2CuS_3, contains four molecules in an orthorhombic unit of the dimensions:

$$a_0 = 6.46 \text{ A.}; \quad b_0 = 11.117 \text{ A.}; \quad c_0 = 6.233 \text{ A.}$$

The space group is V_h^{16} ($Pcmn$), with atoms in the positions:

$(4c)$ $\quad \pm(u\ ^1/_4\ v;\ ^1/_2-u,^1/_4,v+^1/_2)$

$(8d)$ $\quad \pm(xyz;\ x+^1/_2,y+^1/_2,^1/_2-z;\ x,^1/_2-y,z;\ ^1/_2-x,y,z+^1/_2)$

The chosen parameters are listed below.

Atom	Position	x	y	z
Cu	$(4c)$	0.583	$^1/_4$	0.127
Fe	$(8d)$	0.0875	0.088	0.134
S(1)	$(4c)$	0.913	$^1/_4$	0.2625
S(2)	$(8d)$	0.413	0.0835	0.274

This gives a structure (Fig. VIIB,33) which can be considered as a sort of superlattice on the wurtzite arrangement (**III,c2**), with slabs of metal-sulfur tetrahedra alternating in their orientations in layers along the b_0 axis. Within an Fe–S slab the tetrahedra share edges and have Fe–S = 2.25–2.29 A. The CuS$_4$ tetrahedra with Cu–S = 2.27–2.34 A. share corners only.

It is probable that *sternbergite*, Fe$_2$AgS$_3$, has this structure, with

$$a_0 = 6.62\ \text{A.;} \quad b_0 = 11.66\ \text{A.;} \quad c_0 = 6.34\ \text{A.}$$

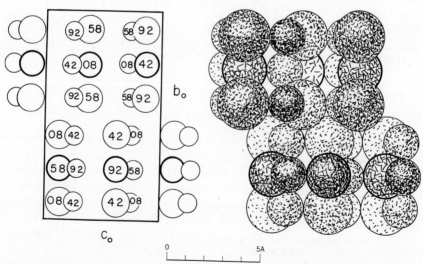

Fig. VIIB,33a (left). The orthorhombic Fe$_2$CuS$_3$ structure projected along its a_0 axis. The sulfur atoms are the smallest circles; the copper atoms are larger and heavily ringed. Origin in lower left.

Fig. VIIB,33b (right). The orthorhombic Fe$_2$CuS$_3$ structure viewed along its a_0 axis. The sulfur are the small, the iron the large dotted circles.

VII,b35. The mineral *bournonite*, $CuPbSbS_3$, is orthorhombic, with a tetramolecular cell of the dimensions:

$$a_0 = 8.162 \text{ A.}; \quad b_0 = 8.7105 \text{ A.}; \quad c_0 = 7.8105 \text{ A.}$$

The space group is C_{2v}^7 $(Pn2m)$, with atoms in the positions:

(2a) $uv0; \bar{u},v+^1/_2,^1/_2$
(4b) $xyz; xy\bar{z}; \bar{x},y+^1/_2,^1/_2-z; \bar{x},y+^1/_2,z+^1/_2$

The chosen positions and parameters are listed in Table VIIB,15.

TABLE VIIB,15
Positions and Parameters of the Atoms in $CuPbSbS_3$ and $CuPbAsS_3$[a]

Atom	Position	x	y	z
Cu	(4b)	0.269 (0.272)	0.440 (0.440)	0.249 (0.247)
Pb(1)	(2a)	0.079 (0.080)	0 (0)	0 (0)
Pb(2)	(2a)	0.440 (0.450)	0.670 (0.675)	0 (0)
Sb(1) [As]	(2a)	−0.073 (−0.067)	0.550 (0.610)	0 (0)
Sb(2) [As]	(2a)	0.515 (0.510)	0.125 (0.110)	0 (0)
S(1)	(2a)	0.225 (0.236)	0.310 (0.300)	0 (0)
S(2)	(2a)	−0.225 (−0.223)	0.810 (0.810)	0 (0)
S(3)	(4b)	0.084 (0.096)	0.706 (0.695)	0.248 (0.238)
S(4)	(4b)	0.564 (0.550)	0.425 (0.440)	0.267 (0.266)

[a] Parameters for $CuPbAsS_3$ are given in parentheses.

For *seligmannite*, $CuPbAsS_3$:

$$a_0 = 8.081 \text{ A.}; \quad b_0 = 8.747 \text{ A.}; \quad c_0 = 7.636 \text{ A.}$$

Atoms are in the positions of Table VIIB,15 with the parameters stated in parentheses.

This structure is illustrated in Figure VIIB,34. In these crystals the coordination of the metallic atoms is somewhat indefinite, depending on the maximum atomic separation considered significant. Shortest interatomic distances are: Cu–S = 2.29 A.; Pb–S = ca. Sb–S = ca. 2.90 A.

This is an arrangement that bears definite relationships to those of aikinite, $CuPbBiS_3$, (**VII,b33**) and antimonite, Sb_2S_3.

VII,b36. The sulfide *pyrargyrite*, Ag_3SbS_3, is rhombohedral, with a bimolecular cell of the dimensions:

$$a_0 = 7.00 \text{ A.}, \qquad \alpha = 103°57'$$

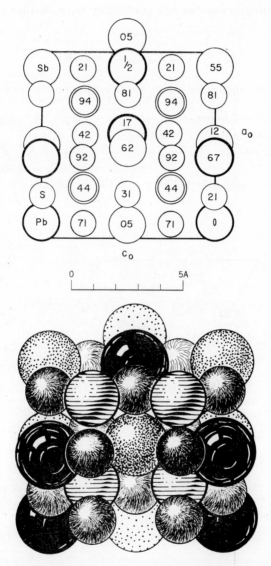

Fig. VIIB,34a (top). The orthorhombic structure of CuPbSbS₃ viewed along its b_0 axis. The copper atoms are doubly ringed. Origin in lower right.

Fig. VIIB,34b (bottom). A packing drawing of the CuPbSbS₃ arrangement seen along its b_0 axis. The lead atoms are large and black, the sulfur atoms are fine-line shaded and the antimony atoms are dotted.

The corresponding hexagonal cell containing six molecules has the edge lengths:

$$a_0' = 11.02 \text{ A.}, \qquad c_0' = 8.73 \text{ A.}$$

Atoms of these six molecules are in the following positions of C_{3v}^6 ($R3c$):

Sb: (6a) $00u;\ 0,0,u+\frac{1}{2}$; rh   with $u = 0$ (arbitrary)
Ag: (18b) $xyz;$ $\bar{y},x-y,z;$ $y-x,\bar{x},z;$
        $\bar{y},\bar{x},z+\frac{1}{2};$ $x,x-y,z+\frac{1}{2};$ $y-x,y,z+\frac{1}{2}$; rh
                              with $x = 0.250,\ y = 0.305,\ z = 0.210$
S: (18b)   with $x = 0.220,\ y = 0.105,\ z = 0.355$

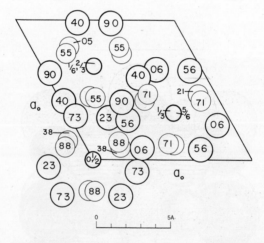

Fig. VIIB,35a. The hexagonal (rhombohedral) structure of Ag_3AsS_3 projected along its principal axis. The arsenic atoms are the small ringed circles; the silver atoms are the largest, heavily ringed circles.

The corresponding arsenide, *proustite*, Ag_3AsS_3, has this structure. For it,

$$a_0 = 6.87 \text{ A.}, \quad \alpha = 103°31' \quad \text{for the unit rhombohedron}$$
$$a_0' = 10.80 \text{ A.}, \; c_0' = 8.69 \text{ A.} \quad \text{for the hexagonal cell}$$

Atoms, in positions corresponding to those for the antimonide, have the parameters: $u(As) = 0$ (arbitrary); $x(Ag) = 0.246$, $y(Ag) = 0.298$, $z(Ag) = 0.235$; $x(S) = 0.220$, $y(S) = 0.095$, $z(S) = 0.385$.

In this arrangement, as shown in Figure VIIB,35, each arsenic atom has three sulfur neighbors at a distance of 2.25 A.; in Ag_3SbS_3 the corresponding distance is Sb–S = 2.45 A. Each sulfur atom in the arsenic mineral is distant from two silver atoms at Ag–S = 2.40 A. and one arsenic atom at 2.25 A.

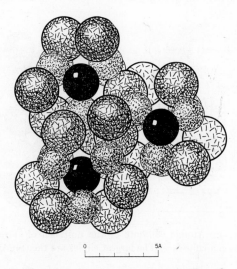

0   5A

Fig. VIIB,35b. A view along its c_0 axis of the hexagonal structure of Ag_3AsS_3. The arsenic atoms are large and black, the sulfur atoms small and dotted.

VII,b37. The *tetrahedrites*, as a family of minerals having the composition $(Cu,Ag)_3(Sb,As)S_3$ are cubic with a value of a_0 that varies between ca. 10.2 and 10.6 A.; a rather pure sample had the edge 10.21 A. There are eight molecules in this unit and atoms have been found to be in the following special positions of T_d^3 ($I\bar{4}3m$):

Ag,Cu(1):   (12e) $\pm(u00; 0u0; 00u)$; B.C.   with $u = 0.25$
Ag,Cu(2):   (12d) $\pm(^1/_2\,0\,^1/_4; \,^1/_4\,^1/_2\,0; \,0\,^1/_4\,^1/_2)$; B.C.
   Sb,As: (8c) $uuu;\ \bar{u}\bar{u}u;\ \bar{u}u\bar{u};\ u\bar{u}\bar{u}$; B.C.   with $u = 0.28$
      S: (24g) $uuv;\ vuu;\ uvu;\ u\bar{u}\bar{v};\ \bar{v}u\bar{u};\ \bar{u}\bar{v}u;$
                  $\bar{u}u\bar{v};\ \bar{v}\bar{u}u;\ u\bar{v}\bar{u};\ \bar{u}\bar{u}v;\ v\bar{u}\bar{u};\ \bar{u}v\bar{u}$; B.C.

with $u = 0.125,\ v = 0.375$

The structure that results, as shown in Figure VIIB,36, resembles that of ZnS (**III,c1**), but with one-quarter of the atoms of sulfur missing.

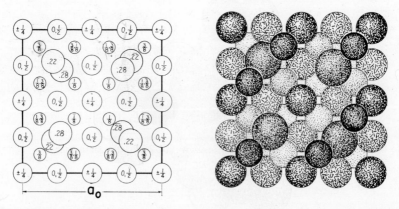

Fig. VIIB,36a (left).   Atomic contents of the unit cell of tetrahedrite, $(Cu,Ag)_3$-$(Sb,As)S_3$, projected on a cube face. The largest circles are the (Sb,As) and the smallest are the sulfur atoms.

Fig. VIIB,36b (right). A packing drawing of atoms in the structure of tetrahedrite if they are given their neutral radii. Line-shaded spheres are the (Cu,Ag) atoms. The Sb atoms are the large, the S atoms the smaller dotted circles.

VII,b38. A simple structure has been assigned the compounds KCu_4S_2 and $RbCu_4S_3$. They are tetragonal, with unimolecular cells of the dimensions:

KCu_4S_3: $a_0 = 3.908$ A., $c_0 = 9.28$ A.
$RbCu_4S_3$: $a_0 = 3.928$ A., $c_0 = 9.43$ A.

Atoms have been placed in the following positions of D_{4h}^1 ($P4/mmm$):

$$K(\text{or } Rb): \quad (1b) \quad 0\ 0\ ^1/_2$$
$$Cu: \quad (4i) \quad \pm(0\ ^1/_2\ u;\ ^1/_2\ 0\ u) \qquad \text{with } u = 0.158$$
$$S(1): \quad (1a) \quad 000$$
$$S(2): \quad (2h) \quad \pm(^1/_2\ ^1/_2\ v)$$
$$\text{with } v = 0.297 \text{ for } KCu_4S_3 \text{ and } 0.292 \text{ for } RbCu_4S_3$$

The structure is illustrated in Figure VIIB,37. It is composed of alternate layers along the c_0 axis of Cu_4S_3 groups and alkali atoms. Each copper atom is surrounded by a nearly regular tetrahedron of sulfur atoms with Cu–S(1) = 2.45 A. and Cu–S(2) = 2.34 A.; the S–S edges of these tetrahedra are 3.90 A. in length. Each alkali atom is at the center of a cube of sulfur atoms with K–S(2) = 3.34 A. and Rb–S(2) = 3.39 A.

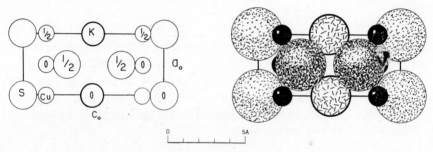

Fig. VIIB,37a (left). The tetragonal structure of KCu_4S_3 projected along an a_0 axis. Origin in lower right.
Fig. VIIB,37b (right). The tetragonal structure of KCu_4S_3 viewed along an a_0 axis. The copper atoms are black; the sulfurs are large and dotted.

Borates

VII,b39. *Boric acid*, H_3BO_3, is triclinic with a tetramolecular unit of the dimensions:

$$a_0 = 7.039 \text{ A.}; \ b_0 = 7.053 \text{ A.}; \ c_0 = 6.578 \text{ A.}$$
$$\alpha = 92°35'; \quad \beta = 101°10'; \quad \gamma = 119°50'$$

All atoms have been put in general positions of C_i^1 ($P\bar{1}$):

$$(2i) \quad \pm(xyz)$$

with the parameters of Table VIIB,16.

TABLE VIIB,16. Parameters of the Atoms in H_3BO_3

Atom	x	y	z
B(1)	0.646	0.427	0.258
B(2)	0.307	0.760	0.242
O(1)	0.424	0.302	0.261
O(2)	0.768	0.328	0.250
O(3)	0.744	0.650	0.261
O(4)	0.532	0.885	0.250
O(5)	0.214	0.540	0.244
O(6)	0.180	0.856	0.233
Hydrogen Positions[a]			
H(1)	0.347	0.361	0.255
H(2)	0.671	0.172	0.250
H(3)	0.890	0.709	0.252
H(4)	0.600	0.817	0.254
H(5)	0.083	0.469	0.246
H(6)	0.297	0.006	0.243

[a] The z parameters for hydrogen have been calculated on the assumption that they lie on lines joining oxygen atoms.

This results in a plate-like structure with sheets of atoms parallel to the a_0b_0 plane. As can be seen from Figure VIIB,38, each sheet consists of planar BO_3 groups which are supposed to be held together by hydrogen bridges between the oxygen atoms.

VII,b40. *Magnesium borate*, $Mg_3(BO_3)_2$, and the corresponding cobaltous compound $Co_3(BO_3)_2$ form orthorhombic crystals with bimolecular cells having the edge lengths:

$$Mg_3(BO_3)_2: \quad a_0 = 5.398 \text{ A.}; \ b_0 = 8.416 \text{ A.}; \ c_0 = 4.497 \text{ A.}$$
$$Co_3(BO_3)_2: \quad a_0 = 5.462 \text{ A.}; \ b_0 = 8.436 \text{ A.}; \ c_0 = 4.529 \text{ A.}$$

Atoms have been found to be in the following positions of V_h^{12} (*Pnmn*):

Mg(1)[or Co]: (2a) $000; \ ^1/_2 \ ^1/_2 \ ^1/_2$

Mg(2)[or Co]: (4f) $\pm (^1/_2 \ u \ 0; \ 0, u+^1/_2, ^1/_2)$ with $u = 0.821$

B: (4g) $\pm (u0v; \ u+^1/_2, ^1/_2, ^1/_2-v)$ with $u = 0.25$ and $v = 0.56$

O(1): (4g) with $u = 0.316$ and $v = 0.258$

O(2): (8h) $\pm (xyz; \ \bar{x}y\bar{z}; \ ^1/_2-x, ^1/_2-y, z+^1/_2; \ x+^1/_2, ^1/_2-y, ^1/_2-z)$ with $x = 0.218, \ y = 0.139$ and $z = 0.705$

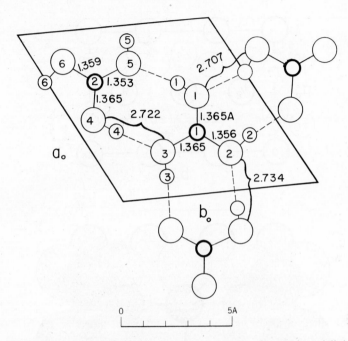

Fig. VIIB,38. A projection on the a_0b_0 plane of molecules of the triclinic H_3BO_3 structure. The heavy small circles are the boron atoms; the light small circles are hydrogen. Numbers within circles refer to the atomic designations of Table VIIB,16. The other numbers are important separations within and between molecules.

In contrast to the situation prevailing in certain other borates, this structure (Fig. VIIB,39) contains discrete borate (BO_3) ions. They are nearly equilateral triangles in which the B–O distances are 1.34 and 1.42 A. Each metal atom is octahedrally surrounded by six oxygen atoms at distances which lie between 2.06 and 2.16 A. for both the magnesium and cobalt salts.

VII,b41. The mineral *hambergite*, $Be_2BO_3(OH)$, has an eight-molecule orthorhombic unit of the dimensions:

$$a_0 = 9.75 \text{ A.}; \quad b_0 = 12.20 \text{ A.}; \quad c_0 = 4.43 \text{ A.}$$

The chosen structure, based on V_h^{15} (*Pbca*), has all atoms in general positions:

$$(8c) \quad \pm(xyz; \ x+\tfrac{1}{2},\tfrac{1}{2}-y,\bar{z}; \ \bar{x},y+\tfrac{1}{2},\tfrac{1}{2}-z; \ \tfrac{1}{2}-x,\bar{y},z+\tfrac{1}{2})$$

with the parameters of Table VIIB,17.

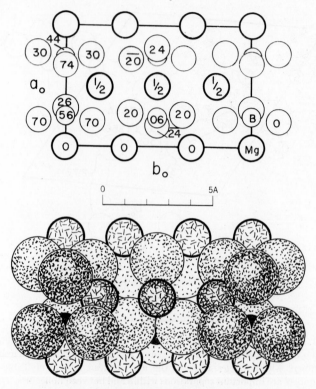

Fig. VIIB,39a (top). The orthorhombic structure of $Mg_3(BO_3)_2$ viewed along its c_0 axis. Origin in lower left.

Fig. VIIB,39b (bottom). The orthorhombic $Mg_3(B_3O)_2$ structure viewed along its c_0 axis. The magnesium atoms are line-shaded, the boron atoms are black.

TABLE VIIB,17
Parameters of the Atoms in Hambergite

Atom	x	y	z
Be(1)	−0.031	0.183	0.458
Be(2)	0.236	0.069	0.458
B	0.117	0.103	−0.028
O(1)	0.031	0.183	−0.167
O(2)	0.097	0.103	0.278
O(3)	0.194	0.038	−0.167
OH	−0.167	0.183	0.167

The resulting structure is shown in Figure VIIB,40. Boron atoms are at the centers of oxygen triangles, with B–O = 1.28–1.42 A. Beryllium atoms are at the centers of tetrahedra of oxygen atoms, one of which is a hydroxyl, with Be–O = 1.56–1.81 A. and Be–OH = 1.74 or 1.85 A. Each oxygen has two beryllium and one boron neighbors, while each hydroxyl has two beryllium atoms nearby.

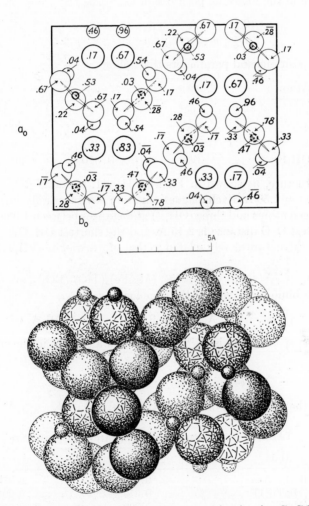

Fig. VIIB,40a (top). The orthorhombic structure of hambergite, Be₂BO₃(OH), projected along its c_0 axis. The smallest, heavily outlined circles are boron, the heavily outlined large circles are hydroxyls. The other large circles are oxygen. Origin in lower left.

Fig. VIIB,40b (bottom). A packing drawing of the orthorhombic structure of Be₂BO₃-(OH) seen along its c_0 axis. The hydroxyls are line-shaded. Small circles are beryllium; the borons do not show.

VII,b42. The mineral *fluoborite*, $Mg_3(OH,F)_3BO_3$, has a bimolecular hexagonal unit of the dimensions:

$$a_0 = 9.06 \text{ A.}, \qquad c_0 = 3.06 \text{ A.}$$

Atoms are in the following positions of C_{6h}^2 $(C6_3/m)$:

$$(2c) \quad \pm(^1/_3 \ ^2/_3 \ ^1/_4)$$
$$(6h) \quad \pm(u \ v \ ^1/_4; \ \bar{v}, u-v, ^1/_4; \ v-u, \bar{u}, ^1/_4)$$

with the positions and parameters that follow:

Atom	Position	u	v
B	$(2c)$	$^1/_3$	$^2/_3$
Mg	$(6h)$	0.381	0.038
O	$(6h)$	0.381	0.537
(OH,F)	$(6h)$	0.310	0.218

The structure, as shown in Figure VIIB,41, contains BO_3 triangles in which B–O = 1.50 A. Each magnesium atom is octahedrally surrounded by three oxygen atoms and three (OH,F) at distances between 1.99 and 2.35 A. The shortest O–O distance is 2.45 A. and the shortest OH–OH = 3.06 A.

This is a structure closely related to that of warwickite (**VII,b43**).

VII,b43. Crystals of *warwickite*, $(Mg,Fe)_3(TiO_2)(BO_3)_2$, are orthorhombic, with a bimolecular unit of the edge lengths:

$$a_0 = 9.20 \text{ A.}; \quad b_0 = 9.45 \text{ A.}; \quad c_0 = 3.01 \text{ A.}$$

The chosen space group is V_h^{16} $(Pnam)$, with all atoms in the special positions:

$$(4c) \quad \pm(u \ v \ ^1/_4; \ u+^1/_2, ^1/_2-v, ^1/_4)$$

and with the parameters of Table VIIB,18.

TABLE VIIB,18. Parameters of the Atoms in Warwickite

Atom	u	v
B	0.340	0.875
[Mg,Ti](1)	0.370	0.201
[Mg,Ti](2)	0.390	0.556
O(1)	0.000	0.125
O(2)	0.250	0.000
O(3)	0.264	0.750
O(4)	0.484	0.875

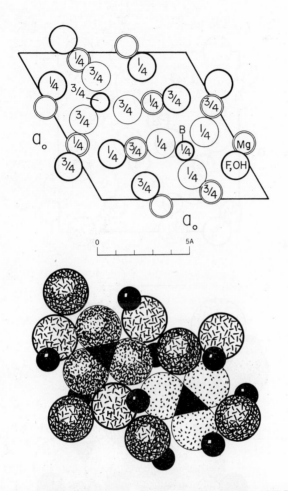

Fig. VIIB,41a (top). The hexagonal structure of fluoborite, $3Mg(OH,F)BO_3$, viewed along its c_0 axis. Origin in lower left.

Fig. VIIB,41b (bottom). A packing drawing of the hexagonal structure of $3Mg(OH,F)$-BO_3 seen along its c_0 axis. The black triangles are boron, the black circles magnesium atoms. The fluorines (and hydroxyls) are heavily outlined and line-shaded.

The resulting structure is shown in Figure VIIB,42. Distances between boron and oxygen in the BO_3 triangles are 1.30, 1.35, and 1.45 A. The titanium and magnesium atoms are not distinguished from one another; each has six octahedrally distributed oxygen atoms at distances between 1.95 and 2.25 A.

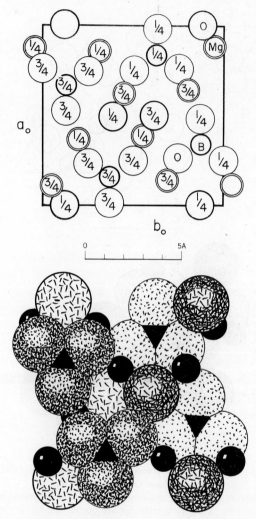

Fig. VIIB,42a (top). The orthorhombic structure of warwickite, $(Mg,Fe)_3TiB_2O_8$, seen along its c_0 axis. Origin in lower left. In this structure no distinction is made between the metallic atoms.

Fig. VIIB,42b (bottom). A packing drawing of the orthorhombic structure of warwickite seen along its c_0 axis. Both the metallic atoms and the borons are black, the borons being shown as triangles. Oxygen atoms not forming part of BO_3 groups have been line-shaded.

The following compounds also with this structure have the cell dimensions:

$Fe_2(CoO)_2(BO_3)_2$: $a_0 = 9.243$ A.; $b_0 = 9.39$ A.; $c_0 = 3.135$ A.
$Fe_2(FeO)_2(BO_3)_2$: $a_0 = 9.243$ A.; $b_0 = 9.468$ A.; $c_0 = 3.158$ A.
$Fe_2(MgO)_2(BO_3)_2$
   ludwigite[II]: $a_0 = 9.258$ A.; $b_0 = 9.427$ A.; $c_0 = 3.104$ A.
$Fe_2(NiO)_2(BO_3)_2$: $a_0 = 9.141$ A., $b_0 = 9.351$ A., $c_0 = 3.047$ A.

VII,b44.   The boron-containing mineral *ludwigite*, $(Mg,Fe)_2(MgO)_4$-$(BO_3)_2$, has a structure related to those of fluoborite (**VII,b42**), and warwickite (**VII,b43**). Its bimolecular orthorhombic unit has the dimensions:

$$a_0 = 9.14 \text{ A.}; \quad b_0 = 12.45 \text{ A.}; \quad c_0 = 3.05 \text{ A.}$$

All atoms are in the following special positions of V_h^9 (*Pbam*):

(2c) $^1/_2\,0\,0; 0\,^1/_2\,0$
(2b) $0\,0\,^1/_2; \,^1/_2\,^1/_2\,^1/_2$
(4g) $\pm(uv0; \,^1/_2-u,v+^1/_2,0)$
(4h) $\pm(u\,v\,^1/_2; \,^1/_2-u,v+^1/_2,^1/_2)$

with the positions and parameters of Table VIIB,19.

TABLE VIIB,19
Positions and Parameters of the Atoms in Ludwigite

Atom	Position	x	y	z
Mg(1)	(2b)	0	0	$^1/_2$
Mg(2)	(2c)	$^1/_2$	0	0
Mg(3)	(4h)	0.000	0.275	$^1/_2$
Fe	(4g)	0.250	0.114	0
B	(4g)	0.288	0.352	0
O(1)	(4g)	0.375	0.444	0
O(2)	(4g)	0.361	0.250	0
O(3)	(4g)	0.136	0.349	0
O(4)	(4h)	0.375	0.058	$^1/_2$
O(5)	(4h)	0.112	0.145	$^1/_2$

As can be seen from Figure VIIB,43, the resulting structure has triangular BO_3 groups, with B–O = 1.40 and 1.50 A. The metal atoms, both bivalent and trivalent, are octahedrally surrounded by oxygens, with R–O lying between 1.98 and 2.45 A. The closest approach of oxygen atoms to one another is 2.40 A.

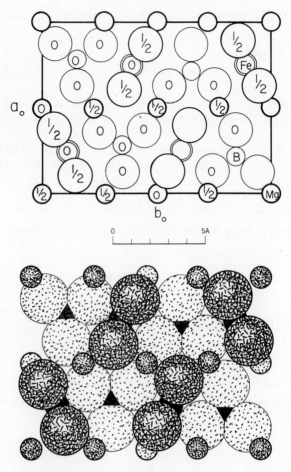

Fig. VIIB,43a (top). The orthorhombic structure of ludwigite, (Mg,Fe₄)B₂O₈(MgO)₂, seen along its c_0 axis. Origin in lower left.

Fig. VIIB,43b (bottom). A packing drawing of the orthorhombic structure of ludwigite viewed along its c_0 axis. Both the iron and the boron are black, the boron being triangular in shape and the irons scarcely showing. Magnesium atoms are small, heavily ringed and dotted. Oxygen atoms not part of BO₃ groups are line-shaded.

Other compounds with this structure have the cell dimensions:

$$Fe_2(CoO)_4(BO_3)_2: \quad a_0 = 9.35 \text{ A.}; \quad b_0 = 12.28 \text{ A.}; c_0 = 3.03 \text{ A.}$$
$$Fe_2(CuO)_4(BO_3)_2: \quad a_0 = 9.397 \text{ A.}; \quad b_0 = 12.02 \text{ A.}; c_0 = 3.13 \text{ A.}$$
$$Fe_2(FeO)_4(BO_3)_2: \quad a_0 = 9.44 \text{ A.}; \quad b_0 = 12.26 \text{ A.}; c_0 = 3.065 \text{ A.}$$
$$Fe_2(NiO)_4(BO_3)_2: \quad a_0 = 9.248 \text{ A.}; \quad b_0 = 12.26 \text{ A.}; c_0 = 3.01 \text{ A.}$$

VII,b45.   The symmetry of *pinakiolite*, $Mn_3Mg(MgO)_2(BO_3)_2$, originally described as orthorhombic, is monoclinic. Its bimolecular cell has the dimensions:

$$a_0 = 5.36 \text{ A.}; \quad b_0 = 5.98 \text{ A.}; \quad c_0 = 12.73 \text{ A.}; \quad \beta = 120°34'$$

Atoms have been placed in the following special and general positions of C_{2h}^2 $(P2_1/m)$:

$$
\begin{array}{ll}
(2a) & 000; 0\,^1/_2\,0 \\
(2d) & ^1/_2\,0\,^1/_2; \,^1/_2\,^1/_2\,^1/_2 \\
(2b) & ^1/_2\,0\,0; \,^1/_2\,^1/_2\,0 \\
(2e) & u\,^1/_4\,v; \,\bar{u}\,^3/_4\,\bar{v} \\
(4f) & \pm(xyz; \,x,^1/_2-y,z)
\end{array}
$$

with the positions and parameters of Table VIIB,20.

TABLE VIIB,20
Positions and Parameters of the Atoms in Pinakiolite

Atom	Position	x	y	z
B	$(4f)$	0.500	0.000	0.250
$Mn^{3+}(1)$	$(2b)$	$^1/_2$	0	0
$Mn^{3+}(2)$	$(2d)$	$^1/_2$	0	$^1/_2$
Mn^{2+}	$(2e)$	0.053	$^1/_4$	−0.197
$Mg(1)$	$(2a)$	0	0	0
$Mg(2)$	$(2e)$	−0.053	$^1/_4$	0.197
$Mg(3)$	$(2e)$	0.00	$^1/_4$	0.50
$O(1)$	$(2e)$	0.239	$^1/_4$	−0.011
$O(2)$	$(2e)$	−0.239	$^1/_4$	0.011
$O(3)$	$(4f)$	0.175	0	0.175
$O(4)$	$(4f)$	0.638	0	0.366
$O(5)$	$(4f)$	0.643	0	0.180
$O(6)$	$(2e)$	0.202	$^1/_4$	0.396
$O(7)$	$(2e)$	−0.202	$^1/_4$	−0.396

This structure (Fig. VIIB,44) is, especially in the distribution of its oxygen atoms, close to that of ludwigite. In the BO_3 triangles that are present, B–O = 1.29, 1.37, and 1.58 A. All metal atoms are octahedrally surrounded by six oxygen atoms at distances that lie between 1.94 and 2.46 A. Thus, in this arrangement, bivalent and trivalent metal atoms cannot be distinguished from one another by their oxygen environments.

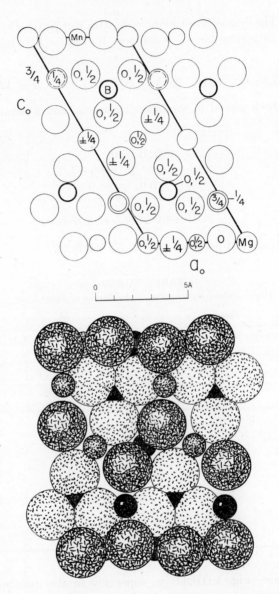

Fig. VIIB,44a (top). The monoclinic structure of pinakiolite, $Mg_3Mn_3B_2O_8$, seen along its b_0 axis. Origin in lower left.

Fig. VIIB,44b (bottom). A packing drawing of the monoclinic structure of Mg_3Mn_3-B_2O_8 seen along its b_0 axis. The boron atoms are black triangles, the manganese black circles. The magnesium atoms are small and dotted. Oxygen atoms not forming part of BO_3 groups are line-shaded.

Phosphites and Arsenites

VII,b46. *Phosphorous acid,* H_3PO_3, is orthorhombic, with a unit containing eight molecules. Its cell has the dimensions:

$$a_0 = 7.257 \text{ A.}; \quad b_0 = 12.044 \text{ A.}; \quad c_0 = 6.845 \text{ A.}$$

The space group is C_{2v}^9 $(Pna2_1)$, with all atoms in the positions:

$$(4a) \quad xyz; \; \bar{x},\bar{y},z+\tfrac{1}{2}; \; \tfrac{1}{2}-x,y+\tfrac{1}{2},z+\tfrac{1}{2}; \; x+\tfrac{1}{2},\tfrac{1}{2}-y,z$$

The parameters as initially determined with x-rays for the phosphorus and oxygen atoms are given in parentheses in Table VIIB,21. Neutron dif-

TABLE VIIB,21
Parameters of the Atoms in H_3PO_3

Atom	x	y	z
P(1)	0.001 (0.002)	0.1265 (0.126)	0.001 (0.003)
P(2)	0.044 (0.046)	0.378 (0.378)	0.4665 (0.461)
O(1)	0.211 (0.208)	0.1195 (0.116)	0.971 (0.951)
O(2)	0.905 (0.910)	0.209 (0.209)	0.871 (0.874)
O(3)	0.9425 (0.940)	0.0035 (0.004)	0.962 (0.985)
O(4)	0.9905 (0.994)	0.258 (0.259)	0.524 (0.518)
O(5)	0.082 (0.075)	0.369 (0.369)	0.245 (0.241)
O(6)	0.905 (0.909)	0.463 (0.459)	0.520 (0.518)
H(1)	0.208	0.401	0.568
H(2)	0.470	0.355	0.200
H(3)	0.075	0.437	0.168
H(4)	0.963	0.244	0.660
H(5)	0.782	0.314	0.948
H(6)	0.312	0.515	0.983

fraction data published somewhat more recently lead to the values without parentheses as well as to hydrogen parameters. In this neutron work a different origin was chosen within the unit cell; the parameters in terms of it have been transformed to those of the table by the relations: $x = x'$ (used in the neutron study), $y = \tfrac{1}{2}-y'$, $z = \tfrac{3}{4}-z'$.

In this structure (Fig. VIIB,45), the two crystallographically different H_3PO_3 molecules are of the same dimensions. The generally pyramidal shape and the approximate interatomic distances and bond angles are indicated in Figure VIIB,46. As this shows, one hydrogen atom is directly attached to the phosphorus, the other two to oxygens. The third, free, oxygen, definitely closer to the phosphorus atom, is connected through two

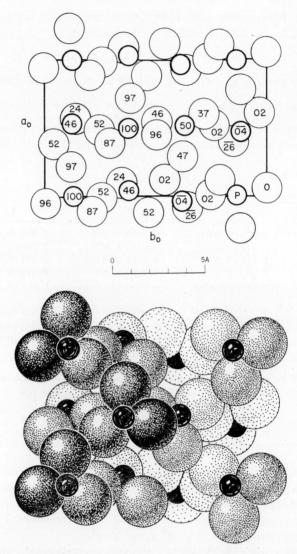

Fig. VIIB,45a (top). A projection of the orthorhombic structure of H_3PO_3 along its c_0 axis. Hydrogen positions are not shown. Origin in lower left.
Fig. VIIB,45b (bottom). A packing drawing of the H_3PO_3 arrangement viewed along its c_0 axis. The phosphorus atoms are black.

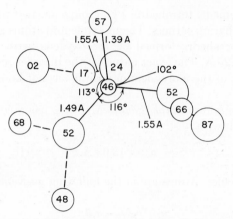

Fig. VIIB,46. One of the H_3PO_3 molecules of Fig. VIIB,45a. (left, middle) showing the positions of its hydrogen atoms and its bond dimensions.

hydrogen bonds to neighboring molecules. Each molecule has four such bonds, with O–H–O = 2.52–2.60 A. Such a molecule agrees with the fact that this acid is dibasic and not tribasic as its gross formula suggests.

VII,b47. The *molybdenum polyphosphate*, $MoO_2(PO_3)_2$, forms ortho-rhombic crystals with the bimolecular unit:

$$a_0 = 7.683 \text{ A.}; \quad b_0 = 7.426 \text{ A.}; \quad c_0 = 4.889 \text{ A.}$$

The space group appears to be V_h^{11} (*Pbcm*), but since there are no two-fold positions in this space group, this would require some randomness in the distribution of the molybdenum atoms. The following special positions are occupied:

$$(4c) \quad \pm(u \, ^1/_4 \, 0; \, u \, ^1/_4 \, ^1/_2)$$
$$(4d) \quad \pm(u \, v \, ^1/_4; \, \bar{u}, v + ^1/_2, ^1/_4)$$

It is considered that the two molybdenum atoms are distributed among the four equivalent positions of (4c) with $u = 0.219$. The other atoms have been placed as follows:

Atom	Position	x	y	z
P	(4d)	0.642	0.139	$^1/_4$
O(1)	(4c)	0.717	$^1/_4$	0
O(2)	(4d)	0.090	0.141	$^1/_4$
O(3)	(4d)	0.440	0.137	$^1/_4$
O(4)	(4d)	0.269	0.463	$^1/_4$

In this arrangement, tetrahedral PO_4 groups are tied together in chains parallel to c_0 by sharing corners. The molybdenum atoms are at the centers of some of the octahedra formed by the oxygens, with Mo–O lying between 1.77 and 2.25 A. The exact nature of the disorder resulting from this incomplete filling of the molybdenum positions is not known.

VII,b48. *Aluminum metaphosphate*, $Al(PO_3)_3$, has a complicated structure. Its cubic unit with

$$a_0 = 13.63 \text{ A.}$$

contains 16 molecules. Atoms are in the following positions of $T_d{}^6$ ($I\bar{4}3d$):

(16f)　　uuu;　　　　　　　　$u+{}^1/_2, {}^1/_2-u, \bar{u}$;
　　　　　$\bar{u}, u+{}^1/_2, {}^1/_2-u$;　　${}^1/_2-u, \bar{u}, u+{}^1/_2$;
　　　　　$u+{}^1/_4, u+{}^1/_4, u+{}^1/_4$; $u+{}^3/_4, {}^1/_4-u, {}^3/_4-u$;
　　　　　${}^3/_4-u, u+{}^3/_4, {}^1/_4-u$; ${}^1/_4-u, {}^3/_4-u, u+{}^3/_4$; B.C.

(48e)　　xyz;　　　　　　　　zxy;　　　　　　　　yzx;
　　　　　$x+{}^1/_2, {}^1/_2-y, \bar{z}$;　$z+{}^1/_2, {}^1/_2-x, \bar{y}$;　$y+{}^1/_2, {}^1/_2-z, \bar{x}$;
　　　　　$\bar{x}, y+{}^1/_2, {}^1/_2-z$;　$\bar{z}, x+{}^1/_2, {}^1/_2-y$;　$\bar{y}, z+{}^1/_2, {}^1/_2-x$;
　　　　　${}^1/_2-x, \bar{y}, z+{}^1/_2$;　${}^1/_2-z, \bar{x}, y+{}^1/_2$;　${}^1/_2-y, \bar{z}, x+{}^1/_2$;
　　　　　$y+{}^1/_4, x+{}^1/_4, z+{}^1/_4$; $z+{}^1/_4, y+{}^1/_4, x+{}^1/_4$; $x+{}^1/_4, z+{}^1/_4, y+{}^1/_4$;
　　　　　$y+{}^3/_4, {}^1/_4-x, {}^3/_4-z$; $z+{}^3/_4, {}^1/_4-y, {}^3/_4-x$; $x+{}^3/_4, {}^1/_4-z, {}^3/_4-y$;
　　　　　${}^3/_4-y, x+{}^3/_4, {}^1/_4-z$; ${}^3/_4-z, y+{}^3/_4, {}^1/_4-x$; ${}^3/_4-x, z+{}^3/_4, {}^1/_4-y$;
　　　　　${}^1/_4-y, {}^3/_4-x, z+{}^3/_4$; ${}^1/_4-z, {}^3/_4-y, x+{}^3/_4$; ${}^1/_4-x, {}^3/_4-z, y+{}^3/_4$; B.C.

with the parameters stated below:

Atom	Position	x	y	z
Al	(16f)	0.117	0.117	0.117
P	(48e)	0.340	0.063	0.124
O(1)	(48e)	0.090	0.110	0.800
O(2)	(48e)	0.095	0.141	0.245
O(3)	(48e)	0.137	0.096	0.986

The phosphorus and oxygen atoms in the arrangement that results form a system of linked PO_4 tetrahedra which associate themselves to build P_4O_{12} complexes. The P–O distances within a tetrahedron vary between 1.39 and 1.60 A. O(1) atoms are equidistant from two phosphorus atoms; the others, besides forming part of the PO_4 tetrahedra, provide octahedra about the aluminum atoms at distances of 1.80 and 1.83 A.

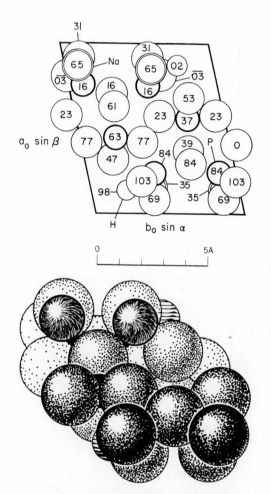

Fig. VIIB,47a (top). The triclinic structure of $Na_2H(PO_3)_3$ projected along its c_0 axis Origin in lower left.

Fig. VIIB,47b (bottom). A packing drawing of the $Na_2H(PO_3)_3$ structure seen along its c_0 axis. The phosphorus atoms are black but scarcely show. Sodium atoms are fine-line shaded; the oxygens are dotted.

VII,b49. The *acid sodium polyphosphate*, $[Na_2H(PO_3)_3]_n$, is triclinic, with the bimolecular cell:

$$a_0 = 7.72 \text{ A.}; \quad b_0 = 6.76 \text{ A.}; \quad c_0 = 7.11 \text{ A.}$$
$$\alpha = 90°36'; \quad \beta = 92°24'; \quad \gamma = 103°6'$$

TABLE VIIB,22
Parameters of the Atoms in $[Na_2H(PO_3)_3]_n$

Atom	x	y	z
Na(1)	0.155	0.418	0.351
Na(2)	0.139	0.902	0.351
P(1)	0.231	0.903	0.837
P(2)	0.228	0 461	0.840
P(3)	0.446	0.236	0.635
O(1)	0.082	0.880	0.691
O(2)	0.082	0.416	0.686
O(3)	0.196	0.353	0.027
O(4)	0.197	0.992	0 028
O(5)	0.300	0.707	0.886
O(6)	0.409	0.037	0.770
O(7)	0.410	0.421	0.774
O(8)	0.317	0.197	0.470
O(9)	0.358	0.720	0.387

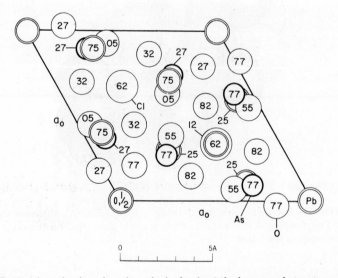

Fig. VIIB,48a. A projection along its principal axis of the hexagonal structure of Pb_5Cl-$(AsO_3)_3$.

All atoms are in the general positions $\pm(xyz)$ of C_i^1 ($P\bar{1}$) with the parameters of Table VIIB,22.

In this arrangement (Fig. VIIB,47) there are endless chains of PO_4 tetrahedra held together by sharing corners and running in the direction of the b_0 axis. Within tetrahedra, P–O = 1.49–1.66 A. Each sodium atom is surrounded by six oxygens at distances between 2.28 and 2.50 A.

The positions of the hydrogen atoms could not be established with certainty, but there are indications that they may be approximately defined by the parameters: $x(H) = 0.126$, $y(H) = 0.235$, $z(H) = 0.980$.

It is thought that this is the substance previously described as monoclinic with the composition $Na_2P_4O_{11}\cdot H_2O$.

VII,b50. The mineral *finnemanite*, $Pb_5Cl(AsO_3)_3$, though it resembles apatite both in composition and in dimensions of the hexagonal unit cell, has been assigned a very different atomic arrangement. Its unit containing, like apatite, two molecules has the edges:

$$a_0 = 10.28 \text{ A.}, \qquad c_0 = 7.00 \text{ A.}$$

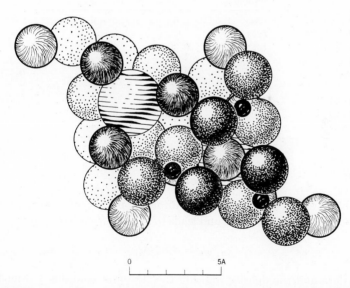

Fig. VIIB,48b. A packing drawing of the finnemanite arrangement viewed along its c_0 axis. The arsenic atoms are black, the lead atoms fine-line shaded. Oxygens are dotted.

The selected space group is C_6^6 ($P6_3$), with atoms in the positions:

(2a) $00u;\ 0,0,u+{}^1/_2$

(2b) ${}^1/_3\ {}^2/_3\ u;\ {}^2/_3,{}^1/_3,u+{}^1/_2$

(6c) $xyz;\qquad \bar{y},x-y,z;\qquad y-x,\bar{x},z;$
       $\bar{x},\bar{y},z+{}^1/_2;\ y,y-x,z+{}^1/_2;\ x-y,x,z+{}^1/_2$

The selected parameters are those of Table VIIB,23.

In this structure (Fig. VIIB,48) the AsO_3 group has its usual pyramidal form, with As–O = 1.80–1.83 A. The three kinds of lead atoms that are present are differently coordinated: Pb(1) has six octahedrally coordinated oxygens with Pb–O = 2.63 A.; Pb(2) has six oxygen neighbors at 2.56–2.58 A. and two rather distant chlorines at Pb–Cl = 3.50 A.; Pb(3) has three oxygens (Pb–O = 2.21–2.45 A.) and one chlorine (Pb–Cl = 2.75 A.) tetrahedrally arranged around it.

TABLE VIIB,23

Positions and Parameters of the Atoms in $Pb_5Cl(AsO_3)_3$

Atom	Position	x	y	z
Pb(1)	(2a)	0	0	0
Pb(2)	(2b)	${}^1/_3$	${}^2/_3$	0.620
Pb(3)	(6c)	0.300	0.400	0.250
Cl	(2b)	${}^1/_3$	${}^2/_3$	0.120
As	(6c)	0.270	0.370	0.770
O(1)	(6c)	0.205	0.170	0.750
O(2)	(6c)	0.145	0.440	0.820
O(3)	(6c)	0.375	0.445	0.550

Miscellaneous

VII,b51.   The very simple x-ray diffraction pattern of *sodium ceric oxide,* Na_2CeO_3, is like that of NaCl and leads to a unit cube of the edge:

$$a_0 = 4.82 \text{ A.}$$

These data and the density indicate that the unit contains 4/3 molecules, i.e., four oxygen atoms plus four metallic atoms. It would appear that the oxygen atoms are in close-packed positions [such as 000; F.C.], while the metallic atoms are distributed among the other close-packed positions occupied in sodium chloride, namely, ${}^1/_2\,{}^1/_2\,{}^1/_2$; F.C. The intimate relation between this structure and that found for $LiFeO_2$ (**VI,23**) is obvious.

Four other oxides have been found to have this arrangement. They are:

$$Li_2TiO_3: \quad a_0 = 4.1355 \text{ A.}$$
$$Na_2PbO_3: \quad a_0 = 4.690 \text{ A.}$$
$$Na_2PrO_3: \quad a_0 = 4.84 \text{ A.}$$
$$Na_2TbO_3: \quad a_0 = 4.740 \text{ A.}$$

Two mixed oxides of the following compositions also have this simple structure. Their cells have the dimensions:

$$Li_2Fe_{0.5}Ta_{0.5}O_3: \quad a_0 = 4.192 \text{ A.}$$
$$Li_2Zn_{0.5}Ta_{0.5}O_3: \quad a_0 = 4.214 \text{ A.}$$

A number of sulfides, selenides, and tellurides give correspondingly simple patterns. They have, however, been described in terms of disordered ZnS (**IV,c1**) rather than NaCl arrangements.
Their unit cubes have the edges:

$$Cu_2GeS_3: \quad a_0 = 5.30 \text{ A.}$$
$$Cu_2SnS_3: \quad a_0 = 5.43 \text{ A.}$$
$$Cu_2GeSe_3: \quad a_0 = 5.55 \text{ A.}$$
$$Cu_2SnSe_3: \quad a_0 = 5.69 \text{ A.}$$
$$Cu_2GeTe_3: \quad a_0 = 5.95 \text{ A.}$$
$$Cu_2SnTe_3: \quad a_0 = 6.04 \text{ A.}$$

VII,b52. *Lithium stannate*, Li_2SnO_3, has been given a structure based on a large monoclinic unit containing eight molecules and having the dimensions:

$$a_0 = 5.29 \text{ A.}; \quad b_0 = 9.19 \text{ A.}; \quad c_0 = 29.61 \text{ A.}; \quad \beta = 100°6'$$

Atoms have been placed in the following positions of C_{2h}^6 ($C2/c$):

(4d) $^1/_4 \, ^1/_4 \, ^1/_2; \, ^3/_4 \, ^1/_4 \, 0; \, ^3/_4 \, ^3/_4 \, ^1/_2; \, ^1/_4 \, ^3/_4 \, 0$
(4e) $\pm (0 \; u \; ^1/_4; \, ^1/_2, u + ^1/_2, ^1/_4)$
(8f) $\pm (xyz; \, \bar{x}, y, ^1/_2 - z; \, x + ^1/_2, y + ^1/_2, z; \, ^1/_2 - x, y + ^1/_2, ^1/_2 - z)$

In the original article the parameters were stated to be those of Table VIIB,24. It was also stated there that in the arrangement thus described Li–O and Sn–O = 2.07 A. This is scarcely possible, however, if the z parameters are those of the table. In the article, mention is made of a shift of origin to a point $0 \; ^1/_{12} \; ^1/_4$, but there is no clear statement of what is meant and it would appear that additional work is needed.

TABLE VIIB,24

Positions and Parameters of the Atoms in Li_2SnO_3

Atom	Position	x	y	z
Li(1)	(4d)	$1/4$	$1/4$	$1/2$
Li(2)	(4e)	0	0.92	$1/4$
Li(3)	(8f)	0.25	0.08	0.00
Sn(1)	(4e)	0	0.58	$1/4$
Sn(2)	(4e)	0	0.25	$1/4$
O(1)	(8f)	0.37	0.25	0.108
O(2)	(8f)	0.37	0.58	0.108
O(3)	(8f)	0.37	0.92	0.108

Several other substances found to possess this structure have the following cell dimensions:

Li_2TiO_3: $a_0 = 5.05$ A.; $b_0 = 8.76$ A.; $c_0 = 28.60$ A.; $\beta = 100°0'$

Na_2SnO_3: $a_0 = 5.50$ A.; $b_0 = 9.53$ A.; $c_0 = 32.51$ A.; $\beta = 99°36'$

Na_2ZrO_3: $a_0 = 5.60$ A.; $b_0 = 9.70$ A.; $c_0 = 32.75$ A.; $\beta = 99°42'$

$\beta\text{-}Na_2PbO_3$: $a_0 = 5.68$ A.; $b_0 = 9.84$ A.; $c_0 = 32.90$ A.; $\beta = 99°48'$

VII,b53. Similar but not identical structures have been described for *sodium metagermanate*, Na_2GeO_3, and the corresponding silicate (Chapter XII). There are four molecules in an orthorhombic unit which for the germanate has the dimensions:

$$a_0 = 10.88 \text{ A.}; \quad b_0 = 6.22 \text{ A.}; \quad c_0 = 4.92 \text{ A.}$$

The space group has been chosen as C_{2v}^{12} (*Cmc*), with atoms in the positions:

Na: (8b) $xyz;$ $\bar{x}yz;$
$\bar{x},\bar{y},z+1/2;$ $x,\bar{y},z+1/2;$
$x+1/2,y+1/2,z;$ $1/2-x,y+1/2,z;$
$1/2-x,1/2-y,z+1/2; x+1/2,1/2-y,z+1/2$
with $x = 0.170$, $y = 0.333$, $z = 0.555$

Ge: (4a) $0uv; 0,\bar{u},v+1/2; 1/2,u+1/2,v; 1/2,1/2-u,v+1/2$
with $u = 0.140$, $v = 0.125$

O(1): (8b) with $x = 0.137$, $y = 0.284$, $z = 0.00$

O(2): (4a) with $u = 0.140$, $v = 0.50$

For the silicate Na_2SiO_3,

$$a_0 = 10.43 \text{ A.}; \quad b_0 = 6.02 \text{ A.}; \quad c_0 = 4.81 \text{ A.}$$

Atoms have been given the corresponding positions, with

$$
\begin{array}{lll}
\text{Na:} & (8b) & \text{with } x = 0.166,\ y = 0.339,\ z = 0.563 \\
\text{Si:} & (4a) & \text{with } u = 0.166,\ v = 0.563 \\
\text{O}(1): & (8b) & \text{with } x = 0.130,\ y = 0.286,\ z = 0.500 \\
\text{O}(2): & (4a) & \text{with } u = 0.077,\ v = 0.895.
\end{array}
$$

In these two arrangements (Figs. VIIB,49 and VIIB,50) the sodium atoms have been identically placed and the O(2) atoms similarly placed. The Si(Ge) and O(1) parameters differ in the two compounds by a shift of ca. $1/2$ along c_0.

Both arrangements contain $Ge(Si)O_4$ tetrahedra linked by sharing corners. In the germanate, these form chains running along c_0; in the silicate they form double Si_2O_6 chains. In the silicate the sodium atoms have fivefold coordination to oxygen with Na–O = 2.27–2.45 A.; in the germanate the coordination is fourfold, with Na–O = 2.23–2.46 A.

It will be important through further work to be certain that these two compounds actually have these two differing atomic distributions. The corresponding lithium compounds have similar unit cells and the same space group C_{2v}^{12} (*Cmc*):

$$
\begin{array}{llll}
Li_2GeO_3: & a_0 = 9.625 \text{ A.}; & b_0 = 5.557 \text{ A.}; & c_0 = 4.815 \text{ A.} \\
Li_2SiO_3: & a_0 = 9.41 \text{ A.}; & b_0 = 5.43 \text{ A.}; & c_0 = 4.660 \text{ A.}
\end{array}
$$

VII,b54. The three compounds Li_5GeN_3, Li_5SiN_3, and Li_5TiN_3 are cubic with structures described as superlattices on the fluorite arrangement. The unit cubes, containing 32/3 of a formula weight, have the edges:

$$
\begin{array}{ll}
Li_5GeN_3: & a_0 = 9.61 \text{ A.} \\
Li_5SiN_3: & a_0 = 9.44 \text{ A.} \\
Li_5TiN_3: & a_0 = 9.70 \text{ A.}
\end{array}
$$

Atoms have been placed in the following positions of T_h^7 (*Ia3*):

$$
\begin{array}{lll}
\text{N}(1): & (8a) & 000;\ \text{F.C., B.C.} \\
\text{N}(2): & (24d) & \pm(u\,0\,1/4;\ u\,1/2\,3/4);\ \text{tr; B.C.}
\end{array}
$$

with $u = 0.205$, 0.215, and 0.230 for Li_5GeN_3, Li_5SiN_3, and Li_5TiN_3, respectively.

Li(1): (48e) $\pm(xyz;\ x,\bar{y},1/2-z;\ z,\bar{x},1/2-y;\ y,\bar{z},1/2-x)$; tr; B.C.

with $x = z = 1/8$, $y = 3/8$ in all cases

$^{16}/_3$ Li(2) + $^{32}/_3$ M: (16c) $\pm(uuu;\ u+1/2,1/2-u,\bar{u})$; B.C.

with $u = 0.115$ for Li_5GeN_3, 0.117 for Li_5SiN_3, and 0.120 for Li_5TiN_3.

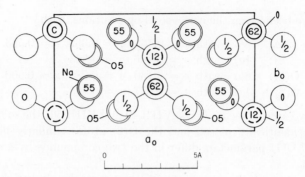

Fig. VIIB,49. The orthorhombic arrangement proposed for Na_2GeO_3 as projected along its c_0 axis. Origin in lower right.

Similar phosphides and arsenides have been described as being related to fluorite but with the metallic atoms statistically distributed. Their cubic cells of half the lengths of the foregoing are:

$$Li_5GeAs_3: \quad a_0 = 6.09 \text{ A.}$$
$$Li_5GeP_3: \quad a_0 = 5.89 \text{ A.}$$
$$Li_5SiAs_3: \quad a_0 = 6.055 \text{ A.}$$
$$Li_5SiP_3: \quad a_0 = 5.854 \text{ A.}$$
$$Li_5TiAs_3: \quad a_0 = 6.14 \text{ A.}$$
$$Li_5TiP_3: \quad a_0 = 5.96 \text{ A.}$$

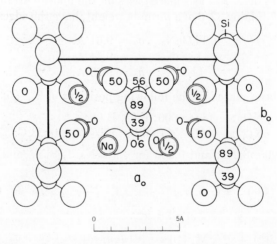

Fig. VIIB,50a. The orthorhombic structure of sodium metasilicate, Na_2SiO_3, seen along its c_0 axis. Origin in lower right.

BIBLIOGRAPHY TABLE, CHAPTER VIIB

Compound	Paragraph	Literature
Ag$_2$CsI$_3$	**b29**	1954: B,B&L
Ag$_3$AsS$_3$ (proustite)	**b36**	1936: H; 1937: H
Ag$_3$SbS$_3$ (pyrargyrite)	**b36**	1928: G&M; 1936: H; 1937: H
Al(PO$_3$)$_3$	**b48**	1927: H&W; 1937: P&S
Al$_3$NO$_3$	[VIII,**b1**]	1959: Y&Y
BaCO$_3$·2CeFCO$_3$ (cordylite)	**b11**	1931: O
BaCa(CO$_3$)$_2$	**a2, b8**	1930: G&M; 1960: A
Ba(NO$_3$)$_2$	**b17**	1917: N&H; 1922: V; 1927: V&D; 1928: J&vM; 1931: V&B; 1942: V&R
Ba$_2$ZnS$_3$	**b33**	1961: S&H
Be$_2$(OH)BO$_3$ (hambergite)	**b41**	1931: Z
CaCO$_3$·CeFCO$_3$ (synchisite)	**b11**	1931: O; 1953: D&D; 1961: I&S
CaCO$_3$·2CeFCO$_3$ (parisite)	**b11**	1931: O; 1953: D&D
2CaCO$_3$·3CeFCO$_3$ (roentgenite)	**b11**	1953: D&D

(continued)

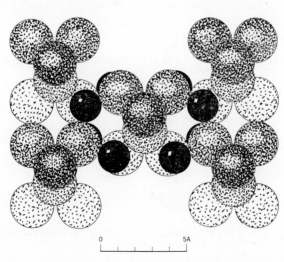

Fig. VIIB,50b. A packing drawing of the orthorhombic structure of Na$_2$SiO$_3$ seen along its c_0 axis. The sodium atoms are black, the silicons are hidden.

BIBLIOGRAPHY TABLE, CHAPTER VIIB (*continued*)

Compound	Paragraph	Literature
$CaMg(CO_3)_2$ (dolomite)	b7	1914: B; 1923: S; 1924: W&M; 1925: B&B; 1926: G; 1930: H; 1935: S; 1958: H&B; 1959: S&S; 1961: G
$CaMn(CO_3)_2$	b7	1961: G
$Ca(NO_3)_2$	b17	1922: V; 1928: J&vM; 1931: K,H&P; V&B; 1955: M
$CaSn(BO_3)_2$ (nordenskioldite)	b7	1934: R; 1935: E&R
$CdMg(CO_3)_2$	b7	1961: G
$Ce(IO_3)_4$	b27	1956: C&L
$(Ce,La...)FCO_3$ (bastnäsite)	b11	1929: O; 1931: O; 1953: D&D; 1961: I&S
$Co_3(BO_3)_2$	b40	1949: B
$Cs(UO_2)(NO_3)_3$	b19	1961: M&M
Cs_2AgCl_3	b33	1949: B&MG
Cs_2AgI_3	b33	1949: B&MG
$Cu(OH)IO_3$ (salesite)	b26	1962: G
$CuPbAsS_3$ (seligmannite)	b35	1956: H&L; L
$CuPbBiS_3$ (aikinite)	b33	1953: W; 1954: W
$CuPbSbS_3$ (bournonite)	b35	1932: O; 1956: H&L; L
Cu_2CsCl_3	b31	1954: B,B&L
Cu_2GeS_3	b51	1961: P,K,B&A
Cu_2GeSe_3	b51	1961: P,K,B&A
Cu_2GeTe_3	b51	1961: P,K,B&A
$Cu_2(OH)_2CO_3$ (malachite)	b14	1932: B; 1933: B; 1950: R&W; 1951: W
$Cu_2(OH)_3Br$	b20	1948: A; 1949: W; 1961: O,I,L&L
$Cu_2(OH)_3Cl$	b20	1950: A; F
$Cu_2(OH)_3I$	b20	1961: O,I,L&L
$Cu_2(OH)_3NO_3$	b20	1950: N&S; 1951: N&S; 1952: N&S
Cu_2SnS_3	b51	1961: P,K,B&A
Cu_2SnSe_3	b51	1961: P,K,B&A
Cu_2SnTe_3	b51	1961: P,K,B&A
$(Cu,Ag)_3(Sb,As)S_3$ (tetrahedrite)	b37	1927: P; 1928: dJ; M
$Cu_3(OH)_2(CO_3)_2$ (azurite)	b15	1932: B; 1933: B; 1958: G&Z

(*continued*)

BIBLIOGRAPHY TABLE, CHAPTER VIIB (*continued*)

Compound	Paragraph	Literature
Cu_4KS_3	**b38**	1952: R,S&W
Cu_4RbS_3	**b38**	1952: R,S&W
Fe_2AgS_3 (sternbergite)	**b34**	1937: B
Fe_2CuS_3 (cubanite)	**b34**	1936: B; 1937: B; 1945: B; 1947: B; 1955: A&B
$Fe_2(CoO)_2(BO_3)_2$	**b43**	1950: B
$Fe_2(CoO)_4(BO_3)_2$	**b44**	1950: B
$Fe_2(CuO)_4(BO_3)_2$	**b44**	1950: B
$Fe_2(FeO)_2(BO_3)_2$	**b43**	1950: B
$Fe_2(FeO)_4(BO_3)_2$	**b44**	1950: B
$Fe_2(MgO)_2(BO_3)_2$	**b43**	1950: B
$Fe_2(NiO)_2(BO_3)_2$	**b43**	1950: B
$Fe_2(NiO)_4(BO_3)_2$	**b44**	1950: B
Ge_2CuP_3	[VI,43]	1961: F&P; G,V&C
H_2SeO_3	**b24**	1949: W&B
H_3BO_3	**b39**	1934: Z; 1954: Z
H_3PO_3	**b46**	1957: F&L; 1958: L
$KAgCO_3$	**b5**	1944: D&H
$KHCO_3$	**b2**	1936: D; 1952: H; N,T&K; 1954: N,T&K
$K(NHOH)SO_3$	**b22**	1957: B&B
$K(PuO_2)CO_3$	**b6**	1954: E&Z
K_2AgI_3	**b30**	1952: B&K
K_2CuCl_3	**b33**	1949: B&MG
$K_2NH(SO_3)_2$	**b23**	1956: J&J
$LiH_3(SeO_3)_2$	**b25**	1960: V,O&P
$LiNa(BeF_3)_2$ (diopside-like)	[XII]	1953: H
Li_2CO_3	**b4**	1957: Z
$Li_2(Fe_{0.5}Ta_{0.5})O_3$	**b51**	1960: B
Li_2GeO_3	**b53**	1957: H&T
Li_2SiO_3	**b53**	1953: D&D
Li_2SnO_3	**b52**	1951: H&L; 1954: L
Li_2TiO_3	**b51, b52**	1933: K; 1935: K; 1943: B; 1954: L
$Li_2(Zn_{0.5}Ta_{0.5})O_3$	**b51**	1960: B
Li_5GeAs_3	**b54**	1954: J&S

(*continued*)

BIBLIOGRAPHY TABLE, CHAPTER VIIB (*continued*)

Compound	Paragraph	Literature
Li_5GeN_3	b54	1953: J,W&MS
Li_5GeP_3	b54	1954: J&S
Li_5SiAs_3	b54	1954: J&S
Li_5SiN_3	b54	1953: J,W&MS
Li_5SiP_3	b54	1954: J&S
Li_5TiAs_3	b54	1954: J&S
Li_5TiN_3	b54	1953: J,W&MS
Li_5TiP_3	b54	1954: J&S
$Mg_3(BO_3)_2$	b40	1949: B
$Mg_3Ca(CO_3)_4$ (huntite)	b10	1953: F; 1962: G&B
$Mg_3(OH,F)_3BO_3$ (fluoborite)	b42	1950: T
$(Mg,Fe)_3(TiO_2)(BO_3)_2$ (warwickite)	b43	1950: B; T,W&I
$(Mg,Fe)_2(MgO)_4(BO_3)_2$ (ludwigite)	b44	1950: B; T,W&I
$Mn_3Mg(MgO)_2(BO_3)_2$ (pinakiolite)	b45	1950: B; T,W&I
$MoO_2(PO_3)_2$	b47	1962: K
NH_4HCO_3	b3	1932: M; 1950: B&A
$NH_4H_2(NO_3)_3$	b18	1950: D&L
$NH_4(PuO_2)CO_3$	b6	1954: E&Z
$(NH_4)_2AgI_3$	b30	1952: B&K
$(NH_4)_2CuBr_3$	b33	1952: B&vA
$(NH_4)_2CuCl_3$	b32	1952: B&vA
$NaHCO_3$	b1	1933: Z; 1962: S&S
$Na_2Ca_2(CO_3)_3$ (shortite)	b9	1949: W
Na_2CeO_3	b51	1940: Z&M
Na_2GeO_3	b53	1954: G
$Na_2H(PO_3)_3$	b49	1961: T&J; 1962: J
$Na_2Mg(CO_3)_2.NaBr$	b12	1931: S&W
$Na_2Mg(CO_3)_2.NaCl$ (northupite)	b12	1931: G&K; S&W; 1933: W
Na_2PbO_3	b51, b52	1954: L
Na_2PrO_3	b51	1940: Z&M
Na_2SO_3	b21	1931: Z&B; 1952: NBS; 1956: T&K
Na_2SiO_3	b53	1952: G&P

(*continued*)

BIBLIOGRAPHY TABLE, CHAPTER VIIB (*continued*)

Compound	Paragraph	Literature
Na$_2$SnO$_3$	**b52**	1941: H&L; 1954: L
Na$_2$TbO$_3$	**b51**	1962: H&L
Na$_2$ZrO$_3$	**b52**	1954: L
Na$_4$Mg$_2$(CO$_3$)$_4$. Na$_2$SO$_4$ (tychite)	**b12**	1931: S&W; 1933: W
Pb(NO$_3$)$_2$	**b17**	1917: N&H; 1922: V; 1927: V&D; 1928: J&vM; 1931: V&B; 1942: V&R; 1955: M; 1957: H
Pb$_2$Br$_2$CO$_3$	**b13**	1944: S&P; 1945: O; 1946: S&P
Pb$_2$Cl$_2$CO$_3$ (phosgenite)	**b13**	1934: O; 1944: S&P; 1945: O; 1946: S&P
Pb$_3$(OH)$_2$(CO$_3$)$_2$	**b16**	1956: C
Pb$_5$Cl(AsO$_3$)$_3$ (finnemanite)	**b50**	1927: A&P; 1955: G
Rb(AmO$_2$)CO$_3$	**b6**	1954: E&Z
Rb(UO$_2$)(NO$_3$)$_3$	**b19**	1958: H&S
Rb$_2$AgI$_3$	**b30**	1952: B&K
Si$_2$CuP$_3$	**[VI,43]**	1961: F&P
Sr(NO$_3$)$_2$	**b17**	1917: N&H; 1922: V; 1928: J&vM; 1931: V&B; 1942: V&R; 1959: D,S&M
Zr(IO$_3$)$_4$	**b28**	1960: NBS; 1961: L&C

BIBLIOGRAPHY, CHAPTER VII

1914

Bragg, W. L., "The Analysis of Crystals by the X-Ray Spectrometer," *Proc. Roy. Soc.* (*London*), **89A**, 468.

1915

Bragg, W. H., "X-Rays and Crystal Structure," *Phil. Trans. Roy. Soc.*, **215**, 253.

1917

Nishikawa, S., and Hudinuki, K., "The Structure of the Nitrates of Lead, Barium and Strontium," *Proc. Math. Phys. Soc. Tokyo*, **9**, 197.

1919

Schiebold, E., "The Application of the Laue Diagram to the Determination of the Structure of Calcite," *Abhandl. Math.-Phys. Kl. Sächs. Akad. Wiss. Leipzig*, **36**, 65.

1920

Wyckoff, R. W. G., "The Crystal Structure of Sodium Nitrate," *Phys. Rev.*, **16**, 149.

Wyckoff, R. W. G., "The Crystal Structures of Some Carbonates of the Calcite Group," *Am. J. Sci.*, **50**, 317.

1921

Dickinson, R. G., and Goodhue, E. A., "The Crystal Structures of Sodium Chlorate and Sodium Bromate," *J. Am. Chem. Soc.*, **43**, 2045.

Kolkmeijer, N. H., Bijvoet, J. M., and Karssen, A., "Investigation by means of X-Rays of the Crystal Structure of Sodium Chlorate and Sodium Bromate," *Proc. Acad. Sci. Amsterdam*, **23**, 644.

1922

Vegard, L., "The Structure of the Isomorphous Group; the Nitrates of Ca, Sr, Ba and Pb," *Z. Physik*, **9**, 395; *Videnskaps. Skrifter I. Mat.-Naturv. Kl.*, **1922**, No. 3.

Vegard, L., "The Atomic Arrangement in the Optically Active Crystals Sodium Bromate and Sodium Chlorate," *Z. Physik*, **12**, 289; *Videnskaps. Skrifter I. Mat.-Naturv. Kl.*, **1922**, No. 16.

Wulff, G., "The Structure of Sodium Chlorate," *Z. Krist.*, **57**, 190.

1923

Karssen, A., "X-Ray Investigation of the Structure of Crystals of Sodium Bromate and Sodium Chlorate," *Rec. Trav. Chim.*, **42**, 904.

Kiby, W., "The Crystal Structure of Sodium Chlorate," *Z. Physik*, **17**, 213; cf. Kolkmeijer, N. H., Bijvoet, J. M., and Karssen, A., *ibid.*, **20**, 82.

Kolkmeijer, N. H., Bijvoet, J. M., and Karssen, A., "The Structure of Crystals of Sodium Chlorate and Bromate," *Z. Physik*, **14**, 291; cf. Vegard, L., *ibid.*, **18**, 379.

Mauguin, C., "The Reflection of Certain Unusual Lattice Planes of Calcite," *Compt. Rend.*, **176**, 1331.

Schiebold, E., "Experiments with X-Ray Rotation Diagrams," *Z. Krist.*, **57**, 579.

1924

Bragg, W. L., "The Structure of Aragonite," *Proc. Roy. Soc. (London)*, **105A**, 16.

Heide, F., "Vaterite," *Central. Mineral. Geol.*, **1924**, 641.

Levi, G. R., and Ferrari, A., "The Crystalline Lattices of Rhombohedral Carbonates of Bivalent Metals," *Rend. Accad. Lincei*, **33**, 516.

Rinne, F., "An X-Ray Investigation of Some Finely Divided Minerals, Artificial Products and Dense Rocks," *Z. Krist.*, **60**, 55.

Wyckoff, R. W. G., and Merwin, H. E., "The Crystal Structure of Dolomite," *Am. J. Sci.*, **8**, 447.

1925

Barth, T., "The Crystal Structure of Perovskite and Related Compounds," *Norsk Geol. Tidssk.*, **8**, 201.

Beets, H. N., "A Determination of the Angle Between the Cleavage Faces of Calcite by Use of X-Rays," *Phys. Rev.*, **25**, 621.

Bragg, W. H., and Bragg, W. L., *X-Rays and Crystal Structure*, 5th ed., G. Bell & Son, London.

Compton, A. H., Beets, H. N., and DeFoe, O. K., "The Grating Space of Calcite and Rock Salt," *Phys. Rev.*, **25**, 625.

Gibson, R. E., Wyckoff, R. W. G., and Merwin, H. E., "Vaterite and μ-Calcium Carbonate," *Am. J. Sci.*, **10**, 325.

Levi, G. R., and Natta, G., "The Crystalline Structure of Perovskite," *Rend. Accad. Lincei*, **2**, 39; **4**, 54 (1926).

Smith J. H., "Molecular Symmetry in Crystal Structure," *Nature*, **115**, 334.

Tomkeieff, S. I., "The Structure of Aragonite," *Mineral. Mag.*, **20**, 408.

Wyckoff, R. W. G., "Orthorhombic Space Group Criteria and Their Application to Aragonite," *Am. J. Sci.*, **9**, 145; *Z. Krist.*, **61**, 425.

1926

Garrabos, L., "X-Ray Study of Symmetry and Twins of Dolomite," *Bull. Soc. Franc. Mineral*, **49**, 110.

Goldschmidt, V. M., "The Laws of Crystal Chemistry," *Skrifter Norske Videnskaps-Akad. Oslo I. Mat.-Naturv. Kl.*, **1926**, No. 2; *Naturwiss.*, **14**, 477.

Goldschmidt, V. M., "Researches on the Structure and Properties of Crystals," *Skrifter Norske Videnskaps-Akad. Oslo I. Mat.-Naturv. Kl.*, **1926**, No. 8.

1927

Aminoff, G., and Parsons, A. L., "The Symmetry and Lattice Dimensions of Finnemanite and Mimetite," *Geol. Foren. Stockholm Forh.*, **49**, 438.

Brentano, J., and Dawson, W. E., "Determination of the Lattice Spacing and of the Rhombohedral Angle of Magnesium Carbonate from a Microcrystalline Powder," *Phil. Mag.*, **3**, 411.

Ferrari, A., and Baroni, A., "On the Crystalline Structure of the Double Chloride of Cadmium and Cesium, $CsCdCl_3$. Observations on the Monometric Structure of the Type $A(BX_3)$," *Rend. Accad. Lincei*, **6**, 418.

Ferrari, A., and Fontana, C. G., "The Structure of Silver Chlorate," *Rend. Accad. Lincei*, **6**, 312.

Hendricks, S. B., and Wyckoff, R. W. G., "The Space Group of Aluminum Metaphosphate," *Am. J. Sci.*, **13**, 491.

Natta, G., "The Crystalline Structure of Cesium Trichloro-mercurate," *Rend. Accad. Lincei*, **5**, 1003.

Palacios, J., "The Crystalline Structure of Tetrahedrite," *Anales Real Soc. Espan. Fis. Quim.*, **25**, 246.

Tsuboi, C., "On the Effect of Temperature upon the Crystal Structure of Calcite," *Proc. Imp. Acad. (Japan)*, **3**, 17.

Vegard, L., and Dale, H., "Investigations on Mixed Crystals and Alloys," *Skrifter Norske Videnskaps-Akad. Oslo I. Mat-Naturv. Kl.*, **1927**, No. 14; *Z. Krist.*, **67**, 148 (1928).

1928

Gossner, B., and Mussgnug, F., "The Crystal Structure of Pyrargyrite," *Central. Mineral. Geol.*, **1928A**, 65.

Harang, L., "The Crystal Structure of the Tetragonal Compounds $AgClO_3$ and $AgBrO_3$," *Z. Krist.*, **66**, 399.

Jaeger, F. M., and van Melle, F. A., "On the Symmetry and Structure of the Cubic Nitrates of Calcium, Strontium, Barium and Lead," *Verslag Akad. Wetenschap. Amsterdam*, **37**, 528; *Proc. Acad. Sci. Amsterdam*, **31**, 651.

Jong, W. F. de, "The Crystal Structures of Arsenopyrite, Bornite and Tetrahedrite," Thesis, Delft.

Machatschki, F., "Formula and Crystal Structure of Tetrahedrite," *Norsk Geol. Tidssk.*, **10**, No. 1; *Z. Krist.*, **68**, 204.

Natta, G., and Passerini, L., "Isomorphism, Polymorphism and Morphotropy I. Compounds of the ABX_3 Type," *Gazz. Chim. Ital.*, **58**, 472.

Skaliks, W., "Some Double Compounds of the Alkali Carbonates with the Alkaline Earth Carbonates," *Schrift. Königsberger Gelehrt. Ges. Naturwiss. Kl.*, **5**, 93.

Wilson, T. A., "The Lattice Constants and the Space Groups of $BaCO_3$ and $SrCO_3$," *Phys. Rev.*, **31**, 305.

Zachariasen, W. H., "The Crystal Structure of the Sesquioxides and Compounds of the Type ABO_3," *Skrifter Norske Videnskaps-Akad. Oslo I. Mat.-Naturv. Kl.*, **1928**, No. 4.

1929

Brentano, J., and Adamson, J., "Precision Measurements of X-Ray Reflections from Crystal Powders. The Lattice Constants of Zinc Carbonate, Manganese Carbonate and Cadmium Oxide," *Phil. Mag.*, **7**, 507.

Ferrari, A., and Colla, C., "The Crystal Structure of the Neutral Carbonates of Cobalt and Nickel," *Rend. Accad. Lincei*, **10**, 594.

Mizgier, S., "On the Structure of Lublinite," *Z. Krist.*, **70**, 160.

Oftedal, I., "The Crystal Structure of Bastnäsite," *Z. Krist.*, **72**, 239.

Ruff, O., Ebert, F., and Stephan, E., "The System ZrO_2–CaO," *Z. Anorg. Chem.*, **180**, 215.

Zachariasen, W. H., "The Crystal Structure of Potassium Chlorate," *Z. Krist.*, **71**, 501.

Zachariasen, W. H., "The Crystal Structure of Sodium Chlorate," *Z. Krist.*, **71**, 517.

1930

O'Daniel, H., "A New Occurrence of Tarnowitzite in Tsumeb-Otavi and the Question of Mutual Isomorphous Replacement of Calcium and Lead," *Z. Krist.*, **74**, 333.

Gaertner, H. R. v., "The Crystal Structure of Loparite and Pyrochlore," *Neues Jahrb. Mineral. Geol.*, *Beilage-Bd.*, **61A**, 1.

Gossner, B., and Mussgnug, F., "Krokoite, Lautarite and Dietzeite and their Crystallographic Relationships," *Z. Krist.*, **75**, 410.

Gossner, B., and Mussgnug, F., "Alstonite and Milarite—A Contribution to the Study of Complex Crystals," *Centr. Mineral. Geol.*, **1930A**, 220.

Gossner, B., and Mussgnug, F., "Barytocalcite and Its Structural Relations to Other Materials," *Centr. Mineral. Geol.*, **1930A**, 321.

Halla, F., "X-Ray Distinctions Between Magnesite and Dolomite," *Sitzungsber. Akad. Wiss. Wien, Math.-Naturv. Kl. Abt. IIb*, **139**, 683; *Monatsh. Chem.*, **57**, 1.

Krieger, P., "X-Ray Diffraction Study of the Series Calcite-Rhodochrosite," *Am. Mineralogist*, **15**, 23.

Sirkar, S. C., "On the Laue Photographs of Iridescent Crystals of Potassium Chlorate," *Indian J. Phys.*, **5**, 337.

Taylor, N. W., "The Crystal Structures of the Compounds Zn_2TiO_4, Zn_2SnO_4, Ni_2SiO_4 and $NiTiO_3$," *Z. Physik. Chem.*, **9B**, 241.

1931

Bearden, J. A., "Variations in the Grating Constant of Calcite Crystals," *Phys. Rev.*, **38**, 1389.

Bearden, J. A., "Grating Constant of Calcite Crystals," *Phys. Rev.*, **38**, 2089.

Cork, J. M., and Gerhard, S. L., "Crystal Structure of the Series of Barium and Strontium Carbonates," *Am. Mineralogist*, **16**, 71.

Edwards, D. A., "A Determination of the Complete Crystal Structure of Potassium Nitrate," *Z. Krist.*, **80**, 154.

Fukusima, E., "Determination of the Parameter of Calcite by the Temperature Effect," *J. Sci. Hiroshima Univ.*, **1A**, 195.

Gossner, B., and Koch, I., "The Crystal Lattice of Langbeinite, Northupite and Hanksite," *Z. Krist.*, **80**, 455.

Kracek, F. C., Hendricks, S. B., and Posnjak, E., "Group Rotation in Solid Ammonium and Calcium Nitrates," *Nature*, **128**, 410.

Kracek, F. C., Posnjak, E., and Hendricks, S. B., "Gradual Transition in Sodium Nitrate II. The Structure at Various Temperatures and Its Bearing on Molecular Rotation," *J. Am. Chem. Soc.*, **53**, 3339.

Oftedal, I., "The Crystal Structure of Bastnäsite (Ce,La,···)FCO_3," *Z. Krist.*, **78**, 462.

Oftedal, I., "Parasite, Synchisite and Cordylite, X-Ray Investigations," *Z. Krist.*, **79**, 437.

Shiba, H., and Watanabe, T., "The Crystal Structure of Northupite, Brominated Northupite and Tychite," *Compt. Rend.*, **193**, 1421.

Vegard, L., and Bilberg, L., "The Crystal Structure of Nitrates of Calcium, Strontium, Barium and Lead," *Norske Videnskaps-Akad. Oslo I. Mat. Naturv. Kl.*, **1931**, No. 12.

Zachariasen, W. H., "The Crystalline Structure of Hambergite, $Be_2BO_3(OH)$," *Z. Krist.*, **76**, 289.

Zachariasen, W. H., and Barta, F. A., "Crystal Structure of Lithium Iodate," *Phys. Rev.*, **37**, 1626.

Zachariasen, W. H., and Buckley, H. E., "Crystal Lattice of Anhydrous Sodium Sulfite," *Phys. Rev.*, **37**, 1295.

1932

Baccaredda, M., "The Structure of Spherocobaltite," *Rend. Accad. Lincei*, 16, 248.

Brasseur, H., "Contribution to the Structure of Malachite," *Z. Krist.*, 82, 111.

Brasseur, H., "The Structure of Azurite," *Z. Krist.*, 82, 195.

Bijvoet, J. M., and Ketelaar, J. A. A., "Molecular Rotation in Solid Sodium Nitrate," *J. Am. Chem. Soc.*, 54, 625.

Garrido, J., "Crystalline Structure of Ammonium Iodate," *Anales Soc. Real Espan. Fis. Quim.*, 30, 811.

Goldschmidt, V. M., and Hauptmann, H., "Isomorphism of Borates and Carbonates," *Nachr. Ges. Wiss. Gottingen, Math.-Phys. Kl.*, 1932, 53.

Hendricks, S. B., Posnjak, E., and Kracek, F. C., "Molecular Rotation in the Solid State. The Variation of the Crystal Structure of Ammonium Nitrate with Temperature," *J. Am. Chem. Soc.*, 54, 2766.

Jong, W. F. de, "The Crystal Structure of the Cubic Na-W-Bronzes," *Z. Krist.*, 81, 314.

Mooney, R. C. L., "The Crystal Structure of Ammonium Bicarbonate," *Phys. Rev.*, 39, 861.

Oftedal, I., "The Space Group of Bournonite," *Z. Krist.*, 83, 157.

Quill, L. L., "The Lattice Constants of Columbium, Tantalum and Several Columbates and Tantalates," *Z. Anorg. Allgem. Chem.*, 208, 257.

West, C. D., "The Crystal Structure of Rhombic Ammonium Nitrate," *J. Am. Chem. Soc.*, 54, 2256.

1933

Brasseur, H., "The Structure of Azurite and Malachite," Thesis, Liège.

Colby, M. Y., and LaCoste, L. J. B., "The Crystal Structure of Cerussite," *Z. Krist.*, 84, 299.

Dennis, L. M., and Rochow, E. G., "Oxyacids of Fluorine II," *J. Am. Chem. Soc.*, 55, 2431.

Hoffman, A., "Difference in Size of the Ions of Zirconium and Hafnium," *Naturwiss.*, 21, 676.

Kordes, E., "On the Structure of Li_2TiO_3," *Fortschr. Mineral. Krist.*, 18, 27.

Pauling, L., and Sherman, J., "Note on the Crystal Structure of Rubidium Nitrate," *Z. Krist.*, 84, 213.

Watanabe, T., "The Crystalline Structure of Northupite and Tychite," *Sci. Papers Inst. Phys. Chem. Res. (Tokyo)*, 21, 40.

Weigle, J., "Precision Measurements of Rhombohedral Crystal Lattices: Sodium Nitrate," *Helv. Phys. Acta*, 7, 46.

Zachariasen, W. H., "The Crystal Lattice of Sodium Bicarbonate," *J. Chem. Phys.*, 1, 634.

1934

Barth, T. F. W., and Posnjak, E., "The Crystal Structure of Ilmenite," *Z. Krist.*, 88, 265.

Barth, T. F. W., and Posnjak, E., "Notes on Some Structures of the Ilmenite Type," *Z. Krist.*, 88, 271.

Elliott, N., "The Crystal Structure and Magnetic Susceptibility of Cesium Argentous Auric Chloride and Cesium Aurous Auric Chloride," *J. Chem. Phys.*, 2, 419.

Gossner, B., and Kraus, O., "On the Structure of Jeremejewite, BAlO₃," *Zentr. Mineral. Geol.*, **1934A**, 348.

Hoffman, A., "Oxygen Acids of Quadrivalent Cerium and Thorium," *Naturwiss.*, **22**, 206.

Onorato, E., "The Structure of Phosgenite," *Periodico Mineral.*, **5**, 1.

Posnjak, E., and Barth, T. F. W., "Notes on Some Structures of the Ilmenite Type," *Z. Krist.*, **88**, 271.

Ramdohr, P., "Nordenskioldine from a Cassiterite Deposit," *Neues Jahrb. Mineral. Geol. Beil.-Bd.*, **68A**, 288.

Saini, H., and Mercier, A., "Thermal Expansion of Sodium Nitrate Determined with the Help of X-Rays," *Helv. Phys. Acta*, **7**, 267.

Waldbauer, L., and McCann, D. C., "Cesium Nitrate and the Perovskite Structure," *J. Chem. Phys.*, **2**, 615.

Weigle, J., and Saini, H., "The Thermal Expansion of Calcite," *Helv. Phys. Acta*, **7**, 257.

Zachariasen, W. H., "The Crystal Lattice of Boric Acid," *Z. Krist.*, **88**, 150.

1935

Colby, M. Y., and Lacoste, L. J. B., "The Crystal Structure of Witherite, BaCO₃," *Z. Krist.*, **90A**, 1.

Ehrenberg, W., and Ramdohr, P., "The Structure of Nordenskioldite," *Neues J. Mineral. Geol.*, **69A**, 1.

Ferrari, A., and Colla, C., "Cadmium Nitrites of Monovalent Metals," *Gazz. Chim. Ital.*, **65**, 797.

Ferrari, A., and Colla, C., "Mercuric Nitrites of Monovalent Metals," *Gazz. Chim. Ital.*, **65**, 789.

Hägg, G., "The Cubic Sodium-Tungsten Bronzes," *Z. Physik. Chem.*, **29B**, 192.

Hoffman, A., "Compounds with Perowskite Structure," *Z. Physik. Chem.*, **28B**, 65.

Kordes, E., "Li₂TiO₃ and Mixed Crystals," *Z. Krist.*, **92A**, 139.

Schoklitsch, K., "On the Structures of Several Carbonates," *Z. Krist.*, **90A**, 433.

1936

Buerger, M. J., "The Crystal Structure of Cubanite, CuFe₂S₃," *Am. Mineralogist*, **21**, 205.

Dhar, J., "The Crystal Structure of KHCO₃," *Current Sci. (India)*, **4**, 867.

Ferrari, A., and Colla, C., "Solid Solutions of Divalent Metal Carbonates," *Gazz. Chim. Ital.*, **66**, 571.

Harker, D., "The Application of the Three-Dimensional Patterson Method and the Crystal Structures of Proustite, Ag₃AsS₃, and Pyrargyrite, Ag₃SbS₃," *J. Chem. Phys.*, **4**, 381.

Natta, G., and Baccaredda, M., "The Structure of Sb₂O₅·H₂O," *Gazz. Chim. Ital.*, **66**, 308.

Zedlitz, O., "On Perowskite, etc.," *Fortschr. Mineral. Krist.*, **22**, 66.

1937

Buerger, N. W., "The Unit Cell and Space Group of Sternbergite, Fe₂AgS₃," *Am. Mineralogist*, **22**, 847.

Buerger, M. J., "The Crystal Structure of Cubanite," *Am. Mineralogist*, **22**, 1117.

548

CRYSTAL STRUCTURES

Cole, S. S., and Espenschied, H., "On the Crystal Structure of PbTiO₃," *J. Phys. Chem.*, **41B**, 445.

Ferrari, A., "On the Structure of Some Mixed Hexachloraurates," *Gazz. Chim. Ital.*, **67**, 94.

Finbak, C., and Hassel, O., "The Structure of CsNO₃," *Z. Physik*, **5**, 460.

Finback, C., and Hassel, O., "The Rotation of Anion Polyhedra in Cubic Structures, III. Nitrates," *Z. Physik. Chem.*, **35B**, 25.

Hocart, R., "The Structure of the Proustite and Pyrargyrite," *Compt. Rend.*, **205**, 68.

Pauling, L., and Sherman, J., "The Crystal Structure of Al(PO₃)₃," *Z. Krist.*, **96A**, 481.

1938

Brasseur, H., and Pauling, L., "The Crystal Structure of NH₄CdCl₃," *J. Am. Chem. Soc.*, **60**, 2886.

Elliott, N., and Pauling, L., "The Crystal Structure of Cs₂Au₂Cl₆ and Cs₂AuAgCl₆," *J. Am. Chem. Soc.*, **60**, 1846.

Hamilton, J. E., "The Crystal Structure of NaBrO₃," *Z. Krist.*, **100A**, 104.

Harmsen, E. J., "The Crystal Structure of NH₄HgCl₃," *Z. Krist.*, **100A**, 208.

Lindsay, G. A., and Hoyt, H. C., "The Constants of Cerrusite," *Z. Krist.*, **100A**, 360.

Schrewelius, N., "X-Ray Investigation of the Compounds NaSb(OH)₆, NaSbF₆, NaSbO₃, and Analogous Substances," *Z. Anorg. Allgem. Chem.*, **238**, 241.

Siegl, W., "On the Structure of Plumbocalcite," *Z. Krist.*, **99A**, 95.

Strunz, H., "On Rhodizite and Jeremejewite," *Naturwiss.*, **26**, 217.

1939

Barth, T. F. W., "The Crystal Structure of the Pressure Modification of KNO₃," *Z. Physik. Chem.*, **43B**, 448.

MacGillavry, C. H., Nijveld, H., Dierdorp, S., and Karsten, J., "The Crystal Structure of NH₄CdCl₃ and RbCdCl₃," *Rec. Trav. Chim.*, **58**, 193

Silberstein, A., "X-Ray Structure of KCuBr₃," *Compt. Rend.*, **209**, 540.

Zedlitz, O., "Perewskite," *Neues Jahrb. Mineral. Geol.*, **75**, 245.

1940

Huber, K., "The Formation of Anomalous Solid Solutions of Alkali and Lead Halogen Compounds," *Helv. Chim. Acta*, **23**, 302.

Ievins, A., and Straumanis, M., "Lattice Constant of Calcite Determined by the Rotating-Crystal Method," *Z. Physik*, **116**, 194.

Pocza, J., "The Structure of AgClO₃," *Magyar Kem. Folyoirat*, **46**, 141.

Spiegelberg, P., "X-Ray Studies on Potassium Antimonates," *Arkiv Kemi*, **14A**, No. 5, 12 pp.

Tyren, F., "Precision Studies of Soft X-Rays with the Concave Grating," *Nova Acta Regiae Soc. Sci. Upsaliensis*, **12**, #1.

Zintl, E., and Morawietz, W., "Double Oxides with the NaCl Structure," *Z. Anorg. Allgem. Chem.*, **245**, 26.

1941

Rogers, M. T., and Helmholz, L., "The Structure of Iodic Acid," *J. Am. Chem. Soc.*, **63**, 278.

1942

Duyn, D. van, "Tungsten Bronze," *Rec. Trav. Chim.*, **61**, 669.

Naráy-Szabó, I., and Pocza, J., "The Crystal Structure of $AgClO_3$," *Z. Krist.*, **104A**, 28.

Vegard, L., and Roer, K. I., "Method of X-Ray Examination of the Effect of Temperature on the Crystal Lattice and Its Application to the Nitrates of the Bivalent Metals," *Avhandl. Norske Vidensk. Akad. Oslo, I. Math. Naturw. Kl.*, **1941**, 3.

1943

Barblan, F. F., "The Crystal Chemistry of Fe_2O_3 and TiO_2 and Their Alkali Compounds," *Schweiz. Mineral. Petrog. Mitt.*, **23**, 295.

Byström, A., "X-Ray Investigation of the System $MgO–Al_2O_3–SiO_2$," *Ber. Deut. Keram. Ges.*, **24**, 2.

MacGillavry, C. H., and Panthaleon van Eck, C. L., "The Crystal Structure of $NaIO_3$ and NH_4IO_3," *Rec. Trav. Chim.*, **62**, 729.

Naráy-Szabó, I., "The Structure of Compounds ABO_3, 'Sister Structures,' " *Naturwiss.*, **31**, 466.

Naráy-Szabó, I., "The Structural Type of Perovskite," *Naturwiss.*, **31**, 202.

O'Daniel, H., and Tscheischwili, L., "A Model Substance for Silicates," *Naturwiss.*, **31**, 209.

Rivoir, L., and Abbad, M., "The Structure of Rhombic $TlNO_3$," *Anales Fis. Quim. (Madrid)*, **39**, 306.

Sørum, H., "Crystal Structure of $NaVO_3$," *Kgl. Norske Videnskab. Selskabs, Forh.*, **15**, 39; *Chem. Zentr.*, **1944**, I, 206.

Zedlitz, O., "The Structure of Perovskite," *Naturwiss.*, **31**, 369.

1944

Donohue, J., and Helmholz, L., "The Crystal Structure of $KAgCO_3$," *J. Am. Chem. Soc.*, **66**, 295.

Sillen, L. G., and Pettersson, R., "The Crystal Structure of Phosgenite and $Pb_2Br_2CO_3$," *Naturwiss.*, **32**, 41.

1945

Buerger, M. J., "Structure of Cubanite, $CuFe_2S_3$, and the Coordination of Ferromagnetic Iron," *J. Am. Chem. Soc.*, **67**, 2056.

Forrester, W. F., and Hinde, R. M., "Crystal Structure of Barium Titanate," *Nature*, **156**, 177.

Ketelaar, J. A. A., and Strijk, B., "The Atomic Arrangement in Solid Sodium Nitrate at High Temperatures," *Rec. Trav. Chim.*, **64**, 174.

Megaw, H. D., "Crystal Structure of $BaTiO_3$," *Nature*, **155**, 484.

O'Daniel, H., and Tscheischwili, L., "The Structure of $NaBeFe_3$ and β-$CaSiO_3$," *Neues Jahrb. Mineral., Geol. Monatsh.*, **1945–48A**, 56.

Oftedal, I., "On the Crystal Structure of Phosgenite $Pb_2Cl_2CO_3$ and Synthetic Compounds," *Norsk. Geol. Tidsskr.*, **24**, 79.

Rooksby, H. P., "Compounds of the Structural Type of $CaTiO_3$," *Nature*, **155**, 484.

1946

Klug, H. P., and Sears, G. W., Jr., "Crystal-Chemical Study of Cesium Trichlorocuprate," *J. Am. Chem. Soc.*, **68**, 1133.

Megaw, H. D., "Crystal Structure of Barium Titanium Oxide at Different Temperatures," *Experientia*, **2**, 183.

Megaw, H. D., "Changes in Polycrystalline Barium-Strontium Titanate at Its Transition Temperature," *Nature*, **157**, 20.

Megaw, H. D., "Crystal Structure of Double Oxides of the Perovskite Type," *Proc. Phys. Soc. (London)*, **58**, 133.

Sillen, L. G., and Pettersson, R., "On the Crystal Structure of Phosgenite and Pb_2Br_2-CO_3," *Arkiv Kemi*, **21A**, No. 13.

1947

Borchert, W., and Schroeder, R., "Crystallographic and X-Ray Data on Wittichenite from Sadisdorf," *Heidelberger Beitr. Mineral. Petrog.*, **1**, 112.

Brandenberger, E., "Crystal Structure of $K(CdCl_3)$," *Experientia*, **3**, 149.

Buerger, M. J., "The Crystal Structure of Cubanite," *Am. Mineralogist*, **32**, 415.

Bretteville, A. P. de, and Levin, S. B., "The Lattice Constants of a Single Crystal of Barium Titanate," *Am. Mineralogist*, **32**, 686.

Goodwin, T. H., and Whetstone, J., "The Crystal Structure of Ammonium Nitrate III, and Atomic Scattering Factors in Ionic Crystals," *J. Chem. Soc.*, **1947**, 1455.

Kay, H. F., and Rhodes, R. G., "Barium Titanate Crystals," *Nature*, **160**, 126.

Megaw, H. D., "Temperature Changes in the Crystal Structure of Barium Titanium Oxide," *Proc. Roy. Soc. (London)*, **A189**, 261.

Naráy-Szabó, I., "The Perovskite-Structure Family," *Muegyet. Kozlemen.*, **1947**, No. 1, 30.

Naráy-Szabó, I., and Neugebauer, J., "The Crystal Structure of Sodium Iodate," *J. Am. Chem. Soc.*, **69**, 1280.

Rivoir, L., and Abbad, M., "The Structure of Thallous Iodate," *Anales Fis. Quim. (Madrid)*, **43**, 1051.

Tahvonen, P. E., "The Crystal Structure of Sodium Nitrate and the Atomic Scattering Factors of the Atoms in the Nitrate Group," *Ann. Acad. Sci. Fennicae, Ser. A, I, Math.-Phys.*, No. 42, 24 pp.

Tahvonen, P. E., "X-Ray Investigation of Molecular Rotation in Potassium Nitrate Crystals," *Ann. Acad. Sci. Fennicae, Ser. A, I, Math.-Phys.*, No. 44, 20 pp.

Wells, A. F., "The Crystal Structure of $CsCuCl_3$ and the Crystal Chemistry of Complex Halides ABX_3," *J. Chem. Soc. London*, **1947**, 1662.

1948

Aebi, F., "The Crystal Structure of the Basic Copper Bromide, $CuBr_2 \cdot 3Cu(OH)_2$," *Helv. Chim. Acta*, **31**, 369.

Burbank, R. D., and Evans, H. T., Jr., "The Crystal Structure of Hexagonal Barium Titanate," *Acta Cryst.*, **1**, 330.

Evans, H. T., Jr., and Burbank, R. D., "The Crystal Structure of Hexagonal Barium Titanate," *J. Chem. Phys.*, **16**, 634.

Michel, A., and Pouillard, E., "Preparation of Synthetic Neutral Ferrous Titanate or Ilmenite," *Bull. Soc. Chim. France*, **1948**, 962.

Rivoir, L., and Abbad, M., "The Structure of Thallous Bromate," *Anales Real. Soc. Espan. Fis. Quim (Madrid), Ser. A*, **44**, 5.

Santana, D., "The Structure of Thallous Iodate," *Anales Real Soc. Espan. Fis. Quim. (Madrid), Ser. A,* **44,** 557.

Vul, B. M., and Gol'dman, I. M., "A New Form of Barium Titanate," *Dokl. Akad. Nauk USSR,* **60,** 41.

1949

Berger, S. V., "The Crystal Structure of the Isomorphous Orthoborates of Cobalt and Magnesium," *Acta Chem. Scand.,* **3,** 660.

Brink, C., and MacGillavry, C. H., "The Crystal Structure of K_2CuCl_3 and Isomorphous Substances," *Acta Cryst.,* **2,** 158.

Danielson, G. C., "Domain Orientation in Polycrystalline Barium Titanate," *Acta Cryst.,* **2,** 90.

Kay, H. F., Wellard, H. J., and Vousden, P., "Atomic Positions and Optical Properties of Barium Titanate," *Nature,* **163,** 636.

Luzzati, V., "Crystal Structure of Anhydrous Nitric Acid," *Compt. Rend.,* **229,** 1349.

Magnéli, A., "Crystal-Structure Studies on Tetragonal Sodium Tungsten Bronze," *Arkiv Kemi,* **1,** 269.

Rhodes, R. G., "Structure of Barium Titanate at Low Temperatures," *Acta Cryst.,* **2,** 417.

Romers, C., Ketelaar, J. A. A., and MacGillavry, C. H., "Crystal Structure of Ammonium Tetrametaphosphate," *Nature,* **164,** 960.

Wells, A. F., "The Crystal Structure of Atacamite and the Crystal Chemistry of Cupric Compounds," *Acta Cryst.,* **2,** 175.

Wells, A. F., and Bailey, M., "The Structure of Inorganic Oxy Acids. The Crystal Structure of Selenious Acid," *J. Chem. Soc.,* **1949,** 1282.

Wickman, F. E., "The Crystal Structure of Shortite, $Ca_2Na_2(CO_3)_3$," *Arkiv Mineral. Geol.,* **1,** 95.

Zachariasen, W. H., "Crystal Chemical Studies of the 5f Series of Elements. XIII. The Crystal Structure of U_2F_9 and $NaTh_2F_9$," *Acta Cryst.,* **2,** 390.

Zachariasen, W. H., "Crystal Chemical Studies of the 5f Series of Elements. XII. New Compounds Representing Known Structure Types," *Acta Cryst.,* **2,** 388.

1950

Aebi, F., "The Structure of Basic Salts with Pseudohexagonal Layer Lattices," *Acta Cryst.,* **3,** 370.

Andress, K. R., and Fischer, K., "Remarks on the Structure of Ammonium Tetrametaphosphates," *Acta Cryst.,* **3,** 399.

Askham, F., Fankuchen, I., and Ward, R., "The Preparation and Structure of Lanthanum Cobaltic Oxide," *J. Am. Chem. Soc.,* **72,** 3799.

Bertaut, E.-F., "The Structure of Boroferrites," *Acta Cryst.,* **3,** 473.

Brooks, R., and Alcock, T. C., "Crystal Structure of Ammonium Bicarbonate and a Possible Relationship with Ammonium Hypophosphate," *Nature,* **166,** 435.

Duke, J. R. C., and Llewellyn, F. J., "Crystal Structure of Ammonium Trinitrate," *Acta Cryst.,* **3,** 305.

Ferrari, A., Cavalca, L., and Tonelli, M. G., "The Structure of Rhombic Thallous Nitrate," *Gazz. Chim. Ital.,* **80,** 199.

Frondel, C., "Paratacamite and Some Related Copper Chlorides," *Mineral. Mag.,* **29,** 34.

Jonker, G. H., and Santen, J. H. van, "Ferromagnetic Compounds of Manganese with the Perewskite Structure," *Physica*, 16, 337.

Lukesh, J. S., "The Unit Cell and Space Group of Ammonium Metavanadate," *Acta Cryst.*, 3, 476.

Luzzati, V., "Crystal Structure of Nitric Acid," *Mem. Serv. Chim. Etat (Paris)*, 35, No. 3, 7.

Magnéli, A., and Nilsson, R., "Lithium Tungsten Bronze of the Perovskite Type," *Acta Chem. Scand.*, 4, 398.

Nowacki, W., and Scheidegger, R., "The Crystallography of Monoclinic Basic Copper Nitrate, $Cu(NO_3)_2 \cdot 3Cu(OH)_2$," *Acta Cryst.*, 3, 472.

Peacock, M. A., "Mineral Thio Salts. XV. Xanthoconite and Pyrostilpnite," *Mineral. Mag.*, 29, 346.

Ramsdell, L. S., and Wolfe, C. W., "The Unit Cell of Malachite," *Am. Mineralogist*, 35, 119.

Shirane, G., Hoshino, S., and Suzuki, K., "Crystal Structures of Lead Titanate and Lead Barium Titanate," *J. Phys. Soc. Japan*, 5, 453.

Takeuchi, Y., "The Structure of Fluoborite," *Acta Cryst.*, 3, 209.

Takeuchi, Y., Watanabe, T., and Ito, T., "Crystal Structures of Warwickite, Ludwigite and Pinakiolite," *Acta Cryst.*, 3, 98.

Zemann, J., "The Crystal Chemistry of Bismuth," *Tschermaks Mineral. Petrog. Mitt.*, 1, 361.

1951

Brimm, E. O., Brantley, J. C., Lorenz, J. H., and Jellinek, M. H., "Sodium and Potassium Tungsten Bronzes," *J. Am. Chem. Soc.*, 73, 5427.

Dymon, J. J., and King, A. J., "Structure Studies of the Two Forms of Sodium Tripolyphosphate," *Acta Cryst.*, 4, 378.

Edwards, J. W., Speiser, R., and Johnston, H. L., "Structure of Barium Titanate at Elevated Temperatures," *J. Am. Chem. Soc.*, 73, 2934.

Evans, H. T., Jr., "The Crystal Structure of Tetragonal Barium Titanate," *Acta Cryst.*, 4, 377.

Gomis, V., and García-Blanco, S., "Structure of Rubidium Chlorate," *Anales Real Soc. Espan. Fis. Quim. (Madrid)*, Ser. A., 47, 95.

Guiot-Guillain, G., "Crystal Structure of Ferrites of Lanthanum and Praseodymium," *Compt. Rend.*, 232, 1832.

Hund, F., and Lang, G., "Lithium Metastannate," *Naturwiss.*, 38, 502.

Korhonen, U., "The Crystal Structure of Cubic $RbNO_3$," *Ann. Acad. Sci. Fennicae*, Ser. A, I, No. 102, 37 pp.

Lander, J. J., "The Crystal Structures of $NiO \cdot 3BaO$, $NiO \cdot BaO$, $BaNiO_3$, and Intermediate Phases with Composition near $Ba_2Ni_2O_5$," *Acta Cryst.*, 4, 148.

Luzzati, V., "Crystal Structure of Anhydrous Nitric Acid," *Acta Cryst.*, 4, 120.

Nikitina, E. A., and Kokurina, A. S., "Reduction of Silicotungstates with Hydrogen. IV. Sodium Bronze and Reduction of Sodium Silicotungstate," *Zh. Obshch. Khim.*, 21, 1940.

Nowacki, W., and Scheidegger, R., "The Crystal Structure of Monoclinic Basic Copper Nitrate," *Experientia*, 7, 454; *Chimia (Aarau)* 5, 103; *Angew. Chem.*, 42, 194.

Plieth, K., and Wurster, C., "The Structure of Metaphosphates. X-Ray Investigation of a Fibrous Sodium Metaphosphate—$NaPO_3$ (Kurrol Salt)," *Z. Anorg. Allgem. Chem.*, 267, 49.

Rhodes, R. G., "Barium Titanate Twinning at Low Temperatures," *Acta Cryst.*, **4**, 105.

Romers, C., Ketelaar, J. A. A., and MacGillavry, C. H., "Crystal Structure of Ammonium Tetrametaphosphate," *Acta Cryst.*, **4**, 114.

Sawaguchi, E., Maniwa, H., and Hoshino, S., "Antiferroelectric Structure of Lead Zirconate," *Phys. Rev.*, **83**, 1078.

Shirane, G., and Hoshino, S., "Phase Transition in Lead Titanate," *J. Phys. Soc. Japan*, **6**, 265.

Smith, P., and Martínez Carrera, S., "Structure of Thallium Chlorate," *Anal. Real Soc. Espan. Fis. Quim. (Madrid)*, Ser. A, **47**, 89.

Ueda, R., and Shirane, G., "X-Ray Study on Phase Transition of Lead Zirconate, $PbZrO_3$," *J. Phys. Soc. Japan*, **6**, 209.

Vousden, P., "The Structure of the Ferroelectric Niobates and Tantalates," *Acta Cryst.*, **4**, 68.

Vousden, P., "Unit Cell Dimensions and Symmetry of Certain Ferroelectric Compounds of Niobium and Tantalum at Room Temperatures," *Acta Cryst.*, **4**, 373.

Vousden, P., "Structure of Ferroelectric Sodium Niobate at Room Temperature" *Acta Cryst.*, **4**, 545.

Wells, A. F., "Malachite: Re-Examination of Crystal Structure," *Acta Cryst.*, **4**, 200.

Wood, E. A., "Polymorphism in Potassium Niobate, Sodium Niobate and other ABO_3 Compounds," *Acta Cryst.*, **4**, 353.

Wood, E. A., "Evidence for the Non-Cubic High Temperature Phase of $BaTiO_3$," *J. Chem. Phys.*, **19**, 976.

1952

Brink, C., and Arkel, A. E. van, "The Crystal Structures of $(NH_4)_2CuCl_3$, and $(NH_4)_2CuBr_3$," *Acta Cryst.*, **5**, 506.

Brink, C., and Kroese, H. A., "The Crystal Structure of K_2AgI_3 and Isomorphous Substances," *Acta Cryst.*, **5**, 433.

Brisi, C., "Crystal Structures of $KCdF_3$ and $KCaF_3$," *Ann. Chim. (Rome)*, **42**, 356.

Dungan, R. H., Kane, D. F., and Bickford, L. R., Jr., "Lattice Constants and Dielectric Properties of Barium Titanate-Barium Stannate-Strontium Titanate Bodies," *J. Am. Ceram. Soc.*, **35**, 318.

Grund, A., and Pizy, M., "Crystal Structure of Anhydrous Sodium Metasilicate," *Acta Cryst.*, **5**, 837.

Herpin, P., "Structure of Potassium Acid Carbonate," *Compt. Rend.*, **234**, 2205.

Langles, R. de SL., "Preparation and Structure of Anhydrous Nickel Carbonate," *Ann. Chim. (Paris)*, **7**, 568.

Lapitskii, A. V., "Anhydrous Potassium Metacolumbate," *Zh. Obshch. Khim.*, **22**, 379.

Ludekens, W. L. W., and Welch, A. J. E., "Reactions Between Metal Oxides and Fluorides: Some New Double-Fluoride Structures of the Type ABF_3," *Acta Cryst.*, **5**, 841.

Nitta, I., Tomiie, Y., and Koo, C. H., "Crystal Structure of Potassium Bicarbonate," *Acta Cryst.*, **5**, 292.

Nowacki, W., and Scheidegger, R., "Crystal Structure Investigation of Monoclinic Basic Copper Nitrate, $Cu_4(NO_3)_2(OH)_6$," *Helv. Chim. Acta*, **35**, 375.

Pepinsky, R., McCarty, C. M., Zemyan, E., and Drenck, K., "A Ferroelectric Ammonium Metaphosphate," *Phys. Rev.*, **86**, 793.

Rüdorff, W., Schwarz, H. G., and Walter, M., "Structure Investigation of the Alkali Thiocuprates," *Z. Anorg. Allgem. Chem.*, **269**, 141.

Sawaguchi, E., "Lattice Constant of PbZrO₃," *J. Phys. Soc. Japan*, **7**, 110.

Ward. R., Wold, A., Gushee, B., and Ridgley, D. H., "Ternary Compounds: First Technical Report for the Period June 1, 1951 to May 31, 1952," U.S. At. Energy Comm. NP-3915, U22515, 46 pp.

Zhdanov, G. S., and Sanadze, V. V., "X-Ray Analysis of Hg(SCN)₂·ASCN, Where A = K or NH₄," *Zh. Fiz. Khim.*, **26**, 469.

1953

Brogren, G., "The Possibility of Using Ground Calcite Crystals as X-Ray Gratings," *Arkiv Fysik*, **6**, 479.

Brous, J., Fankuchen, I., and Banks, E., "Rare Earth Titanates as a Perovskite Structure," *Acta Cryst.*, **6**, 67.

Donnay, G., and Donnay, J. D. H., "Crystal Geometry of Some Alkali Silicates," *Am. Mineralogist*, **38**, 163.

Donnay, G., and Donnay, J. D. H., "The Crystallography of Bastnäsite, Parisite, Roentgenite, and Synchisite," *Am. Mineralogist*, **38**, 932.

Evans, H. T., Jr., "Use of a Geiger Counter for the Measurement of X-Ray Intensities from Small Single Crystals," *Rev. Sci. Instr.*, **24**, 156.

Faust, G. T., "Huntite, Mg₃Ca(CO₃)₄, a New Mineral," *Am. Mineralogist*, **38**, 4.

Fousek, J., "Structural Changes in Barium Titanate," *Czech. J. Phys.*, **3**, 315.

Hahn, T., "Model Relations between Silicates and Fluorine Beryllates," *Neues Jahrb. Mineral., Abhandl.*, **86**, 1.

Ievins, A., and Ozols, J., "The Precise Determination of the Parameters of the Elementary Cell for Monoclinic Crystals," *Dokl. Akad. Nauk SSSR*, **91**, 527; *Latvijas PSR Zinatnu Akad. Vestis*, **1953**, No. 6, 105.

Juza, R., Weber, H. H., and Meyer-Simon, E., "Ternary Nitrides and Oxynitrides of Elements of the Fourth Group," *Z. Anorg. Allgem. Chem.*, **273**, 48.

Korhonen, U., "The Crystal Structure of Cesium Nitrate above 161°C.," *Ann. Acad. Sci. Fennicae, Ser. A., I*, No. 150, 16 pp.

Pecora, W. T., and Kerr, J. H., "Burbankite and Calkensite, Two New Carbonate Minerals from Montana," *Am. Mineralogist*, **38**, 1169.

Vries, R. C. de, and Roy, R., "The Systems KF–MgF₂ and AgF–ZnF₂," *J. Am. Chem. Soc.*, **75**, 2479.

Wickman, F. E., "The Crystal Structure of Aikinite, CuPbBiS₃," *Arkiv Mineral. Geol.*, **1**, 501.

1954

Borodin, L. S., and Kazakova, M. E., "Irinite, a New Mineral of the Perewskite Group," *Dokl. Akad. Nauk SSSR*, **97**, 725.

Brink, C., Binnendijk, N. F., and Linde, J. van de, "The Crystal Structures of CsCu₂Cl₃ and CsAg₂I₃," *Acta Cryst.*, **7**, 176.

Brown, B. W., and Banks, E., "The Sodium Tungsten Bronzes," *J. Am. Chem. Soc.*, **76**, 963.

Dornberger-Schiff, K., Liebau, F., and Thilo, E., "Crystal Structure of (NaAsO₃)ₓ, of Madrell Salt and of β-Wollastonite," *Naturwiss.*, **41**, 551; *Acta Cryst.*, **8**, 752 (1955).

Edstrand, M., and Ingri, N., "Crystal Structure of the Double Lithium Antimony(V) Oxide LiSbO₃," *Acta Chem. Scand.*, **8**, 1021.

Ellinger, F. H., and Zachariasen, W. H., "The Crystal Structure of $KPuO_2CO_3$, NH_4-PuO_2CO_3 and $RbAmO_2CO_3$," *J. Phys. Chem.*, **58**, 405.

Evans, H. T., Jr., and Block, S., "Crystal Structure of KVO_3," *Am. Mineralogist*, **39**, 326.

Garrett, B. S., "Crystal Structures of Oxalic Acid Dihydrate and α-Iodic Acid by Neutron Diffraction," U. S. At. Energy Comm. Rept. ORNL-1745, 149 pp.

Ginetti, Y., "Crystal Structure of Sodium Metagermanate," *Bull. Soc. Chim. Belg.*, **63**, 460.

Ginetti, Y., "Crystal Structure of Copper Metagermanate," *Bull. Soc. Chim. Belg.*, **63**, 209.

Juza, R., and Schulz, W., "Ternary Phosphides and Arsenides of Lithium with Elements of Groups III and IV," *Z. Anorg. Allgem. Chem.*, **275**, 65.

Keith, M. L., and Roy R., "Structural Relations among Double Oxides of Trivalent Elements," *Am. Mineralogist*, **39**, 1.

Kestigian, M., and Ward, R., "The Preparation of Lanthanum Titanium Oxide, $LaTiO_3$," *J. Am. Chem. Soc.*, **76**, 6027.

Lang, G., "Crystal Structure of Some Examples of the Compound Class $M_2^IM^{IV}O_3$ as Contributions to the Clarification of the Classification of Li_2TiO_3," *Z. Anorg. Allgem. Chem.*, **276**, 77.

Looby, J. T., and Katz, L., "Yttrium Chromium Oxide, a New Compound of the Perowskite Type," *J. Am. Chem. Soc.*, **76**, 6029.

Malinofsky, W. W., and Kedesdy, H., "Barium Iron Oxide Isomorphs of Hexagonal and Tetragonal $BaTiO_3$," *J. Am. Chem. Soc.*, **76**, 3090.

Nitta, I., Tomiie, Y., and Koo, C. H., "The Relation Among the Results of Various Structure Investigations of Potassium Bicarbonate," *Acta Cryst.*, **7**, 140.

Roy, R., "Multiple Ion Substitution in the Perowskite Lattice," *J. Am. Ceram. Soc.*, **37**, 581.

Rüdorff, W., and Becker, H., "Interaction of Vanadium(III) Oxide and Vanadium(IV) Oxide with Some Metal Oxides," *Z. Naturforsch.*, **9b**, 613.

Rüdorff, W., and Pfister, F., "Alkaline Earth Uranates(VI) and Their Reduction Products," *Z. Naturforsch.*, **9b**, 568.

Ruggiero, A., and Ferro, R., "Orthogalliates of Rare Earth Elements," *Atti Accad. Nazl. Lincei, Rend., Classe Sci. Fis. Mat. Nat.*, **17**, 48.

Ruggiero, A., and Ferro, R., "Orthocobaltites of the Rare Earth Elements," *Atti Accad. Nazl. Lincei, Rend., Classe Sci. Fis. Mat. Nat.*, **17**, 254.

Shirane, G., Danner, H., Pavlovic, A., and Pepinsky, R., "Phase Transitions in Ferroelectric $KNbO_3$," *Phys. Rev.*, **93**, 672.

Syneček, V., and Hanic, F., "The Crystal Structure of Ammonium Metavanadate," *Czech. J. Phys.*, **4**, 120.

Weiss, A., and Damm. K., "Sodium Trichloromercurate(II), $Na(HgCl_3)$. Mercury Halogenides IV," *Z. Naturforsch.*, **9b**, 82.

Wickman, F. E., "The Crystal Structure of Aikinite, $CuPbBiS_3$," *Arkiv Mineral.*, **1**, 501.

Wold, A., and Ward, R., "Perovskite-Type Oxides of Cobalt, Chromium and Vanadium with Some Rare Earth Elements," *J. Am. Chem. Soc.*, **76**, 1029.

Zachariasen, W. H., "The Precise Structure of Orthoboric Acid," *Acta Cryst.*, **7**, 305.

1955

Azaroff, L. V., and Buerger, M. J., "Refinement of the Structure of Cubanite, $CuFe_2S_3$," *Am. Mineralogist*, **40**, 213.

Christ, C. L., Clark, J. R., and Evans, H. T., Jr., "Crystal Structure of Rutherfordine, UO_2CO_3," *Science*, **121**, 472.

Frazer, B. C., Danner, H. R., and Pepinsky, R., "Single-Crystal Neutron Analysis of Tetragonal $BaTiO_3$," *Phys. Rev.*, **100**, 745.

Gabrielson, O., "The Crystal Structure of Finnemanite, $PbCl(AsO_3)_3$," *Arkiv Mineral. Geol.*, **2**, 1.

Gasperin, M., "Synthesis and Identification of Two Double Oxides of Tantalum and Tin," *Compt. Rend.*, **240**, 2340.

Harwood, M. G., "The Crystal Structure of Lanthanum-Strontium Manganites," *Proc. Phys. Soc. (London)*, **68B**, 586.

Ismailzade, I. G., "X-Ray Study of the Structures of the Solid Solutions of the Titanates of Barium and of Lead," *Dokl. Akad. Nauk Azerb. SSR*, **11**, 527.

Lapitskii, A. V., and Simanov, Y. P., "Lithium Metaniobate and Metatantalate," *Zh. Fiz. Khim.*, **29**, 1201.

Menary, J. W., "Some Lattice Constants," *Acta Cryst.*, **8**, 840.

Ruggiero, A., and Ferro, R., "Orthoaluminates, Orthochromites and Orthoferrites of the Rare Earth Elements," *Gazz. Chim. Ital.*, **85**, 892.

Shirane, G., Pepinsky, R., and Frazer, B. C., "X-Ray and Neutron Diffraction Study of Ferroelectric $PbTiO_3$," *Phys. Rev.*, **97**, 1179.

Tishchenko, G. N., "Electron Diffraction Investigation of the Structure of $CsNiCl_3$," *Tr. Inst. Kristallogr., Akad. Nauk SSSR*, **1955**, 93.

Yakel, H. L., Jr., "The Structures of Some Compounds of the Perovskite Type," *Acta Cryst.*, **8**, 394.

1956

Candlin, R., "Thermal Changes in the Structure of Sodium Sesquicarbonate," *Acta Cryst.*, **9**, 545.

Corbridge, D. E. C., "The Crystal Structure of Rubidium Metaphosphate," *Acta Cryst.*, **9**, 308.

Cowley, J. M., "Electron Diffraction Study of the Structure of Basic Lead Carbonate, $2PbCO_3 \cdot Pb(OH)_2$," *Acta Cryst.*, **9**, 391.

Cromer, D. T., and Larson, A. C., "The Crystal Structure of $Ce(IO_3)_4$," *Acta Cryst.*, **9**, 1015.

Fischmeister, H. F., "X-Ray Measurements of the Thermal Expansion of Trigonal Potassium, Lithium and Silver Nitrates," *J. Inorg. Nucl. Chem.*, **3**, 182.

Francombe, M. H., "High-Temperature Structure Transitions in Sodium Niobate," *Acta Cryst.*, **9**, 256.

Geller, S., "Crystal Structure of Gadolinium Orthoferrite, $GdFeO_3$," *J. Chem. Phys.*, **24**, 1236.

Geller, S., and Bala, V. B., "Crystallographic Studies of Perovskite-Like Compounds II. Rare Earth Aluminates," *Acta Cryst.*, **9**, 1019.

Geller, S., and Wood, E. A., "Crystallographic Studies of Perovskite-Like Compounds. I. Rare Earth Orthoferrites and $YFeO_3$, $YCrO_3$, $YAlO_3$," *Acta Cryst.*, **9**, 563.

Hahn, H., and Mutschke, U., "Ternary Chalcogenides. XI. Experiments on the Preparation of Thioperovskites," *Z. Anorg. Allgem. Chem.*, **288**, 269.

Hellner, E., and Leineweber, G., "Sulfidic Ores of Complex Composition. I. The Structure of Bournonite, CuPbSbS$_3$, and Seligmannite, CuPbAsS$_3$," *Z. Krist.*, **107**, 150.

Hilmer, W., and Dornberger-Schiff, K., "Crystal Structure of Lithium Polyarsenate (LiAsO$_3$)$_x$," *Acta Cryst.*, **9**, 87.

Jeffrey, G. A., and Jones, D. W., "The Crystal Structure of Potassium Aminedisulphonate," *Acta Cryst.*, **9**, 283.

Leineweber, G., "Structure Analysis of Bournonite and Seligmannite by Means of Superposition Methods," *Z. Krist.*, **108**, 161.

Liebau, F., "The Crystal Structure of Sodium Polyarsenate (NaAsO$_3$)$_x$," *Acta Cryst.*, **9**, 811; *Abhandl. Deut. Akad. Wiss. Berlin, Kl. Chem. Geol. Biol.*, **1955**, No. 7, 117 (1957).

McCarroll, W. H., Ward, R., and Katz, L., "Ternary Oxides of Tetravalent Molybdenum," *J. Am. Chem. Soc.*, **78**, 2909.

Rabenau, A., "Perovskite and Fluorite Phases in ZrO$_2$–LaO$_{1.5}$–MgO and ZrO$_2$–LaO$_{1.5}$–CaO Systems," *Z. Anorg. Allgem. Chem.*, **288**, 221.

Remy, H., and Hansen, F., "X-Ray Investigation of the System KF–MgF$_2$," *Z. Anorg. Allgem. Chem.*, **283**, 277.

Schmitz-Dumont, O., Bergerhoff, G., and Hartert, E., "Effect of the Cation Radius on the Energy of Formation of Addition Compounds. VII. The Systems Alkali Fluorides/Lead Fluoride," *Z. Anorg. Allgem. Chem.*, **283**, 314.

Shirane, G., Pepinsky, R., and Frazer, B. C., "X-Ray- and Neutron-Diffraction Study of Ferroelectric PbTiO$_3$," *Acta Cryst.*, **9**, 131.

Tang, Y.-C., and Kuei, L.-L., "The Crystal Structure of Anhydrous Sodium Sulfite," Hua Hsüeh Hsüeh Pao, **22**, 572.

1957

Belt, R. F., and Baenziger, N. C., "The Crystal Structure of Potassium Hydroxylamine-*N*-Sulfonate," *J. Am. Chem. Soc.*, **79**, 316.

Ferroni, E., Sabatini, A., and Orioli, P., "Structure of Cesium Nitrate," *Gazz. Chim. Ital.*, **87**, 630.

Ferroni, E., Sabatini, A., and Orioli, P., "Structural Relations between Cubic Cesium Nitrate and Trigonal Cesium Nitrate," *Ricerca Sci.*, **27**, 1557.

Furberg, S., and Landmark, P., "The Crystal Structure of Phosphorous Acid," *Acta Chem. Scand.*, **11**, 1505.

Geller, S., "Crystallographic Studies of Perovskite-Like Compounds IV. Rare Earth Scandates, Vanadites, Galliates, Orthochromites," *Acta Cryst.*, **10**, 243.

Gilleo, M. A., "Crystallographic Studies of Perovskite-Like Compounds. III. La(M$_x$, Mn$_{1-x}$)O$_3$ with M = Co, Fe, and Cr," *Acta Cryst.*, **10**, 161.

Goldsmith, J. R., and Graf, D. L., "The System CaO–MnO–Co$_2$: Solid-Solution and Decomposition Relations," *Geochim. Cosmochim. Acta*, **11**, 310.

Hahn, H., and Theune, U., "On the Crystal Structure of Li$_2$GeO$_3$ and LiGa$_5$O$_8$," *Naturwiss.*, **44**, 33.

Hamilton, W. C., "A Neutron Crystallographic Study of Lead Nitrate," *Acta Cryst.*, **10**, 103.

Hilmer, W., "Determination of the Crystal Structure of Lithium Polyarsenate, (LiAsO$_3$)$_x$," *Abhandl. Deut. Akad. Wiss. Berlin, Kl. Chem. Geol. Biol.*, **1955**, No. 7, 125.

Hinde, R. M., and Kellett, E. A., "Unit Cell and Space Group of Thallous Nitrate, TlNO$_3$," *Acta Cryst.*, **10**, 383.

Jona, F., Shirane, G., Mazzi, F., and Pepinsky, R., "X-Ray and Neutron Diffraction Study of Antiferroelectric Lead Zirconate, PbZrO₃," *Phys. Rev.*, **105**, 849.

Kay, H. F., and Bailey, P. C., "Structure and Properties of CaTiO₃," *Acta Cryst.*, **10**, 219.

Kay, H. F., and Miles, J. L., "The Structure of Cadmium Titanate and Sodium Tantalate," *Acta Cryst.*, **10**, 213.

Kestigan, M., Dickinson, J. G., and Ward, R., "Ion-Deficient Phases in Titanium and Vanadium Compounds of the Perovskite Type," *J. Am. Chem. Soc.*, **79**, 5598.

Koehler, W. C., and Wollan, E. O., "Neutron-Diffraction Study of the Magnetic Properties of Perovskite-Like Compounds, LaBO₃," *Phys. Chem. Solids*, **2**, 100.

Møller, C. K., "A Phase Transition in Caesium Plumbochloride," *Nature*, **180**, 981.

Ramachandran, G. N., and Chandrasekaran, K. S., "The Absolute Configuration of Sodium Chlorate," *Acta Cryst.*, **10**, 671.

Ramachandran, G. N., and Lonappan, M. A., "The Structure of High-Temperature Potassium Chlorate," *Acta Cryst.*, **10**, 281.

Roth, R. S., "Classification of Perovskite and Other ABO₃-Type Compounds," *J. Res. Natl. Bur. Std.*, **58**, 75, *Research Paper* 2736.

Sass, R. L., Vidale, R., and Donohue, J., "Interatomic Distances and Thermal Anisotropy in Sodium Nitrate and Calcite," *Acta Cryst.*, **10**, 567.

Shirane, G., Danner, H., and Pepinsky, R., "Neutron Diffraction Study of Orthorhombic BaTiO₃," *Phys. Rev.*, **105**, 856.

Simanov, Y. P., Batsanova, L. R., and Kovba, L. M., "X-Ray Investigation of Double Fluorides of Bivalent Manganese," *Zh. Neorgan. Khim.*, **2**, 2410.

Vickery, R. C., and Klann, A., "Crystallographic and Magnetochemical Studies of ABO₃ Group Compounds of Lanthanon and Manganese Oxides," *J. Chem. Phys.*, **27**, 1161.

Wold, A., Post, B., and Banks, E., "Lanthanum Rhodium and Lanthanum Cobalt Oxides," *J. Am. Chem. Soc.*, **79**, 6365.

Zemann, J., "The Crystal Structure of Li₂CO₃," *Acta Cryst.*, **10**, 664.

1958

Amorós Portolés, J. L., Alonso, P., and Canut, M. L., "Polymorphic Transitions in Single Crystals. I. Superstructure Formation in the IV-V (−18°) Transition of Ammonium Nitrate," *Publ. Dept. Crist. Mineral. (Madrid)*, **4**, 30.

Amorós Portolés, J. L., Alonso, P., and Canut, M. L., "Polymorphic Transitions in Single Crystals. II. IV–II (84°) Transition and II (55°) Metastable Form of Ammonium Nitrate," *Publ. Dept. Crist. Mineral. (Madrid)*, **4**, 38.

Aravindakshan, C., "An Accurate Redetermination of the Structure of Potassium Chlorate, KClO₃," *Z. Krist.*, **111**, 35.

Cloud, W. H., "Crystal Structure and Ferrimagnetism in NiMnO₃, and CoMnO₃," *Phys. Rev.*, **111**, 1046.

Donohue, J., Miller, S. J., and Cline, R. F., "The Effect of Various Substituents on the Lattice Constants of Tetragonal Barium Titanate," *Acta Cryst.*, **11**, 693.

Gattow, G., and Zemann, J., "Redetermination of the Crystal Structure of Azurite, Cu₃(OH)₂(CO₃)₂," *Acta Cryst.*, **11**, 866; *Naturwiss.*, **45**, 208.

Goldsmith, J. R., and Graf, D. L., "Relation between Lattice Constants and Composition of the Ca-Mg Carbonates," *Am. Mineralogist*, **43**, 84.

Hoard, J. L., and Stroupe, J. D., "Structure of Crystalline Rubidium Uranyl Nitrate," *U. S. At. Energy Comm.* TID-5290, Bk. 1, 323.

Howie, R. A., and Broadhurst, F. M., "X-Ray Data for Dolomite and Ankerite," *Am. Mineralogist*, **43**, 1210.

Jost, K. H., "The Structure of Kurrol's Silver Salt, $(AgPO_3)_x$," *Z. Anorg. Allgem. Chem.*, **296**, 154.

Loopstra, B. O., "X-Ray and Neutron Diffraction Study of Calcium Hypophosphite and Phosphorous Acid," *JENER (Joint Establ. Nucl. Energy Res.)*, Publ. No. **15**, 64 pp.

Petrášova, M., Madǎr, J., and Hanic, F., "The Crystal Structure of Potassium Metavanadate," *Chem. Zvesti*, **12**, 410.

Swoboda, T. J., Toole, R. C., and Vaughan, J. D., "New Magnetic Compounds of the Ilmenite-Type Structure," *Phys. Chem. Solids*, **5**, 293.

Verbitskaya, T. N., Zhdanov, G. S., Venevtsev, Y. N., and Soloviev, S. P., "Electrical and X-Ray Diffraction Studies of the $BaTiO_3$–$BaZrO_3$ System," *Kristallografiya*, **3**, 186.

Weiss, R., "Preparation and Structure of Lead and Strontium Metaplumbates," *Compt. Rend.*, **246**, 3073.

1959

Aravindakshan, C., "An Accurate Redetermination of the Structure of Sodium Chlorate, $NaClO_3$," *Z. Krist.*, **111**, 241.

Bergerhoff, G., "Metallocomplexes. The Crystal Structure of Argentosulfonium Nitrate, $[Ag_3S]$ $[NO_3]$," *Z. Anorg. Allgem. Chem.*, **299**, 328.

Bower, J. G., Sparks, R. A., and Trueblood, K. N., "Refinement of the Crystal Strutures of Sodium Chlorate and Potassium Chlorate," *U. S. Govt. Res. Rept.*, **32**, 119, *U. S. Dept. Comm. Off. Tech. Serv. P.B. Rept.* **139**,994, 31 pp.

Deshpande, V. T., Sirdeshmukh, D. B., and Mudholker, V. M., "The Lattice Constant of Strontium Nitrate, $Sr(NO_3)_2$," *Acta Cryst.*, **12**, 257.

Edwards, A. J., and Peacock, R. D., "The Structures of Potassium Trifluorocuprate(II) and Potassium Trifluorochromate(II)," *J. Chem. Soc.*, **1959**, 4126.

Fresia, E. J., Katz, L., and Ward, R., "Cation Substitution in Perovskite-Like Phases," *J. Am. Chem. Soc.*, **81**, 4783.

Ismailzade, I. G., "X-Ray Study of the Structure of the Metatantalates of Strontium, Lead, and Barium, and of the Systems $(Pb,Ba,Sr,Ca)Nb_2O_6$ and $(Pb,Sr,Ba)Nb_2O_6$," *Kristallografiya*, **4**, 658.

Jost, K. H., "The Structure of the *a* Form of Kurrol's Sodium Salt," *Chem. Zvesti*, **13**, 738.

Meyer, H.-J., "On Vaterite and Its Structure," *Angew. Chem.*, **71**, 678.

Møller, C. K., "Structure of Perovskite-Like Caesium Plumbo Trihalides," *Kgl. Danske Videnskab. Selskab, Mat. Fys. Medd.*, **32**, No. 2, 1.

Niggli, A., "X-Ray Crystallographic Investigation of Silver Nitrate," *Z. Krist.*, **111**, 269.

Okazaki, A., Suemune, Y., and Fuchikami, T., "The Crystal Structures of $KMnF_3$, $KFeF_3$, $KCoF_3$, $KNiF_3$ and $KCuF_3$," *J. Phys. Soc. Japan*, **14**, 1823.

Pistorius, C. W. F. T., "High Pressure Preparation and Structure of Crystalline Nickelous Carbonate," *Experientia*, **15**, 328.

Randall, J. J., and Ward, R., "The Preparation of Some Ternary Oxides of the Platinum Metals," *J. Am. Chem. Soc.*, **81**, 2629.

Robbins, C. R., and Levin, E. M., "The System Magnesium Oxide–Germanium Dioxide," *Am. J. Sci.*, **257**, 63.

Rüdorff, W., Kändler, J., Lincke, G., and Babel, D., "Double Fluorides of Nickel and Cobalt," *Angew. Chem.*, **71**, 672.

Seifert, H. J., and Ehrlich, P., "The Systems NaCl/VCl₂, KCl/VCl₂ and CsCl/VCl₂," *Z. Anorg. Allgem. Chem.*, **302**, 284.

Shinnaka, Y., "X-Ray Study of Molecular Rotation in Cubic Ammonium Nitrate," *J. Phys. Soc. Japan*, **14**, 1073.

Shinnaka, Y., "X-Ray-Study on Molecular Rotation in Tetragonal Ammonium Nitrate," *J. Phys. Soc. Japan*, **14**, 1707.

Shirane, G., Pickart, S. J., and Ishikawa, Y., "Neutron Diffraction Study of Antiferromagnetic MnTiO₃ and NiTiO₃," *J. Phys. Soc. Japan*, **14**, 1352.

Shirane, G., Pickart, S. J., Nathans, R., and Ishikawa, Y., "Neutron Diffraction Study of Antiferromagnetic FeTiO₃ and its Solid Solutions with α-Fe₂O₃," *Phys. Chem. Solids*, **10**, 35.

Steinfink, H., and Sans, F. J., "Refinement of the Crystal Structure of Dolomite," *Am. Mineralogist*, **44**, 679.

Wagner, G., and Binder, H., "The Binary Systems BaO–SnO₂ and BaO–PbO₂. II. Crystal Structure Determinations," *Z. Anorg. Allgem. Chem.*, **298**, 12.

Wise, W. S., "An Occurrence of Geikielite," *Am. Mineralogist*, **44**, 879.

Yamaguchi, G., and Yanagida, H., "Study on the Reductive Spinel—a New Spinel Formula AlN–Al₂O₃ instead of the Previous One Al₂O₃," *Bull. Chem. Soc. Japan*, **32**, 1264.

1960

Agranovskaya, A. I., "Physical-Chemical Investigation of the Formation of Complex Ferroelectrics with the Perovskite Structure," *Izv. Akad. Nauk SSSR, Ser. Fiz.*, **24**, 1275.

Alm, K.-F., "The Crystal Structure of Barytocalcite BaCa(CO₃)₂," *Arkiv Mineral. Geol.*, **2**, 399.

Amorós Portolés, J. L., "Polymorphism in Single Crystals," *U. S. Dept. Comm., Off. Tech. Serv. P. B. Rept.*, **147**, 253, 45 pp.

Amorós Portolés, J. L., and Banerjee, R. L., "Mechanism of the IV→II Transition in Ammonium Nitrate," *Bol. Real Soc. Españ. Hist. Nat., Secc. Geol.*, **58**, 165.

Atoji, M., and Rundle, R. E., "Neutron Diffraction Study on Sodium Tungsten Bronzes NaₓWO₃ (x = 0.9 ∼ 0.6)," *J. Chem. Phys.*, **32**, 627.

Berry, C. R., and Combs, C. M., "Antiferromagnetic FeVO₃," *J. Appl. Phys.*, **31**, 1130.

Brixner, L. H., "Preparation and Structure Determination of Some New Cubic and Tetragonally-Distorted Perovskites," *J. Phys. Chem.*, **64**, 165.

Brixner, L. H., "X-Ray Study and Electrical Properties of the System BaₓSr₍₁₋ₓ₎MoO₃," *J. Inorg. Nucl. Chem.*, **14**, 225.

Brixner, L. H., "Preparation and Crystallographic Study of Some New Rare Earth Compounds," *J. Inorg. Nucl. Chem.*, **15**, 352.

Brixner, L. H., "Preparation, Structure and Electrical Properties of Some Substituted Lithium-Oxo-Metallates," *J. Inorg. Nucl. Chem.*, **16**, 162.

Dalziel, J. A. W., and Welch, A. J. E., "The Perovskite-Type Structures of DyAlO₃, DyFeO₃ and Some Related Lanthanon Mixed Oxides," *Acta Cryst.*, **13**, 956.

Evans, H. T., Jr., "Crystal Structure Refinement and Vanadium Bonding in the Metavanadates KVO₃, NH₄VO₃ and KVO₃·H₂O," *Z. Krist.*, **114**, 257.

Filip'ev, V. S., Smolyaninov, N. P., Fesenko, E. G., and Belyaev, I. N., "Formation of BiFeO₃ and Determination of Unit Cell," *Kristallografiya*, **5**, 958.

Foëx, M., Mancheron, A., and Liné, M., "The Combination of La$_2$O$_3$ with NiO," *Compt. Rend.*, **250**, 3027.

Galasso, F., and Katz, L., "Crystal Structure of Ba$_{0.5x}$TaO$_{3-x}$," *Nature*, **188**, 1099.

Inkinen, O., "Experimental Structure Amplitudes of Trigonal Sodium Nitrate and the Atomic Form Amplitudes of Its Sodium Ions," *Ann. Acad. Sci. Fennicae Ser. A, IV*, No. 55, 1.

Ismailzade, I. G., "X-Ray Investigation of BaNb$_2$O$_6$–CaNb$_2$O$_6$ and BaNb$_2$O$_6$–SrNb$_2$O$_6$ Systems," *Kristallografiya*, **5**, 268.

Ismailzade, I. G., "X-Ray Diffraction Investigation of the Pb$_3$NiNb$_2$O$_9$–Pb$_3$MgNb$_2$O$_9$ System," *Kristallografiya*, **5**, 316.

Kantola, M., and Vilhonen, E., "X-Ray Measurements of the Thermal Expansion of NaNO$_3$," *Ann. Acad. Sci. Fennicae Ser. A VI*, No. 54, 1.

Kovba, L. M., and Golubenko, A. N., "Lithium Uranate, LiUO$_3$," *Zh. Strukt. Khim.*, **1**, 390.

Liebau, F., "Crystal Chemistry of Silicates, Germanates and Fluoberyllates of the Formula Type ABX$_3$," *Neues Jahrb. Mineral., Abhandl.*, **94**, 1209.

Russell, L. E., Harrison, J. D. L., and Brett, N. H., "Perovskite-Type Compounds Based on Plutonium," *J. Nucl. Mater.*, **2**, 310.

Sharp, W. E., "The Cell Constants of Artificial Siderite," *Am. Mineralogist*, **45**, 241.

Smith, A. J., "The System Cadmium Oxide–Stannic Oxide," *Acta Cryst.*, **13**, 749.

Smith, A. J., and Welch, A. J. E., "Some Mixed Metal Oxides of Perovskite Structure," *Acta Cryst.*, **13**, 653.

Vedam, K., Okaya, Y., and Pepinsky, R., "Crystal Structure of Ferroelectric LiH$_3$-(SeO$_3$)$_2$," *Phys. Rev.*, **119**, 1252.

Zaslavskii, A. I., and Tutov, A. G., "The Structure of a New Anti-Ferromagnetic BiFeO$_3$," *Dokl. Akad. Nauk SSSR*, **135**, 815.

1961

Beckman, O., and Knox, K., "Magnetic Properties of KMnF$_3$. I. Crystallographic Studies," *Phys. Rev.*, **121**, 376.

Chao, E. C. T., Evans, H. T., Jr., Skinner, B. J., and Milton, C., "Neighborite, NaMgF$_3$, a New Mineral from the Green River Formation, South Ouray, Utah," *Am. Mineralogist*, **46**, 379.

Derbyshire, S. W., Fraker, A. C., and Stadelmaier, H. H., "A Barium Iron Oxide with the Perovskite Structure," *Acta Cryst.*, **14**, 1293.

Dickson, J. G., Katz, L., and Ward, R., "Compounds with the Hexagonal Barium Titanate Structure," *J. Am. Chem. Soc.*, **83**, 3026.

Evans, H. T., Jr., "An X-Ray Diffraction Study of Tetragonal Barium Titanate," *Acta Cryst.*, **14**, 1019.

Fedulov, S. A., Venevtsev, Y. N., Zhdanov, G. S., and Smazhevskaya, E. G., "High-temperature X-Ray and Thermal-Analysis Studies of Bismuth Ferrite," *Kristallografiya*, **6**, 795.

Folberth, O. G., and Pfister, H., "New Ternary Semiconducting Phosphides, MgGeP$_2$, CuSi$_2$P$_3$ and CuGe$_2$P$_3$," *Acta Cryst.*, **14**, 325.

Goryunova, N. A., Vaipolin, A. A., and Chiang, P.-H., "Solubility of Germanium in Certain Ternary Compounds with a Tetrahedral Structure," *Fiz. Khim., Leningrad*, **1961**, 26.

Graf, D. L., "Crystallographic Tables for the Rhombohedral Carbonates," *Am. Mineralogist*, **46**, 1283.

Hoppe, R., Liebe, W., and Dähne, W., "Fluomanganates of the Alkali Metals," Z. Anorg. Allgem. Chem., **307**, 276.

Iitaka, Y., and Stalder, H. A., "Synchisite and Bastnasite," Schweiz. Mineral. Petrog. Mitt., **41**, 485.

Johnston, W. D., and Sestrich, D., "La$_x$Ba$_{1-x}$TiO$_3$ System," J. Inorg. Nucl. Chem., **20**, 32.

Jost, K. H., "The Structure of Silver Polyphosphate, (AgPO$_3$)$_x$," Acta Cryst., **14**, 779.

Jost, K. H., "The Structure of the Kurrol Salt (NaPO$_3$)$_x$, Type A," Acta Cryst., **14**, 844.

Kennedy, S. W., and Patterson, J. H., "Structural Study of Thallous Nitrate III," Z. Krist., **116**, 143.

Khanolkar, D. D., "Crystal Structure Data of some Rhodites and Ruthenites," Current Sci. (India), **30**, 52.

Knox, K., "Perovskite-Like Fluorides. I. Structures of KMnF$_3$, KFeF$_3$, KCoF$_3$, KNiF$_3$ and KZnF$_3$. Crystal Field Effects in the Series and in KCrF$_3$ and KCuF$_3$," Acta Cryst., **14**, 583.

Larson, A. C., and Cromer, D. T., "The Crystal Structure of Zr(IO$_3$)$_4$," Acta Cryst., **14**, 128.

Levin, E. M., Roth, R. S., and Martin, J. B., "Polymorphism of ABO$_3$ Type Rare Earth Borates," Am. Mineralogist, **46**, 1030.

Malčić, S. S., and Manojlović, L. M., "Crystal Structure of Cesium Uranyl Nitrate," Bull. Inst. Nucl. Sci., "Boris Kidrich," (Belgrade), **11**, 135.

Náray-Szabó, I., and Kálmán, A., "On the Structure and Polymorphism of Potassium Iodate, KIO$_3$," Acta Cryst., **14**, 791.

Okazaki, A., and Suemune, Y., "The Crystal Structure of KCuF$_3$," J. Phys. Soc. Japan, **16**, 176.

Okazaki, A., and Suemune, Y., "The Crystal Structures of KMnF$_3$, KFeF$_3$, KCoF$_3$, KNiF$_3$ and KCuF$_3$ above and below their Néel Temperatures," J. Phys. Soc. Japan, **16**, 671.

Oswald, H. R., Iitaka, Y., Locchi, S., and Ludi, A., "The Crystal Structure of Cu$_2$(OH)$_3$Br and Cu$_2$(OH)$_3$I," Helv. Chim. Acta, **44**, 2103.

Palatnik, L. S., Komnik, Y. F., Belova, E. K., and Atroschenko, L. V., "Ternary Semiconducting Compounds Containing Copper and Elements from Groups IV and VI," Kristallografiya, **6**, 960.

Pistorius, C. W. F. T., "Melting Curve to 15,000 Bars and Lattice Constants of Bridgman's AgNO$_3$ II," Z. Krist., **115**, 291.

Schnering, H. G., v., "Crystal Structure of Na[Zn(OH)$_3$)]," Naturwiss., **48**, 665.

Schnering, H. G., and Hoppe, R., "Ba$_2$ZnS$_3$," Z. Anorg. Allgem. Chem., **312**, 99.

Sleight, A. W., and Ward, R., "Compounds of Heptavalent Rhenium with the Perovskite Structure," J. Am. Chem. Soc., **83**, 1088.

Thilo, E., and Jost, K. H., "Structures of the Anion Chains in Insoluble Crystalline Macromolecular Alkali Polyphosphates. Preliminary Results for [Na$_2$H(PO$_3$)$_3$]$_x$ and the β-Form of Na$_n$H$_2$P$_n$O$_{3n+1}$," Kristallografiya, **6**, 828.

Wells, M., and Megaw, D., "The Structures of NaNbO$_3$ and Na$_{0.975}$K$_{0.025}$NbO$_3$," Proc. Phys. Soc. (London), **78**, 1258.

1962

Amorós Portolés, J. L., Arrese, F., and Canut, M., "The Crystal Structure of the Low-Temperature Phase of NH$_4$NO$_3$(V) at −150°C," Z. Krist., **117**, 92.

Burns, J. H., "Unit Cell and Space Group of LiBrO$_3$," Acta Cryst., **15**, 89.

Galasso, F., and Darby, W., "Ordering of the Octahedrally Coordinated Cation Position in the Perovskite Structure," *J. Phys. Chem.*, **66**, 131.

Ghose, S., "The Crystal Structure of Salesite, CuIO₃(OH)," *Naturwiss.*, **49**, 102; *Acta Cryst.*, **15**, 1105.

Gillespie, R. B., Gantzel, P. K., and Trueblood, K. N., "The Crystal Structure of Ammonium Chlorate," *Acta Cryst.*, **15**, 1271.

Graf, D. L., and Bradley, W. F., "The Crystal Structure of Huntite, Mg₃Ca(CO₃)₄," *Acta Cryst.*, **15**, 238.

Hardy, A., "Crystal Structures of Two Allotropic Varieties of Barium Manganite. New Structure ABO₃," *Acta Cryst.*, **15**, 179.

Hilmer, W., "The Structure of the High Temperature Form of Barium Germanate BaGeO₃(h)," *Acta Cryst.*, **15**, 1101.

Hoppe, R., and Lidecke, W., "The Terbate(IV) Na₂TbO₃," *Naturwiss.*, **49**, 255.

Ivanova, V. V., Kapyshev, A. G., Venevtsev, Y. N., and Zhdanov, G. S., "X-Ray Determination of the Symmetry of Elementary Cells of the Ferroelectric Materials (K₀.₅Bi₀.₅)TiO₃ and (Na₀.₅Bi₀.₅)TiO₃ and of High Temperature Phase Transitions in (K₀.₅Bi₀.₅)TiO₃," *Izv. Akad. Nauk SSSR, Ser. Fiz.*, **26**, 354.

Jost, K. H., "The Structure of an Acid Sodium Polyphosphate," *Acta Cryst.*, **15**, 951.

Kierkegaard, P., "On the Crystal Structure of MoO₂(PO₃)₂," *Arkiv Kemi*, **18**, 521.

Kupriyanov, M. F., and Fesenko, E. G., "X-Ray Structural Studies of Phase Transitions in Perovskite-Type Compounds," *Kristallografiya*, **7**, 451.

Kurki-Suonio, K., "Nonapproximate Analysis of Experimental Structure Amplitudes. II. The Correct Model for NaNO₃," *Ann. Acad. Sci. Fennicae, Ser. A. VI*, **94**, 34 pp.

Mil'kova, L. P., and Porai-Koshits, M. A., "Lattice Parameters, Crystal Symmetry, and Principal Structure Features of Some Metafluoroberyllates," *Izv. Akad. Nauk SSSR, Ser. Fiz.*, **26**, 368.

Okazaki, A., and Suemune, Y., "The Crystal Structure of KMF₃," *J. Phys. Soc. Japan*, **17**, Suppl. B-I, 204.

Ozerov, R. P., Rannev, N. V., Pakhomov, V. I., Rez, I. S., and Zhdanov, G. S., "Structure of KIO₃ at Room Temperature," *Kristallografiya*, **7**, 620.

Ringwood, A. E., and Seabrook, M., "High-Pressure Transition of MgGeO₃ from Pyroxene to Corundum Structure," *J. Geophys. Res.*, **67**, 1690.

Sass, R. L., and Scheuerman, R. F., "The Crystal Structure of Sodium Bicarbonate," *Acta Cryst.*, **15**, 77.

Schröder, J., "The Mixed-Crystal Systems La(Sr)CoO₃ and La(Th)CoO₃," *Z. Naturforsch.*, **17b**, 346.

Srikanta, S., Tare, V. B., Sinha, A. P. B., and Biswas, A. B., "Structural Properties of (Ba,Pb)₁₋ₐ(Ti,Nb)O₃ Systems," *Acta Cryst.*, **15**, 255.

Zalkin, A., Lee, K., and Templeton, D. H., "Crystal Structure of CsMnF₃," *J. Chem. Phys.*, **37**, 697.

NAME INDEX*

A

Actinium sesquioxide, 3, 40
Actinium sesquisulfide, 40, 163
Actinium tribromide, 78, 121
Actinium trichloride, 78, 121
Actinium trifluoride, 62, 121
Aikinite, 505–506, 508, 538
Alstonite, 367, 537
Aluminum bismuth trioxide, 398, 446
Aluminum bromide, 57–59, 121
Aluminum carbide, 163–164, 168
Aluminum carbonitride, 197–198, 201
Aluminum chloride, 55–56, 121
Aluminum fluoride, 46–47, 50, 121
Aluminum hydroxide, 78–80, 121
Aluminum metaphosphate, 528, 537
Aluminum oxide, 7–8, 40
Aluminum sesquisulfide, 2–3, 8, 40
tri-Aluminum tetroxide, 168 [**3**, Ch. VIII]
Americium sesquioxide, 3, 5, 40
Americium sesquisulfide, 40, 163
Americium tetrafluoride, 129, 147
Americium tribromide, 64, 121
Americium trichloride, 78, 121
Americium trifluoride, 62, 121
Americium triiodide, 64, 121
Ammonia, 117, 119, 124
Ammonium acid fluoride, 279, 347
Ammonium acid nitrate, 483–484, 540
Ammonium azide, 280, 347
Ammonium beryllium trifluoride, 432, 453
Ammonium bicarbonate, 462–463, 540
Ammonium cadmium trichloride, 428–429, 453
Ammonium cadmium trinitrite, 396, 453
Ammonium chlorate, 381, 453
Ammonium chlorite, 306–307, 347
Ammonium chlorobromoiodide, 297, 299–300, 346

Ammonium cobalt trifluoride, 393, 453
di-Ammonium cuprous bromide, 505, 540
di-Ammonium cuprous chloride, 503–504, 540
Ammonium hypophosphite, 308, 310, 347
Ammonium iodate, 389–390, 454
Ammonium isothiocyanate, 284, 286, 347
Ammonium manganese trifluoride, 393, 454
Ammonium mercuric chloride, 426, 453
Ammonium mercury isothiocyanate, 426, 428, 454
Ammonium metavanadate, 430–431, 454
Ammonium nickel trifluoride, 393, 454
Ammonium nitrate, 368–374, 454
Ammonium plutonyl carbonate, 465, 540
di-Ammonium silver iodide, 501, 540
Ammonium silver isothiocyanate, 290–291, 346
Ammonium tetrametaphosphate, 432–435, 454
Ammonium triiodide, 297, 299, 347
Ammonium trinitrate, 483–484, 540
Ammonium trinitride, 280, 347
Ankerite, 468
Antimony pentachloride, 174–175, 201
Antimony sesquioxide, 17–18, 20–21, 43
Antimony tribromide, 66, 125
Antimony trichloride, 65–67, 125
Antimony trifluoride, 64–65, 125
Antimony triiodide, 45, 125
Antimony triselenide, 27–28, 43
Antimony trisulfide, 27, 43, 508
Antimony tritelluride, 30, 43
Aragonite, 364–366, 448
Argentic. *See* Silver.
Argentous. *See* Silver.
Arsenic triiodide, 45, 50, 121
Arsenic trisulfide, 26–27, 40
Arsenious oxide, 17–20, 40

* In addition to those elements and compounds reported in this volume, this index contains citations of future volumes of this work. Such citations are enclosed in brackets and give both the volume and the chapter number (e.g., [**4**, Ch. XII]).

FORMULA INDEX*

A

AcBr$_3$, 78, 121
AcCl$_3$, 78, 121
AcF$_3$, 62, 121
Ac$_2$O$_3$, 3, 40
Ac$_2$S$_3$, 40, 163
AgAlS$_2$, 338, 343
AgAlSe$_2$, 338, 343
AgAlTe$_2$, 338, 343
AgBiS$_2$, 312, 343
AgBiSe$_2$, 292, 312, 334–335, 343
AgBiTe$_2$, 312, 343
AgBrO$_3$, 386, 446
AgClO$_2$, 307–309, 343
AgClO$_3$, 386, 446
AgCrO$_2$, 292, 343
AgCrS$_2$, 295, 343
AgCrSe$_2$, 296, 343
AgFeO$_2$, 292, 343
AgFeS$_2$, 338, 343
AgGaS$_2$, 338, 343
AgGaSe$_2$, 338, 343
AgGaTe$_2$, 338, 344
AgInS$_2$, 338, 344
AgInSe$_2$, 338, 344
AgInTe$_2$, 338, 344
AgNCO, 284, 344
AgNO$_2$, 303, 344
AgNO$_3$, 362, 446
AgN$_3$, 280–282, 344
(AgPO$_3$)$_n$, 438–439, 446
AgSCN, 284, 286–287, 344
AgSbO$_3$, 445–446
AgSbS$_2$, 312, 344
AgSbSe$_2$, 312, 344
AgSbTe$_2$, 312, 344
AgZnF$_3$, 391, 446
Ag$_2$CsI$_3$, 497, 499, 537

Ag$_2$O$_3$, 15, 40
Ag$_2$PbO$_2$, 318, 320, 344
Ag$_3$AsS$_2$, 511, 537
Ag$_3$AuTe$_2$, 342–343
Ag$_3$SNO$_3$, 377–378, 446
Ag$_3$SbS$_3$, 508, 510–511, 537
AlBiO$_3$, 398, 446
AlBr$_3$, 57–59, 121
AlCl$_3$, 55–56, 121
AlF$_3$, 46–47, 50, 121
Al(OH)$_3$, 78–80, 121
Al(PO$_3$)$_3$, 528, 537
Al$_2$O$_3$, 7–8, 40
Al$_2$S$_3$, 2–3, 8, 40
Al$_3$O$_4$, 168 [**3**, Ch. VIII]
Al$_4$C$_3$, 163–164, 168
Al$_5$C$_3$N, 197–198, 201
AmBr$_3$, 64, 121
AmCl$_3$, 78, 121
AmF$_3$, 62, 121
AmF$_4$, 129, 147
AmI$_3$, 64, 121
Am$_2$O$_3$, 3, 5, 40
Am$_2$S$_3$, 40, 163
AsH$_3$, 119, 121
AsI$_3$, 45, 50, 121
As$_2$B$_{13}$, 140, 246
As$_2$NiO$_4$, 149 [**3**, Ch. VIII]
As$_2$O$_3$, 17–20, 40
As$_2$S$_3$, 26–27, 40
AuCl$_3$, 69–70, 121

B

BBr$_3$, 71, 121
BCl$_3$, 70, 121
BI$_3$, 71, 121
B$_2$H$_6$, 121
B$_2$O$_3$, 16–17, 40

B$_4$C, 138–139, 147
B$_4$H$_{10}$, 235–237, 245
B$_5$H$_9$, 237–238, 245
B$_5$H$_{11}$, 237–239, 246
B$_9$H$_{15}$, 240–241, 246
B$_{10}$H$_{14}$, 242–243, 246
BaB$_6$, 203, 244
BaCO$_3$, 367, 446
BaCO$_3$.2CeFCO$_3$, 473, 537
BaCa$_{0.5}$W$_{0.5}$O$_3$, 395, 446
BaCa(CO$_3$)$_2$, 367, 468–469, 537
BaCaZrGeO$_6$, 396, 447
BaCdO$_2$, 317, 344
BaCe$_{0.5}$Nb$_{0.5}$O$_3$, 395, 447
BaCeO$_3$, 391, 447
BaCo$_{0.33}$Ta$_{0.67}$O$_3$, 395, 447
BaCo$_{0.5}$W$_{0.5}$O$_3$, 395, 447
BaCoO$_2$, 315, 344
BaDy$_{0.5}$Nb$_{0.5}$O$_3$, 395, 398, 447
BaEr$_{0.5}$Nb$_{0.5}$O$_3$, 395, 398, 447
BaEu$_{0.5}$Nb$_{0.5}$O$_3$, 395, 398, 447
BaFe$_{0.5}$Nb$_{0.5}$O$_3$, 395, 447
BaFe$_{0.5}$Ta$_{0.5}$O$_3$, 395, 447
BaFe$_{0.5}$W$_{0.5}$O$_3$, 395, 447
BaFeO$_3$, 391, 398, 447
BaGd$_{0.5}$Nb$_{0.5}$O$_3$, 395, 398, 447
BaGeO$_3$, 430, 442–443, 447
BaHo$_{0.5}$Nb$_{0.5}$O$_3$, 395, 398, 447
BaIn$_{0.5}$Nb$_{0.5}$O$_3$, 395, 399, 447
BaIr$_{0.33}$Ti$_{0.67}$O$_3$, 416, 447
BaLa$_{0.5}$Nb$_{0.5}$O$_3$, 399, 447

* In addition to those elements and compounds reported in this volume, this index contains citations of other volumes of this work. These are enclosed in brackets and give both the volume number, and page number (e. g., [**1**, 333]), or the volume number and chapter number (e.g., [**3**, Ch. IX]).

UB$_{12}$, 230–231, 246
UBr$_3$, 78, 126
UCl$_3$, 78, 126
UCl$_4$, 130, 148
UCl$_6$, 205–206, 244
UD$_3$, 119–120, 126
UF$_3$, 62, 126
UF$_4$, 129, 148
UF$_5$, 175–177, 201
UF$_6$, 206–207, 244
UH$_3$, 119–121, 126
UI$_3$, 64, 126
UO$_2$CO$_3$, 378–379, 458
UO$_2$F$_2$ [1, 288, 407]
UO$_3$, 53, 95–96, 126
USe$_3$, 102, 126
UTa$_2$O$_8$, 220, 245
U$_2$C$_3$, 38–39, 44
U$_2$F$_9$, 225–226, 245
U$_2$N$_3$, 5, 44
U$_2$S$_3$, 29, 44
U$_2$Se$_3$, 29, 44
U$_3$As$_4$, 162, 169
U$_3$Bi$_4$, 162, 169
U$_3$O$_7$, 221, 245
U$_3$O$_8$, 220–221, 245
U$_3$P$_4$, 162, 169
U$_3$Sb$_4$, 162, 169
U$_3$Si$_2$, 32, 44
U$_3$Te$_4$, 162, 169
U$_4$O$_9$, 221, 245

V

VCl$_3$, 50, 126
VF$_3$, 48–49, 126
V$_2$O$_3$, 7–8, 44
V$_2$O$_5$, 184, 186–187, 201
V$_2$TiO$_5$, 193 [3, Ch. IX]
V$_3$B$_2$, 32, 44
V$_3$B$_4$, 168, 169
V$_3$O$_5$, 192–193, 201
V$_3$S, 105–107, 126
V$_3$Se$_4$, 151, 169
V$_3$Te$_4$, 151, 169
V$_6$O$_{13}$, 232–233, 246

W

WCl$_6$, 203–205, 244
WO$_3$, 82–83, 126
W$_2$B$_5$, 188–189, 201
W$_4$O$_{11}$, 246
W$_{10}$O$_{29}$, 246
W$_{18}$O$_{49}$, 90–92, 246
W$_{20}$O$_{58}$, 91, 93, 246

Y

YAlO$_3$, 395, 409, 458
YBO$_3$, 362, 369, 458
YB$_6$, 203, 244
YB$_{12}$, 230, 246
YCl$_3$, 56–57, 126
YCrO$_3$, 395, 399, 409, 458
YF$_3$, 58, 60, 126
YFeO$_3$, 395, 409, 458
YH$_3$, 62, 126

Y(OH)$_3$, 77, 126
YScO$_3$, 409, 458
Y$_2$O$_2$S, 3, 44
Y$_2$O$_3$, 4–5, 44
YbBO$_3$, 369, 458
YbB$_6$, 203, 244
YbCl$_3$, 57, 126
YbF$_3$, 61, 126
Yb(OH)$_3$, 78, 126
Yb$_2$O$_2$S, 3, 44
Yb$_2$O$_2$Se, 3, 44
Yb$_2$O$_3$, 4–5, 44

Z

ZnCO$_3$, 362, 458
ZnGeAs$_2$, 338, 348
ZnGeP$_2$, 338, 348
ZnSiAs$_2$, 338
ZnSiP$_2$, 338
ZnSnAs$_2$, 339, 348
Zn$_3$As$_2$, 36, 44
Zn$_3$N$_2$, 5, 44
Zn$_3$P$_2$, 33–35, 44
Zn$_4$O(BO$_2$)$_6$, 329–331, 348
ZrB$_{12}$, 230, 246
ZrBr$_4$, 132, 148
ZrCl$_4$, 132, 148
ZrF$_4$, 127–128, 148
Zr(IO$_3$)$_4$, 495, 497–498, 541
ZrSe$_3$, 102, 126
ZrTaNO, 334, 348

9 1236

BRA

642

1-MONTH

NON-CIRCULATING